GOD IN THE AGE OF SCIENCE?

God in the Age of Science?

A Critique of Religious Reason

HERMAN PHILIPSE

OXFORD
UNIVERSITY PRESS

Great Clarendon Street, Oxford, OX2 6DP,
United Kingdom

Oxford University Press is a department of the University of Oxford.
It furthers the University's objective of excellence in research, scholarship,
and education by publishing worldwide. Oxford is a registered trade mark of
Oxford University Press in the UK and in certain other countries

© Herman Philipse 2012

The moral rights of the author have been asserted

First published in 2012
First published in paperback 2014

All rights reserved. No part of this publication may be reproduced, stored in
a retrieval system, or transmitted, in any form or by any means, without the
prior permission in writing of Oxford University Press, or as expressly permitted
by law, by licence or under terms agreed with the appropriate reprographics
rights organization. Enquiries concerning reproduction outside the scope of the
above should be sent to the Rights Department, Oxford University Press, at the
address above

You must not circulate this work in any other form
and you must impose this same condition on any acquirer

Published in the United States of America by Oxford University Press
198 Madison Avenue, New York, NY 10016, United States of America

British Library Cataloguing in Publication Data
Data available

Library of Congress Cataloguing in Publication Data
Data available

ISBN 978–0–19–969753–3 (Hbk)
ISBN 978–0–19–870152–1 (Pbk)

Links to third party websites are provided by Oxford in good faith and
for information only. Oxford disclaims any responsibility for the materials
contained in any third party website referenced in this work.

For
PETER M. S. HACKER
il miglior fabbro

I think it may not be amiss to take notice, that however *Faith* be opposed to Reason, *Faith* is nothing but a firm assent of the mind: which if it be regulated, as is our duty, cannot be afforded to any thing, but upon good reason; and so cannot be opposite to it. He that believes, without having any reason for believing, may be in love with his own fancies; but neither seeks truth as he ought, nor pays the obedience due to his Maker, who would have him use those discerning faculties he has given him, to keep him out of mistake and error. (John Locke, *An Essay Concerning Human Understanding* Bk. IV, Ch. xvii, § 24)

It seemed to Brother Juniper that it was high time for theology to take its place among the exact sciences and he had long intended putting it there. What he had lacked hitherto was a laboratory. (Thornton Wilder, *The Bridge of San Luis Rey*, Part One)

Contents

Preface   xi
Acknowledgements   xvii

PART I. NATURAL THEOLOGY

1. The Priority of Natural Theology   3
2. The Rise, Fall, and Resurrection of Natural Theology   19
3. The Reformed Objection to Natural Theology   31
4. Refutation of the Reformed Objection   43
5. The Rationality of Natural Theology   65
6. A Grand Strategy   76

PART II. THEISM AS A THEORY

7. Analogy, Metaphor, and Coherence   95
8. God's Necessity   120
9. The Predictive Power of Theism   140
10. The Immunization of Theism   161

PART III. THE PROBABILITY OF THEISM

11. Ultimate Explanation and Prior Probability   191
12. Cosmological Arguments   221
13. Arguments from Order to Design   256
14. Other Inductive Arguments   279
15. Religious Experience and the Burden of Proof   310

Conclusion   338

References   347
Index   363

Preface

At some phase in their life, many people who are educated in a particular religious tradition start to ask questions about the value of their practices and the truth of their beliefs, such as those typical of Christianity, Hinduism, Islam, Judaism, Mormonism, Shinto, Sikhism, or any of the countless minor living religions and sects. 'Suppose that I had been born into another religious tradition,' they may wonder, 'would I have endorsed the creed of that religion as wholeheartedly as I am now accepting my own faith? If so, are there good reasons for holding the religious beliefs that I now hold, or, perhaps, compelling arguments for rejecting them?' Most people do not find it easy to think about such issues, because there are emotional and intellectual obstacles. Religious faith satisfies deep longings of the human heart; some religions and sects strongly discourage or even punish critical thought about their central tenets; and the issues involved are complex and difficult to grasp.

This book is intended primarily for those readers who have raised such questions several times, and who want to penetrate deeper into the challenging problems involved. These problems belong to what is usually called the philosophy of religion, which deals with issues concerning the meaningfulness, justification, and truth of religious creeds. Apart from creeds or clusters of religious beliefs, religions have many other aspects, such as social organizations, rituals, architecture, dietary prescriptions, forms of art, and moral norms. Since this book focuses on religious beliefs, however, and because the relative importance of religious doctrine as against the prominence of rituals or social practices is perhaps greatest in the mature monotheistic traditions of Christianity, Islam, and Judaism, most of my arguments will be concerned explicitly with monotheism. Yet the reader can easily adjust these arguments in order to apply them to polytheistic religions as well.

I have written this volume also for colleagues and students in university departments of philosophy and theology. Indeed, the scope of the book is such that it may be used as a textbook for courses in the philosophy of religion. Instead of giving a neutral overview of the field, however, I argue for my own opinions on the issues involved and against those of other authors. It is my experience that often this approach is more stimulating for students than that of textbooks in which the author attempts to remain uncommitted. I trust that many readers will relish attempting to refute my views.

In our times, even hardened secularists and stubborn unbelievers have two good reasons for becoming interested in the philosophy of religion, an external and an internal one. The external reason is that religion is becoming increasingly important in the world at large. One cause of this development is purely demographic. Since 1804, when the human population of the world reached one billion, population growth has exploded. In 1927 there were two billion humans alive, in 1960 three billion, in 1975 four billion, and between 2012 and 2050 the human population is expected to grow from seven billion to a predicted number of roughly 9.3 billion. Since this population growth takes place predominantly in religious regions, the more secularized countries have a dwindling share in the population of the world at large.

Another cause of the increasing importance of religion in the world is what is called globalization. Because of air travel, media of mass communication, extensive migration, worldwide economy, and the Internet, world views belonging to different cultural traditions clash with each other more often than ever before in world history. Within individuals, such clashes may produce what psychologists call 'cognitive dissonance', the unpleasant awareness of conflicting opinions. No wonder that this cognitive dissonance, if added to other feelings of frustration, sometimes leads to anger and aggression, which may then be justified by religious doctrines, such as varieties of Islamism. As a result, conflicts in which religious motivations play a role now figure frequently in our news media.

Of the many issues that can be raised with regard to religions, one central question is whether there are good reasons for thinking that some specific subset of religious beliefs makes sense and is true. This question is the core issue of the philosophy of religion. For most of the twentieth century, the philosophy of religion (at least as practised on the European continent) had a somewhat dubious intellectual reputation among philosophers, because it was a field contaminated by obscure rhetoric, sloppy arguments, and meaningless jargon. During the last decades, however, a number of analytic philosophers, who are well trained in conceptual analysis, logic, probability theory, and the philosophy of science, have developed new approaches to the philosophy of religion, which are sufficiently sophisticated to deserve ample attention. In other words, the philosophy of religion has become an academic discipline that is intellectually fascinating in its own right. This is the second, internal reason for becoming interested in the philosophy of religion as practised by analytic philosophers, a reason which holds for both believers and secularists.

* * *

One might distinguish between two different objectives in the philosophy of religion. The *apologetic* objective is to develop arguments, and overall strategies of arguing, that purport to legitimize belief in one specific religious creed (*positive* apologetics), and to refute objections against that creed (*negative* apologetics). Philosophers of religion who pursue positive apologetic objectives by developing arguments, not based upon revelation or faith, to the effect that a specific creed is true, are usually called *rational* or *natural theologians*. The *critical* objective is to investigate whether apologetic arguments or strategies are convincing, and to develop objections against specific religious creeds or against all religions.

Apologetic philosophers of religion must also be critical ones, since each religious doctrine being to some extent logically incompatible with many others, in arguing for the truth of one creed they must argue against its religious rivals. Indeed, most religious believers are atheists with regard to deities of other religions, since they deny the existence of these deities, although typically polytheists will be more inclusive and tolerant than monotheists as regards this point. People who are usually called 'atheists' are in fact *universal* atheists, in contrast with the particular and selective atheism of religious believers. The objective of the present book is a critical one, and its spirit is purely Socratic. Or rather, its aim is to discover what is meaningful and true in the religious domain.

Critical treatises on the philosophy of religion are sometimes rejected by religious-minded reviewers who argue that such works merely chase after paper tigers instead of analysing the best and most worthy instances of religious belief. For example, it has been objected to *The God Delusion*, published by Richard Dawkins in 2006, that its author imagines 'like a bumptious young barrister that you can defeat the opposition while being complacently ignorant of its toughest case.'[1] In order to avoid this snare, I discuss the views of some main apologetic philosophers of religion in the analytic tradition, such as D. Z. Phillips and Alvin Plantinga. For several reasons, a pivotal role in my book is granted to the argumentative strategy in defence of monotheism or *theism* developed by Richard Swinburne, who was Nolloth Professor of the Philosophy of the Christian Religion at the University of Oxford from 1985 to 2002, and who is one of the most prominent natural theologians of our time.

First, the doctrine of monotheism that Swinburne defends is common to the three great Abrahamic traditions, Judaism, Christianity, and Islam. As a consequence, his apologetic strategy is of greater religious relevance than, say, the quite different apologetic policy of Alvin Plantinga, the import of which will turn out to be limited to a particular monotheistic tradition. Swinburne distinguishes clearly between his arguments for monotheism *tout court*, or 'bare theism' as he calls it, and his more specific apologetic arguments for the 'ramified theism' of Christianity, which presuppose the defence of monotheism as such. Hence, all monotheists might endorse Swinburne's apologetic arguments for bare theism, to the extent that these arguments are sound, whereas agnostics, atheists, polytheists, or pantheists should attempt to refute at least some of them. In this book, I shall amply investigate Swinburne's case for bare theism.

A second reason for selecting Richard Swinburne as an important protagonist is that his argumentative strategy for bare theism seems to me the most promising one from an apologetic point of view. Many traditional philosophers of religion have attempted to develop deductive arguments for the existence of their god(s) akin to the types of argument we find in mathematics, claiming certainty for their conclusions. But the existence of the monotheistic god, as this god is traditionally conceived, is not at all like the 'existence' of numbers or other mathematical 'entities'. If he exists, that god is a divine person with many kinds of causal powers, whose existence can best be argued for on the basis of his alleged empirically detectable doings.

Furthermore, deductive varieties of the arguments for the existence of God typically rely on at least one premise that the unbeliever will reject, such as a version of Leibniz's principle of sufficient reason, or the impossibility of an infinite causal regress. Swinburne claims plausibly, then, that the logical form of the arguments for bare theism should be that of inductive arguments to the best explanation, and that Bayesian confirmation theory is a useful tool for presenting such arguments. The question is not whether we can *prove* that God exists or that he does not exist, but rather how *probable* it is that he exists, given the total evidence pro and contra, if at least the very notion of God makes sense.

In analytic philosophy, it is not very common to focus on the writings of one author to this extent. Should we not concentrate on arguments as such, and not on the works of a particular philosopher? But as Swinburne stresses, one should not treat arguments

---

[1] Quoted from Terry Eagleton's review of *The God Delusion* (2006) by Richard Dawkins: Eagleton (2006), p. 32.

in support of theism in isolation from each other, since the force of one inductive argument may be increased or diminished by that of another. In order to do justice to the 'toughest case' in apologetic philosophy of religion, then, one should study each argument within the context of an overall argumentative strategy, and in philosophy such a strategy usually is best elaborated in the writings of one author. The first editions of Swinburne's works have been amply discussed in philosophy journals. My critical analyses are based upon later revised editions, such as the second edition of *The Existence of God* (2004).

* * *

Those who write on analytic philosophy of religion often dedicate each chapter of a book to one of the (kinds of) arguments for or against theism, such as ontological arguments, cosmological arguments, the arguments from or to design, arguments from consciousness, arguments from providence, the problem of evil, or arguments from religious experience and miracles. This is also the case in classics of critical philosophy of religion, such as J. L. Mackie's *The Miracle of Theism* (1982), Michael Martin's *Atheism: A Philosophical Justification* (1990), *The Non-existence of God* by Nicholas Everitt (2004), Jordan Howard Sobel's massive volume called *Logic and Theism* (2004), or *Arguing about Gods* (2006a) by Graham Oppy. The present book has a somewhat different structure, because I pay more attention than is usual to the strategic choices and dilemmas with which religious believers and apologetic philosophers of religion are faced in our age of science.

The scientific progress of the last four centuries has severely limited the number of tenable religious options open to believers and apologetic philosophers of religion. For example, since many traditional religious doctrines expressed in allegedly divine revelations, such as the idea that the world was created a few thousand years before the birth of Christ or Mohammed, have been refuted by scientific discoveries, an intellectually responsible believer cannot rely any more on the text of a holy book without further ado. Globally speaking, there are only four strategic options (b, d, e, f, see below) open to the believer and the religious apologist, which may be presented as the horns of three interlocked dilemmas, or as the end nodes of a decision tree. Let me briefly hint at these options here; they will be elaborated in chapters of this book.

The first dilemma is that between cognitive and non-cognitive interpretations of religious statements, such as 'God exists', 'Allah is the only god', or 'there are many gods'. According to (a) cognitive interpretations, a religious believer who says that God exists is making a factual claim, which is either true or false, whereas according to (b) non-cognitive interpretations, such sayings have very different functions, and an evaluation in terms of truth or falsity is inappropriate. Clearly, option (b) of a non-cognitive interpretation, as ascribed to Wittgenstein, for example, and defended by some of his followers, may seem to have a great benefit to the believer. For if it is correct, traditional criticisms of religion to the effect that religious credal statements are false, simply do not apply. Such criticisms are diagnosed as misguided, because allegedly they are based upon a misunderstanding of 'the religious language game'. This huge benefit risks being cancelled out, however, by the costs of the non-cognitive option. Clearly, most believers of the past have thought, and most present religious believers think, that whether God exists or not is a factual issue, and that his actual existence is of the highest importance to their lives.

If one opts for (a) the cognitive interpretation of the religious statement that God exists, a second dilemma emerges. Either (c) a religious belief has to be backed up by evidence or reasons in order to be legitimate or reasonable or justified, or (d) such a backup is not or may not be needed. This latter option (d), according to which no positive supporting reasons are necessarily required for a religious belief to be legitimate or 'warranted', has been most ably defended by Alvin Plantinga under the banner of the *Reformed objection to natural theology*. The basis of Plantinga's argument is his epistemological theory of warrant, which is an *externalist* doctrine in the technical sense of 'externalism' used by writers on the theory of knowledge. My discussion of Plantinga's work in this book will be relatively brief. I shall argue that religious beliefs, and indeed most modern religious believers, cannot be called justified or reasonable or warranted unless the latter, or at least some experts in their community, can support their beliefs by adducing good positive reasons to the effect that these beliefs are true.

Let us suppose, then, that (c) a religious belief can be reasonable or justified or warranted only if it is shown to be (probably) true by positive reasons and evidence. As I shall argue, these reasons should not derive from, or not derive merely from, a revelation or from faith. In other words, rational or natural theology is indispensable to modern religious believers, if at least they want to be called reasonable or rational. But now a third dilemma arises, which concerns the methods that the natural theologian should use in arguing for the thesis that God, Yahweh, or Allah exists. Either (e) these methods are quite unlike the methods used by scientists and scholars when they investigate a factual hypothesis of existence, or (f) they are like these methods. Both horns of this third dilemma imply great perils for religious believers. If they opt for the second horn (f), there is the risk that religious belief is open to refutation or disconfirmation by empirical research, such as investigations of the effectiveness of prayers said in the past. If, however, they opt for horn (e), there is the risk that their methods and arguments lack credibility in our scientific age, since we commonly require a public and persuasive validation of methods of research. Why should methods of religious investigation be an exception?

A final reason for singling out Richard Swinburne's works as the 'toughest case' for a critical philosopher of religion is that to my mind, he presents the most sophisticated solution available in the literature to this third dilemma. If I am correct in this respect, the critical results of my book will be pertinent to the prospects of rational or natural theology as such.

* * *

The set-up of the book is as follows. In the six relatively short chapters of Part One, which is introductory, it is argued that for modern believers religious faith is not an epistemic virtue but a vice, unless there are convincing arguments for the truth of its contents. Recourse to an alleged divine revelation can only be a ground for faith if this revealed theology is backed up by arguments of rational or natural theology (Chapter 1). The history of natural theology in Western culture is summarized in a nutshell (Chapter 2), and the so-called *Reformed objection* to natural theology is shown to be invalid for intellectually responsible modern believers (Chapters 3 and 4). Which kind of rationality should be aimed at by natural or rational theology is briefly investigated (Chapter 5). Concerning the objective rationality aspired to by this discipline, it is argued that believers who endorse the existence of one or more invisible spirits, such as the Christian god, are faced with a methodological dilemma. Do they have at their disposal reliable and validated methods of religious research? If so, it seems that the content of their faith can be refuted, at least in principle. If not, how can a religious creed be credible at all in our scientific era (Chapter 6)?

Part Two discusses some difficulties of formulating meaningfully the contents of a religious doctrine such as (mono)theism. Is it possible to eliminate the many apparent contradictions in the traditional theistic conception of God without introducing analogy or metaphor? And if analogy is unavoidable, can theism be a 'theory' that is susceptible of empirical confirmation? Indeed, if analogy pervades all religious language, does it not follow that believers simply do not know what they mean when they say that God or Allah exists (Chapter 7)? I discuss in detail some of the properties that monotheistic theologians traditionally attribute to their god, such as being a bodiless person, or being necessary in some sense (Chapter 8). Most arguments for the existence of this god, such as the various arguments from design, presuppose that theism is a theory or hypothesis with some predictive power. But how can theism have predictive power at all (Chapter 9)? And if it has such power, how can the theologian immunize theism against empirical refutations? In other words, how can natural theology avoid the pitfall of what is commonly called God-of-the-gaps (Chapter 10)?

Part Three is concerned with the arguments for and against (mono)theism. The logic of these arguments should be a logic of probability, and according to many writers, Bayesian confirmation theory is its best formalization. Yet a Bayesian such as Richard Swinburne is faced with many difficulties. Are the estimates of prior probabilities not purely subjective? Is the existence of a bodiless god not extremely improbable, given our background knowledge of intelligent beings on Earth, if this idea is intelligible at all? And how can the theist solve what philosophers of science call the problem of under-determination? Monotheists such as Swinburne will stress the importance of simplicity as a criterion of theory-choice. But is this criterion truth-conducive or merely pragmatic? And does it favour theism as the best theory (Chapter 11)?

In Chapters 12–14 of Part Three, probabilistic versions of the traditional arguments for the existence of a monotheistic god are developed and assessed. By adding up all these arguments in a cumulative case, the natural theologian wants to establish that the thesis that God exists has a logical or epistemic probability sufficient for full endorsement. But how can this be shown convincingly if we cannot determine the relevant probabilities quantitatively? As some readers may expect, natural theologians will attempt to shift the burden of proof to the opponent, and one prominent tactic for doing so by means of the argument from religious experience is discussed in detail (Chapter 15).

In the Conclusion, the many threads spun out in this book will be knotted together in order to provide an answer to the crucial question: which view concerning religious matters has the best credentials: theism, agnosticism, some version of atheism, or perhaps a polytheistic creed?

Herman Philipse

Acknowledgements

I am particularly grateful to Richard Swinburne for generously commenting on various versions of my manuscript, and to Alvin Plantinga for reading and criticizing a draft of Chapters 3 and 4. These prominent philosophers of religion taught me much, and they corrected quite a few misunderstandings of their views. As always, discussions with my dear friend Peter Hacker have been a great source of inspiration, and his castigations concerning my English were stimulating indeed. Many other friends and colleagues have commented on drafts of one or more chapters of this book. I would like to thank warmly Tjitze Baarda, Bob Becking, Anthony Booth, Irene Conradie, Igor Douven, Dirk-Martin Grube, Willem van Hoorn, John Hyman, Theo Kuipers, Joop Leo, Renate Loll, Tim Mawson, Annette Merz, Thomas Müller, Fred Muller, Katherine Munn, Rik Peels, Jeroen de Ridder, Jan Willem Romeijn, Marcel Sarot, Eric Schliesser, Bas Van Fraassen, Eddie van de Kraats, and René van Woudenberg for their critical comments and helpful suggestions. I also profited greatly from the critical observations of the two anonymous referees of Oxford University Press, and of a third referee, who did not object to revealing his name after the Press decided to publish the book, John Schellenberg.

My assistant Nick Boerma tirelessly traced the numerous articles written in recent times on the topics discussed, and critically commented on most chapters. Parts of the manuscript were presented at conferences on the philosophy of religion, such as those of the British Society for the Philosophy of Religion at Oxford in 2007 and 2009, and the conference on Formal Perspectives on the Epistemology of Religion at Leuven, Belgium, in June 2009. I am grateful to the audiences of my talks for their instructive and critical comments. I am also much obliged to participants in seminars on the philosophy of religion, such as the Oxford reading group at St Peter's College, and the seminar on the philosophy of religion I directed at the University of Utrecht, The Netherlands, which is my home university. Some of the chapters were presented to the Dutch National Seminar on Analytic Philosophy at the University of Utrecht, convened by Rik Peels and chaired by myself. I am deeply indebted to the governing body of this university for appointing me in 2003 as a distinguished professor, liberating me from my recurrent managerial duties as a chairman of the Faculty of Philosophy at the University of Leiden, so that I now have ample time for research.

Finally, it is a great pleasure to thank the Delegates and officers of Oxford University Press and, in particular, Peter Momtchiloff, whose expertise, efficiency, and charm are a boon to authors working with him.

PART I
NATURAL THEOLOGY

1

The Priority of Natural Theology

In a strict sense, the term 'theology' covers all systematic ways of talking and writing about God or gods.[1] In Western academic traditions, it is customary to distinguish between two different kinds of theology. The one, called *revealed* theology, is the endeavour by believers to interpret and systematize the contents of texts that are considered divine revelations, such as the Bible, the Koran, the Vedas, or the Book of Mormon. The other, called *natural* or *rational* theology, consists in attempts to prove or show to be probable the existence of God or gods, and to acquire knowledge about them, on the basis of evidence or premises that can also be accepted by non-believers, such as empirical knowledge about the natural world.

It is the aim of this first chapter to establish provisionally the thesis that natural theology has a priority over revealed theology (the 'reformed' objection that we might have properly basic religious knowledge, which does not stand in need of inferential justification, will be discussed in Chapters 3 and 4). But which kind of priority is meant? In the context of discovery, there usually is no temporal priority of natural theology. Individuals do not discover and endorse religious beliefs for the first time because they have studied its intricate arguments. Living religions are shared forms of life, into which most believers were born. Typically, religious education starts early in infancy, long before children develop their mature intellectual capacities, and it often proceeds by means of ritualised repetitions. Critical thought about the doctrinal tenets of a religion is discouraged in many traditional cultures. Indeed, it is often considered indecent or sinful, and is sometimes punished by social exclusion or worse.

Even the relatively rare adult converts to an established religion are usually not attracted to it because they have become convinced by the arguments of natural theology that its core beliefs are true. There are other aspects of religions or sects that may be appealing to outsiders, such as ravishing rituals, the sense of belonging provided by close-knit communities, the elevated feeling of human dignity conferred on the believers by the conviction that they belong to the elect, the appeal of high-minded ethical ideals, financial and social interests, or the longing for a life after death, which only the faithful will enjoy. Adults may also convert because they want to marry

---

[1] In a loose sense, 'theology' is also used for the empirical sciences of religion, such as the sociology and the history of religions, and for strictly historical interpretations of religious texts. But this loose sense is not relevant to my argument.

a believing partner. In some cases, converts have had exceptional experiences, which they interpret within the framework of an established religion.

In the great monotheistic religions and in Hinduism, the doctrinal contents of faith are contained in and derived from a corpus of ancient texts, which are often considered eternally valid divine revelations. It is to such a revelation, and, indeed, to revealed theology, that the faithful and their pastors or priests are ultimately referred whenever they seek enlightenment and stand in need of reasons for justifying their religious beliefs or practices. Yet, these revelations and the numerous problems that they raise provide the believer with sufficient reasons to engage in natural theology and critical philosophy of religion.

In the present chapter I shall argue briefly that natural or rational theology is indispensable for the conscientious religious believer. It may be that believers have come to endorse a religious creed because they were raised within a religious community, or in a quest for personal significance, and not because they know of good arguments or evidence for its truth. Historically speaking, belief in a revelation may precede rational deliberation. Yet, rational or natural theology has an epistemological priority over revelation in the context of justification. What I mean by this is that, at least in our modern, science-informed culture, merely referring to the text of a revelation and to revealed theology as a justification for what one believes will never be sufficient. If one aims at being a rational or reasonable person in endorsing a creed as true, one will also need the arguments of natural theology in order to justify one's reliance on a specific religious revelation in the first place.

1.1 ISSUES OF TRUTH AND INTERPRETATION

Somewhat schematically, we might distinguish between, on the one hand, reasons for engaging in natural theology that are put forward by authors of holy or revealed texts, and, on the other hand, reasons resulting from problems about these texts. An example of the first kind of reason is contained in Paul's letter to the Romans, I.18–20:

> For the wrath of God is revealed from heaven against all ungodliness and wickedness of men who by their wickedness suppress the truth. For what can be known about God is plain to them, because God has shown it to them. Ever since the creation of the world his invisible nature, namely, his eternal power and deity, has been clearly perceived in the things that have been made. So they are without excuse...[2]

According to the standard interpretation of this passage, 'what can be known' refers to what one can know about God on the basis of our empirical knowledge of the world, and not to God's special revelation to Israel and in Christ. Clearly, the idea is that God is not a despotic autocrat, who punishes people merely because they do not believe in him, but that he has a righteous nature, so that his wrath must be deserved. Allegedly it is deserved by unbelievers who have not received some direct propositional revelation from God, if at least the whole world is an unmistakable sign or a *revelatio generalis* of his existence, so that the unbeliever should have known better.

---

[2] All quotes from the Bible are taken from the Revised Standard Version, as published in The New Oxford Annotated Bible.

However, since Paul is somewhat over-optimistic about the clarity with which the world testifies to the existence of God – if that were really clear, nearly all polytheistic Hindus, Buddhists, and atheists would immediately convert to monotheism – at least some clever monotheists ought to engage in natural theology, and produce convincing evidence or arguments for the truth of their monotheism on the basis of empirical and public phenomena. Here is a clear case, then, in which the text of a revelation incites believers to engage in natural theology, that is, in apologetic philosophy of religion.

By far the greatest number of reasons for practising natural theology is of the second kind, however, and in this and the next section I shall discuss briefly six types of arguments for doing so (A–F). A first reason (A) is that many contradictions are discovered within allegedly revealed texts, and that it is not always easy to explain them away by the accommodating interpretations of revealed theology.

For example, there is at first sight an embarrassing contradiction within the New Testament concerning what has been regarded traditionally as the great central fact of Christianity: the resurrection of Jesus. In his first letter to the Corinthians (15.35–50), probably written between 53–57 CE, Paul seems to deny that Jesus was resurrected with his earthly or physical body, arguing that he was raised with a new, spiritual and heavenly body (*sooma pneumatikon*), since 'flesh and blood cannot inherit the kingdom of God'. But the author(s) of the gospel according to Mark, who possibly wrote around 70 CE, and the authors of the later gospels that were incorporated into the New Testament, tell the story of the empty grave. This latter account seems to contradict the older view of Paul, since it implies that Christ was resurrected with his earthly physique. The unknown authors of the four canonical gospels, which were later attributed to Mark, Matthew, Luke, and John in order to give them more authority, wrote their texts thirty-five to sixty-five years after Jesus' death.[3] They had not known Jesus and probably had never been in Jerusalem. Did they invent the story of the empty tomb in order to assimilate Jesus to great heroes of the Greco-Roman world within which they were living, such as Hercules, Romulus, and Aeneas? There still is a lively debate among experts on this issue.[4]

Another important contradiction within the New Testament is concerned with the attitude Christians should adopt in order to have a right standing before God. Paul argued that keeping the Jewish law can have no role whatsoever in salvation (even though, of course, people should obey the law anyway). For if people could be justified before God by doing what God prescribed in the law, there would have been no reason to crucify Jesus as a sacrifice for the sins of humans. Paul concluded that the only way to be justified is by having faith in our atonement by the crucifixion and resurrection of Jesus.[5] However, the author(s) of the gospel according to Matthew, writing some

---

[3] Whereas Jesus' disciples were illiterate peasants from Galilee, who spoke Aramaic, the authors of the gospels attributed to disciples of Jesus were well-educated Greek-speaking Christians who very probably lived outside Palestine. Cf. Ehrman (2009), p. 106.

[4] Cf. for a recent overview of the vast literature: Allison (2005), Chapter 6. According to Allison, who is a believing Christian and who declares that he 'should very much like to believe in the literal resurrection of Jesus' (p. 214), there is no contradiction, since 'Paul believed in "some sort of continuity between the present physical body and the totally transformed resurrection body – in spite of all discontinuity"' (p. 314). But the textual evidence Allison adduces for this view, such as 1 *Cor.* 6.12–20, is insufficient. For a more ample discussion of this issue, see Chapter 10.3–5, below.

[5] Cf. *Romans* (3.21ff.), and *Galatians* (2.16; 3.10–14; 3.24–25; 5.1–6; 6.12–15).

thirty years after Paul, clearly disagreed with him on this crucial issue. In *Matthew* we read that followers of Jesus who do not keep the law, and, indeed, who do not keep it better than most religious Jews, will never attain salvation: 'For I tell you, unless your righteousness exceeds that of the scribes and Pharisees, you will never enter the kingdom of heaven' (*Matthew* 5.17–20).[6]

The Koran also contains numerous contradictions, which are not easily resolved by the interpretative methods of revealed theology. One of them concerns the attitude that Muslims should adopt with regard to those who do not believe in the god of Islam, such as polytheists. According to a first text, usually called 'the sword verse', Muslims must kill unbelievers unless they convert:

Then, when the sacred months are drawn away, slay the polytheists wherever you find them, and take them, and confine them, and lie in wait for them at every place of ambush. But if they repent, and perform the prayer, and pay the alms, then let them go their way; God is all-forgiving, all-compassionate (Q9:5).

According to the 'tribute verse', however, unbelievers need not be fought or killed if they do not convert, as long as they pay some kind of tax and endure some kind of humiliation:

Fight those who believe not in God and the Last Day and do not forbid what God and His Messenger have forbidden – such men as practise not the religion of truth, being of those who have been given the Book – until they pay the tribute out of hand and have been humbled (Q9:29).

It may be that this second text does not contradict the first, because *lex specialis derogat lege generali*, and it is often interpreted as saying that Jews and Christians, who have been given the Book, need not convert to Islam if they pay tribute and are humbled. But this solution is not available for a third text, which seems to preach unconditional tolerance with regard to unbelievers:

No compulsion is there in religion. Rectitude has become clear from error (Q2:256).

Whereas this last text was an embarrassment for many Muslims in earlier ages, it is stressed much by those modern Muslims who want to avoid the disgrace that Islam is a religion propagated by the sword or by more technologically advanced forms of violence. But it is not easy to interpret away the contradiction between the third text and the other, more violent ones.

Of two contradictory propositions, at least one must be false, and if both propositions are part of a monotheistic divine revelation, this raises the question of how an omniscient veracious god can reveal to us something that is not true. There seem to be only three possibilities here, each of which casts serious doubts upon the supernatural inspiration of an allegedly revealed text. Either the relevant god really did not tell the truth at one moment, or the receiver of the revelation misinterpreted it, or, finally, the supposed revelation was not a real revelation at all, but has some natural explanation. For example, modern psychiatric research has shown that some patients who suffer from temporal lobe epilepsy or schizophrenia have a tendency to be fanatically

---

[6] For a popular account of such contradictions in the New Testament, cf. Ehrman (2009), pp. 85–92 and *passim*.

religious, and hear voices and the like. In the ancient world, these diseases were generally interpreted as interventions by demons or gods. Should we not suppose that the very intelligent founders of religions, such as St Paul or Muhammad, were afflicted sometimes by such mental disorders, the symptoms of which they in all sincerity attributed to divine interventions?[7] Surely believers need arguments of natural theology in order to dispel these sceptical doubts, so that this discipline has an epistemological priority over revealed theology in the context of justification.

One might object to this conclusion that resolving the apparent contradictions between passages in a revelation is the proper task of revealed theology. Should not all contradictions be interpreted away on the assumption that the texts of the Bible or the Koran, though written down by humans, are ultimately inspired by an omniscient and veracious god? If, for example, the Bible is a communication from God to humankind, it might be no wonder that its texts require deep and perceptive reflection in order for us to understand what they mean.[8] But there are two reasons why this objection is not very convincing. One is that with regard to many contradictions it turns out to be extremely difficult to explain them away by a religious interpretation. Take, for example, the contradiction between Paul's letters to the Romans and to the Galatians on the one hand, and the gospel according to Matthew on the other hand concerning the proper attitude believers should adopt to be justified before God. It is difficult to imagine a clearer contradiction: either obedience to the law is necessary for our salvation (Matthew) or it is not (Paul), even though we should obey the law for other reasons. These contentions cannot both be true.

The second reason why the objection fails to be convincing is that the interpretative results of the so-called *historical-critical* method of interpretation, or *historical biblical criticism*, are much more credible than those of revealed theology. Whereas the Christian variety of the latter discipline assumes that God is the ultimate author of the Bible, so that all biblical passages are written, ultimately, by one and the same person, who is omniscient, veracious, and eternal, the former method does not use this assumption of a supernatural inspiration. The critical method focuses instead on the historical situation and possible intentions of the human authors.[9] It turns out that most contradictions can be understood very well by assuming that the texts concerned were written by different human authors who were inspired by different oral traditions in different parts of the Roman Empire, and who had different beliefs and objectives in writing their texts. But if this is so, is it not a very compelling reason to reject the presupposition of Christian revealed theology that God is the ultimate author of the Bible (*mutatis mutandis* for other revelations)? Believers need to practise natural theology in order to refute this plausible conclusion.

---

[7] This was suggested by a number of nineteenth-century authors, such as Nietzsche (1881, §68). For modern research on temporal lobe epilepsy and extreme forms of religiosity, see: Carrazana et al. (1999), Dewhurst and Beard (1970), and Landsborough (1987). For the interpretation of epileptic attacks in traditional cultures, see: Jilek-Aall (1999), and Wohlers (1999).

[8] Cf. Plantinga (2000), pp. 381–90.

[9] Plantinga (2000), Chapter 12, nicely spells out the philosophical differences between these two approaches, and distinguishes three varieties of historical biblical criticism. But since he does not compare in detail the results of revealed theology with those of historical bible scholarship, his argument that the latter cannot be a defeater of Christian belief is unconvincing.

A second type of argument (B) for engaging in natural or rational theology is provided by empirical discoveries and advances in science and scholarship over the ages, which have shown that many passages in revelations are not true, at least if taken in their traditional interpretations.[10] Indeed, a book such as the Bible is so full of incredible stories that quite often no sophisticated investigations are needed in order to establish that it contains falsehoods, such as the claim that Adam, Seth, Enosh, and Kenan all lived for more than nine hundred years (*Genesis* 5.5–14). Should we reinterpret these passages in order to make them compatible with ordinary experience or with the results of modern science and other empirical data, which we now have?

Again, in such cases of conflict there seem to be three possibilities only, and each is equally problematic for a religious believer. We may map these possibilities as the horns of two interlocked dilemmas, the first of which is as follows. Either (a) we stick to an originalist and strictly historical interpretation of the text, or (b) we develop a modernizing interpretation, which makes it compatible with contemporary science and scholarship.

In favour of the first option (a), traditional theologians put forward the argument from divine authority. If an old holy text is revealed by God, we should understand this text as it was originally intended, since the authority of God is absolute, and humans have no right to change the meaning of a revelation on their own account. But this first option immediately triggers a second dilemma, given the incompatibility between the holy text as originally intended and results of modern science and scholarship. Should we (c) endorse the former or (d) endorse the latter?

Preference for a holy text given its incompatibility with scientific results (c) was well expressed in a book on astronomy published in 1873 at the publishing house of the Lutheran Synod of Missouri, in which the author squarely rejects all the astronomical discoveries made in modern times that conflict with biblical texts:

Let no one understand me as inquiring first where truth is to be found – in the Bible or with the astronomers. No; I know that beforehand – that my God never lies, never makes a mistake; out of his mouth comes only truth, when he speaks of the structure of the universe, of the earth, sun, moon, and stars ...[11]

Although some contemporary creationists prefer this option as well, it is unattractive in view of the high reliability and technological fruitfulness of many scientific procedures and results.

Since few religious believers who endorse a revelation will be able to accept horn (d), that the relevant passage in the revelation is simply false, on the grounds I mentioned above, most modern believers resort to horn (b) of the first dilemma. Either they embrace a doctrine of 'the living scripture', which claims that the meaning of a revelation changes over time, or they hold that the true meaning gradually dawns upon humanity in the course of history. In Islam, we find a well-funded industry of reading the discoveries of modern science back into the Koran. For example, the

---

[10] A classic overview is White (1896). Although more recent historical research has corrected White's results at many points and has criticized his warfare-metaphor, White's survey is still of great value. For criticisms of White, see, for example, Lindberg and Numbers (1986).

[11] *Astronomische Unterredung zwischen einem Liebhaber der Astronomie und mehreren berühmten Astronomern der Neuzeit*, by 'J.C.W.L.' (St. Louis, 1873), quoted by White (1896), Vol. I, p. 151.

verse in which God says of heaven 'We extend it wide' (Q51: 47), is now translated as 'We are expanding it', so that it refers to the modern cosmological discovery of an expanding universe.[12] In Hinduism, similar attempts are made to substantiate the claim that quite some modern science can be found in the Vedas.

Yet from a religious point of view this second horn (b) of the first dilemma is not very attractive either. It implies that the authority of science and historical scholarship has precedence over the authority of a divine revelation, and that the external advances in secular knowledge set the agenda for the religious reinterpretation of that revelation. Moreover, whereas there is a fairly reliable methodology for the historical or originalist interpretation of texts, there is no well-established method for modernizing interpretations, which is agreed upon by all factions within a religion. As a consequence, it seems that, religiously speaking, nearly 'anything goes' if one accepts the doctrine of the living scripture or another principle that justifies modernizing interpretations, as is shown by the many irresolvable disagreements between such interpretations. Finally, we may wonder why theologians have not proposed the interpretations of their holy texts, which allegedly contain modern scientific insights, before scientists discovered these insights. If these interpretations are the correct ones and divinely inspired, should theologians not have developed them independently from science on the basis of the holy texts, their theological background knowledge, and promptings by the Holy Spirit?

Summarizing these difficulties, we may say that the religious believer who relies on a revelation is faced with the following trilemma of options in the face of scientific and scholarly progress. Either (c) reject modern scientific and scholarly results if they contradict the revelation in its originalist interpretation, or (d) accept that this revelation contains falsehoods, or, finally, (b) accept that the progress of science and scholarship sets the agenda for reinterpreting the revelation, so that the authority of science overrules religious authority.

One might try to escape between the horns of this trilemma by arguing that a revelation such as the Christian one, the Koran, or the Vedic scriptures consists of two parts, one part containing the essential and eternal truths of Christianity or Islam or Hinduism, and another part containing the *Weltanschauung* of an ancient culture. One might argue further that God (or the gods) had good grounds to speak to the original receivers of the revelation in the vocabulary of their world-view, which is now outdated, and to give them a 'culture-relative revelation' instead of an absolute one, because they simply would not understand it otherwise. This is one of the solutions that Richard Swinburne offers in his book *Revelation*, following to some extent the tradition of so-called doctrines of accommodation.[13] As he claims:

---

[12] My examples of the Koran in this section are borrowed from Cook (2000).
[13] Richard Swinburne (RMA), pp. 75–84. According to traditional doctrines of accommodation, biblical passages such as *Joshua* 10.12–14, where it is said that God made the Sun and the Moon stand still in order to enable Joshua to win a battle, are written in a language 'accommodated' to the understanding of the common man, so that they are not incompatible with the Copernican view of the diurnal rotation of the Earth, for example. Such a doctrine of accommodation was proposed already by Copernicus's pupil Georg Joachim Rheticus (1514–74), without whose assistance Copernicus's book *De revolutionibus orbium coelestium Libri VI* would not have been published during his lifetime (in 1543, the year of Copernicus's death), and a doctrine of accommodation was also endorsed by Copernicans such as Giordano Bruno (1548–1600), Johannes Kepler (1571–1630),

False scientific presuppositions would make no difference to the religious content of the message, that is, to the kind of life and worship which it sought to encourage. A mistaken view of what God had created, or where Heaven was, would not affect the praiseworthiness of God, or the desirability of Heaven. It therefore follows [...] that, so long as context allows a clear distinction between statement and presupposition, false scientific presuppositions would not render the revelation false.[14]

However, this view is confronted by difficulties of its own. First of all, the kind of life recommended to the earliest followers of Christ, for example, depended partly on the factual presupposition that the utopian kingdom of God on Earth would arrive soon during the time of their life, so that no investments in a long-term future were needed.[15] But the kingdom of God did not arrive, which created the so-called problem of the *Postponed Parousia*.[16] This shows that scientific and other factual presuppositions are not irrelevant to what is considered the core Christian message. A scrupulous history of what Christians throughout the ages regarded as essential revelations in the Bible will reveal that this essential content has shifted over time and, at least in developed countries, has gradually dwindled during the last centuries. For example, a good many contemporary Christians do not believe in an afterlife any more, because they cannot reconcile the idea of spiritual survival after bodily death with the results of modern brain research, which show in ever greater detail the extent to which our mental life depends upon specific bodily processes.

Another difficulty for the doctrine of accommodation consists in a new dilemma. Were the false scientific presuppositions of the outdated *Weltanschauung* in terms of which the revelation is formulated, part of what God communicated to the original receivers of the revelation or not? In the first case, the omniscient god deceived his audience in a somewhat patronizing manner. Instead of revealing to early believers the true view of the universe, with its trillions of galaxies and super-massive black holes, he communicated to them a false but consolingly comfortable picture of the world, in which humans play a central role. How can one trust such a patronizing deceiver with regard to the other things he is saying, which constitute the 'religious content' of the message? In the second case, much of what is claimed to be revealed is in fact not revealed by God. Those who regarded themselves as 'eyewitnesses and ministers of the word', the word being God's revelation in Christ, are in fact telling us many things from hearsay.[17] But how can one trust witnesses who have from hearsay what they claim to have witnessed? In both cases, our confidence in the text of a revelation should be seriously undermined, and we have to engage in natural theology in order to discover whether, how, and where we can restore it.

---

and Galileo Galilei (1564–1642). Cf. for Rheticus' tract on the Holy Scripture and the motion of the Earth: Hooykaas (1984).

[14] Swinburne (RMA), p. 77.

[15] Cf., for example, *Mark* 9.1: 'Truly, I say to you, there are some standing here who will not taste death before they see that the kingdom of God has come with power'; *Mark* 13.30: 'Truly, I say to you, this generation will not pass away before all these things take place'; *Matthew* 24.44: 'Therefore you also must be ready; for the Son of man is coming at an hour you do not expect'.

[16] Cf. Sanders (1993), Chapter 11.

[17] Cf. *Luke* I.2–3: 'those who from the beginning were eyewitnesses and ministers of the word'. Of course, historical Bible scholarship has established that 'the gospels as we have them were not written by eyewitnesses on the basis of first-hand knowledge of Jesus'. Cf. Sanders (1993), p. 63.

1.2 FOUR FURTHER REASONS

This brings me to a third type of argument (C) for practising natural theology or critical philosophy of religion. Historical research done during the nineteenth and twentieth centuries has demonstrated to what extent the content of revelations is influenced by earlier sources and cultures, which were not at all regarded as divinely inspired by the alleged receivers of these revelations. As Andrew D. White already wrote in his classic *A History of the Warfare of Science with Theology in Christendom* of 1896:

> [i]t has now become perfectly clear that from the same sources which inspired the accounts of the creation of the universe among the Chaldeo-Babylonian, the Assyrian, the Phoenician, and other ancient civilizations came the ideas which hold so prominent a place in the sacred books of the Hebrews.[18]

The same is true of other ingredients of Christianity, such as the belief that a god dies in order to save his people, and it also holds for the Koran. Again, the question is how a believer can reliably distinguish between those contents of revelations that were really inspired by God and contents that the authors had from hearsay, or which are false. Since revelations do not provide us with the intellectual instruments for doing so, the believer will have to resort to cultural history, historical bible criticism, and natural theology.

When we radicalise this third reason for engaging in natural theology, we obtain a fourth one (D). If by historical research large parts of alleged revelations can be traced back to older sources which are not considered divine revelations by present-day religious authorities, it will seem far-fetched to claim that a god or gods played a role in the genesis of specific texts at all. Is it not much more plausible to assume that full explanations of the origin of all alleged revelations will merely refer to human and all-too-human causal factors? These texts never contain pieces of knowledge or moral insights that humanity did not already possess before the alleged revelatory communication by a god, or which humans could not have acquired without divine assistance. Hence, there is no good reason to postulate a god in order to explain the origins of these texts.

But if there is no convincing argument from the text of an alleged revelation for positing a god as one of its sources, the existence of a god should be argued for independently of revelations. Indeed, since a revelation is by definition a direct communication from a god to a human being, the claim that a specific text contains a revelation presupposes that this god exists. Hence, one should first establish the existence of that god by arguments of natural theology, before one can believe in a specific revelation and engage in revealed theology. Or at least one should argue that the hypothesis of a divine origin makes the existence of this text more likely than it is made by the rival hypothesis of a purely secular origin.

A fifth reason (E) for engaging in natural theology is the following. As we have seen above, alleged revelations contain factual falsehoods, which cannot always be removed

---

[18] White (1896), Vol. I, p. 2.

by accommodating interpretations. What holds for the factual contents of revelations is also true for their moral doctrines. On the one hand, holy texts contain moral norms, officially revealed by a god, which many of us now find unacceptable and even wicked, or at least barbaric. We read for example in *Deuteronomy* 21.18–21:

If a man has a stubborn and rebellious son, who will not obey the voice of his father or the voice of his mother, and, though they chastise him, will not give heed to them, then his father and his mother shall take hold of him and bring him out to the elders of his city at the gate of the place where he lives, and they shall say to the elders of his city, "This our son is stubborn and rebellious, he will not obey our voice; he is a glutton and a drunkard." Then all the men of the city shall stone him to death with stones; so you shall purge the evil from your midst; and all Israel shall hear, and fear.

Stoning to death is a punishment that was popular at the time *Deuteronomy* was written, and it is prescribed for many other sins, such as not being a virgin when you marry, or committing adultery (22.20–24).

Christians will perhaps object that I quote from the Old Testament and not from the New. But in the New Testament, we also encounter many problematic moral evaluations, such as Paul's view that homosexuality is sinful and a punishment by God for unbelief. As Paul says in his letter to the Romans 1.27: 'and the men [...] gave up natural relations with women and were consumed with passion for one another, men committing shameless acts with men and receiving in their own persons the due penalty for their error'. If Christian believers are not convinced by this passage, because they endorse the biblical pronouncements on homosexuality, other passages from the New Testament may be more compelling. According to the Revelation to John, those who worship 'the beast' will 'drink the wine of God's wrath, poured unmixed into the cup of his anger' and will be 'tormented with fire and sulphur [...] for ever and ever'.[19] But is it morally acceptable that those who sinned intermittently by worshipping another god, or committed evil deeds, will burn eternally in the Lake of Fire? Should punishments not be proportional to the crimes committed both in severity and in duration?

On the other hand, values that many of us now hold in high esteem are not found in most revelations. The Bible does not recommend the value of intellectual curiosity, for example, which is essential to scientific and scholarly research, or the value of religious tolerance, or equal rights for men and women. However, if revelations are not a reliable guide for ethics, religious believers have to resort to natural theology in order to find out to what extent their revelation can be trusted in this respect, and, indeed, whether it can be trusted at all.

Finally, a sixth reason (F) for engaging in natural theology derives from what is usually called the problem of the diversity of religions. Believers who ground their religious beliefs on a revelation may be subjectively justified in holding these beliefs as long as they have never heard about other revelations that are incompatible with their own. However, as soon as they become aware of such rival revelations, which is likely in our globalized world, these incompatible revelations are potential defeaters of the religious beliefs they endorse. Why should one prefer, for example, the Christian belief that Jesus is God incarnate to the Muslim belief that God was never humanly

---

[19] *Revelation* 14.9–11, and *passim*.

incarnated because there is only one god? And why should one prefer the monotheism of the Abrahamic religions to Hindu polytheism (cf. Chapter 4.2–4, below)?

With a small leap of the imagination, believers will fancy that if they had been born into another religious tradition, they would have been equally convinced of a religious creed incompatible with the one they happen to endorse now. However, if the religion to which one adheres is selected by the accident of birth, and if one also has the conviction that salvation and eternal life depend upon accepting the creed of the only true religion, one has a powerful motive for engaging in a comparative research of religions. This means engaging in natural theology and attempting to show that one religious revelation is more likely to be true than the competing revelations, given the available evidence.

1.3 A PYRRHONIAN CRISIS?

The problem of the diversity of religions can be further elucidated with reference to the idea of a Pyrrhonian crisis. One might say that an authoritative revelation such as the New Testament, the Koran, or the Book of Mormon aims at functioning as a criterion for religious truth.[20] What is written in such a holy text, if interpreted adequately, must be true, so the traditional believer argues, because it is revealed, or at least inspired, by an omniscient and veracious god.

If revelations function as criteria of religious truth in this manner, incompatibilities between the texts of different revelations can be seen as raising a dispute about the correct criterion of religious truth. But according to the Greek sceptic Sextus Empiricus, a dispute about the criterion of truth can never be resolved, because in order to argue for one's criterion of truth one has to use premises of which one claims that they are true. Hence, one has to presuppose one's criterion of truth in order to substantiate the claim that it is the correct criterion.[21] In other words, as soon as a dispute about the criterion of truth arises, it is likely to degenerate into a mutual bombardment with circular arguments. This is called a 'Pyrrhonian crisis' in honour of Pyrrho of Elis (around 365–275 BCE), the originator of a school of Greek sceptics.

A similar Pyrrhonian crisis may emerge *within* one of the revealed religions, if a dispute arises about the criterion for the religious truth of interpretations of the revelation. This happened within Christianity, for example, when Luther claimed that the individual conscience of a believer could function as the (procedural) criterion for establishing the true interpretation, whereas the Church of Rome held that the authority of Pope and Councils had to function as the procedural criterion.[22] The Vatican could easily point out which disastrous consequences the Lutheran criterion would have, for it would lead to an ever-increasing fragmentation of Christianity. But arguments to the effect that its own institutional criterion of truth is correct turned out to be either circular or very weak. Because of this dispute on the criterion for establishing the true interpretation of Scripture and other factors, the schism of the Reformation within Christianity has not been overcome to this very day.

---

[20] Cf., for example, Pope John Paul II (1998), §23.
[21] Sextus Empiricus, *Outlines of Pyrrhonism* (1933), II, Chapter iv.
[22] Cf. Popkin (1979), Chapter 1.

However, there is in principle a non-circular way of resolving a dispute about a criterion of truth for a given limited domain, such as religion. One might attempt to argue that quite probably a specific revelation is the true one, whereas conflicting revelations are less likely to be genuine revelations, on the basis of premises that all parties, including the non-believers, may accept as true. In order to do so, one should first show or make probable on the basis of such premises that the god of the relevant revelation exists. This is the very project of natural theology.

1.4 NATURAL THEOLOGY DEFINED

The six reasons (A–F) for engaging in natural theology summarized in the preceding two sections show two things, which I call the indispensability and the epistemological priority of natural or rational theology. Rational theology (*theologia rationalis*) is commonly defined in opposition to theological articulations of revealed religion (*theologia revelata*) as the attempt to develop a coherent conception of one god, or of more than one deity, and to produce arguments, a priori or empirical, which establish or make probable the existence of this one god or of these deities. Rational theology is also called natural theology, because the premises of its arguments do not rely on supernatural sources such as revelations, although some authors restrict the application of the label 'natural theology' to those forms of rational theology that derive their arguments solely from more or less detailed knowledge of nature.[23] I shall ignore such restrictions and use the two labels as synonyms. In contrast to revealed theology, then, we may define rational or natural theology as the attempt to argue for the truth of a specific religious view on the basis of premises that non-believers will be able to endorse, that is, without appealing to the alleged authority of a revelation.

We may conclude on the six grounds which I have given, that rational or natural theology is indispensable to believers, because revelations are in themselves insufficient to justify or warrant religious belief in view of the many difficulties with which they are beset. Although the alleged revelations have a crucial role in the so-called context of discovery, since without them the adherents of revealed religions would never have come across the contents of their creeds, their value in the context of justification is limited indeed. For example, particular religions are confronted by the problem of religious diversity, which is liable to function as a defeater of each revealed religion, unless one can show that one's own alleged revelation is reliable and real, whereas conflicting alleged revelations are not. The traditional religious standards for validating a revelation, such as the internal criteria of its truth, the sublime nature of its moral doctrine and the fact that it contains verified predictions, or the external criteria of miracles and the noble character of a prophet, are invoked by most major religions in favour of their own revelation, so that they cannot settle the problem of the diversity of religions without further rational argument.

What is more, if a religion such as Roman Catholicism regards its own revelation as the only true and final one, so that it rejects the alleged revelations of Muslims or

---

[23] Cf. Kant, *Kritik der reinen Vernunft* (KdrV), B 659ff.

Mormons as pseudo revelations, it has to conclude that, in general, the subjective experience of receiving a revelation is a highly unreliable source of knowledge, because the Catholic doctrine of a revelation-monopoly implies that the vast majority of alleged religious revelations are not genuine.[24] But how can one convincingly argue that one's own alleged revelation can be trusted if one also claims implicitly that receiving revelations is not a reliable epistemic source or method? We can only conclude that, if the problem of the diversity of religions can be solved at all for believers, this must be done by natural theology, which has to develop arguments that are logically independent of revelations, and that plead in favour of one religious view and against the others.[25] This is why in the context of justification, natural theology has epistemological priority over revealed theology.

1.5 FOUR CONDITIONS

In this first chapter, I have focused on reasons for engaging in natural theology that should be endorsed by believers in revealed religions. But it can also be rational for unbelievers to spend some time on the philosophy of religion if the following four conditions hold.[26] First, we should not (yet) know for certain whether one god exists, more than one god exists, or no gods exist, and which religious creed is true, if any. If we already knew for certain that universal atheism or some specific religious doctrine were true, it would be irrational to spend more time on research, because it would not make a difference. But in fact most of us think that we do not know this for certain, and intelligent people disagree on the issue, so that this first condition is satisfied.

Second, acquiring true beliefs on the religious issue should have some importance for us. This condition seems to be satisfied as well, for a number of reasons. In general, having true beliefs is necessary for achieving our purposes, and all religions claim that some human purposes, such as enjoying deep contentment during this life or earning an eternal blissful afterlife, can only be achieved if their creed is true and if one follows the way of that religion. If no religious creed is true, however, we should be ascetic with regard to these lofty ambitions and learn to accept that the ideal of achieving them is a chimera. Clearly, then, it makes a difference to our way of life whether we are religious or not. Moreover, if there is a god on whom we depend for our happiness, this may create moral obligations with regard to this god, such as the obligation to worship her

---

[24] Cf. the *Declaratio Dominus Iesus*, published by the Congregatio pro Doctrina Fidei of the Vatican in 2000.
[25] Religious believers may attempt to solve the problem by arguing that all religions of the world are diverse ways of conceptualizing 'the transcendent', and that they are all valid or true to some extent, because the 'noumenal' transcendent may appear to humans in many different ways. But of course, this pluralist solution, defended by John Hick and others, is nothing but a new religion, which is rejected by many of the established religions, so that it does not really solve the problem of religious diversity.
[26] More precisely, it would be rational to spend time on it in proportion to the degree in which these four conditions are satisfied. Cf. Swinburne (FR), Chapter 3, for an extensive discussion of this issue.

or him, whereas prayer and worship are a waste of time if there is no god. Finally, having true beliefs may be important to us in itself, apart from its instrumental value for achieving other purposes. '[I]f history and physics are of importance for this reason, religious knowledge is obviously of far greater importance', Richard Swinburne rightly argues, since 'a true belief here, whether theistic or atheistic, is of enormous importance for our whole world-view'.[27] Summarizing these reasons, we may conclude that having true beliefs about the religious issue is important for most of us, so that the second condition is satisfied as well.

Even so, it would be irrational to engage in the philosophy of religion if it were unlikely that we can find out by research or critical reflection whether any religious view is true. Although most great religious traditions and most atheists have always held that rational enquiry can establish the (probable) truth or falsity of religious propositions about the existence of God or gods, one may have serious doubts at this point for three kinds of reasons.

Some theologians have put forward a religious argument to the effect that religious belief is altogether beyond the jurisdiction of rational arguments, saying, for example, that faith radically transcends our rational faculties, because the latter are too limited for this sublime task, or because they are contaminated by the Fall of man. If this is the case, so these theologians argue *à la* Kierkegaard, we should reject rational considerations in the religious domain and engage in a religion by a blind leap of faith. But in which direction should we leap blindly, and which religion should we embrace? We rarely think it wise to engage blindly in some course of life. The theological reasons for doing so in religious matters presuppose that we already accept a specific irrational version of a given religion, whereas the issue was whether we should endorse one specific religious creed in the first place.

The second kind of reason for thinking that it is futile to engage in the philosophy of religion is epistemological. Philosophers of the Enlightenment, such as Hume and Kant, have argued that no rational investigation into the truth of religious doctrines can yield knowledge, because of the nature of the inductive method or in view of the essential limitations of our epistemic faculties. Although these arguments of Hume and Kant were very influential in Western culture, often pushing religious believers into irrationalism concerning religion, I shall argue briefly in the next chapter that they are based on an outdated philosophy of science.[28]

Finally, a third kind of reason for regarding the philosophy of religion as pointless is the opinion that rational debates about the truth of religious doctrines have gone on for many centuries without leading to any progress or consensus. It seems sensible to conclude by a pessimistic induction that engaging in the philosophy of religion will not contribute much to forming our opinion on religions, and that whether one adheres to a specific religion or not cannot be more than a matter of irrational decision or unjustifiable habit. As Daniel Dennett says: 'I decided some time ago that diminishing returns had set in on the arguments about God's existence, and I doubt that any break-throughs are in the offing, from either side'.[29] But again, it may be that these

---

[27] Swinburne (FRa), p. 80. In (FRb), p. 85, Swinburne substituted 'true religious belief' for 'religious knowledge'.
[28] Cf. also Swinburne (FRb), pp. 103–6.   [29] Dennett (2006), p. 27.

debates in the past were not fruitful because they were not informed by a correct view of the methods to be used. Are breakthroughs in the sciences not often achieved because of new methodological insights after many ages of fruitless discussions? If we provisionally accept this argument, which I shall develop in the next chapter, we may conclude that the third condition for it being rational to spend some time on the philosophy of religion is also satisfied, to wit, that this investment may yield convincing grounds for endorsing one religious creed rather than another, or for becoming a universal atheist, or for embracing agnosticism.

About the fourth and final condition that has to be satisfied I can be brief, since each individual reader should judge for her- or himself to what extent it obtains. This condition is that, given one's overall aims and capacities, and the limited time of one's life, engaging in the philosophy of religion is preferable to doing other things at that moment, such as earning money, playing soccer, or going to the movies. In our complex culture, there is a fine-grained division of labour, which has progressed very far in the intellectual domain. We rely on the authority of experts in many areas of knowledge, and we have good reasons for doing so if we can trust that the methods used in those areas are reliable. It is precisely because we cannot have this type of trust in the area of revealed religion that believers need to engage in natural theology.

But even here there is a division of intellectual labour. Objectively speaking, natural theology has an epistemological priority over revealed theology in the context of justification. But this does not imply that, subjectively speaking, it is rational for all believers and unbelievers to engage in the philosophy of religion to the same extent. Some specialists devote their entire life to this discipline, whereas non-specialists may read one book only. Since the first three conditions are satisfied at least to some extent, it seems to be rational for every educated person to spend some time on the philosophy of religion. Perhaps the reader will draw some inspiration from the following passage quoted from the works of René Descartes.

In the opening of his first *Metaphysical Meditation*, Descartes wrote:

Some years ago I was struck by the large number of falsehoods that I had accepted as true in my childhood, and by the highly doubtful nature of the whole edifice that I had subsequently based on them. I realized that it was necessary, once in the course of my life, to demolish everything completely and start again right from the foundations if I wanted to establish anything at all in the sciences that was stable and likely to last. But the task looked an enormous one, and I began to wait until I should reach a mature enough age to ensure that no subsequent time of life would be more suitable for tackling such inquiries. This led me to put the project off for so long that I would now be to blame if by pondering over it any further I wasted the time still left for carrying it out. So today I have expressly rid my mind of all worries and arranged for myself a clear stretch of free time. I am here quite alone, and at last I will devote myself sincerely and without reservation to the general demolition of my opinions.[30]

Descartes' project was a Herculean one, and present-day philosophers of science agree that it was misconceived. If we want to establish something in the sciences that is stable and likely to last, there is no good reason whatsoever to begin by demolishing

---

[30] Descartes, AT IX, p. 17. Quoted from Cottingham et al. (1988), p. 17.

everything completely and start again from the foundations. This is because absolutely secure foundations of the kind Descartes was looking for do not exist, and because there is a division of labour in the sciences. But in the far more limited domain of religion, it is a good advice to suspend judgement provisionally once in the course of one's life, in order to enquire to what extent the beliefs one always held can survive critical scrutiny.

2

The Rise, Fall, and Resurrection of Natural Theology

Readers who are convinced by the arguments in the previous chapter and decide to spend some time on natural theology or on the critical philosophy of religion, are immediately confronted by an *embarras de choix*: where should they begin? In the sciences of nature, which are clearly progressive, one will always start by studying a recent textbook published by a prestigious press. But in philosophy it is less obvious that there is intellectual progress, and as a result much time in philosophical curricula is spent on authors of the distant past such as Plato, Aristotle, Augustine, Aquinas, Descartes, Spinoza, Hume, Kant, and Hegel. Why, then, should we focus on recent analytic philosophy of religion, and particularly on the works of Richard Swinburne, for example, instead of studying the celebrated five ways of Thomas Aquinas? Let me try to answer this question in the present chapter by giving a bird's-eye view of the history of Western rational or natural theology.

2.1 FROM XENOPHANES TO AQUINAS

Natural theology started in Europe long before Christianity appeared upon the scene, when Greek philosophers began to criticize and modify traditional polytheistic religions. Some of them, such as Xenophanes (*c.*570 – *c.*475 BC), rejected the anthropomorphism and immorality of the Greek gods as depicted by the poets, whereas others proposed physical explanations for phenomena that were earlier attributed to divine intervention or design. These two tendencies led to many conflicting natural theologies, the most influential of which were developed by Plato, Aristotle, and the Stoics, whereas the Epicureans serenely criticized as a debilitating illusion the idea that gods intervene in human affairs.

Plato remained a polytheist of sorts, defending traditional mythology and arguing in the *Timaeus* that a divine artificer had assembled the cosmos, but Aristotle may be considered a monotheist. In book 8 of his *Physics* and book 12 of his *Metaphysics*, he argued that there must be one First Mover, who causes all motions in the universe by being the universal object of desire, because an infinite regress in the causes of motion is impossible. Since this First Mover is a fully actualised being that enjoys perfect blessedness by eternally reflecting on itself, it is self-sufficient and would never create the world, which in fact exists eternally. The Stoics, finally, were primarily interested in mastering misfortune and in physical theory, but they also held that all processes in the universe are controlled by a divine mind, so that they endorsed a theory of absolute divine providence.

It was inevitable that when Christianity started its long march to supremacy within the Greco-Roman world, it had to develop its theology partly in terms of Greek philosophy, which it sought to overcome at the same time. Neglecting here the baffling complexities of this intellectual history, we might say that from the cultural catastrophe of the fall of the Western Roman Empire onwards until the Renaissance, that is, during the millennium of the Dark Ages, natural theology was slowly on the rise within Christianity, because most speculative talent in the Western world was absorbed by issues about the Christian god. Many scholars agree that natural theology reached its acme in the thirteenth century, when Thomas Aquinas adapted Aristotelian metaphysics to the ends of Christian theology, and, inspired by Arabic philosophers, used Aristotelian arguments in order to show that the Christian creator-god exists. Should we then turn to Thomas if we are interested in natural theology today?

There are three features of the Thomist marriage between Aristotle and Christianity by reference to which one might argue that natural theology as developed by Aquinas is superseded by later intellectual developments. First of all, Thomas' arguments depend substantially on the conceptual resources of Aristotelian science and metaphysics. But Aristotelian science and metaphysics were rejected during the Scientific Revolution, and no Thomist proof of the existence of God is sound as it stands according to modern standards. Anthony Kenny concluded his examination of Thomas' celebrated Five Ways by saying that they 'fail [...] principally because it is much more difficult than at first appears to separate them from their background in medieval cosmology'.[1]

Second, the Thomist arguments for the existence of God were informed by the ideal of demonstrative scientific knowledge as developed in Aristotle's *Posterior Analytics*. Although Thomas correctly supposed that arguments for the existence of God have to start from a premise concerning undeniable empirical facts, such as the fact that there is movement in the universe, or that there is efficient causality and contingency, he also assumed that ideally they should be demonstrative arguments, that is, deductively valid arguments, the premises of which are necessarily true. Unfortunately, however, none of Thomas' deductive arguments can be considered a sound demonstrative argument. Moreover, we now reject the Aristotelian conception of demonstrative science as an inadequate ideal for scientific knowledge.[2]

Let me illustrate this negative verdict on Thomist natural theology by the fourth of Thomas' five ways, as exposed in *Summa theologiae* Ia.2.3, which runs as follows:

---

[1] Kenny (1969), p. 3. However, Oderberg (2007) argues that some of Thomas' arguments are 'ripe for reinvestigation' (p. 349).

[2] Thomas does not restrict himself completely to demonstrative arguments, however. Citing Aristotle as having 'very well said' that 'it belongs to an educated man to seek such certitude in each thing as the nature of that thing allows', he acknowledges the need for probabilistic arguments, noting, however, that they do not yield 'knowledge'. Cf. Aristotle, *Nicomachean Ethics* (1934) I1, 1094b: 23–5; Thomas Aquinas, *Summa contra gentiles* (1975) I.3.1 (p. 63), I.9.2: 'Nevertheless, there are certain likely arguments that should be brought forth in order to make divine truth known' (p. 77), and I.9.3; and Kretzmann (1997), p. 52.

(1) We find in the world things that differ in their degree of *perfection*, being, for example, more or less good, noble, intelligent, hot, or rapid.
(2) But 'more' and 'less' are said of different things as they approach in different degrees something that is such to the maximum degree, to wit, the maximally good, noble, intelligent, hot, etc.
(3) The maximum in any genus is the cause of all things in that genus, as fire, which is the maximum heat, is the cause of all hot things in so far as they are hot.
(4) Therefore, there is something that is best, maximally noble, maximally intelligent, etc., which causes everything else, and this is God.

The argument of Thomas' fourth way has a strong neo-Platonist flavour and it is not sound for at least five reasons. It may be that the very idea of a maximum of something is contradictory, such as the notion of a maximum natural number. This implies, secondly, that we can very well speak of 'more' or 'less' without having the notion of a maximum, as indeed we can, so that premise (2) is false. But even if we cannot in certain cases, it does not follow that such a maximum should exist in reality, since we may have the notion of a maximum without there being anything in reality that corresponds to this notion. Premise (3), which aims at ruling out this possibility, is not true in general, because one cannot say that the less beautiful is always caused by the more beautiful – Rembrandt painted supremely beautiful pictures, whereas heat is not always caused by fire. And supposing that it did follow that a maximum exists in all cases, it is not at all clear that this maximum is God, for should we really suppose that God is maximally hot or rapid? If the argument were sound, and if we refused to accept that the Christian God incorporates all conceivable maxima, it seems that a proliferation of really existing maxima would result, such as the maximally wicked (the Devil), the maximally dirty (Dirt itself), and the maximal miser, which is surely a surprising version of polytheism.

A third and final feature of Thomas' natural theology, which makes it less attractive for modern readers, is that Thomas did not sufficiently face the problem of the diversity of religions, so that he never seriously considered the possibility that the Christian creed is false. Indeed, he subscribed to the traditional scholastic view that revealed Christianity or the *doctrina sacra* is simply true, if interpreted adequately, and that the task of natural theology is primarily that of 'making known [...] the truth that the Catholic faith professes, and of setting aside the errors that are opposed to it'.[3] In our modern globalized world, in which religious cultures easily clash because of the advanced means of telecommunication, massive migration, and rapid transportation, the problem of the diversity of religions is of primary importance. It should be addressed conscientiously by a natural theology that is up to its task. So let us leave Aquinas behind, and see whether the later history of Western natural theology offers more promising avenues.

2.2 FROM THE REFORMATION TO KANT

In spite of their deep divergences concerning natural theology, the scholastic philosophers of medieval Christianity had one conviction in common. Whereas natural

---

[3] Aquinas (1975), I.2.2 (p. 62).

theology could freely use the arguments developed by pagan philosophers such as Plato and Aristotle, they all held that the objectives of ancient philosophy had been met and transcended by the Christian revelation. As a consequence, philosophy, including natural theology, should and could merely play a secondary role as a supporter of revealed theology, that is, as an *ancilla theologiae*. Although most philosophers agreed that natural theology is indispensable, it could never take precedence over revealed religion, and in fact there was no intellectual space for a philosophy apart from Christianity. After the Renaissance, however, this scholastic consensus concerning the relation between revealed religion and natural theology was undermined by two cultural forces, which initially pulled in opposite directions.

On the one hand, the founding fathers of the Reformation such as Calvin and Luther proclaimed that Christians should rely on the Bible only, *sola scriptura*, arguing that Christianity had been contaminated by styles of thinking derived from the pagan philosophers of the Ancient world. They radicalized criticisms of Aristotelian arguments in natural theology put forward by Ockham, Petrarch, or Erasmus, and claimed that there is an incompatibility between scholastic philosophy and true Christianity. It was said, for example, that human reason had been corrupted by original sin, and that God transfers his teachings by direct inspiration to uneducated people who read the Bible with an open mind, rather than to sophisticated philosophers. In short, the Reformation heralded a downfall of natural theology after its continuous rise during the Middle Ages, although reformed writers such as Melanchton or Voetius felt obliged to develop their own type of scholasticism in order to rebut arguments against their dogmatic views.

On the other hand, the new discoveries made during the Scientific Revolution at first tended to increase the relative importance of natural theology. By contradicting the text of the Bible as it had been interpreted traditionally, scientific advances cast doubt on the reliability of the revelation. Furthermore, by comparing the text of the Vulgate with the Greek and Hebrew originals, philologists started to question the authority of some traditional dogmas, such as Mary's virginity. Should the philosopher not develop a rational reconstruction of religion, which would show how much of the Christian revelation was really tenable in the light of modern science and historical scholarship? Many philosophers of the Scientific Revolution held that the Bible was primarily intended for practical guidance of the uneducated, whereas religious truth had to be discovered by using human reason in the study of nature. During the seventeenth century, Spinoza went furthest in this direction, arguing in his *Theological–Political Treatise* that the Bible should be interpreted as a purely historical document, and developing in his *Ethics* a radically pan(en)theistic natural theology, which leaves no room for a transcendent god as distinct from the world. At the end of the eighteenth century, deistic authors such as Thomas Paine openly rejected the idea that the Bible is a revelation, claiming that God would never have revealed himself merely in one of the thousands of human languages. In *The Age of Reason* of 1794, Paine argued that the only real divine revelation is to be found in 'the book of nature', which is written in a universal language available to all mankind, and that the only true theology is a natural or rational theology of deism, developed by studying the natural world.[4]

---

[4] Cf. Paine (1987), pp. 420–1.

Although the Reformation and the Scientific Revolution initially had opposite impacts on the status of natural theology, the one heralding its final fall and the other raising it to supremacy over revealed religion, the latter effect was short-lived, at least among leading philosophers. The reason was that many different systems of natural theology or theological metaphysics were developed, whereas there seemed to be no reliable method for deciding between them. Put in a Kantian terminology, this deadlock meant that natural or rational theology had not succeeded in reaching the 'secure path of a scientific discipline', so that a critical investigation was called for in order to decide whether natural theology is possible at all.

David Hume published a negative verdict on this issue in 1748, which was informed by his analysis of causality and induction. Although we can conclude that an unobserved entity exists if it is a probable cause of effects that we experience, Hume argued that such an argument from effect to cause is justified only if it is based upon an inductively established causal regularity. For example, we legitimately conclude that a human being walked upon the beach if we see human footprints, because we have established by repeated experience the causal regularity that humans walking on sand cause human footprints. Similarly, if we see a beautiful ruin on an abandoned site, we may conclude that once there were humans who constructed it, because we have often perceived that humans construct buildings of stone, whereas no other animals that we know can do such things.

If, however, we conclude from certain aspects of the universe that it has been created by an omnipotent, omniscient, and perfectly good god, this conclusion can never be justified. In order to establish a causal regularity between A-s and B-s by induction, we must have repeated experiences of both A-s and B-s, but we lack any experience of God, and the alleged creation of the world supposedly was a unique event. For these two reasons, Hume argued, the traditional physico-teleological argument for the existence of God from features of the world is at best an argument by analogy, based upon our experience of human creative activities. As such, however, the argument is excessively weak, because the differences between humans and the putative incorporeal Christian god are much greater than the resemblances. Also, Christian philosophers ascribe properties to God as the alleged cause of the world that are not justified by our knowledge of the effect, since we should not suppose more in a cause than is necessary for explaining the effects, if the cause is known by the effect only. For example, since the world is a rather sorry place, we cannot conclude that its alleged Maker, if any, is perfectly good.[5]

Kant's verdict on the possibility of rational theology as a scientific discipline (*Wissenschaft*) was as negative as Hume's, although he relied on a conception of scientific method very different from Hume's empiricist doctrine. Like many scientist-

---

[5] Hume (1748), section xi. Of course, Hume had to be cautious in publishing this critique of Christian natural theology, and he dressed it up as a speech of Epicurus to the Athenians, invented by a friend, who talked about Jupiter instead of the Christian god. But it is interesting to note that the most fundamental point of criticism is put forward by Hume himself in the last paragraph of the section. Shortly after 1748, Hume started to work on the *Dialogues Concerning Natural Religion*, which were published only in 1779, after Hume's death in 1776.

philosophers in the eighteenth century, Kant was convinced that Newton's classical mechanics realized the ancient ideal of a demonstrative science, an ideal which Kant endorsed. This could not be the case, however, if the laws of mechanics were merely empirical laws, justified by induction on the basis of particular experiences, as Newton and Hume had held.[6] Kant concluded that some fundamental laws of science, including the propositions of Euclidean geometry, are necessarily true and a priori, that is, they can be known to be true independently of experience, whereas they are also *synthetic* in the sense that they provide information about reality instead of being merely conceptual truths.

In his *Critique of Pure Reason* of 1781, Kant solved the problem of how such synthetic a priori knowledge is possible in physics and mathematics by proposing his transcendental philosophy. According to this theory, we can have a priori knowledge because it is produced by epistemic structures that are inherent in our cognitive apparatus, whereas this knowledge is applicable to the world-as-we-perceive-it because the input of our senses is processed by these same epistemic structures, so that all things of which we can be perceptually aware must conform to our synthetic a priori knowledge. It is a counterintuitive consequence of this transcendental theory that we can know nothing whatsoever about the world-as-it-is-in-itself. Yet this latter world has to be posited as the ultimate cause of our perceptual input, since otherwise the world-as-we-perceive-it would be the only world there is, so that our perceptions would create the world.

Applying this transcendental epistemology to the main problem of his first *Critique*, the problem of whether metaphysics is possible as a scientific discipline, Kant concluded that metaphysics is possible only to the extent that it aims at mapping the synthetic a priori knowledge of the world-as-we-perceive-it, that is, of the so-called phenomenal world. This is done, for example, by Kant's book on *The Metaphysical Foundations of Natural Science* (1786), which contains the basic principles of Newtonian mechanics. However, a 'special' metaphysical discipline such as rational or natural theology is ruled out by Kant's transcendental philosophy, since it aims at synthetic a priori knowledge of something (God) that belongs to the world-as-it-is-in-itself, the so-called noumenal world, of which we cannot know anything. Kant famously observed in the preface to the second edition of the first *Critique* that by limiting the possibility of theological *knowledge*, he intended to make room for religious *faith*.[7] Indeed, his distinction between a phenomenal world and a noumenal world seemed to offer an elegant solution to the contradictions between science and religion. Whereas religion posits an almighty person in the noumenal world, of which real knowledge is impossible, science is merely concerned with the world of transcendentally constructed phenomena. Many philosophers concluded from the writings of Hume and Kant that some form of fideism was the only option that remained for religion.

---

[6] The differences between Newton's and Hume's views on scientific method do not matter here.
[7] Kant (KdrV), B xxx: 'Ich mußte also das *Wissen* aufheben, um zum *Glauben* Platz zu bekommen...' ('I therefore had to annul *knowledge* in order to make room for *faith*').

2.3 DOWNFALL AND RESURRECTION

Hume's and Kant's negative verdicts on the very possibility of natural theology precipitated its fall during the nineteenth century, and although Hegel attempted to restore a modified Christian metaphysics by claiming an obscure sort of super-rationality for it, the Romantic period favoured an irrational or anti-rational interpretation of religion. This demise of natural theology in the nineteenth century was further reinforced by the gradual elimination of theological notions from the natural sciences. Whereas Newton assumed that the instability of the solar system, caused by the complex gravitational interactions between the Sun and the major planets, had to be corrected periodically by God, the superior calculations of Laplace and others eliminated this theological element from cosmology. Even more importantly, the traditional doctrine that the complex functional constitution of organisms could only be explained by supposing that the first individuals of each species had been designed by God, which William Paley still defended in his *Natural Theology* of 1802, was refuted by Darwin's theory of evolution through natural selection as propounded in *The Origin of Species* of 1859, which offers a vastly superior explanation of the factual adaptations of organisms, and of the diversity and evolution of species.

It is no wonder, then, that the philosophy of religion after Darwin became ever more dominated by two tendencies. On the one hand, many philosophers became atheists in the wake of Baron d'Holbach, Feuerbach, and Karl Marx, since both biblical revelation and natural theology had now been intellectually discredited. In 1882 Nietzsche certified the 'death of God' without even feeling the need to argue for it. On the other hand, those philosophers who remained religious tended to defend irrational, a-rational, or even anti-rational views of religion, according to which the very attempt to engage in natural theology is misguided. Søren Kierkegaard, who had argued during the second quarter of the nineteenth century that reason has to be crucified if we want to relate to God, became hugely influential shortly after 1900, and he was the main source of inspiration for the religious decisionism of the Existentialists in the twentieth century. Another important author was William James, who defended in his essay 'The Will to Believe' of 1896 the thesis that 'our passional nature [...] must decide an option between propositions, whenever it is a genuine option that cannot by its nature be decided on intellectual grounds'.[8]

The linguistic turn in twentieth-century philosophy radicalized even further the elimination of natural theology by both atheists and anti-rational religious believers, although neo-Thomism flourished in traditional Catholic circles. Relying on a verification theory of meaning, logical positivists such as Carnap and Ayer argued that religious claims concerning the existence of God are not true or false, but cognitively meaningless. Accordingly, there is no point in trying to find out by engaging in natural theology whether such claims are true or false. The speech acts traditionally interpreted as making religious claims should rather be seen as expressing

---

[8] James (1896), 'The Will to Believe', section iv.

attitudes or commitments to specific kinds of behaviour, as R. B. Braithwaite argued in his celebrated Eddington lecture.[9]

A somewhat similar conclusion has been defended on different grounds by religious philosophers such as Rush Rhees and D. Z. Phillips, who endorsed Wittgenstein's later conception of philosophy. The task of the philosopher would not be to build theories or discover truths, as empirical scientists do, but rather to analyse the logical grammar of language games in order to resolve deep conceptual confusions which lie at the root of the paradigmatic philosophical problems. Applying this conception to the philosophy of religion, they argued that although a statement such as 'God exists' superficially resembles factual existence claims, its logical grammar is very different, because it has the function of putting human life into the completely new perspective of religious faith. As a consequence, engaging in natural theology is not only beyond the proper task of the philosopher. It is also misguided in itself, since it is based upon a mistaken interpretation of religious language.

Having arrived at this point, the reader will be inclined to conclude that natural theology is as dead as a doornail, because its downfall is final in view of the epistemological critiques of Hume and Kant, and the linguistic criticisms of logical positivists and Wittgensteinean philosophers of religion. Such a conclusion would be in agreement with what many lay persons tend to think about religious affiliations anyway, to wit, that they are a matter of habit or feeling or blind decision or grace rather than of rational deliberation, not only in fact but also *de jure*. Yet the philosophy of religion, both apologetic and critical, is today a flowering field of academic research within analytic philosophy, so much so that we may speak of a surprising resurrection of natural theology.

This resurrection began when in the 1960s and 1970s analytically trained philosophers such as William Alston, Alvin Plantinga, Richard Swinburne, and Nicholas Wolterstorff started to apply the tools of analytic philosophy to natural theology, thereby radically transforming the field. The spirit of this renewal of natural theology was well expressed by Swinburne, when he wrote in 1977:

It is one of the intellectual tragedies of our age that when philosophy in English-speaking countries has developed high standards of argument and clear thinking, the style of theological writing has been largely influenced by the continental philosophy of Existentialism, which, despite its considerable other merits, has been distinguished by a very loose and sloppy style of argument. If argument has a place in theology, large-scale theology needs clear and rigorous argument. That point was very well grasped by Thomas Aquinas and Duns Scotus, by Berkeley, Butler, and Paley. It is high time for theology to return to their standards.[10]

Three years later, the general public was informed about the resurrection of natural theology within analytic philosophy when *Time* ran a story on 7 April 1980 called 'Modernizing the Case for God', which contains the following passage:

In a quiet revolution in thought and argument that hardly anybody could have foreseen only two decades ago, God is making a comeback. Most intriguingly, this is happening not among

---

[9] Braithwaite (1955).   [10] Swinburne, (CT), p. 7.

theologians or ordinary believers, but in the crisp intellectual circles of academic philosophers, where the consensus had long banished the Almighty from fruitful discourse.[11]

2.4 THE RESURRECTION OF NATURAL THEOLOGY VALIDATED

One may wonder how this comeback of God in philosophy can be justified in view of the earlier verdicts on natural theology by Hume, Kant, the logical positivists, and the Wittgensteinean philosophers of religion. Let me conclude this chapter by explaining very briefly why their entombment of natural theology should be considered premature in the light of more recent intellectual developments, mainly in the philosophy of science.

In his analysis of causality and induction, Hume presupposed that the only causes of things that we can know are precedent observable phenomena. He concluded that all inductive arguments from effects to causes must be based upon enumerative inductions concerning the relation between observable A-s and B-s. But clearly, this view of inductive argument as used in science is much too restrictive. Quite often, scientists explain observable phenomena by positing unobservables, such as sub-atomic particles, super-massive black holes, or a big bang, and the theories that posit such unobservables are considered to be justified at a given time if they provide us with the best explanation available of the observed phenomena. The strengths of such *abductive arguments* or *arguments to the best explanation* can be assessed by the various models of modern confirmation theory, and there seems to be no good a priori reason for excluding the possibility that an existence claim about a god can be justified in this way. Consequently, Hume's global negative verdict on natural theology was informed by an outdated philosophy of science, and we have to investigate in detail the specific arguments put forward by natural theologians in order to see to what extent they confirm their theological theory.[12]

Although Kant's philosophy of science was very different from Hume's, it is outdated as well. Kant's assumption that Euclidean geometry and some fundamental principles of Newtonian natural science, such as a deterministic principle of causality, are synthetic a priori truths (that is, necessarily true, known independently of experience, and informative about the world), was discredited during the first quarter of the twentieth century by quantum mechanics, which is not deterministic, and by the theory of general relativity, which uses Riemannian geometry. The very idea of synthetic a priori truths was discredited further by the invention of non-Euclidean geometries, which led to the distinction between pure and applied mathematics, and by Wittgenstein's reinterpretation of principles such as 'all events have a determining cause' as norms of representation instead of a priori truths about nature. As a result,

---

[11] 'Modernizing the Case for God', *Time*, 7 April 1980, pp. 65–6, also quoted by Craig (2002), p. 2.
[12] Hume's rejection of abductive arguments to the best explanation has been updated by Van Fraassen's influential book *The Scientific Image*, which renewed a discussion on realism in the philosophy of science (Van Fraassen, 1980). But practising scientists use abductive arguments all the time.

the very problem of how such synthetic a priori propositions are possible, which Kant's transcendental epistemology was supposed to solve, turns out to be otiose, and the transcendental theory with its radical distinction between a noumenal and a phenomenal world is unjustified. But it was precisely this radical distinction between a noumenal world and a phenomenal world that informed Kant's rejection of natural theology as a possible scientific discipline, since God, who supposedly resides in the noumenal world, cannot be known by scientific disciplines, which are concerned with the phenomenal world only. Again, we come to the conclusion that there are no good a priori reasons for excluding the possibility of a theological hypothesis being confirmed by a convincing argument to the best explanation on the basis of empirical evidence.

This conclusion *eo ipso* refutes the logical positivist view that all talk of gods is cognitively meaningless. Admittedly, many theologians do not utter meaningful sentences when they pretend to speak of something referred to by the name 'God'. For example, Karl Barth claimed in his early works that God is so radically different (*totaliter aliter*) from all mundane things that we cannot describe him by using our language, not even analogically.[13] Anglo-Saxon religious philosophers such as John Hick have endorsed this doctrine of *divine ineffability*.[14] If this were correct, however, we could give no referential meaning whatsoever to the name 'God', for in God's case we can do so only by providing some meaningful description of what the name is supposed to stand for. It follows that Karl Barth has not stated anything when he says 'God exists', or 'God is radically different', for the name 'God' is used idly, without any possible means of determining its reference. The logical positivists tended to claim, however, that *no* talk of God could be cognitively meaningful, invoking their verification theory of meaning. But if one cannot exclude a priori that a theological hypothesis may be confirmed by arguments to the best explanation on the basis of empirical evidence, such a hypothesis will be meaningful even according to a somewhat sophisticated verification theory of meaning. Incidentally, formulating a satisfactory version of the verification theory of meaning has turned out to be very difficult, and most philosophers have abandoned the theory altogether.

Finally, against the negative verdict on natural theology by Wittgensteinean philosophers such as D. Z. Phillips, we should raise two objections. The first is that their analysis of religious language is not at all an impartial description, which they claim it is. An unprejudiced analysis of the many forms of religious language, which we encounter in studying traditions such as the Christian or Islamic ones, reveals that religious language is not understood in a uniform way by all believers, and that the logical grammar of the name 'God' is laid out differently by various authors. Given this diversity, it is surprising that a Wittgensteinean philosopher of religion such as D. Z. Phillips claims that *the* logical grammar of the word 'God' must be understood in the way he prefers, and that all other ways of understanding it are symptoms of superstition. One may suspect that this essentialist conception of the logical grammar of 'God' is not the result of an attempt to analyse religious

---

[13] Barth (1957), p. 76. [14] Cf. Plantinga (2000), pp. 43–63.

language as it is in fact used. Rather, it aims at introducing a more or less radical revision of the logical grammar of 'God', which would immunize religious claims against all imaginable challenges.

The second objection is that the logical grammar of 'God' as revised by Phillips does not seem to make much sense. For example, we might come to think that given the precariousness of life, we are living beings in need of divine grace, and we might even think that we have received grace (whatever conception of grace we may have). But to someone who raises the question whether in fact there is a god who can bestow grace upon us, and reproaches Phillips for not allowing this question to be asked, he answers as follows:

so far from omitting the notion of divine reality, I am endeavouring to elucidate its grammar. It is a misunderstanding to try to get 'behind' grace to God, since 'grace' is a synonym for 'God'. As with 'generosity is good', so with 'the grace of God' we are not attributing a predicate to an indefinable subject. We are being given a rule for one use of 'good' and 'God', respectively. God's reality and God's divinity, that is, his grace and love, come to the same thing. God is not 'real' in any other sense.[15]

Surely, according to the ordinary grammar of the relevant words, 'God' functions (more or less) like a proper name, whereas 'generosity' is an abstract noun and 'good' in this context an adjective of recommendation, so that it is highly implausible to model the logical depth grammar of 'the grace of God' on 'generosity is good'. We may conclude that quite probably natural theologians who raise the question as to whether God in fact exists have a better grasp of the logical grammar of the word 'God' as it is used by most believers than Wittgensteinean philosophers of religion such as D. Z. Phillips.[16]

I started this chapter by sketching the *embarras de choix* that confronts someone who has decided to engage in the philosophy of religion. Where should one begin? A bird's-eye view of the history of natural theology has shown, I hope, that the best starting point is contemporary analytic philosophy of religion. Many scholars in the field hold that if one wants to focus on one philosopher who has developed a promising global strategy in natural theology, the best choice is the work of Richard Swinburne. For example, Norman Kretzmann wrote that:

The single most accomplished contemporary practitioner and advocate of natural theology is Richard Swinburne, whose series of books constitutes a monumental achievement in this field.[17]

Even higher praise is bestowed by William Abraham, who claims that:

In fact Swinburne is rightly seen as a kind of new Aquinas,[18]

whereas Alvin Plantinga declares:

In my opinion, Swinburne's arguments [...to the effect] that Christian belief is probable with respect to public evidence [...] are the best on offer.[19]

---

[15] Phillips (2005), p. 461.
[16] Cf. Bloemendaal (2006), *passim*.
[17] Kretzmann (1997), p. 3.
[18] Abraham (1997), p. 587.
[19] Plantinga (2001), p. 219.

But this does not mean that Swinburne's attempt to justify the belief that God in fact exists by developing an integrated series of empirical arguments to the best explanation, is the only possible approach in apologetic philosophy of religion. Indeed, other philosophers claim that someone may be fully warranted in believing that God exists *without providing any positive arguments whatsoever*. If this contention were correct, natural theology would not be indispensable to the religious believer, after all. Whether it is indeed correct, is the topic of the next two chapters.

3

The Reformed Objection to Natural Theology

In the first chapter of this book I have argued that natural theology is epistemologically prior to revealed theology in the context of justification. No religious belief, such as the belief that a specific god exists, is justified or warranted unless it is sufficiently supported by reasons or evidence. This conclusion is often called the *evidentialist challenge* to religious doctrines, and it is primarily concerned with the credentials of a belief or proposition, rather than with the rationality of individual believers.[1] Indeed, it may be subjectively reasonable for a believer who has grown up in an isolated and homogeneous religious community, and who has never encountered dissidents, to endorse without qualms the religious doctrines proclaimed by its spiritual leader, even though these doctrines are not based on convincing arguments or evidence.

But in our modern societies such isolated religious communities are rare. We all know that many different religions exist, the creeds of which are mutually incompatible on several points, and most modern believers are aware of at least some of the defects of revelations pointed out in Chapter 1. It seems, then, that it cannot be rational or reasonable for modern believers to endorse a specific religious creed, unless they rely on convincing results of natural theology. And if, as some philosophers have argued, we have a moral obligation to form our opinions responsibly, we might say that having a religious belief is not a virtue but a vice, unless one knows of, or unless someone in one's religious community knows of, convincing arguments for the truth of its contents.[2]

As we have seen in Chapter 2, however, there are religious authors who claim that human reason on its own is either too sinful or too weak to arrive at true beliefs about their deity. These authors hold that faith *cannot* be acquired by believers merely on the basis of evidence and argumentative deliberations: it can only be given to them by God as a saving grace. Moreover, many religious thinkers have argued that religious beliefs

---

[1] Cf. Parsons (1989), p. 37.
[2] Cf. Kenny (2006), pp. 59–60: 'In my view, faith is not a virtue, but a vice, unless certain conditions are fulfilled. One is that the existence of God can be rationally established without appeal to faith. Accepting something as a matter of faith is taking God's word for its truth: but one cannot take God's word for it that He exists'. What kind of obligations we have in forming our opinions is hotly debated by present-day virtue epistemologists.

should not be based upon the arguments of natural theology, since a belief so based is likely to be unstable and wavering, whereas religious faith allegedly requires an absolute commitment. Such purely religious objections to natural theology and its indispensability will not impress the philosopher, however, because they presuppose a particular version of a particular religious creed, which has to be justified by natural theology in the first place. But some Christian philosophers have attempted to transform them into an epistemological objection that can stand on its own feet. Without any doubt, the most prominent author in this area is Alvin Plantinga. In the present and in the next chapter, I briefly explain and assess his *reformed objection to natural theology*.[3]

3.1 PROPERLY BASIC BELIEF

Inspired by Calvin and the Dutch theologian Herman Bavinck (1854–1921), Plantinga argued in his 1982 paper 'The Reformed Objection to Natural Theology' that 'a person is entirely within his epistemic rights, entirely rational, in believing in God, even if he has no argument for his belief and does not believe it on the basis of any other beliefs he holds'.[4] In subsequent papers, such as 'Reason and Belief in God' of 1983, Plantinga has further developed this view. Applying his theory of epistemic warrant as explained in *Warrant and Proper Function* of 1993, he has argued in his voluminous *summa* of 2000, entitled *Warranted Christian Belief*, that a Christian may even *know* that God exists, and, indeed, may know that 'the full panoply of Christian belief' is true, without being able to provide any arguments or evidence for this belief.[5] In other words, Plantinga squarely rejects the thesis of the indispensability and epistemological priority of natural theology, at least if natural theology is understood as an argumentative enterprise, which attempts to demonstrate or make probable the existence of a specific god or gods by inferences from public evidence, that is, from premises which both believers and non-believers endorse.[6]

How can Plantinga defend his thesis that religious believers may know that their god(s) exist(s), for example, without having any arguments or evidence for such a belief? Plantinga's global argumentative strategy is as follows. Plausibly supposing that some kind of foundationalist epistemology is more credible than all versions of coherentist epistemology, he assumes that not all our beliefs can stand in need of inferential backing, since this would lead either to an infinite regress or to a viciously circular chain of justifying arguments. In fact, many of our beliefs must be *basic* beliefs in the sense that we hold them without inferential backing or argument unless we are confronted by contrary evidence. Quite often, holding such a basic belief will amount

---

[3] Cf. for a still instructive critique of Plantinga's early (1983) reformed objection to natural theology: Parsons (1989).
[4] Plantinga (1982), p. 333.
[5] Plantinga (2000), pp. 241, 357, 499, and *passim*.
[6] Plantinga will admit that religious believers need something like 'negative apologetics', which aims at refuting the various objections raised against particular religious beliefs, like the problem of evil. But he denies that religious believers also need 'positive apologetics' in order to be rational or justified in endorsing their beliefs, that is, arguments or evidence that make the contents of these beliefs probable or plausible. Cf. for discussion Chapter 4.4, below.

to having knowledge, so that we can know that *p* without having any inferential justification for our belief that *p*. For example, I know that there is a tree outside my window because I see there is, and I know that yesterday I dined with a friend because I remember it. If we apply this foundationalist epistemology to religious belief, we can say that we may be entitled to believe that a specific god exists without having any arguments for it if (and only if) this belief is *properly* basic, that is, if we not only *in fact* believe that this god exists without having any arguments for it, but also are somehow *entitled* to hold this belief without arguments.[7]

Clearly, the main challenge for Plantinga's global strategy is to show how a belief in a monotheistic god, or in the Christian god (called 'God'), and indeed the 'full panoply' of Christian belief, can be properly basic, since at first sight such beliefs do not seem to be plausible candidates for having that epistemic status.[8] We usually consider as properly basic, for example, perceptual beliefs, such as the belief that there now is a computer in front of me, or memory beliefs about things we experienced, such as the belief that I had dinner yesterday, or self-evident beliefs such as that $1 + 1 = 2$. The belief that Allah or God or Zeus exists does not obviously belong to any of the paradigmatic sets of properly basic beliefs. In order to meet this challenge, Plantinga deploys both a negative and a positive tactic. *Negatively*, he argues that there are no good reasons to exclude a priori that a belief in God *can* be properly basic. In particular, he holds that a mistaken epistemology, which he calls *classical foundationalism*, has been the main motive for the conviction that belief in God *cannot* be properly basic.

Plantinga defines classical foundationalism as the disjunction of ancient and medieval foundationalism on the one hand and modern foundationalism as defended by Descartes and the empiricist tradition on the other hand. In general, foundationalism is the epistemological view that whereas many of our beliefs can only be justified, entirely or in part, by inferences on the basis of other beliefs, there must also be beliefs that are properly *basic* or *foundational* in the sense that such an inferential justification is not needed.[9] What the *classical* types of foundationalism have in common is that they set very demanding requirements for beliefs to be regarded as properly basic. For example, following Aristotle's *Posterior Analytics*, many medieval philosophers, as well as Descartes, required that the basic propositions of scientific knowledge have to be

---

[7] According to Plantinga's early definition, 'a belief is *properly basic* for me if it is basic for me and I am justified, violating no epistemic duties, in accepting it in the basic way'. Cf. Plantinga (1993a), pp. 19–20, footnote 35; cf. pp. 70–2. This definition is concerned with proper basicality with regard to justification. In his (2000), Plantinga prefers to define proper basicality with regard to warrant and rationality. Cf. pp. 175–98 and 343–9.

[8] Of course we should distinguish between (a) a belief in God (where I use the word 'God' as a proper name, the alleged referent of which is identified by a definite description), and (b) the belief that God exists. Belief in sense (a), that is, believing in someone, is having trust in someone, and it presupposes belief in sense (b). For reasons of economy, however, I often use 'belief in *x*' as short for 'belief that *x* exists'.

[9] If one does not define 'knowledge' as justified true belief, one should distinguish between foundationalism concerning knowledge and foundationalism concerning justification. Cf. for instance, Audi (1998) for an introduction to these concepts.

self-evident or even indubitable. And in the empiricist tradition, it was often said that all basic beliefs must be evident to the senses or even incorrigible.[10]

Plantinga is right to reject these demanding requirements, because we accept many perceptual and memory beliefs, for example, as properly basic although they do not meet the extravagant standards of these strong versions of foundationalism. To put this point differently, classical foundationalism would imply a devastating scepticism with regard to many beliefs that we commonly accept as properly basic, and this is a good reason for rejecting that view. Furthermore, classical foundationalism seems to be self-referentially incoherent, for this doctrine is not properly basic according to its own criteria, and attempts of philosophers to argue for it convincingly, starting from properly basic beliefs as premises, have all failed. Hence, if the doctrine is true, it must be unjustified, for it stipulates standards for justified belief that it does not itself meet.[11]

As a consequence, Plantinga is also right in rejecting arguments to the effect that since a belief in God does not meet the demanding standards of classical foundationalism, such a belief cannot be properly basic. But it does not follow, of course, that if one settles for a weaker version of foundationalism, as he does, belief in the existence of God *can* be properly basic, or that we *can* know that God exists in the basic way, or that the full panoply of Christian beliefs *can* be known to be true in the basic way, that is, without any inferential backing. In order to show this, Plantinga employs a *positive* tactic, which involves his theory of warrant, the centrepiece of his theory of knowledge.[12]

3.2 WARRANT AND PROPER FUNCTION

We can use the verb 'to know' in a number of different constructions, such as with a direct object (as in 'I know quite well Paris, Spanish, John'), with a wh-pronoun (as in 'I know what time it is, where I left my keys, when she will leave'), combined with 'how' (as in 'I know how to solve this problem'), or with a that-clause (as in 'I know that God exists'). Although the interrelations between these uses are complex and a worthy object of analysis, philosophical theorists of knowledge usually focus on the uses of 'to know' that can be expressed with a that-clause, that is, on so-called propositional knowledge.[13] They then ask the age-old question raised already by Plato in his *Theaetetus*: what distinguishes knowledge that p from the mere true belief that p (where 'p' stands for any proposition)?

Typically, having propositional knowledge requires more than merely having a true belief, because one might hit on a true belief accidentally, as when one guesses what is

---

[10] Cf. Plantinga (1982), pp. 333–8; Plantinga (1983), pp. 39–63, 71–3; Plantinga (1993a), *passim*; and Plantinga (2000), Chapter 3. According to Ayer (1936), propositions that 'refer solely to the content of a single experience' can be 'incorrigible' when verified by the occurrence of this experience, in the sense that 'it is impossible to be mistaken about them except in a verbal sense' (p. 13).

[11] Cf. Plantinga (1993a), pp. 84–6; Plantinga (1993b), pp. 182–3; Plantinga (2000), pp. 93–9.

[12] At least, this is what Plantinga does in the mature version of his reformed objection. I abstract here from the historical development of this objection in Plantinga's oeuvre.

[13] Cf., for example, Alan White (1982).

the case. We would not call such an accidentally true belief 'knowledge'. Many philosophers have tried to develop a general view of this 'more' that turns a true belief into knowledge, which holds for all imaginable cases of knowing that p, even very far-fetched ones, assuming that it is somehow useful and feasible to do so. Plantinga is one of these philosophers, although he is sophisticated about method, admitting that we must distinguish between paradigm cases of knowledge and more marginal cases.[14] Plantinga introduced the technical term 'warrant' to denote:

that further quality or quantity (perhaps it comes in degrees), whatever precisely it may be, enough of which distinguishes knowledge from mere true belief.

He also developed a general theory of warrant, which may be summarized as follows.[15] A belief that p has warrant for a subject S if and only if four conditions are satisfied:

(a) S's having the belief is the result of S's cognitive faculties functioning properly,
(b) in a cognitive environment sufficiently similar to that for which the faculties were designed,
(c) according to a design plan aimed at the production of true beliefs, when
(d) there is a high statistical probability of beliefs so produced being true.

Plantinga adds that when a belief meets these four conditions and thus does enjoy warrant, the degree of warrant that the belief enjoys depends upon the firmness with which one holds it.[16] For example, my belief that there is a keyboard in front of me on which I am typing has warrant for me because (a) it is the result of my perceptual faculties working properly in (b) the typical earthly environment they were 'designed' for, whereas (c, d) their 'design plan' is such that using my perceptual faculties mostly results in my having true beliefs about my proximal perceptible environment. In fact, my belief has a sufficiently high degree of warrant, because I hold it very firmly, and when a true belief that p has a sufficiently high degree of warrant for S, S knows that p.

Plantinga's four conditions need some elucidation. Plantinga holds that just as organs like our heart or lungs may function properly, subject to no disorder or dysfunction, so perceptual organs or cognitive faculties may function properly. This is required by condition (a) for warrant. He further holds that the concept of a proper function is 'inextricably bound' with another, that of a 'design plan': a faculty functions properly if it functions as it is 'designed' or 'meant' to work, and each organ is 'designed' to work in a specific environment. For example, we can breathe neither under water nor (most of us) at the top of Mount Everest, and we cannot see very well in the dark. According to condition (b) for warrant, our belief should be the result of a cognitive faculty functioning properly in an appropriate environment. Furthermore, the notion of a design plan implies that faculties and organs have specific aims, functions, or purposes. For example, the function or purpose of the heart is to pump blood, and when one is at rest, the human heart functions properly if the

---

[14] Cf. Plantinga (1993a), pp. 212–13, Plantinga (1993b), p. ix, and Plantinga (2000), p. 156.
[15] Plantinga (2000), p. 153. Cf. also Plantinga (1993b), Chapters 1 and 2.
[16] Plantinga (2000), pp. 153–6, and Plantinga (1993b), pp. 7–9 and *passim*. In his (2000), Plantinga added a fifth condition in order to exclude so-called Gettier cases, the 'resolution condition' (pp. 156–61), which is not relevant to my argument in this and the next chapter. For a short summary of Plantinga's theory of warrant, cf. also Plantinga and Tooley (2008), pp. 1–14.

pulse rate is about 50 to 80 beats per minute, etc. Conditions (c) and (d) require that the properly functioning faculty, the use of which yields the belief that *p*, be designed for the purpose of producing true beliefs, and that it be successfully designed in the sense that it usually does so.

In his positive tactic, which aims at showing that religious beliefs can be properly basic, Plantinga applies this theory of warrant as follows. Suppose that God exists. Then it could very well be, and is even quite likely, that he has implanted in us humans a specific religious cognitive faculty or module, a *sensus divinitatis* as conceived in terms loosely reminiscent of Calvin, for example, well-designed with the purpose of producing in us on specific occasions (such as when we read the Bible, or are enthralled by a landscape, or feel remorse) true non-inferential religious beliefs, such as the belief that God exists and is worthy of worship. If so, it may also be the case that this faculty is now functioning properly, and is functioning in the environment for which it was designed, namely human beings on Earth. Then, if I believe that God exists without having any arguments or evidence for it, this belief will be properly basic, because it is basic and warranted in Plantinga's sense. Furthermore, if I believe quite strongly that God exists, if it is true that he exists, and if conditions (a–d) are in fact satisfied, I *know* that God exists, even though I do not have any argument or evidence for this view. In other words, a non-inferential belief that God exists can be justified or rational or warranted in the sense of being properly basic, and if sufficiently strong it can amount to knowledge that God exists.

3.3 ORIGINAL SIN AND THE EXTENDED AQUINAS/CALVIN MODEL

Alvin Plantinga calls the idea that God may have endowed human beings with the cognitive mechanism of a *sensus divinitatis* the 'Aquinas/Calvin Model' of religious knowledge, or the 'A/C model' for short. He uses the term 'model' roughly in the sense in which giving a model of a state of affairs means showing how it can be that it actually obtains.[17] According to the A/C model, the non-inferential belief that God exists may be properly basic and warranted if it is true. If sufficiently strong in a given situation, this belief may even amount to knowledge. As a consequence, believers can be fully rational and warranted in holding this theistic belief, even though they have no arguments whatsoever for thinking that it is true. This may be the case simply because the belief is produced by the reliable cognitive mechanism of the *sensus divinitatis*, even if believers do not suspect in the least that such a mechanism is operative in them.[18] In other words, if Plantinga's positive tactic succeeds, natural theology is not indispensable to those who believe that God exists. And this is part of what Plantinga wanted to show.

Plantinga claims four merits for his A/C model of religious knowledge, two of which are important here. First, the model is both *logically possible*, in the sense that its

---

[17] Plantinga (2000), p. 168.
[18] Cf. Plantinga (2000), p. 179: 'It is not the case, of course, that a person who acquires belief by way of the *sensus divinitatis* need have any well-formed ideas about the source or origin of the belief, or any idea that there is such a faculty as the *sensus divinitatis*'.

description does not contain or entail a contradiction, and also *epistemically possible*, in the sense that its description is consistent with everything we know. Second, Plantinga claims that there are no cogent objections to the model that are not *also* cogent objections to monotheism or Christian belief. This second point is crucial to the main message of Plantinga's book *Warranted Christian Belief.* Many critics of religion, such as Marx, Freud, or Nietzsche, hold that religious beliefs are unjustified or irrational, because there are no convincing arguments for their truth and these beliefs have no warrant. In Plantinga's jargon, this is the *de jure* objection to religious belief. Typically, such critics do not even attempt to argue that religious beliefs are in fact false, which would be the *de facto* objection. Critics of religion usually assume that *de jure* objections are logically independent of the *de facto* objection, so that they do not need to argue that religious beliefs are false in order to argue that holding them is irrational.[19]

However, if the A/C model of religious belief is plausible on the assumption that God exists, so that religious believers do not need to support this belief (that God exists) by arguments or evidence in order to be rational, warranted, and fully within their epistemic rights, one cannot convincingly argue that religious belief is irrational without also arguing that it is not true.[20] In other words, if Plantinga's positive tactic succeeds, one cannot successfully put forward *de jure* objections against religious believers without also mounting convincing *de facto* objections against religious beliefs.[21] Clearly, if this positive tactic succeeds, the burden of proof for critics of religion becomes much heavier. The naive non-intellectual religious believer does not need to fear any more critics of religion such as Marx, Freud, Nietzsche, or Daniel Dennett, who do not bother to argue that religious beliefs are not true. Suppose, however, that there are cogent objections to the A/C model that are *not* also cogent objections to theism. In that case, Plantinga's positive tactic will fail, and his global strategy of arguing that *de jure* objections against believers logically depend upon *de facto* objections against religious beliefs will be shipwrecked.

But to the extent that I have spelled out Plantinga's A/C model of religious belief or knowledge, it still lacks one of the essential merits he claims for it. Indeed, this restricted A/C model is open to a massive empirical objection, which is logically independent of any cogent *de facto* objection to theism in general. For suppose that God exists and that he has endowed all humans with a *sensus divinitatis*. Since God is omnipotent and perfectly good, the design plan of this cognitive mechanism will be aimed at the production of true beliefs, which is condition (c) for warrant, and it will be well designed in that it usually produces true beliefs, which is condition (d). Furthermore, we may assume that humans are living in the cognitive environment for which the *sensus divinitatis* was designed, which is condition (b). But if these things are the case, and if there are abundant triggers for the *sensus divinitatis* in our environment, as Plantinga assumes, we should expect that all humans in the past

---

[19] Plantinga (2000), pp. 168–9 and *passim*. For an elaborate exegesis of the *de jure* objection, cf. Plantinga (2000), Chapters 3–5.
[20] That a model like the A/C model is quite likely an adequate one if theism or classical Christian belief is true, is the fourth merit that Plantinga claims for the model. Cf. Plantinga (2000), pp. 169–70 and 188–90.
[21] Plantinga (2000), pp. 190–1.

had, and all humans now living have a strong belief that the god of Christianity exists. Indeed, if the *sensus divinitatis* is functioning properly (condition (a)), 'God's presence and glory would be as obvious and uncontroversial to us all as the presence of other minds, physical objects, and the past', Plantinga says.[22]

But clearly this is not the case, neither diachronically nor synchronically. According to *Genesis*, if taken literally, God created Adam and Eve on the sixth day of creation, and if we believe the chronology of the Bible, this happened a few thousand years before the Common Era. It follows that according to *Genesis* the first humans must have been theists or monotheists, so that, diachronically speaking, the first religion of humanity has been monotheism. We now know on the basis of scientific and scholarly research in many disciplines, however, that this creation story is no more than a myth, and nowadays most theologians interpret it as such. Whereas today cosmologists estimate the age of the universe to be some 13.7 billion years, and the age of the Earth has been discovered to be around 4.5 billion years, the hominid lineage probably originated by continuous evolution between 7 and 5 million years ago, and the earliest fossils of anatomically modern humans are over 100,000 years old.[23] Although the prehistory of religion is a matter of speculation, it is clear that the earliest documented religions, such as those of the Indus Valley civilization between 2800 and 2000 BCE or of the Nile Valley from 5000 BCE onwards, were all polytheistic, so that very probably monotheism appeared upon the scene very late in history.[24] Indeed, the first recorded monotheistic experiment took place in ancient Egypt only in the fourteenth century BCE under the reign of Pharaoh Akhenaton, and probably the Israelite religion became monotheistic only in the seventh and sixth centuries BCE.[25]

It seems, then, that the history of mankind and the history of religion refute the restricted A/C model as explained up to this point, according to which God has endowed all humans with a (properly functioning, etc.) *sensus divinitatis*. The same conclusion holds synchronically. According to current estimates, around 55% of the present human population of the Earth is monotheistic. Officially there are about 2.1 billion Christians and 1.5 billion Muslims, to which we have to add monotheistic Hindus, 23 million Sikhs, 7 million Baha'is, 2.6 million Zoroastrians, etc., on a world population of over 6.5 billion.[26] So we may construct the following empirical refutation of the hypothesis that God has endowed all humans with a (properly functioning) *sensus divinitatis*. If this hypothesis were true, nearly all humans of the past and the present would have a distinct knowledge of the monotheistic god and would, as a consequence, be monotheists. But a great many human beings were and are not monotheists. This is the case even if my estimates of the dissemination of monotheism in the past and present are too pessimistic. Hence the hypothesis is false. This objection shows that the restricted A/C hypothesis as developed up to this point is not epistemically possible. Moreover, the objection is independent of cogent

---

[22] Plantinga (2000), p. 214.   [23] Cf. Ridley (2004), §18.7.2.
[24] Not everyone agrees that this is the most probable scenario. Cf., for example, Stark (2007), a book that defends monotheistic views apart from being a sociological analysis of the history of religions.
[25] About the origins of Israelite monotheism, see Smith (2001).
[26] See www.adherents.com, consulted on 14 July 2009. According to this site, there are now 1.1 billion people who are secular/nonreligious/agnostic/atheistic. In July 2009 the human population of the world was estimated at 6.771 billion, and by now it is estimated above 7 billion.

objections against monotheism as such, for the idea of a *sensus divinitatis* is not an essential ingredient of monotheism.

How can Plantinga avoid this refutation without endangering his positive tactic of showing that religious belief may be warranted as properly basic? He cannot argue that God may have designed the human *sensus divinitatis* badly, so that the statistical probability that it yields true beliefs is low, for in that case condition (d) for warrant would not be satisfied. Nor can he argue that the *sensus divinitatis* was not designed to produce true beliefs in the first place, for then condition (c) for warrant is violated. Similarly, Plantinga has to assume that condition (b) is satisfied, according to which the *sensus divinitatis* of monotheistic believers is functioning in a cognitive environment sufficiently similar to that for which God designed it.[27] It follows that there is only one logical possibility left for Plantinga if he wants to protect his A/C model from empirical refutation: he has to argue that very often the *sensus divinitatis* is not functioning properly, and that this is why many people do not have monotheistic and Christian convictions. In other words, he has to argue that in both humanity's past and its present, condition (a) often is not satisfied. To explain why this is so, and why the *sensus divinitatis* nevertheless functions properly in the case of believing Christians, is, I suppose, an important objective of Chapters 7–9 of *Warranted Christian Belief*. This is my exegetical hypothesis, however, for officially the aim of these chapters is to show how *all* specific beliefs of *Christian* theism may be warranted as properly basic.[28]

Very briefly, Plantinga's defence runs as follows. When God created the first human beings, their *sensus divinitatis* worked as it was designed to. As Plantinga says, 'they had extensive and intimate knowledge of God, and *sound* affections, including gratitude for God's goodness'.[29] But then, *original sin* was committed, which had 'ruinous *cognitive* consequences', because 'the *sensus divinitatis* was damaged and deformed'. Moreover, 'sin induces in us a *resistance* to the deliverances of the *sensus divinitatis*'.[30] By incorporating into his A/C model this Christian doctrine of original sin, according to which original sin contaminates our cognitive apparatus and is somehow inherited by all later humans, Plantinga can accommodate the facts that we do not find monotheism in the early documented history of mankind, and that many of our contemporaries are not monotheists.

But now a new explanatory gap opens up, which is plugged with further extensions of the A/C model.[31] How can it be that after humanity lapsed into this doomed state of original sin, some of us have monotheistic and indeed Christian beliefs at all? And how can these beliefs be warranted in the basic way, so that believers do not need the arguments of natural theology in order to be warranted, justified and fully within their epistemic rights when they claim that God exists, for example? Plantinga fills this gap

---

[27] Unless, of course, one assumes that the *sensus divinitatis* was designed for the environment of Paradise only. But then no Christian believer on Earth would have warrant for Christian beliefs.
[28] Cf. Plantinga (2000), pp. 199–203, 241, and *passim*.
[29] Plantinga (2000), p. 204 (Plantinga's italics). There is no empirical evidence whatsoever supporting these claims.
[30] Plantinga (2000), p. 205 (Plantinga's italics).
[31] In Plantinga's (2000) exposition, sin is included in the 'simple' A/C model (pp. 184ff), but this model including sin is not epistemically possible either if one assumes that Christian beliefs of present-day believers can be warranted in the basic way.

with some other Christian doctrines.[32] God himself has provided a remedy for original sin by the incarnation, crucifixion, and resurrection of Christ. In order to inform humanity of this remedy, He has inspired the authors of the Bible to write their holy books, so that God is the ultimate author of the Old and the New Testament. Furthermore, God has sent the Holy Spirit, who produces in some humans the supernatural gift of faith and seals it upon their hearts. Whenever people read the Scripture and become convinced that what they read is true, this belief is produced in them directly or non-inferentially by the internal instigation of the Holy Spirit as a grace of God. In such cases, they have warrant in the basic way not only for the belief that God exists, but also for the 'full panoply of Christian belief in all its particularity', since 'such a process that consists in direct divine activity cannot fail to function properly'.[33]

It follows that if God – as Christians conceive of him – exists, Christians may be warranted in believing 'the full panoply of Christian belief in all its particularity' without having to support their beliefs by arguments. In other words, according to this extended A/C model, *all* Christian beliefs can be properly basic. And if the believer holds these beliefs sufficiently firmly, holding them will amount to having knowledge. As Plantinga says:

> My Christian belief can have warrant, and warrant sufficient for knowledge, even if I don't know of and cannot make a good historical case for the reliability of the biblical writers or for what they teach. I don't *need* a good historical case for the truth of the central teachings of the gospel to be warranted in accepting them. I needn't be able to find a good argument, historical or otherwise, for the resurrection of Jesus Christ, or for his being the divine Son of God, or for the Christian claim that his suffering and death constitute an atoning sacrifice whereby we can be restored to the right relationship with God. On the model, the warrant for Christian belief doesn't require that I or anyone else have this kind of historical information; the warrant floats free of such questions.[34]

3.4 THE POWER OF PLANTINGA'S POSITIVE TACTIC

It is crucial for a book on natural theology to investigate whether Plantinga's positive tactic succeeds. If the tactic is effective, natural theology is not indispensable for the Christian believer, and the faithful may legitimately and rationally persist in holding

---

[32] Plantinga (2000), Chapters 8 and 9.

[33] Plantinga (2000), pp. 241 and 246, note 10, cf. p. 257. As Plantinga notes, however, his four conditions for warrant have to be adjusted somewhat to the case of the internal instigation of the Holy Spirit, and this supernatural process 'functions properly in a limiting sense of the term' (ibidem). It is an interesting question *how much* of what the Bible says receives warrant for a Christian reader by the Internal Instigation of the Holy Spirit (IIHS). A Christian fundamentalist will answer: 'Everything!' But Plantinga has to be more cautious at this point, since many passages in the Bible are refuted by scientific research and scholarship if interpreted in the sense in which the authors probably intended them. How much of the Christian faith can be considered properly basic because of the IIHS if the extended model is to be epistemically possible? How much is contained in the 'full panoply of Christian belief in all its particularity'? The answer to this question will vary in relation to scientific and scholarly progress. According to Plantinga, it at least includes trinity, incarnation, atonement, and resurrection. Cf. Plantinga (2000), pp. 241, 357, 499.

[34] Plantinga (2000), p. 259.

their religious beliefs firmly while being blissfully ignorant about the philosophy of religion.[35] Or so it seems, since the logical power of Plantinga's positive tactic can be overestimated easily. For this reason, I conclude the present chapter by briefly elucidating the logical force of Plantinga's positive tactic on the assumption that it is successful, before investigating in the next chapter whether it can succeed.

The logical power of Plantinga's positive tactic is limited for two reasons: the tactic is conditional, and it works only for certain types of religious believers. Plantinga's positive tactic is conditional in at least two respects. First, we have seen that a simple A/C model merely including the *sensus divinitatis* is open to massive empirical objections unless it is expanded with the doctrines of original sin, God as the author of Scripture, the divine incarnation including the crucifixion and resurrection of Christ, the Holy Spirit, and non-inferential Faith as a Grace of God. In other words, the positive tactic can succeed *only if* one incorporates into it a number of typically Christian (or some other) doctrines, so that the soothing import of Plantinga's philosophy of religion is limited to the community of Christian believers who endorse these doctrines (or to communities that endorse these other doctrines). Indeed, it is one of the two official aims of *Warranted Christian Belief* to provide 'a good way for *Christians* to think about the epistemology of Christian belief'.[36] In this respect, the import of Plantinga's philosophy of religion is more restricted than the import of Richard Swinburne's arguments for (bare) theism, for example, which will be appealing to all monotheists.

Second, Plantinga's positive tactic is conditional in that the (extended) A/C model is a likely model of religious belief and knowledge *only if God exists* as Christians conceive of him. In other words, only if God exists might a non-inferential belief in God, and indeed the full panoply of Christian belief, be warranted in the basic way, so that one can endorse these beliefs rationally without giving any argument. If God does not exist, we humans will not have a *sensus divinitatis*, the deficiencies of which are due to original sin, and the Holy Spirit will not instigate Christian believers to endorse the full panoply of Christian beliefs as true in the basic way. If we nevertheless have non-inferential Christian beliefs, these beliefs are both false and lack the warrant of being properly basic.

However, intellectually responsible Christian readers of Plantinga's books will not only want to be reassured that non-inferential religious beliefs *may be* warranted *if* God exists. They will also want to discover *whether* these beliefs have warrant, since, having encountered objections against the Christian creed, such as those spelled out in Chapter 1, they will have legitimate doubts as to whether the Christian beliefs are really true. And in order to discover whether these beliefs have warrant, they have to find out first whether, or at least to be reassured that, God exists. Yet in this core issue of the philosophy of religion, the issue of finding out whether God exists at all, neither Plantinga's positive tactic nor his theory of warrant are of any help.

It follows that argumentative philosophy of religion or natural theology is indispensable, after all, except to religious believers who never doubt the truth of their Christian creed. It is this type of unwavering Christian only who will find comfort in

---

[35] According to Plantinga, this holds at least for positive apologetics. The believer may need negative apologetics, however. Cf. Chapter 4.4, below, for further discussion.
[36] Plantinga (2000), p. xiii (my italics).

Plantinga's assurance that *if* God exists, Christian beliefs are likely to be warranted even though they are not backed up by arguments or evidence. As enlightened believers may unkindly observe, Plantinga's positive tactic does the very dubious service to humanity of reinforcing religious dogmatists in their dogmatism, and of bolstering up bigots in their bigotry. More reflective Christians, however, who are beset by doubts concerning the truth of their religious creed, cannot profit from Plantinga's positive tactic. They cannot reassure themselves that their beliefs may be warranted even if they have no arguments to support them. For their beliefs may be warranted in the basic way *only if* God exists, and that was precisely what they were calling into question.[37]

Since Plantinga's positive tactic is conditional and will entirely reassure the unwavering believer only, its import is limited indeed. In the eyes of non-believers and of Christians who have serious doubts about their religious creed, the positive tactic merely depicts a logical possibility. How probable it is that this possibility is actualized depends upon how probable it is that God exists. In order to assess this latter probability, Plantinga's theory of warrant is not of any service. What we have to do is rather to collect all the evidence that supports the thesis that God exists and all the evidence that disconfirms this hypothesis, in the hope that we may reach a clear conclusion by assessing the total evidence. In other words, we have to engage in the arguments of natural theology in order to discover how likely it is that specific religious beliefs will be warranted without any arguments.[38]

---

[37] Cf. for this type of criticism: Kenny (1992), p. 71: 'The doubting believer in God cannot reassure himself that his belief is warranted; for only if there is a God is his belief warranted, and that is what he was beginning to doubt'. Plantinga will answer (correctly) that his extended A/C model is indeed hypothetical, but that it also does great services both to Christian believers and to those who want to criticize religious beliefs. Christian believers who have doubts concerning the truth of their creed *merely* on the ground that they cannot support it by arguments or evidence, will be comforted by the model, since it shows that such arguments of positive apologetics are not necessarily necessary. And the model teaches critics of religion that *de jure* objections to religious belief must be grounded in *de facto* objections.

[38] Cf. Swinburne (2001) and Plantinga (2001) for a discussion about the import of Plantinga's positive tactic.

4

Refutation of the Reformed Objection

As we have seen in Chapter 3, the logical and persuasive power of the Reformed Objection to natural theology is somewhat more modest than was suggested by Plantinga's 1982 article. He claimed that in general no arguments are needed for the justification of some core religious beliefs, or for endorsing them to be rational. This meant, in Plantinga's words, that:

> the believer is entirely within his epistemic right in believing that God has created the world, even if he has no argument at all for that conclusion. The believer doesn't need natural theology in order to achieve rationality or epistemic propriety in believing; his belief in God can be perfectly rational even if he knows of no cogent argument, deductive or inductive, for the existence of God – indeed, even if there *isn't* any such argument.[1]

Later on, Plantinga even argued that the 'full panoply' of Christian beliefs would be warranted in the basic way, if it is yielded either by a proper functioning *sensus divinitatis* or sealed upon our hearts by the internal instigation of the Holy Spirit.

But because of its conditional nature, Plantinga's positive tactic to the effect that Christian beliefs may be properly basic *if* God exists, turned out to work only for Christians who do not doubt at all whether God exists or, indeed, whether any Christian belief is true. Those who have such doubts, or those who want to proselytise and persuade non-believers that the Christian creed is true, will need the arguments of natural theology.[2] Furthermore, unbelievers will not be impressed by Plantinga's positive tactic. Even if they admit that the Christian creed *might* be warranted as properly basic in Plantinga's sense *if* the creed were true, they will hold that very probably it is not true, and hence that probably humans neither have a *sensus divinitatis* nor are subject to an internal instigation of the Holy Spirit, so that the creed will not be warranted as properly basic. And concerning the *de facto* issue as to whether the Christian creed is true, many unbelievers will say that the burden of proof rests on the Christian believer, since in general it rests on those who put forward positive and problematic contentions, such as the claim that God exists. In short, it is only the unwavering and dogmatic believers to whom natural theology is not indispensable, if at least Plantinga's positive tactic is successful.

---

[1] Plantinga (1982), p. 330 (Plantinga's italics).
[2] Cf. Plantinga (2001), p. 217.

But we now have to investigate whether Plantinga's positive tactic is indeed successful within this limited sphere. Is it true that unwavering believers can be entirely within their epistemic rights, rational, and warranted in believing that God exists, and that the full panoply of the Christian creed is true, without adducing any supporting arguments or evidence for their beliefs? Allowing for a division of intellectual labour, we may formulate this question somewhat differently. Can unwavering believers who are members of a religious community be entirely within their epistemic rights in believing that God exists, even if they are aware of the fact that no member of that community or anyone else can offer any arguments or evidence for the truth of this belief, and indeed, even if they think that there *isn't* any such argument? In the present chapter, I discuss two problems for Plantinga's positive tactic, the second of which is insuperable, and two possible rejoinders by a reformed epistemologist to this second problem.[3] Let me begin by reviewing a traditional worry for externalist theories of warrant, the problem of generality.

4.1 THE PROBLEM OF GENERALITY

Plantinga's theory of warrant for basic beliefs is *externalist* in the sense that the subject who has warrant does not need to know whether conditions (a–d), as stated in Chapter 3.2, are satisfied. For an externalist in epistemology, whether a belief is warranted and amounts to knowledge merely depends on whether the process by which it is produced is of the right kind or type, quite independently of whether the believer is, or can become, aware of whether it is indeed of the right kind. This contrasts with *internalist* views of justified belief and knowledge, according to which what justifies a belief for a subject has to be accessible to this subject.[4] Furthermore, Plantinga's theory of warrant for basic beliefs is *reliabilist* in the broad sense that it proposes as a necessary condition for a belief's being warranted that the process which generates the belief is of a type that generally generates true beliefs (condition d). In other words, the actual token process that produced the belief has to be a reliable one in the sense that, statistically speaking, processes of that kind or type yield true beliefs much more often than false beliefs.

Reliabilist theories of warrant are confronted by a well-known difficulty called the *generality problem*, which arises because one may describe at many different levels of generality the actual token-process by which a belief that p is generated in a subject. As a result, this process may fall both under reliable kinds or types and under unreliable kinds of processes, whereas there may not be a principled manner of choosing between them. For example, if I now seem to remember that I met my beloved for the first time in 1995, the belief that I did so may be unwarranted in the reliabilist sense if it is

---

[3] Cf. Baker (2007), Chapters 3 and 4, for an overview of eight types of criticisms raised with regard to Plantinga's reformed epistemology.

[4] If one wants the dichotomy between internalism and externalism in epistemology to be exhaustive, one should define externalist theories of epistemic justification as theories according to which there is *at least one* condition for justification to the satisfaction of which the cognitive subject has no access. But I shall use the definition of the main text in what follows. This definition allows for mixed theories that have both externalistic and internalistic features.

described as resulting from a process of the type 'human memory of events in the distant past', or 'my memory of events in the distant past', which are unreliable kinds of processes, but it may be warranted if it is described as resulting from 'remembering when one met one's beloved for the first time', or 'my remembering when I met my beloved for the first time', which may be quite reliable kinds of processes. But if there is no principled method of choosing between the various types of processes, reliable and unreliable ones, under which the actual process that produced the belief can be subsumed, the externalist answer to the question whether the belief has warrant is arbitrary.

One may be tempted to solve the problem of generality by stipulating that in order to assess whether a subject S is warranted in having adopted a specific belief b produced by a process P at time t, one should always select the lowest (least abstract and least general) kind of belief-generating processes to which P belongs. But this will not do. Clearly, this lowest kind is precisely the kind characterized as 'the process P-producing-b-in-S-at-t', and this kind or type has only one instance or token, the very process P that took place in S at t. As a consequence, this type of process will be 100% reliable if b is true, and 100 per cent unreliable if b is false. In other words, if one stipulates that the reliability of a process should be evaluated with its lowest kind as a reference set, a belief is always warranted if it is true, and never warranted if it is false, so that warrant collapses into truth. But a tenable theory of the justification or warrant of beliefs should allow that on some occasions having a false belief may be warranted, and that in other cases holding a true belief may not be warranted. It follows that the reliability of a process of belief-acquisition should always be assessed at some higher level of generality, so that the generality problem emerges again. As discussions of the problem in the literature have shown, there is not always an easy escape from it for an externalist theory of justification or warrant.[5]

In Plantinga's theory of warrant, conditions (a–c) aim at eliminating the problem of generality, among other things. If the processes, the reliability of which constitutes a necessary condition (d) for a belief being warranted, are identified as the proper functioning (condition a) or dysfunctioning of epistemic faculties or organs in their proper domains (condition b), there may seem to be in principle a non-arbitrary way of selecting the level of generality at which we describe these processes. One reason is that sense organs such as the eyes are clearly identifiable parts of the human body, so that our cognitive faculties can be identified non-arbitrarily if their use depends on the use of specific sense organs. Another reason is that if our faculties are designed by God with a specific purpose, as Plantinga holds for the hypothetical *sensus divinitatis*, God's design plan determines the proper level of generality.

Compare this with measuring instruments such as thermometers, for example. The type under which specific temperature measurements made with this individual thermometer (call it Boreas) should be subsumed, simply is the type 'measurements made by Boreas', since Boreas is a clearly identifiable instrument made with a specific

---

[5] Cf., for example, Conee and Feldman (1998), and Swinburne (EJ), pp. 13–20. According to Heller (1995), however, the problem of generality is a 'red herring' (p. 503) because '[t]he process has to be as reliable as the evaluator's standards require, and those standards vary from context to context' (p. 501). In other words, the problem 'only arises if we ignore the context-relativity of our use of the term "knowledge"' (ibidem).

purpose. In this case it is not at all arbitrary under which type we should subsume a belief-generating process. If we have carefully calibrated Boreas by doing empirical tests, we will know to what extent its results are reliable, and the problem of generality does not arise. So it may seem at first sight that Plantinga's theory of warrant does not suffer from the generality problem.[6]

Yet one might argue that the problem of generality re-emerges in Plantinga's application of his theory of warrant to the justification of religious beliefs as properly basic. For suppose that the Christian god exists, and suppose that he has implanted in all human beings a specific faculty of religious knowledge, the *sensus divinitatis* in Calvin's sense. Suppose further that this *sensus divinitatis* has been damaged and deformed by original sin but that, in some elected group of humans, the Holy Spirit has restored the organ, and by using it produces in them non-inferential Calvinist beliefs whenever specific triggers are operative, such as seeing a beautiful landscape, reading passages from the Bible, or feeling remorse. Can we now say with regard to the religious beliefs of a specific individual that they are warranted as properly basic? Since we humans cannot calibrate our individual *sensus divinitatis* by tests of its reliability, the critic might argue that the answer to this question merely depends on the level of generality at which one describes the operations of the *sensus divinitatis*.

If we describe the operations of the *sensus divinitatis* very broadly, including all human beings who are living now, for example, the organ as such clearly is unreliable, for many of us are not monotheists, let alone Christians or Calvinists. This is of course admitted by the extended A/C model, since according to this model, the *sensus divinitatis* of most humans is not functioning properly. If we describe the operations of the organ very narrowly, however, including only believing Calvinists who claim that they have specific non-inferential religious beliefs, the organ will be reliable (condition d), since, *ex hypothesi*, these beliefs are true. But note again that externalist justification or warrant should not collapse into truth. We should not say, for example, that the restored *sensus divinitatis* is functioning properly (condition a) in a proper environment (condition b) according to a design plan aimed at the production of true beliefs (condition c) if and only if the resulting beliefs are true. However, in the case of the *sensus divinitatis*, no human being has any empirical access to the workings of this hypothetical mechanism or to its design plan. As a result, we do not have a clue as to when we should say that conditions (a–c) are fulfilled.

If we want to avoid the collapse of justification or warrant into truth, we should assume that in some people or on some occasions these conditions are fulfilled, even though the resulting beliefs are false, whereas in other people or on other occasions the resulting beliefs are true, although one of the necessary conditions for warrant is not satisfied. This means both that non-Calvinists or non-Christians should be included in the reference set for applying condition (d), whereas we should also assume that some people have the true religious basic beliefs although they lack warrant. But how many non-Calvinists should be included? And in how many people are Calvinist basic beliefs produced, if God exists, although one of the necessary conditions (a–c) for warrant is not satisfied? How widely should we define the kind or type of a properly functioning *sensus divinitatis*? There are many alternatives here, and the critic will argue that there

---

[6] Cf. Plantinga (1993b), pp. 28–9.

seems to be no principled way of choosing between them. If there is no principled way of choosing, however, the problem of generality re-emerges. It seems to be arbitrary whether we assert or deny that in a specific case someone's non-inferential Calvinist beliefs are warranted as properly basic.

Plantinga might respond to this first argument that for an externalist theory of warrant it is irrelevant whether *we humans* are able to identify in a non-arbitrary manner the right level of generality for characterizing the operations of the *sensus divinitatis*. Indeed, according to the externalist, we may know that p in the basic way because we have sufficient warrant for p in the basic way even though we cannot discover in principle by calibrating tests of our *sensus divinitatis* that we have sufficient warrant for p. In the present case of the extended A/C model, it would suffice that *God* can solve the problem of generality non-arbitrarily. And of course he can, at least if he exists and if he has endowed human beings with such a faculty of religious knowledge. In that case, he has deliberately designed the faculty and repaired it in some human beings after it had been deformed by original sin, so that he can describe its operations at the right level of generality. Being omniscient, he will also know in each case whether conditions (a–d) are satisfied. Perhaps he has repaired the *sensus divinitatis* of some Calvinists for reasons that escape us, so that only in their case the faculty functions properly and is reliable, whereas the *sensus divinitatis* of other Calvinists is still deformed by original sin. In that case, the religious beliefs of the former Calvinists will have warrant in the basic way, whereas the very same religious beliefs of the latter Calvinists lack this enviable property.

We may conclude that the problem of generality is not a decisive objection against the *logical* and *epistemic possibility* of a warranted basic belief that God exists. What the problem shows is merely that even if God exists, and even if he has implanted a *sensus divinitatis* in us and repaired it in some cases after original sin, *we humans* can never discover by calibrating tests whether our religious beliefs are warranted as properly basic or not. And if this is so, we humans can never *justify* our religious beliefs by adducing the extended A/C model as a *reason* for holding that these beliefs in fact have warrant.[7] There is a second argument, however, inspired by the facts of religious diversity, which purports to show that even if God exists, basic religious beliefs *cannot* be warranted as properly basic, at least not for human beings who are sufficiently aware of the facts of religious diversity.

---

[7] Plantinga will reply (personal correspondence) that even if such a *justification* is impossible, religious believers might have a second-order belief to the effect that their first-order religious beliefs have warrant in the basic way, and that this second-order belief might also have warrant in the basic way (on the extended A/C model). And if this second-order belief is held very strongly, is true, and has warrant in the basic way, it amounts to knowledge according to Plantinga's externalist theory of knowledge. I would reply that in the sense in which we use the word in ordinary contexts, such a second-order belief would not amount to 'knowledge' at all, since we all agree that we can only acquire knowledge concerning the reliability of instruments or epistemic organs by doing empirical calibration tests. Cf. Chapter 5.2 on the issue as to whether externalism is tenable as a theory of knowledge. Furthermore, the fact that on the model of the internal instigation of the Holy Spirit *any* belief could have warrant in the basic way, may be seen as a *reductio ad absurdum* of that model.

4.2 THE PROBLEM OF RELIGIOUS DIVERSITY

Plantinga defines *religious exclusivists* as people who are aware of the existence of other religions, some of the central doctrines of which are incompatible with their own, and yet persist in adhering to their own religious creed, although they do not think that they know of any conclusive arguments in its favour.[8] Plantinga then asks whether it is possible to be a *rational* religious exclusivist. Of course, if having a properly basic belief in the externalist sense of Plantinga's theory of warrant is a sufficient condition for being rational, and if the entire creed of the exclusivist consists of beliefs that are properly basic with respect to warrant, the answer is affirmative, and this conditional claim is what Plantinga maintains. As he writes:

> I've argued that Christian belief – the full panoply of Christian belief, including trinity, incarnation, atonement, resurrection – can, if true, have warrant, can indeed have sufficient warrant for knowledge, and can have that warrant in the basic way.[9]

It is logically possible that God exists and that he has repaired the *sensus divinitatis* after the postulated Fall in Christian believers only. Then, if this organ is functioning properly (condition a) in a suitable environment (condition b) according to a design plan successfully aimed at truth (conditions c and d), and yields some Christian beliefs in the basic way when triggered on specific occasions, the Christian exclusivist has warrant for these beliefs as properly basic. Furthermore, an internal instigation of the Holy Spirit, who is infallible in its instigations, might produce in believers the remainder of the full panoply of Christian beliefs. The problem is, however, that Plantinga's theory of warrant is not, or cannot be, purely externalist.

In order to see why his theory also incorporates internalist elements even for cases of basic belief, we have to introduce the notion of a *defeater*.[10] In general, we may define a defeater as a belief or a piece of evidence that destroys the warrant for another belief, say the belief that p. Suppose, for example, that I believe that I met my beloved for the first time in 1995, and suppose that this is a properly basic belief which is warranted because my memory is functioning properly (condition a) and conditions (b–d) are also satisfied. However, my beloved now tells me with great assurance that in fact we met for the first time only in 1997, so that I am confronted with another basic belief by her testimony, which contradicts the belief that p. Suppose further that my memories and those of my beloved are equally vivacious and detailed, so that initially there is no reason to trust the one more than the other. According to Plantinga, the second belief then *defeats* the warrant of the belief that p, even if in fact my memory was functioning properly and we met in 1995. The notion of a defeater is an internalist notion, because

---

[8] Cf. Plantinga (2000), p. 440, and Plantinga (1995), pp. 174–6.
[9] Plantinga (2000), p. 357.
[10] Cf. Plantinga (2000), pp. 357–63. It is surprising that Plantinga does not discuss in his summa *Warranted Christian Belief* defeaters of biblical passages resulting from empirical and scientific discoveries, such as listed by White (1896), since he claims that one of the Christian beliefs that can be warranted on his model is the belief that 'the Bible is the word of God' (2000, p. 380). I shall come back to this issue in the main text below. Incidentally, there are other internalist elements in Plantinga's theory of warrant for basic beliefs, such as his notion of internal rationality downstream from experience, which I shall not discuss here. Cf. Plantinga (2000), pp. 110–12.

the second belief defeats the warrant for the first belief only if I am aware of the fact that the second belief undermines the first.[11]

John Pollock, who developed the notion of a defeater, distinguishes between rebutting defeaters and undercutting defeaters.[12] A defeater is rebutting if it contradicts the original belief that p, or sufficiently diminishes the probability that it is true, given one's background knowledge, whereas it is an undercutting one if it destroys the warrant for the belief that p without contradicting p or diminishing its probability. Pollock offers the following example of an undercutting defeater.[13] In a factory, the items coming down the assembly line look red, and because of this visual impression you form the belief that they are indeed red. But then a factory supervisor tells you that the items are irradiated by red light in order to enable him to detect certain types of flaws. This new piece of information undermines your belief that the items are red without giving rise to any reason for thinking that it is false. Another example of an undercutting defeater is the case in which one acquires a basic belief that p on the testimony of a friend, and later discovers that this friend is a pathological liar concerning issues such as whether p. Defeaters may be defeated in their turn. In my first example, it may be the case that repeated memory tests teach both my beloved and me that my memory is much better than hers. If I believe on these good grounds that my warrant for the belief that p is much better than her warrant for believing that not-p, my original warrant will be restored by this defeater-defeater. A defeater-defeater that restores the original warrant is called a 'neutralizer' of the defeater.

If we apply this terminology to the case of the religious exclusivist, we can say that awareness of religious beliefs of others that are incompatible with one's own religious beliefs presents one with a rebutting defeater, if one notices that the beliefs of the other persons are held as honestly, with as much conviction, and on as good grounds as one's own.[14] For example, the Christian belief that Jesus of Nazareth is God incarnate is rebutted for Christian believers by their awareness of the sincere Muslim belief that Jesus is not, and indeed cannot be, an incarnation of God because a divine revelation says he is not. Of course, awareness of this contradictory belief, and of the fact that it is held in earnest by intelligent and well-educated human beings on grounds as solid as the grounds for our own belief, does not show that the original belief is false, but it destroys its original warrant. Even if this Christian belief had a purely externalist warrant in Plantinga's sense before the believer became aware of the Muslim defeater,

---

[11] I am using the notion of a defeater here in the internalist sense of a 'purely epistemic rationality defeater'; cf. Plantinga (2000), pp. 359–63. Is it necessary for a defeater d of my belief b that I fully endorse d? This is required by Plantinga's (2000) definition on p. 361 (cf. pp. 359, 366, 368): 'A defeater for a belief b, then, is another belief d such that, given my noetic structure, I cannot rationally hold b, given that I believe d'. I think it is sufficient that I am confronted with d by testimony, for example, and do not have any good reasons or grounds (apart from b) for rejecting d, given my noetic structure.

[12] Cf. Pollock (1974), pp. 42ff., who speaks of *type I* defeaters and *type II* defeaters. My definitions are inspired by Pollock, but differ from his.

[13] Plantinga (1993b), pp. 40–2, embellished Pollock's example (cf. Pollock (1974)), pp. 42–3) and I am using Plantinga's embellishments; cf. Plantinga (2000), pp. 357–73. Cf. also Swinburne (EJ), pp. 28ff.

[14] Internalists may add other requirements for such cases of epistemic peers or 'equal weight', such as the condition that both parties in the dispute have considered the same evidence with the same amount of attention.

the awareness of the contradicting belief annuls the original warrant, so that holding on to the Christian belief is not warranted or rational any more, unless the Christian can neutralize the defeater.

Plantinga rightly stresses that defeaters are relative to the rest of what one knows and believes, that is, to the rest of one's 'noetic structure'.[15] If a Christian believes, for example, that all Muslims are wicked and that God rightly punishes them for their sins by deforming their *sensus divinitatis* and by intentionally withholding from them the redeeming insight in His incarnation, the fact that many Muslims deny God's incarnation in Jesus will not amount to a defeater for the incarnation-belief of such a Christian. In this example, the belief about the wickedness of Muslims functions as a 'defeater deflector'.[16]

Let us suppose, however, that what I shall call *decent* Christian exclusivists reject as deeply immoral and unjustified such traditional Christian denunciations of other religions and their devotees. Indeed, Plantinga suggests that what is new in our times with regard to the problem of religious diversity 'is a more widespread sympathy for other religions, a tendency to see them as more valuable, as containing more by way of truth'.[17] For such decent Christian exclusivists, a rather full awareness of the fact that adherents to other religions sincerely and passionately hold beliefs that are incompatible with (some of) their own will constitute a rebutting defeater. As I said, such a defeater does not show that their beliefs are false. But it destroys the warrant of these beliefs, which can be restored only by defeating or neutralizing the defeater.

One way of defeating the defeater is to engage in natural theology and to develop arguments supporting one's original belief on the basis of premises that all parties can accept. But what Plantinga wanted to show is that the full panoply of Christian belief can be warranted without the positive support of natural theology, where 'can' has the value of a logical and epistemic possibility. How is Plantinga still able to defend this view, if one admits that as soon as Christians become fully aware of the diversity of religions, a purely externalist warrant (if it exists) is no longer sufficient for their belief's being warranted in the basic way, since this warrant is defeated by incompatible rival beliefs and the grounds adduced for them? What is needed in order to restore the warrant is a defeater-defeater, a view or argument that defeats or neutralizes the defeaters. But how can one produce such a neutralizer without engaging in natural theology?

Another way of formulating this problem of the diversity of religions is in terms of arbitrariness. Given that one is aware of the diversity of religions and of the contradictions between their creeds, is it not completely arbitrary to stick to one's own creed, unless one attempts to produce convincing arguments or adduce evidence in its favour and against the probability that some other incompatible and honestly believed creed is true? Surely someone who arbitrarily holds a belief cannot be called 'rational' in holding that belief.

---

[15] Plantinga (2000), pp. 360, 366.
[16] Plantinga defines a 'defeater deflector' *DD* with respect to a potential defeater *D* of a belief *b* roughly as a belief such that given that you have it, *D* does not defeat *b*, whereas if you had not had it, *D* would have defeated *b* (personal correspondence).
[17] Plantinga (2000), pp. 439–40.

Plantinga discusses this objection of arbitrariness at length in *Warranted Christian Belief*, focusing primarily on its moral version, according to which it is somehow arrogant or egoistic to adopt a belief arbitrarily. This is unfortunate, however, because the most serious problem for Plantinga is not the moral variety but the epistemological version of the objection, according to which religious exclusivists cannot hold on to their beliefs *rationally* or *reasonably* in the light of contradicting beliefs, unless they produce a defeater or neutralizer of these defeaters.[18]

Yet Plantinga also proposes a solution to this epistemological problem, albeit *ambulando*, as he says somewhat light-heartedly.[19] It reads as follows:

> If the believer concedes that she *doesn't* have any special source of knowledge or true belief with respect to Christian belief – no *sensus divinitatis*, no internal instigation of the Holy Spirit, no teaching by a church inspired and protected from error by the Holy Spirit, nothing not available to those who disagree with her – *then*, perhaps, she can properly be charged with an arbitrary egoism, and *then*, perhaps, she will have a defeater for her Christian belief. But why should she concede these things?[20]

There is a delicate problem of interpretation concerning this quote. If by 'not conceding that you do not have a special source of knowledge' is meant that you do not have any beliefs whatsoever about this issue, this is not enough to defeat the defeater. In order to neutralize the defeater, believers should have the positive belief that they *do have* a special source of knowledge, such as an internal instigation of the Holy Spirit, which adherents of incompatible creeds lack. So let us interpret the passage in this latter sense, which is the correct one, as is clear from the context.[21]

Then the idea seems to be that the awareness of other religions, to the extent that their creeds are incompatible with Christian beliefs, would indeed defeat a Christian belief (call it p) by destroying its original warrant, unless Christian believers also have another belief (call it j), namely *that* they have a warrant which believers in other religions lack, because, for example, their *sensus divinitatis* is functioning properly since it has been restored after the Fall, whereas believers in other religions have not been blessed with such a regenerating restoration. While the original warrant for the Christian belief (p) was a purely externalist one (the believer has a *sensus divinitatis* which is in fact functioning properly, etc.), this warrant is now backed up by an internalist justifying argument (j), namely *that* the believer has the original warrant, and that such a warrant is lacking in the case of those who endorse incompatible beliefs. The crucial question is, of course, whether this second belief can really function as an adequate defeater-defeater or neutralizer. For three reasons it seems clear that it cannot.

First, religious exclusivists who are really aware of the details of other religions know not only that these other religions contradict their own religious beliefs at many points. They also know that many other religions back up their beliefs by epistemological claims to the effect that these latter beliefs are *uniquely* warranted by revelations, or by well-functioning divinely implanted epistemic faculties, or by internal

---

[18] Plantinga (2000), pp. 437–57. [19] Plantinga (2000), p. 443.
[20] Plantinga (2000), p. 453 (Plantinga's italics), jo. p. 443.
[21] Cf. Plantinga (2000), p. 453: 'she must think that she has access to a source of warranted belief the other lacks'.

instigations of infallible spirits and deities. In other words, a rather full awareness of other religions not only defeats the religious belief that p; it also defeats the epistemological belief that j, since the epistemological claims of other religions, such as Islam or Hinduism, contradict the Calvinist or Christian view that the only existing deity, called God, has implanted a *sensus divinitatis* in us, the proper functioning of which warrants the Christian beliefs only, etc.[22] But if the belief that j is itself defeated by an awareness of other religions, it cannot function as a defeater-defeater or neutralizer that restores the original warrant for Christian beliefs.[23] The epistemological aporia here is structurally similar to the one created by the fact that many conflicting religions used to appeal to miracles in order to validate their revelations. As Hume once observed, since each appeal to a miracle aims at validating one's own religion only, it 'has [...] the same force, though more indirectly, to overthrow every other system', and '[i]n destroying a rival system, it likewise destroys the credit of those miracles, on which that system was established'.[24] Of course, this holds vice versa also for the appeal to miracles of believers in rival religions: that appeal destroys the credit of the miracles *we* invoke.

The second reason is that a defeater-defeater (j), which aims at restoring the warrant for the belief that p, should in most cases rely on independent evidence, which, by definition, is not available for the religious exclusivist. For example, in the memory story related above, the original warrant for my belief that I met my beloved for the first time in 1995 was the externalist one that my memory was functioning properly. When confronted with the conflicting memory belief that I met her for the first time in 1997, this warrant was destroyed by a rebutting defeater. The original warrant can be restored by independent evidence to the effect that my memory is better than the memory of my beloved – evidence either concerning my superior track record in this respect, or concerning the greater vivacity and specificity of my memory – and such independent evidence neutralizes the defeater.

However, Plantinga seems to think that in the case of Christian religious exclusivists, the *mere belief* that their religious convictions are produced by a properly functioning *sensus divinitatis* or by an internal instigation of the Holy Spirit (= j), would be sufficient to restore the warrant for the belief that p and to neutralize the defeaters of that belief. The analogy in the memory case would be that the *mere claim* that my memory is better than the memory of my beloved would suffice as a

---

[22] Assuming, at least, that the religious exclusivist is 'decent' in the sense defined above and does not endorse defeater deflectors to the effect that the adherents of other religions are all wicked etc. Incidentally, the well-informed religious exclusivist will know that other religions also adopt defeater deflectors to the effect that believers in rival religions are somehow blinded, which amount to undercutting defeaters for the belief that p.

[23] In his (2000), p. 350, Plantinga admits that other religions may adopt equivalents of the A/C model in order to argue that their beliefs may be warranted as properly basic: 'Well, probably something like that *is* true for the other theistic religions: Judaism, Islam, some forms of Hinduism, some forms of Buddhism, some forms of American Indian Religion' (Plantinga's italics). But it may be true for many polytheistic religions as well.

[24] Hume (1748), Sect. X, Part II, §95. The problem of religious diversity differs from the 'Great Pumpkin' objection, discussed by Plantinga (1982), in that the beliefs of other religions are not obviously more fanciful or absurd than those of Christianity. Furthermore, it also differs from the 'Son of Great Pumpkin' objection, discussed in Plantinga (2000), pp. 344–9, because it uses the notion of a defeater.

neutralizer. But that clearly will not do, at least in cases in which both parties contend to rely on the same type of epistemic source. What is needed for such a neutralizer is independent evidence to the effect that my memory is indeed better. Plantinga does not adduce such independent evidence for the corresponding claim that his *sensus divinitatis* is more reliable than the *sensus divinitatis* of Muslims, or that the internal instigations of the Holy Spirit are more reliable than contradictory internal instigations of Allah. Indeed, it is difficult to imagine a plausible non-circular argument to this effect, unless one engages in natural theology and shows convincingly that the Christian trinitarian god exists whereas Allah does not.[25] And if the conclusion of the last section (4.1) is correct, Plantinga cannot adduce such independent evidence, since only God or Allah or some other god can solve the problem of generality.

Finally, the Christian who adduces *j* as a neutralizer of the defeaters of *p* commits the fallacy of begging the question. It is important to stress again that the whole apparatus of defeaters and neutralizers introduces an internalist element into Plantinga's theory of warrant. The claim (*j*) of Christians that they have a properly functioning *sensus divinitatis*, or are subject to an internal instigation of the Holy Spirit, which produces in the basic way their belief that (*p*) the god of Christianity exists, and all other Christian beliefs, is now adduced as a *reason* for holding on to *p* in spite of the defeaters produced by rival religions. But this justification is circular, because Plantinga's main argument was that *j* is probably true *if and only if p* is true. In other words, *j* can be a good reason for holding on to *p* only if *p* is known to be true in the first place. As Plantinga would say in similar cases, what we have here is 'a vicious circle in the *receives-its-warrant-from* relation'.[26]

We have to conclude that decent religious exclusivists cannot be rational or reasonable, unless they produce independent evidence for *j*. But since such evidence must include evidence for the hypothesis that God exists, because *j* can be true only if *p* is true, decent religious exclusivists can be rational only if they engage in natural theology. In other words, no decent religious exclusivist can reasonably remain a complete epistemological externalist.[27] As a consequence, the reformed objection to natural theology is invalid for those decent religious believers who are fully aware of the existence of rival religions, that is, for most decent well-educated contemporary believers, however unwavering and dogmatic they may be in their faith.

---

[25] I am indebted to Silver (2001), who presents a carefully elaborated version of this argument against Plantinga. I modified the argument somewhat, however, in order to protect it against the criticisms of Vogelstein (2004). Cf. also Forrest (2002) and Tien (2004).

[26] Plantinga (2000), p. 352. Plantinga denies on p. 456 that the Christian who adduces *j* as a neutralizer for a defeater of *p* is involved in a circularity: 'She isn't putting forward an *argument* for anything; nor is she proposing a *definition*: so how does circularity so much as rear its ugly head? If she were giving an argument for theism and then proposed as a premise that she enjoyed the benefits of one of those special sources of belief, *then* her argument might be circular. But she isn't arguing for that; nor need she be arguing for anything else' (Plantinga's italics). According to Plantinga, Christians who hold that *j* are *merely explaining* their belief that *p* and not arguing for it. But this is not correct if they use *j* as a neutralizer for a defeater of *p*. Although initially they believed that *p* in the basic way, and may have been warranted in the basic way, they now put forward *j* as a *reason* for holding on to *p* in spite of a defeater. So there clearly is a circularity here, as I pointed out in the main text.

[27] Cf. Willard (2003) for other arguments to this effect.

But it was precisely for such well-educated Christian believers that Plantinga designed his positive tactic in the first place. As he said in the Preface to *Warranted Christian Belief*, that book is about the 'intellectual or rational acceptability of Christian belief [...] for *us, now*', that is, for 'educated and intelligent people living in the twenty-first century, with all that has happened over the last four or five hundred years'.[28] In other words, the positive tactic should work not only for 'ignorant fundamentalists', who are unaware of the existence of rival religions and of the many difficulties in their own revelations as pointed out in my first chapter, but, Plantinga says, in particular for 'sophisticated, aware, educated, turn-of-the-millennium people, who have read their Freud and Nietzsche, their Hume and Mackie (their Dennett and Dawkins)'.[29] I hope to have shown, however, that these sophisticated believers cannot have warrant for their Christian beliefs without supporting them by arguments or evidence. Pace Plantinga, natural theology turns out to be indispensable, after all.

4.3 INTRINSIC NEUTRALIZERS

The debate on Reformed Epistemology of Religion has gone on in analytic philosophy for some thirty years now, and it is not difficult to imagine how supporters might attempt to answer the refutation of the reformed objection to natural theology presented in section 4.2. Let me single out for discussion two possible reformed rejoinders. According to the first one, well-educated modern Christian religious exclusivists simply do not need to neutralize potential defeaters by arguments in order to be fully rational, intellectually respectable, and completely within their epistemic rights. The second rejoinder starts by admitting that religious exclusivists have to rebut, or at least to undercut, some defeaters by arguments, but it stresses that this argumentative enterprise is very limited, and does not amount to natural theology in the sense conceived of by those who endorse the evidentialist challenge to religious belief. As a consequence, the reformed objection would still stand against natural theology in this evidentialist sense.

The first rejoinder will employ the notion of an intrinsic defeater-defeater or an intrinsic neutralizer. By definition, a properly basic belief is an intrinsic neutralizer of one of its potential defeaters if and only if it has more warrant than this potential defeater.[30] Suppose, for example, that my wife accuses me of having committed adultery with a lady L after I dined with the latter in a restaurant. My wife has some evidence for her accusation, such as reports from friends in the restaurant who said to her that I was rather intimate with L during the evening, and me telling her that I accompanied L to her house after dinner, which explains the fact that I came home very late. Even worse, I have a reputation for womanizing, and L confidentially intimated to my wife that I did make love to her during the evening of that very day.

But as a matter of fact, after having accompanied L to her house, I kissed her goodbye (literally) at the doorstep and walked home, taking a long detour. Since

---

[28] Plantinga (2000), pp. vii–viii (Plantinga's italics).
[29] Plantinga (2000), p. 200; cf. p. 242.
[30] Plantinga (1986), the section on 'Intellectual Sophistication and Basic Belief in God', and Plantinga (2000), pp. 450–1.

I remember all of this very clearly, my belief (p) that I did not commit adultery with L that day is properly warranted in the basic way, if we assume that Plantinga's four conditions for warrant are satisfied. However, the evidence against me is very strong. This potential rebutting defeater ($\sim p$) of my belief that p is presented in detail to me during an unpleasant conversation with my wife, so that I now have strong evidence for the denial of p. I cannot refute $\sim p$ by any arguments or evidence, however, so that, from my side, the discussion is embarrassingly repetitive. I merely state again and again that I did not do it. Should we conclude that when presented with this rebutting defeater, I would not be rational in continuing to believe that I did not commit adultery with L that evening, since I can neither adduce any arguments in support of my belief, nor rebut or undercut by argument the potentially defeating evidence?

I suppose that in this case we all agree that, on the contrary, I would be perfectly rational, would flout no intellectual duties, and would be fully warranted in continuing to believe in the basic way that I did not commit adultery. The reason is that my nonpropositional warrant for this basic belief is stronger than the warrant for the potential defeater, so that there is no epistemic parity. The former warrant is based upon personal memory and perception of what I know to be the case, whereas the latter warrant merely stems from circumstantial evidence and testimony. In this example, it is my basic belief that I did not commit adultery (p) *itself* that is my reason for thinking that the defeater ($\sim p$) is false, and not some other belief or argument. Since the warrant for the basic belief that p is stronger than the warrant for $\sim p$, p is an *intrinsic* neutralizer of $\sim p$. By contrast, a belief r may be called an *extrinsic* neutralizer if it defeats a defeater q of a belief p distinct from r.

Using this notion of an intrinsic neutralizer, the reformed epistemologist might argue that if God exists, Christian believers do not need to reply by argument to *any* potential defeater. Their Christian beliefs might simply have such a strong warrant in the basic way that these beliefs are intrinsic neutralizers of *all* possible defeaters. Indeed, in a great many different circumstances, such as when they read the Bible, admire natural scenery, or feel guilty, their *sensus divinitatis*, perhaps assisted by the Holy Spirit, might function properly and produce very strong Christian beliefs, such as the belief that God is speaking truly to them through all biblical texts, the belief that God created all natural scenery, or the belief that God disapproves of what they have done. If God exists, if he has repaired the *sensus divinitatis* of Christians after the Fall, and if the Holy Spirit helps to produce Christian beliefs, whereas Christians do not rely on any inferences when they hold them, these beliefs might have enough warrant in the basic way to neutralize all potential defeaters intrinsically. Indeed, are the internal instigations of the Holy Spirit not infallibly true? Can one imagine a more weighty warrant for a belief in the basic way? For this reason, the reformed epistemologist might stick to the externalist epistemology of religious belief without admitting any internalistic elements, and the reformed objection to natural theology would be valid, after all.

Is this first rejoinder convincing? With his usual sophistication, Plantinga expresses his opinion on the issue as follows:

My opinion (for what it is worth) is that for many theists, the nonpropositional warrant belief in God has for them is indeed greater than that of alleged potential defeaters of theistic belief...[31]

---

[31] Plantinga (1986), reprinted in Craig (2002), p. 55.

The problem is, however, that Plantinga's opinion on this issue may not be worth very much. As we remember from Chapter 3, his positive tactic for showing that belief in God can be properly basic is conditional. Only *if* God exists, can belief in God have warrant in the basic way of God having implanted a properly functioning *sensus divinitatis* in us, etc., so that, only *if* God exists, can this warrant be greater than that of alleged potential defeaters of theistic belief. Let us admit, then, that Christians *might* have intrinsic neutralizers for all defeaters *if* God exists. Would this help them in a discussion with adherents of other religions?

I do not think it would, since polytheistic Hindus might tell us a similar just-so story, for example. When Vishnu created the universe, he installed in all human souls the disposition to know the three main divinities and innumerable lesser gods of polytheistic Hinduism. This disposition is triggered on many, many occasions. Whenever it is triggered, believing Hindus will have warrant in the basic way for the belief that some of their gods are present, or love them, or cause cholera, or provide food.[32] Since Hindu polytheism is a rebutting defeater of theism, and since *if* the former doctrine were true, the warrant for Hindu basic beliefs might be *just as strong* as the warrant for theistic basic beliefs would be *if* theism were true, the notion of an intrinsic neutralizer cannot solve the problem of religious diversity for the reformed epistemologist.

Admittedly, *if* Christian theism were true, the Christian might have an intrinsic neutralizer for all potential defeaters consisting of rival religious creeds. Hence, if the truth of Christianity were an established fact, no defeater-defeater would be necessary in order to remain rational in endorsing Christian beliefs. But the same holds for Hindu polytheists. If their doctrine were true, they might have an intrinsic neutralizer for the potential defeater constituted by Christian monotheism. In order to find out which of these two incompatible hypotheses is correct, both parties will have to engage in natural theology. Christians can no longer be called rational if they simply stick to their own Christian beliefs when confronted with the contention that the polytheistic views held by Hindus are intrinsic neutralizers of these very same Christian beliefs. That would be so even if, in fact, the Christians were the ones who have the intrinsic neutralizers.

There is yet another reason why the notion of an intrinsic neutralizer cannot rescue the reformed objection to natural theology. I illustrated this notion by the example of my basic memory beliefs concerning what happened after I dined with lady *L*. Why is it that these beliefs yielded by personal memory have so much more in the way of warrant than the beliefs constituting and supporting the defeater, so that they can be its intrinsic neutralizer? Let us pause to think about this issue, and then wonder whether the theistic beliefs of a Christian believer can *ever* have so much warrant as well, if they are produced by a well-functioning *sensus divinitatis* or prompted by an internal instigation of the Holy Spirit.

---

[32] Of course, this Vishnu-model should also be extended in several ways in order to make it epistemically possible; cf. Chapter 3.3, above.

With regard to warrant, these memory beliefs are parasitic upon my perceptual and proprioceptual beliefs formed during the evening, such as the (proprio)-perceptual belief that I (literally) kissed lady *L* goodbye at the doorstep of her house and walked home, taking a long detour. Memories of what I perceived can have a strong warrant because they get their warrant from perceptual beliefs, even though the warrant of the former typically is weaker than that of the latter. And why can perceptual beliefs have such a very strong warrant? Plantinga's externalist theory of warrant fails to explain this fact sufficiently. Let us take an example that Plantinga often uses. He looks out into the backyard of his house and sees that the coral tiger lilies are in bloom. Since he has learned the name of these flowers, the belief that the coral tiger lilies are in bloom spontaneously and non-inferentially arises in him.[33] Why does this belief have such a very strong warrant in the basic way, so that it amounts to knowledge?

This is not only, I suggest, because the four conditions of Plantinga's externalist theory of warrant are met and he strongly believes it. The warrant is so strong because he is perceptually aware of the very state of affairs that he believes to be the case. Let us call this state of affairs the *truth-maker* of the belief.[34] Then we might say that in unproblematic cases of sense perception, the truth-maker of the belief is *itself manifest* and *directly present* to him, so that no better evidence for his belief can be conceived of (apart from further perceptual exploration of the truth-maker). Furthermore, when someone asks us how we know that the coral tiger lilies are in bloom while we are perceiving that this is the case, we can always tell this person that we know it because we are perceiving it. Let us express these two aspects of sense perception by saying that when we have perceptual beliefs while perceiving their truth-makers and identifying them as such, we have *transparent access* by using our senses to the very state of affairs of which we believe that it is the case. We believe that the coral tiger lilies are in bloom and we see that they are in bloom. In such cases, the descriptions of what we believe and of what we see (or hear, or smell, etc.) are identical. And because we are aware of the fact that we are perceiving what we believe is the case, such a transparent access is not (merely) an externalist warrant of our belief but (also) an internalist one.[35]

Now contrast this with Plantinga's examples of Christian beliefs produced in the basic way (according to the extended A/C model) by a properly functioning *sensus*

---

[33] Plantinga (2000), p. 175, and *passim*.

[34] The jargon of 'truth-makers' may be misleading to the extent that it suggests that the state of affairs that *p* has some causal relation to the truth of the proposition that *p*. This need not be the case.

[35] I am using here the idiom of naive realism concerning perception that all of us use in everyday life, and I am assuming that in such clear cases of sense-perception, it is always obvious to us that we are not subject to hallucinations or illusions. For these reasons, my view differs from Plantinga's notion of 'internal rationality downstream from experience' (Cf. Plantinga (2000), pp. 110ff). Plantinga presupposes a representational theory of perception, according to which 'perceptual beliefs are formed *in response to* sensuous imagery and *on the basis of* such imagery' (p. 110, Plantinga's italics). But I suggest that perceptual beliefs are formed on the basis of direct and transparent access to their truth-makers. There is no 'imagery' of any kind involved in sense perception. The idea that such imagery is present results from the mistaken belief that a causal and scientific analysis of sense perception *refutes* common-sense direct realism, whereas in fact it merely *micro-reduces* the common-sense conception of sense perception. Cf. Austin (1962) for an early critique of representational theories of perception; cf. also Hacker (1987).

divinitatis or by the internal instigation of the Holy Spirit. When Christians read the Bible, admire natural scenery, or feel guilty, their *sensus divinitatis*, perhaps assisted by the Holy Spirit, produces in them the belief that God is speaking truly to them through all biblical texts, or the belief that God created all natural scenery, or the belief that God disapproves of what they have done. Can these beliefs have the same strong warrant as perceptual beliefs, based on transparent access to the truth-makers? For the following three reasons it will seem obvious that they cannot.

First, what is present in perception and triggers these basic Christian beliefs is not identical with their truth-makers. For example, a page of the Bible that we read, the natural scenery that we perceive, or a feeling of guilt that we have, is not the truth-maker of the religious belief. If, upon reading the Bible, Christians form the belief that God is speaking truly to them through all biblical texts, their belief is merely triggered or occasioned by reading the Bible. The belief is not shown to be true by their readings, even if it is in fact produced by the internal instigation of the Holy Spirit. To put it more colloquially, these Christians are reading the Bible; they are not reading God. It follows, secondly, that Christians who have these basic beliefs do not have transparent access to the state of affairs itself that they believe to be the case. Finally, they cannot become aware just by reflection on their mental life of the fact that their *sensus divinitatis* is producing a true belief in them, if that is a fact, as I argued above in section 4.1.[36] And the same holds for the internal instigation of the Holy Spirit. One cannot become aware by mere reflection on one's mental life of the fact, if it is one, that the Holy Spirit is producing specific beliefs in one, although, of course, on the extended A/C model the Holy Spirit might produce the second-order belief in someone that specific first-order beliefs are produced by the Holy Spirit.

But if Christian beliefs produced by the *sensus divinitatis* or by the internal instigation of the Holy Spirit cannot have the same strong warrant as perceptual beliefs, there is no good reason to think that they can be intrinsic neutralizers of all possible defeaters. On the contrary, it will be likely on the basis of what I argued in Chapter 1, that even if God exists, a basic Christian belief such as that God is speaking truly to Christians through all biblical texts has less by way of warrant than a defeater consisting, for example, in a scientific refutation of one of these texts. Indeed, by arguing that 'the full panoply' of Christian beliefs can be warranted in the basic way on the extended A/C model, including the belief that the Bible is the word of God, Plantinga exposes his A/C model to great risks.[37] Is the extended model still epistemically possible, in the sense that it is consistent with all things we know, as Plantinga claims?[38] Clearly, if the Bible is God's word, so that it cannot contain any falsehoods, all scientific and other refutations of biblical passages also undermine the extended A/C model (cf. Chapter 1, above). Consequently, when confronted with beliefs of other religions or with scientific objections that are potential defeaters of their beliefs, Christians will have to engage in natural theology in order to defeat these defeaters, so that the first reformed rejoinder flunks.

---

[36] Plantinga is inclined to think, and indeed argues, that basic beliefs gained by way of the *sensus divinitatis* of the A/C model are not perceptual beliefs. Cf. Plantinga (2000), pp. 182, 287–8, 335–8. On the possibility of perceiving God, see Chapter 15, below.
[37] Cf. Plantinga (2000), pp. 375–80.
[38] Cf. Chapter 3, above, and Plantinga (2000), pp. 168–9.

4.4 WILL NEGATIVE APOLOGETICS SUFFICE?

Having arrived at this point in our discussion, reformed epistemologists might admit that in order to restore the warrant of their basic religious beliefs, Christians will have to defeat many potential defeaters by using arguments. For example, if someone raises the problem of natural evil as a defeater of the belief that God exists, Christians will have to argue that the presence of natural evil on Earth is neither logically incompatible with God's existence nor yields a good inductive argument against it. Similarly, Christians might rebut the defeaters constituted by incompatible religious creeds by attempting to refute the relevant beliefs of these rival religions. But having admitted this, reformed epistemologists will stage their second rejoinder. Even if Christians have to neutralize defeaters of their religious creed by argument, this does not entail that they also have to produce positive arguments or evidence in support of this creed, as the evidentialist challenge requires. One might rebut or undercut defeaters of the Christian creed without producing any evidence or arguments to the effect that the Christian creed is true. For this latter contention, the theist would have warrant in the basic way, if at least God exists.

In order to clarify this second reformed rejoinder, it will be illuminating to introduce again the distinction between negative apologetics and positive apologetics, which I made in the *Preface*. According to the evidentialist challenge, religious believers should engage in positive apologetics in order to be rational. They should produce arguments and evidence to the effect that it is more probable than not that God exists. Or at least, allowing for a division of intellectual labour, some experts should pursue positive apologetics in order to ensure that other religious believers can be rational in endorsing their religious beliefs. This is not necessary, however, if religious beliefs can be properly basic, as reformed epistemologists argue. Defeaters might destroy the warrant of these basic beliefs. But in order to restore their warrant, it may be sufficient to engage in negative apologetics only. One might defeat the defeaters by showing that probably they have less by way of warrant than the original basic beliefs, if God exists. The reformed epistemologist will argue that this can be achieved without producing positive evidence or arguments supporting these religious beliefs.[39]

The problem is, however, that this can perhaps be done for some potential defeaters, such as arguments from evil, but not for all potential defeaters. Let me develop very briefly one example of a potential defeater that cannot be dealt with by negative apologetics only. If convincing, this example also undermines the global strategy of Plantinga's book *Warranted Christian Belief*. As we remember from Chapter 3.3, the main objective of his strategy is to show that *de jure* criticisms to the effect that Christian beliefs are irrational, such as those advanced by Marx,

---

[39] Cf. Plantinga (2000), pp. 368–9: 'Here is a place, then, where the philosopher can be of service to the Christian community, by pointing out truths which, when added to a Christian noetic structure, can preserve Christian or theistic belief from defeat or provide a defeater-defeater for a defeater of such belief'.

Nietzsche, and Freud, cannot be compelling unless one also advances convincing *de facto* criticisms to the effect that these beliefs are probably false. Because Plantinga holds that such *de facto* criticisms, like the problem of evil, are unconvincing, he can conclude that Christian beliefs are not necessarily irrational.

Somewhat schematically, one might distinguish between two complementary components of the *de jure* criticisms raised by Marx, Nietzsche, and Freud. On the one hand, they presupposed that in order to be rational or justified, religious beliefs have to be argued for on the basis of sufficient evidence, which was assumed to be impossible. This is the evidentialist challenge. On the other hand, they held that religious beliefs result from, and are sustained by, psychological or sociological mechanisms that are either not aimed at truth or not well-functioning, such as class interests of the dominant classes, the will to power of the weak, or mental mechanisms of projection and wish-fulfilment. Let us call this the irrationalist explanation of religion. From each of these two components it seemed to follow that all religious beliefs are irrational. Plantinga implemented his global strategy by proposing his theory of warrant and his extended A/C model, according to which religious beliefs can be properly basic. This yields an alternative to each of the two components of *de jure* criticisms. If religious beliefs can be properly basic, the evidentialist challenge is misguided, and if Christian beliefs result from a properly functioning *sensus divinitatis* or an internal instigation of the Holy Spirit, they are produced by a reliable and truth-conducive faculty or process, and not by mechanisms that are not truth-conducive.

One might think that *de jure* criticisms of religious beliefs are very different and logically independent from *de facto* criticisms. But this need not be the case, and it is not the case with regard to the example of a potential defeater that I shall sketch now. The general idea of this defeater is as follows. On the basis of a consilience of inductions from many disciplines, such as empirical psychology, evolutionary biology, sociology, cultural anthropology, and the history of religions, one might construct a detailed secular explanation of the existing religious beliefs, which is much more advanced and better supported empirically than the speculative accounts of Marx, Nietzsche, and Freud. Why should one focus on these outdated authors instead of relying on the most recent empirical research?

In principle, this secular explanation might be sufficient to account for all existing religious beliefs, that is, the beliefs of all religions of the past and present. It would draw both from empirical studies of the proximate mental mechanisms that produce basic religious beliefs, and propose ultimate explanations of the fact that specific religious ideas arose and proliferated in the course of history. The upshot would be that religious beliefs do not arise and proliferate because they are true and because the corresponding entities exist, such as Zeus, Wodan, Poseidon, Ganesha, or God, but because they have very different social and psychological functions.[40] Whereas usually positing the existence or presence of an entity, like a crocodile or a chimpanzee, is the best explanation of the fact that someone believes that it exists, this is different in the case of gods. Both unbelievers and modern monotheists agree that the existence of the relevant entities should not play a role in explaining the formation or persistence of (nearly) all religious beliefs, although monotheists make an exception for one god: their own.

---

[40] Cf., for example, Wilson (2002).

Since such a secular explanation of the occurring religious beliefs would be sufficient, there is no need to postulate God, or gods, or any other supernatural entity to account for the existence of religious beliefs. On the contrary, such a postulate would be ruled out by Ockham's razor, or the principle of simplicity, according to which one should not postulate entities without necessity.[41] Accordingly, the *de jure* objection and the *de facto* objection would be intimately linked. On the basis of such a secular explanation of religious beliefs, modern and well-informed Christians might come to the conclusion that their Christian beliefs are probably as false as they consider beliefs in Wodan or Zeus to be, and that God does not exist. So we might speak of a *defeating explanation* of religious beliefs, because the secular explanation of religious beliefs is at least an undercutting defeater for all of them.[42]

In order to neutralize this defeater, negative apologetics by the Christian believer will not suffice. The reason is that one might conceive of such a secular explanation of religious beliefs on the one hand, and the hypothetical explanation by the extended A/C model on the other hand, as rivals in a contest for the best explanation of existing religious beliefs. In order to win this contest, Christian believers will have to show that their warranting explanation of Christian beliefs as properly basic is better than the defeating secular explanation. How can this be done?

The secular explanation is supported by massive empirical evidence. For example, psychological research reveals that at an early age, children tend to interpret natural phenomena in an anthropomorphic way.[43] This accounts for the early genesis of polytheistic religious ideas, both individually and historically. That these ideas have many different social and psychological functions, which are not the functions of true beliefs, can be documented by ample empirical research as well.[44] Furthermore, the genesis of monotheism from polytheism may be explained by an update of Hume's natural history of religion, according to which the idea of an infinitely potent, wise, and good God resulted from a flattery competition between or within tribes.

In tribal warfare, each of the competing tribes will hold that their chance of winning a war is greater to the extent that their gods are more powerful and more knowledgeable. Also, in polytheistic religions, one god will be thought of as the ruler of all, and, like the subjects of kings and princes, believers will attempt to win his or her favour by sycophancy. In both cases, there will be a contest of flattery, in which believers attribute to their gods ever more superlative qualities. The natural limit, which cannot

---

[41] Cf. Quinn (1993), the section on 'Defeating Theistic Beliefs', reprinted in Craig (2002), p. 63. Plantinga discusses Quinn's objection at length in his (2000), pp. 367–73.

[42] Cf. for a discussion of this objection, Plantinga (2000), pp. 372–3. Plantinga uses an argument from analogy to refute the idea that such a secular explanation of religious beliefs amounts to an undercutting defeater. Suppose a Berkeleyan philosopher explains our perceptual beliefs by saying that they are caused by God and not by the objects perceived. Would we accept such an explanation as an undercutting defeater for our perceptual beliefs? We would not, of course, but the analogy holds only if one endorses a representational theory of perception, according to which the direct objects of perception are not things in the world but sense data, ideas, or imaginings. As I said in an earlier note to this chapter, I reject such a representational theory.

[43] Cf., for example, Kelemen (2004), Kelemen and Diyanni (2005), and later work by Deborah Kelemen et al.

[44] Cf., for example, Wilson (2002).

be surpassed, is the attribution of maximal or infinite qualities, such as omniscience, omnipotence, and infinite goodness. As Hume wrote:

> In proportion as men's fears or distresses become more urgent, they still invent new strains of adulation; and even he who out-does his predecessors, in swelling up the titles of his divinity, is sure to be out-done by his successors, in newer and more pompous epithets of praise. Thus they proceed; till at last they arrive at infinity itself, beyond which there is no farther progress...[45]

When people are educated from early childhood within a religious framework, they will tend to activate their religious beliefs on many occasions in the basic way. But of course, if the secular explanation of religious beliefs is true, these beliefs will not be properly basic. Very probably, they will not be true either, because they are not produced by truth-conducive mechanisms.

In contrast to the defeating secular explanation of religious beliefs, the reformed explanation of Christian beliefs as being warranted in the basic way is not directly confirmed by any empirical evidence. There is no convincing empirical research showing that we have a *sensus divinitatis* and that this faculty is truth-conducive and reliable, or that there is a Holy Spirit who prompts the full panoply of Christian beliefs in us. As we saw in Chapter 3.4, the extended A/C model is merely a logical possibility. In order to establish that the model is in fact adequate, or at least plausible, and that it is a better explanation of existing Christian beliefs than the secular and defeating explanation, we first have to show that very probably God exists. After all, only if God exists can it be true that he probably created a *sensus divinitatis* in us and restored it after the Fall in some or most Christian believers, and so on. But in order to show that very probably God exists, the Christian believer has to engage in positive apologetics, so that, again, the reformed objection to natural theology turns out to be invalid.

Moreover, if the name of the game is a contest for the best explanation of religious beliefs, reformed epistemologists are landed in some difficult dilemmas. Either (a) they explain religious beliefs by a secular defeating explanation, or (b) they explain them by a religious explanation, which is self-validating in the sense that if the explanation holds, the relevant religious beliefs are probably true and warranted in the basic way. If reformed epistemologists opt for horn (a), they abandon reformed epistemology. If, however, reformed epistemologists choose horn (b), a new dilemma arises. Should they (c) explain *all* religious beliefs by a religious and self-validating explanation, or (d) only some of them? Clearly, horn (c) will lead to contradictions, since it would mean that in order to explain the polytheistic Hindu belief in Ganesha and the innumerable other Hindu gods, one would have to posit all these Hindu gods, whereas in order to explain the belief of Christian monotheists, one would have to hold that only one god exists, that is, God.

Consequently, reformed epistemologists should hold that only their own Christian or monotheistic beliefs have to be explained by a self-validating religious explanation, whereas the religious beliefs of polytheists should be explained by secular defeating explanations. But this would be nothing but a case of special pleading, unless reformed

---

[45] Hume (1757), VI, p. 52.

epistemologists can show that probably Christian theism is true and polytheism is false. So we come again to the conclusion that in order to solve the problem of the plurality of religions, and, indeed, in order to refute an undercutting and rebutting secular explanation of Christian beliefs, natural theology in the sense of positive apologetics is indispensable.[46]

Perhaps some reformed epistemologists are still unconvinced. Surely, they will protest, most Christians do not put forward their Christian beliefs, such as the belief that God exists, as an *explanation* or explanatory *hypothesis* for anything.[47] So how can the Christian be engaged in a contest for the best *explanation* of basic religious beliefs? And '[i]f theistic belief is not proposed as an explanatory hypothesis in the first place, why should its being explanatorily idle, if indeed it is, be held against it?'[48]

But at this stage of the discussion, such an objection is an *ignoratio elenchi*. In order to neutralize the rebutting defeater of an incompatible religious creed, Christians had to hold that the Christian beliefs on the one hand and the incompatible beliefs of other religions on the other hand do not have an equal epistemic weight, or are not on a relevant epistemic par. When faced with potential defeaters, the Christian 'must think that she has access to a source of warranted belief the other lacks'.[49] Suppose, then, that the Christian, who has read her Plantinga, holds that her Christian beliefs result from a well-functioning *sensus divinitatis*, or are produced in her by an internal instigation of the Holy Spirit. As Plantinga admits, this amounts to an *explanation* of her Christian beliefs, which in this argumentative context also functions as a *reason* to hold on to these beliefs in the face of defeaters.[50]

Surely, this explanation can be confronted by another, secular and defeating explanation. Since the Christian explanation is hypothetical, and is credible only if God exists, the existence of God is now put forward as part of the Christian explanation of the existence of Christian beliefs in humans. It follows that at this stage of the discussion, Christians do become engaged in a contest of explanations, which they can win only if they produce evidence or arguments to the effect that God exists. It follows that, if confronted both by incompatible religious creeds and a defeating secular

---

[46] I leave it to the reader to assess the traditional theological strategies for coping with the problem of the diversity of religions, such as the view that the Christian god somehow manifests himself to humanity through the cultural representations of polytheistic religions, or the view that polytheism is due to sin, or, finally, the view that polytheists are like children, who will develop into monotheists when they grow up.

[47] Cf. Plantinga (2000), p. 371: 'The crucial point, here, is that on the [extended A/C] model (and in actuality as well) theistic belief is not ordinarily accepted as an *explanation*. [. . .] On the model, the believer in God ordinarily believes in the basic way, not on the evidential basis of other propositions, and not by way of proposing belief in God as an explanation of something or other' (Plantinga's italics); cf. pp. 330, 370, 386–7, and *passim*.

[48] Plantinga (2000), p. 370.

[49] Plantinga (2000), p. 453.

[50] Plantinga (2000), p. 454: 'She has an explanation: there is the testimony of the Holy Spirit (or of the divinely founded and guided church); the testimony of the Holy Spirit enables us to accept what the Scriptures teach. It is the Holy Spirit who "seals it upon our hearts, so that we may certainly know that God speaks"; it is the work of the Spirit "to convince our hearts that what our ears receive has come from him." She therefore thinks she is in a better epistemic position with respect to this proposition than those who do not share her convictions'.

explanation of religious beliefs in general, Christians (or any other religious believers) have to engage in positive apologetics in order to be rational or justified or warranted in holding on to their beliefs, so that the reformed objection to natural theology is shipwrecked.[51]

---

[51] So it is by combining the problem of religious diversity with the scenario of a secular explanation of all religious beliefs (à la Philip Quinn) that, finally, the reformed objection to natural theology is conclusively refuted. Plantinga (2000), p. 370, correctly objects to Quinn that in order to show that Christian theism is explanatorily idle, it is not sufficient to argue that there is a better explanation *of religious beliefs* than theism, since 'it might still be that theism itself explains lots of *other* things', so that '[t]heistic belief is only *one* of the things that theism can be invoked to explain'. What I am arguing in the main text is, however, that if Christians want to win the competition concerning the best explanation of religious (Christian) beliefs, they have to show that Christian theism is probably true. This might be done by arguing effectively that theism explains many other phenomena, so that these phenomena confirm theism. But in order to do so, Christians must engage in positive apologetics, that is, more precisely, in Richard Swinburne's programme of natural theology.

5

The Rationality of Natural Theology

Engaging in natural or rational theology is indispensable to a religious community, I have argued in the first four chapters of this book, if at least its members want to be rational or justified in holding on to their religious beliefs. But what, precisely, do we mean by 'reason' and 'rationality' in this context? To what type of rationality should the natural theologian aspire? In this chapter, the notion of epistemic rationality will be distinguished from other concepts of rationality, and its various dimensions will be explored briefly.

5.1 CONCEPTS OF RATIONALITY

The notion of rationality most relevant to natural theology is that of *epistemic* rationality, the kind of rationality that is positively related to the probability that a belief is true. In this epistemic sense, people are rational or justified in believing that p if, for example, it is self-evident that p, or if they had a perceptual experience of the state of affairs described by the statement that p, or if they have good reasons or grounds for thinking that p.

We use the term 'rational' in many other senses as well. For instance, if we say that it is rational to do A in order to attain B, we are using the notion of *instrumental* rationality or *Zweckrationalität*, which is concerned with the probability that an action will yield a specific result. In yet another sense of 'rational', we call man a rational animal, meaning that humans, having a language, are endowed with the faculty of reason. On this third sense, many derived senses depend, such as when we call specific persons rational, meaning that they are able to exercise their faculty of reason well and have a sound judgement as a result. In a fourth sense, philosophers speak of 'rational truths' or 'truths of reason', by which expressions they refer to truths that can be discovered without any appeal to experience. Traditionally, scientific disciplines were called 'rational' to the extent that they were concerned with deducing theorems from first principles, so that one spoke of rational mechanics or rational psychology.

Depending on the sense in which we are using the term 'rational', its opposite will be different. Animals are non-rational if they are not endowed with the powers of reason. Human beings are irrational if, being endowed with the faculty of reason, they do not use it properly, whereas all truths that are not truths of reason are empirical.

Accordingly, rational mechanics and rational psychology were traditionally contrasted with empirical mechanics and empirical psychology. I have defined natural or rational theology in contrast with revealed theology, however, so that the former includes empirical arguments for the truth of a specific creed. Acts that clearly do not serve the intended ends may be called irrational, and holding a belief is epistemically irrational if on the available evidence it has a low probability of being true.

The different uses of the term 'rational' are interrelated in many ways, and their interrelations may be misinterpreted easily. It has been argued, for example, that epistemic rationality is always a subspecies of instrumental rationality, the aim being that of having true beliefs and avoiding false ones.[1] But this instrumental conception of epistemic rationality is incorrect for what I shall call *synchronic* or *evidential* epistemic rationality. Unlike deciding to act, having a belief typically is not a matter of choice, whereas instrumental rationality is concerned with the rationality of choosing one course of action rather than another.[2] To put this differently, since having a belief is not an action, it is neither voluntary nor involuntary, whereas the notion of instrumental rationality applies to voluntary actions. But even if having a belief were a voluntary action, the instrumental conception of epistemic rationality differs from the evidential epistemic conception. It may be epistemically rational in the instrumental sense, for example, to believe that God exists, if I also believe that God will reward my belief in him by endowing me with a vast amount of knowledge. But clearly, such a belief that God exists need not be epistemically rational in the evidential sense that, given the total available evidence, the belief is probably true.[3] When I discuss the rationality of religious beliefs, I use the term 'rationality' and its cognates mostly in the evidential epistemic sense.

As a consequence, I shall not be concerned with questions such as whether firmly believing that God or Allah exists will make one happy, in this life or in an afterlife, so that it is rational to try to endorse the belief that he exists. This question is concerned with the instrumental rationality of attempting to acquire religious beliefs, assuming that one wants to be happy, and not with the question of whether it is probable that such beliefs are true. Hence, discussing Blaise Pascal's *wager* falls outside of the scope of this book, as does the *pragmatic test* for religion proposed by William James, who wondered whether religion 'works' for religious believers.

A related issue is whether on balance it is detrimental or beneficial to humanity that people have religious beliefs. Depending upon the answer to this question, one might decide on utilitarian grounds that it is rational to combat and suppress religions, or rather to support and subsidize them. Such a consideration of instrumental rationality may have been one of the reasons why Richard Dawkins wrote his book *The God Delusion*. When in the preface Dawkins imagines 'a world with no religion', what he fancies is a world without many plagues and evils, such as suicide bombers, the attacks of 9/11 and 7/7, crusades, witch-hunts, honour killings, and wars of religion.[4] Dawkins apparently assumes that on balance religions are bad for humanity, and this assumption pervades the rhetoric of his book.

---

[1] Cf., for example, Foley (1987), pp. 6–8, and Laudan (1990).
[2] Cf. Alston (1989), pp. 136–7.
[3] Cf. Mahler (2006), p. 190.   [4] Dawkins (2006), pp. 1–2.

But is the assumption true? If one invites religious believers to imagine a world without religions, they will come up with a very different list of things that would be lacking, such as charities, the beauties of the Notre Dame in Paris or the Dome of the Rock in Jerusalem, moral propriety, Bach's cantatas or the Vedic scriptures, and many other things of great value. Clearly, then, there are conflicting opinions about the issue as to whether on balance religions are detrimental or beneficial to humanity at large. In order to decide who is right here, and whether as a consequence we should fight or foster religions, or perhaps fight some and foster others, we would have to engage in a research project of baffling complexity.

First, we should reach an agreement on the normative issue of what counts as beneficial and what counts as detrimental to humanity. For example, is it beneficial to humanity to conceive of honour in such a way that killing your disobedient wife counts as protecting the honour of your family, or is that detrimental? Second, we should solve difficult problems of definition and causal attribution, such as whether honour killings or Bach's cantatas are to be explained mainly by the influence of religious ideas or mainly by other causes, for we may assume that all such phenomena result from many causal factors. And finally, having resolved these issues, we should engage in a massive project of empirical research in order to sum up all the good and bad effects of specific religions and of religions in general, both in the past and in present times.

Let me confess that such issues of global instrumental rationality are beyond my competence, and, indeed, I think that one must be agnostic about them. As I said above, the (ir)rationality of endorsing religious beliefs discussed in this book is not instrumental but epistemic. If the notion of instrumental rationality is relevant at all to my concerns, this is only so to the extent that epistemic rationality itself is either a means or an aim. It is instrumentally rational to have the highest degree of what I shall call *diachronic* epistemic rationality, if it is an important aim in our life to have true beliefs and to avoid false ones. Being epistemically rational in the diachronic sense to a high degree is the best *means* of attaining this objective of a truth-loving human being. And if being epistemically rational to a high degree is one of our *aims*, it may be instrumentally rational to engage in specific types of research. Unless indicated otherwise, I shall use terms such as 'rational' or 'justified' in the sense of epistemic rationality. Let us now try to get this notion of epistemic rationality and its many dimensions into sharper focus in order to see what the project of rational or natural theology amounts to.

5.2 FOUR DICHOTOMIES

In what epistemic sense, precisely, should engaging in natural theology make it *rational* to hold a religious belief? Philosophers have distinguished many different senses of, and criteria for, epistemic rationality. First, many authors make a distinction between rationality in an *externalist* sense and rationality in an *internalist* sense, which I mentioned in the previous chapter.

Holding a belief is said to be rational or justified in the externalist sense for a subject if the type of process by which it is produced makes it probable that it is true, even though the subject is not, or cannot be, aware of whether the belief has been produced by such a process. Originally, externalist and reliabilist conceptions of rationality or warrant were developed by naturalistic philosophers, who wanted to eliminate as much as possible the notion of awareness or of conscious agents from epistemology. The knowing subject is compared to an input–output machine, such as a scientific instrument for information-gathering, the output of which is said to be warranted if it meets the four conditions of the theory of warrant discussed in Chapters 3 and 4, for example. Ironically, however, a theist such as Alvin Plantinga uses the externalist theory of warrant in order to defend religious believers against the charge of irrationality. As we saw, he argued that subjects can know that the credal statements of Christianity are true although they are not aware of any arguments to this effect, and, indeed, if no such arguments exist. Since the existence and the workings of the *sensus divinitatis* or the internal instigation of the Holy Spirit are processes which occur independently of whether the believer is aware that they occur, the counterintuitive implication of Plantinga's externalism is to remove the possibility of assessing the (ir)rationality of religious belief into a realm that transcends all human capacities.

In the internalist sense, holding a belief that p is rational if and only if the subject is or can be aware of grounds or reasons that make the truth of p probable. These reasons may consist either in arguments that the subject is able to adduce, if the belief is a derived one, or in the fact that the subject has a perceptual experience of the state of affairs described by the statement that p, or remembers to have perceived that p, etc., if the belief is a basic one. When Hume famously observed that 'a wise man, therefore, proportions his belief to the evidence', he was using an internalist notion of epistemic wisdom or rationality, according to which what makes it rational to endorse the belief must be accessible to the subject.[5]

As we have seen in our evaluation of Plantinga's reformed objection to natural theology (Chapter 4), rationality in a purely externalist sense cannot suffice for making it rational to endorse religious beliefs, since modern believers are confronted by the problem of religious diversity, which may function as a defeater of religious beliefs. There are many other defeaters of specific religious beliefs in the modern world, of which educated and intelligent people living in the twenty-first century will be aware. All these defeaters, such as the ones I discussed in Chapter 1, make it indispensable for conscientious religious believers, or at least for some expert in their religious community, to engage in natural theology.

Moreover, a purely externalist conception of epistemic rationality has the inconvenience that subjects cannot adduce the reliability of the processes by which they acquired a belief as evidence for the truth of that belief, since such an argument would transmute their externalist rationality into an internalist rationality. If one thinks, plausibly, that such an argument should always be permitted, there is no reason to exclude the appeal to other types of evidence, so that an at least partially internalist conception of epistemic rationality is preferable to a purely externalist one. Human beings are conscious agents, after all, and they may be called epistemically rational to the extent that they consciously employ their epistemic faculties properly. Engaging in natural theology should make it epistemically rational in an (at least partially) internalist sense to endorse a religious belief.

[5] Hume (1748), Sect. X, Part I, §87.

Let me add some further observations on the externalist/internalist distinction, which support this conclusion. Although it may seem that both the externalist and the internalist philosopher of knowledge aim at answering one and the same question, the question as to whether holding a belief that p is rational, the respective questions they are answering are in fact quite different. The (reliabilist) externalist is wondering whether the belief that p is *caused* by a reliable process, or is the *result* of the workings of a properly functioning epistemic faculty. The internalist is wondering whether there are good *reasons* or *grounds* for thinking that it is true that p, or whether a specific subject is able to adduce such reasons or grounds.

But beliefs in the sense of what can be true or false are not identical with beliefs in the sense of things that can be caused. What can be true or false, right or wrong, correct or incorrect, is the *content* of the belief, *what* is believed, namely *that p*. We may say, accordingly, that it is false that p, or that it is true that p. However, if it makes sense at all to say that a belief is caused, or is the result of a process, what is caused is our believ*ing* that p, and not *what* is believed, namely *that p*. The question 'what is the cause of *that p?*' clearly is a nonsensical question. In general, the set of predicates that we may meaningfully apply to *what* we believe differs from the set of predicates meaningfully applicable to our *believing* something. Whereas *what* we believe may be informative or sterile, correct or incorrect, right or wrong, complicated or simple, justified or not justified, and true or false, our *believing* it may be firm or tentative, obstinate or hesitant, passionate or reluctant, and rational or irrational. It does not make sense to say that what is believed is firm, or obstinate, or reluctant, and it does not make sense either to say that our believing something is true, or complicated, or informative.

For the internalist, the primary question is whether *what* someone believes is *justified* either as a basic belief or inferentially. The question as to whether subjects are *rational* in believ*ing* certain things is a derived one. Subjects are rational in the internalist sense to the extent that they hold beliefs that are justified, have access to the justifications, and hold the beliefs on the basis of those justifications. For the externalist, however, the primary question is concerned with *believings*: how are they caused? In fact, we should distinguish sharply between three different questions concerning the epistemic rationality of the belief that p.

The first question is concerned with the content of the belief without any regard for the subjects who hold it. The question whether, or to what extent, the belief that p is epistemically justified then means: to what extent is it logically or inductively probable that p, given a specified evidence set or collection of premises?[6] This first question merely pertains to the credentials of a belief or proposition given a specified evidence set and specified background assumptions, and not to the rationality of believers. The two other questions are concerned with the epistemic rationality of a specific subject S who holds that p. Internalist epistemologists will wonder whether believing that p is rational *for* S given the evidence available to S and S's background assumptions, or given the grounds on which S bases the belief, since they will hold an evidential

---

[6] Logical or inductive probability may be defined as the probability that a proposition q is true given that another proposition p is true. My notation for this type of probability will be $P(q|p)$.

conception of (synchronic) epistemic rationality.[7] For the internalist, the answer to this second question depends on the answer to the first. Externalist epistemologists wonder whether S's believing that p was caused or produced by a mechanism of a type that makes it probable that the resulting beliefs are true. In a wide sense of 'internalism', the first two of these three questions are internalist, whereas in a narrow sense, only the second question is raised by internalist epistemology.

Natural theologians are interested primarily in the first question as to whether the contents of specific religious beliefs are justified in the sense that they are probably true, given the total available evidence. They also want to discover new evidence and new arguments for or against specific religious doctrines. To what extent a certain subject S is rational in holding these beliefs or doctrines is of derivative interest only, and the natural theologian is not at all interested in the externalist question whether S's believing these things is caused in one way or another. Indeed, the relation of causation is very different from the relations between a proposition and the evidence adduced for it, which are logical relations in a broad sense.

In the definitions of the internalist/externalist dichotomy that one finds in the literature, this causal/logical distinction is often mixed up with other distinctions. One is whether the subject is or is not aware of what makes it rational to believe that p, or whether the subject can or cannot be aware of this. Yet another distinction is that between what is internal in the subject's body or brain and what is not in the subject's body. And a third distinction is between things to which subjects have privileged access, in the sense that they have a way of knowing them that nobody else has, and things to which a subject does not have privileged access. Since all these distinctions are different from each other, whereas some of them are problematic, it would perhaps be better to drop the *terms* 'internalism' and 'externalism' altogether.

In the remainder of this book, I shall use labels such as 'rational' and 'justified' in the wide internalist sense as defined by the dichotomy between reasons and causes. Within the domain of internalist epistemic rationality in this wide sense, we can make many further distinctions. One is between the rationality of holding a belief given the evidence we possess at a specific time on the one hand, and the rationality of holding a belief as a result of an enquiry one has engaged in over time on the other hand. I shall call these two notions *synchronic* and *diachronic* epistemic rationality, respectively.[8] Another distinction is that between the epistemic rationality of *basic* beliefs and the epistemic rationality of *derived* beliefs, assuming that we endorse some version of weak foundationalism. According to a strict definition, a belief is basic if and only if we accept it without supporting it by any other of our beliefs. However, since many of our paradigmatically basic beliefs, such as beliefs based upon perception, are also supported by other beliefs with which they cohere, a broader definition of basic beliefs is perhaps to be preferred. According to this broader definition, a belief is basic if and

---

[7] According to an *evidential* conception of epistemic rationality, a subject S is epistemically rational in believing that p if the belief that p is (sufficiently) supported by S's evidence. One might distinguish this evidential conception of epistemic rationality from an *instrumental* conception of epistemic rationality, according to which having specific beliefs is rational for a subject if endorsing them is conducive to goals such as believing truths and avoiding error (cf. §5.1, above). That these conceptions of epistemic rationality are different is convincingly argued by Mahler (2006), p. 190.

[8] This is the terminology used by Richard Swinburne. Cf. Swinburne (FRb), p. 44, and (EJ), p. 3 and *passim*.

only if we accept it without relying *only* on other beliefs as possible arguments for it, and a belief is derived if we rely *only* on other beliefs.

A basic belief may be called rightly or properly basic in an internalist sense if and only if it satisfies criteria for the rationality of holding basic beliefs, such as being grounded by a perceptual experience, or being self-evident. To what extent it is rational to hold a derived (non-basic) belief will depend on three factors: on the rules of inference that are used, on the question of whether they are used correctly, and on how rational it is to endorse the basic and background beliefs adduced in order to justify the derived belief. The set of basic beliefs adduced by a subject S in order to justify a derived belief b about a factual issue may be called S's 'evidence set' with regard to b.

We may complicate matters even further by using yet a third distinction between types of internalist epistemic rationality. On the one hand, subjects may be rational in holding a belief that p by their own lights or *subjectively*, because the belief seems to them justified either as a basic belief or as a derived belief. On the other hand, subjects may be rational in holding a belief *objectively*, in the sense that their belief is indeed justified.[9] We may refine this rough distinction between *subjective* and *objective* internalist epistemic rationality by introducing intermediate notions, such as a belief being rational according to a community of experts in the light of the evidence set that is available to them at a given time, and of the best set of inferential rules known at that time.

5.3 TYPES OF SYNCHRONIC INTERNALIST RATIONALITY

By combining the three distinctions within the broad domain of internalist epistemic rationality, we may define different subspecies of epistemic rationality, and determine quite precisely in which sense and to what extent a subject S or a community of subjects is epistemically rational in holding a given belief that p. Suppose, for example, that an attractive young girl has been murdered with a knife in a park, and that nearby a drunken man is found with blood of the girl on his hands and clothes. The police may be diachronically and objectively rational in accepting a specific evidence set, because they left no stone unturned to investigate all possible traces in the park and all facts concerning the two people involved. Furthermore, the public prosecutor may be synchronically and subjectively rational in concluding from the evidence set with which he is presented by the police, that quite probably the drunken man committed the atrocity. And yet, the prosecutor may not be diachronically and objectively rational in accepting this derived belief, because, being under time pressure to come up with a suspect for this horrendous crime, he was hampered by tunnel vision and omitted to initiate an investigation of other possible perpetrators. Although he was correct in concluding that, given the evidence set, it was more probable that p (the drunken man committed the crime) than that not-p, he was diachronically and objectively irrational in implicitly assuming that the hypothesis that p was more probable than each of a number of alternative hypotheses.

---

[9] Cf. for this distinction Swinburne (FRb), p. 53, and (EJ), p. 3 and *passim*.

In his book *Faith and Reason*, Richard Swinburne somewhat schematically distinguishes between five specific notions of internalist epistemic rationality, two synchronic and three diachronic, which he numbers by subscripts as rationality$_{1-5}$.[10] I shall now briefly review these notions, although I do not think that the details of Swinburne's account matter too much. As we shall see, he could have distinguished easily between many more notions of internalist epistemic rationality. What is important, however, is to show that such distinctions should be made if we want to be more precise in attributing rationality to a subject who holds a belief. My ultimate aim in the remainder of this chapter is to determine to which type of rationality the rational or natural theologian should aspire. Let me first discuss Swinburne's two notions of synchronic internalist epistemic rationality, which he calls rationality$_1$ and rationality$_2$.

A subject S is synchronically rational in believing that p in the *subjective* sense (rational$_1$) if and only if S's belief is justified by S's own lights at a given moment. In the case of a derived belief, for example, this means that S thinks that the belief is rendered probable, given S's rules of inference, by the total set of what S at a given moment accepts as S's evidence. But of course, S may be wrong on one or more of these three accounts. Beliefs that S accepts as basic may not be properly basic; S's rules of inference may be incorrect, or S may apply these rules in an incorrect manner. If the subject is right in regarding the belief justified on all these accounts at a given moment, S has a synchronically *objectively* rational belief (rationality$_2$). For example, a derived belief is rational$_2$ if and only if the basic beliefs accepted by S are properly basic, and if these basic beliefs in fact render the derived belief as probable as S believes it to be according to the correct rules of inference.

There is some lack of clarity in Swinburne's own definitions of these two kinds of synchronic internalist epistemic rationality, because in formulating them he does not distinguish between the rationality of derived and of basic beliefs.[11] Also, in the case of accepting derived beliefs, subjects may be rational$_2$ in some respects but merely rational$_1$ in other respects. For example, they may be rational$_2$ in accepting their rules of inference, because they are the correct rules, and yet be irrational$_2$ or merely rational$_1$ in applying them, because they misapply them. According to Swinburne's official definition of rationality$_2$, subjects are rational$_2$ in accepting a belief if and only if they are objectively rational in all respects.

One may wonder whether a subject can be irrational$_1$ at all. According to Swinburne, subjects would be synchronically subjectively irrational or irrational$_1$ if there is a failure of internal coherence in their system of beliefs, a failure of which the subjects are unaware.[12] But if the core idea of rationality$_1$ is that one is synchronically rational by one's own lights, an incoherence of which the subject is unaware should not count as a case of irrationality$_1$. Rather, it would be irrational$_2$ only. If there are cases of irrationality$_1$, they must be cases of internalist epistemic bad faith, such as accepting as

---

[10] Swinburne (FRb), pp. 52–7, 66–76. Cf. also (EJ), Chapters 6 and 7.
[11] For example, Swinburne defines the subjective synchronic rationality of a belief as follows (FRb, p. 53): 'A subject S who believes that p will have what I call a synchronically subjectively justified belief or rational$_1$ belief, if and only if his belief that p is based on and rendered probable, given his inductive criteria, by his total actual evidence'. Strictly speaking, this definition applies to derived beliefs only.
[12] Swinburne (FRb), p. 53.

basic a belief which the subject knows is not properly basic, applying a rule of inference which the subject knows is incorrect, or being aware that one misapplies a rule of inference and yet accepting the conclusion one fallaciously arrives at. For example, terminally ill persons may be informed by the doctors that they have only one week left to live, and nevertheless refuse to believe that this is true.

If, however, Swinburne is right in arguing like many epistemologists that, as a matter of logic, we cannot choose at will what to believe, because, for example, we believe that p when we see that p, or when the total set of evidence available to us makes it probable that p, irrationality$_1$ is logically excluded.[13] In other words, if beliefs are 'involuntary' or rather non-voluntary, epistemic bad faith in the synchronic sense is difficult to conceive of. Perhaps we should say of the terminally ill persons that although they in fact believe that they have only one more week to live, since all doctors who examined them told them so, they still hope fervidly that this is not true. The situation is different concerning diachronic internalist epistemic rationality, to which I now turn.

5.4 TYPES OF DIACHRONIC RATIONALITY

In general, we may speak of diachronic internalist epistemic rationality of a belief if that belief is the result of an adequate investigation over time. Engaging in natural or rational theology should aim at some kind of diachronic rationality, because it means doing intellectual research with the purpose of finding out which beliefs in the religious domain are most probably true. Since each investigation will yield some kind of synchronic rationality, that is, a set of basic beliefs and a set of inference rules in the light of which specific beliefs will be rational$_1$ or rational$_2$, we may define in principle four global notions of diachronic internalist epistemic rationality by using again the simple dichotomy of rationality in the subjective or the objective sense. A belief may be rational$_1$ or rational$_2$ as the result of an investigation which was adequate by the subject's own lights, and it may be rational$_1$ or rational$_2$ as the result of an investigation which was objectively adequate.

Somewhat surprisingly, Swinburne defines only three diachronic notions of internalist epistemic rationality instead of these four. He calls a belief rational$_3$ if and only if it results from an investigation *which the subject regards* as adequate according to the subject's own criteria, that is, subjectively adequate in two respects. A belief is rational$_4$ if it is rational$_1$ and results from an investigation that is *in fact* adequate according to the subject's own criteria, hence subjectively adequate in one respect and objectively adequate in another. Swinburne calls this a 'diachronically subjectively justified belief'. Finally, he defines a belief as rational$_5$ if it is rational$_2$ and results from an investigation that is in fact adequate according to the correct criteria for adequacy, that is, objectively adequate in two respects.[14]

---

[13] Swinburne (FRb), pp. 24–6. Cf. (EJ), pp. 39–40. This view is usually called 'doxastic involuntarism'. But this is misleading, since only actions can be voluntary or involuntary, whereas believings are not actions.

[14] Swinburne (FRb), p. 70; cf. (FRa), pp. 49–54.

Swinburne's distinction between rationality$_3$ and rationality$_4$ is somewhat over-sophisticated, since, like a rational$_4$ belief, each rational$_3$ belief must also be a rational$_1$ belief. It follows that the distinction between rationality$_3$ and rationality$_4$ boils down to a difference between the subject's criteria of adequacy being applied correctly according to the subject and the subject's criteria of adequacy in fact being applied correctly. If one takes this subtle distinction into account, however, one should have defined eight different notions of diachronic internalist epistemic rationality instead of three.[15] No doubt it would have been simpler if Swinburne had defined the four notions of diachronic internalist epistemic rationality of beliefs that I indicated above. This does not matter much, however, since it is clear that someone who engages in natural theology should aim at rationality$_5$, that is, at beliefs which are rational$_2$ as a result of an investigation which is in fact adequate according to the correct criteria of adequacy. But what are these correct criteria of adequacy? When is an investigation objectively adequate?

5.5 CRITERIA OF ADEQUACY

There are two interrelated dimensions to this question of the adequacy of research. One dimension is that of rational decision theory or instrumental rationality. Here, the issue is how much research a given person should carry out regarding the truth of a specific set of beliefs, and this may be different for different persons. In Chapter 1.5, I mentioned four criteria that are relevant here: (1) it should not yet be obvious to the person which beliefs are true; (2) whether they are true should have importance for the person; (3) it should be probable that doing research makes a difference in that it yields beliefs that are more likely to be true; and (4) the probable costs of research in terms of time and money should not be too high, given our other purposes and occupations.

Swinburne defines the diachronic rationality of research in terms of these criteria, subsuming criterion (1) under criterion (3), because (1) is among the factors that determine the probability of (3).[16] The specific values of criteria (2) and (4) will be relative to a given person. In Swinburne's own case, they are higher than in the case of most of us, because as a Nolloth Professor of the Philosophy of the Christian Religion he was paid to do research in natural theology. Accordingly, it was both a legal and a moral obligation for him to engage in research of this kind, and it will be much harder for him to live up to these standards of adequacy than it is for other philosophers, let alone for the non-philosophical layperson.

The second dimension to the question of adequacy is a purely internalist and epistemic one, which may be conceived of as pertaining to criterion (3), above. Although criterion (3) is formulated in terms of instrumental rationality concerning the marginal utility of research, the notion of epistemic rationality comes into it as well

---

[15] The reason is that both the criteria of adequacy adopted by the subject and the objectively correct criteria of adequacy can be applied correctly either according to the subject or objectively, whereas the resulting beliefs may be rational$_1$ or rational$_2$ in all these four cases, so that eight different cases result.

[16] Swinburne (FRb), p. 70: 'A belief which is not merely synchronically justified but has been investigated adequately in the light of these considerations, I call a diachronically justified belief'.

for the following reason. It is more probable that doing research will yield beliefs which have a higher probability of being true, to the extent that this research is carried out by using better methods, since a method of research M_1 is better than another method M_2 to the extent that the probability that M_1 yields informative and true results is greater than the probability that M_2 yields such results. It follows that the choice of methods is an important issue for evaluating the adequacy of the research that has been done.

In general, one may say that research is more adequate in this respect, and that endorsing the resulting beliefs is more rational in the diachronic sense (and the beliefs more justified), to the extent that the chosen methodology is a better one. Endorsing the resulting beliefs is rational$_3$ in this respect only if the research is carried out in conformity with the methodology that is best according to the subject, and the methodology is correctly applied by the subject's lights. Endorsing the resulting beliefs is rational$_4$ only if the methodology that is best according to the subject is in fact correctly applied. Finally, endorsing the resulting beliefs is rational$_5$ only if they are yielded by a correct application of the best method, objectively speaking. Clearly, natural or rational theologians should aim at rationality$_5$ in this respect. As a consequence, the choice of the best methods of research is a matter that deserves serious consideration.

6

A Grand Strategy

Which methods of religious investigation should natural theologians choose to employ if they aim at being diachronically rational$_5$ with regard to their religious beliefs, that is, if they aspire at endorsing religious doctrines that are objectively diachronically justified (cf. Chapter 5.4–5)? As I argued, rationality$_5$ can be realized only if one employs methods of research that are objectively best or adequate. The answer to the question as to which particular methods of research are adequate will depend upon the subject matter someone is interested in. If one wants to discover the original meaning of an obscure passage in an old text, one must use the scholarly methods of historical interpretation, and if one wants to know whether there is a super-massive black hole in a galaxy, one should apply various methods of astronomical observation and gravitational calculation. Consequently, what methods natural theologians should choose in order to test and confirm their religious beliefs will depend upon the content of these beliefs.

Historians and anthropologists have discovered an astonishing variety of religious creeds, and each of these creeds is open to many different interpretations. Can there be a general answer, then, to our question concerning the adequate methods of religious research? In this chapter, we shall explore the methodology of natural theology. It will be argued that in our scientific age the natural theologian is faced with a distressing methodological dilemma. This dilemma is so central to the epistemic prospects of religions that the solution which a natural theologian adopts may be called his or her *Grand Strategy*. I shall argue that in his natural theology of bare theism, Richard Swinburne has opted for a grand strategy which, at first sight, is most promising. But even this grand strategy is faced with a double risk, which I call *The Tension*. Let me begin by making some general observations on methodology and the validation of methods.

6.1 METHODS OF RESEARCH

The student of methodology may characterize methods of research at many levels of abstraction or generality. For purposes of review, I roughly distinguish between three levels. At the lowest level of generality (a), we find domain-specific methods, which depend heavily on specialized knowledge about the relevant domain. For example, scientific methods for measuring the temperature in the centre of the Sun, or methods for discovering the masses of black holes in galaxies, cannot be developed or described without using detailed physical knowledge. Similarly, scholarly methods for detecting interpolations in ancient texts presuppose knowledge of stylometry, of the ancient culture, and of the cultural preconceptions and habits of later copyists.

At an intermediate level of abstraction (b), there are methods that do not presuppose domain-specific knowledge and yet cannot be applied everywhere in all disciplines. Statistics provides an example of a set of methods at this level, since in many areas of research, such as the interpretation of philosophical texts, statistical methods are of no or limited use. At the highest level of generality (c), the student of method attempts to develop a methodology that applies to all research in all areas of the pursuit of truth, that is, not only the sciences but also the humanities. Well-known philosophies of science proposed in the twentieth century, such as Karl Popper's falsificationism, Imre Lakatos' methodology of scientific research programmes, or Bayesian confirmation theory, aim at this highest level of generality, although it is a matter of dispute to what extent these methodologies also apply to non-empirical or non-factual disciplines such as mathematics.[1] In all mature areas of empirical or factual research, we find a great wealth of methods at the lower levels (a) and (b), so that experts in these areas rarely take the trouble to get acquainted with the philosophy of science, which is largely situated at level (c).

One may regard the history of the philosophy of science at level (c) as a learning process, in which philosopher-scientists proposed normative models of knowledge and justification, inspired by paradigmatic developments in mathematics and the empirical sciences. During the most revolutionary periods, such as the seventeenth century or the early twentieth century, there was a lively interaction between philosophers and practising scientists or mathematicians, who mutually influenced each other. At the highest level (c) of generality, the major global development in the philosophy of science since Antiquity has been the replacement of the classic Aristotelian conception of *episteme* or *scientia* by a more modest model of theoretical science. According to Aristotle's conception, real knowledge should have the form of an axiomatic-deductive theory, the axioms of which are certain and necessarily true. Modern philosophers of science have dropped these requirements of certainty and necessity, and they also have acknowledged the diversity of styles of research and of theory construction in the sciences and the humanities.

Whereas Aristotle's model of real knowledge was inspired by, and in its turn inspired, the development of axiomatics in mathematics, the invention of non-Euclidean systems of geometry during the nineteenth century decisively refuted the idea that all axioms of mathematics must be necessarily true, and it revolutionized the philosophy of mathematics. In the empirical sciences, the requirement that there had to be necessarily true universal first principles, proposed by Aristotle in his *Posterior Analytics* and endorsed by philosophers such as Aquinas, Descartes, and Kant, was discredited by developments in physics during the early twentieth century, such as quantum mechanics and general relativity theory (cf. Chapter 2.4). These theories incorporated elements, in particular a non-deterministic microphysics or a Riemannian geometry, which are incompatible with principles that were considered necessarily true by Kant

---

[1] With regard to his own methodology, Lakatos (1976) claims that this is the case. In my (1998), I have argued that the hypothetico-deductive method should be applied in the scholarly craft of textual interpretation.

and his later followers around 1900. We may conclude that, although the philosophy of science is a normative discipline in the sense that it recommends rules of method, its models of knowledge and of method can be discredited empirically by developments in the sciences themselves, for example if theories are successful even though they violate these models.

6.2 VALIDATION OF METHODS

At the lowest level of generality (a), empirical assessments of methods of research are of a more direct nature. For example, we may test the reliability of new techniques of measurement or observation by using other techniques, the reliability of which has been established already, or by doing statistical research, or by more specific methods of calibration and triangulation. Such an empirical validation of methods at level (a) is part and parcel of research in the sciences, and a scientific discipline cannot exist without it. The same holds, perhaps in a somewhat mitigated sense, for methods in the humanities, such as methods of textual interpretation and historical investigation. Whereas in some cases the validation of methods and techniques is easy and unproblematic, it may be difficult and a matter of intense controversy in other cases. Let me discuss briefly two well-known examples taken from scientific disciplines in order to illustrate what I mean, and to stress that no discipline of empirical or factual research can be taken seriously if it cannot validate its methods of investigation at level (a).

The first example is concerned with microscopes. If a low-powered electron microscope reveals small dots in red blood platelets, the question arises whether these 'dense bodies' are important elements in blood biology or artefacts of the electron microscope. At least three types of procedures and arguments may be used to show that the dense bodies are real elements instead of optical illusions. First, one may attempt to see the same bodies by using very different physical techniques. In fact it turns out that the dense bodies are also revealed by fluorescent staining and subsequent observation by the fluorescent microscope. Implicit in this first procedure of validation is an *argument from coincidence*. If two physical processes, which have not much in common, produce identical visual configurations, it would be an extremely unlikely coincidence if these configurations were artefacts of the detection methods rather than resulting from real structures in the object under investigation.

Second, we may validate the microscopes by producing macroscopic objects with structures perceptible to the naked eye, which are then reduced in size by a reliable procedure which we know leaves the structures intact, such as photographic reduction. If we discern the very same structures by looking through the microscope, this validates our instrument also for the perception of the dense bodies. Further validations may be obtained, thirdly, by acquiring a theoretical understanding of both the microscopes and of the structures detected. As Ian Hacking observed in his well-known discussion of microscopes,

[w]e become convinced of the reality of bands and interbands on chromosomes not just because we see them, but because we formulate conceptions of what they do, what they are for.[2]

---

[2] Hacking (1983), p. 205. My discussion of this example is based on Chapter 11 of his book.

A second well-known example is Joseph Weber's alleged discovery of large amounts of gravitational radiation coming from space by using a new type of detector of his own design, which he reported in 1969. This discovery was surprising because the amount of radiation he claimed to have detected was far greater than predicted by astronomers and cosmologists on the basis of general relativity theory. Although most scientists agree that according to this theory, moving massive bodies will produce gravity waves, these waves are supposed to be so weak that it is excessively difficult to detect them. Although we cannot generate on Earth discernable amounts of gravitational radiation, it is assumed that violent events in the universe, such as exploding supernovae, black holes, and binary stars, should produce fluxes of gravity waves, which would show themselves on our planet as a tiny oscillation in the value of the constant G related to the gravitational pull of one object on another.

Weber's detector consisted of a massive aluminium alloy bar, weighing several tons, suspended in a vacuum chamber. The gravity waves are supposed to produce changes in the length of the bar by causing changes in the gravitational attraction between its parts. Since such a bar could not be expected to change its dimensions by more than a fraction of the radius of an atom under the influence of pulses of gravitational radiation, it had to be constructed very carefully in such a way that it would oscillate or 'ring' as a bell at the same frequency as the radiation if the latter occurred, which signal then had to be amplified by some device and recorded by another device. Furthermore, since this complex system was built to trace gravity waves by means of detecting vibrations in the metal bar, it had to be insulated very carefully from all other possible causes of vibrations, such as acoustic, electrical, magnetic, or seismic forces. Finally, since there will always be vibrations due to the random movements of the atoms in the bar as long as the temperature is above absolute zero, one has to decide which recordings of oscillations rise above the threshold of background 'thermal noise'.[3]

Although Weber's results were at variance with theoretical predictions, he managed to obtain confirmations of his measurements that looked impressive at first sight. For example, above-threshold peaks of vibrations seemed to be detected simultaneously by two or more detectors which were a thousand miles apart, a coincidence that could be explained plausibly only by some extra-terrestrial disturbance, such as gravity waves. Another coincidence, calling for the same type of explanation, consisted in an apparent 24-hour periodicity of the discovered peaks, related to the astronomical day and not to the Sun, suggesting a source outside the solar system. Initially, the fact that other teams of scientists could not reproduce Weber's discoveries did not seem a decisive argument against his results, because it is excessively difficult to set up such experiments without flaws. But then it was discovered that the first confirmation (simultaneous results) was based upon a confusion of time zones, so that the two sections of the tapes that Weber had compared in fact had been recorded more than four hours apart. Also, the second confirmation of the 'sidereal correlation' faded away over time, whereas Weber did not manage to increase the signal-to-noise ratio of his results over the years when he

---

[3] See for this example Collins and Pinch (1998), Chapter 5. I am summarizing their account very briefly.

improved his apparatus. By 1975, most astrophysicists were convinced that Weber's experiments had not been well designed, and that his method of detection was invalid.

What these examples show is that the validation of methods of research at level (a) is part and parcel of domain-specific research. Validation may be quite difficult, so that much time, energy, and money has to be invested in it. No discipline of factual research can be credible if it is not able to validate some of its domain-specific methods of research and to de-validate others. As I said already, this observation holds not only for the sciences but also for the humanities, although both of my examples were taken from scientific disciplines of factual research.

6.3 THE METHODOLOGICAL DILEMMA

After these introductory remarks on methods and their validation, we may introduce the methodological dilemma for the natural theologian by the following series of considerations. Apologetic natural theologians aim at arguing for the truth of a religious creed of their choice and against rival creeds of other religions, to the extent that these latter doctrines are incompatible with their own. In doing so, they purport to use the very best, or at least adequate, methods of research at all levels, in order to acquire beliefs that are rational$_5$. But how should we characterize religious creeds in general? Can we answer the question as to what methods of research at level (a) natural theologians should choose, irrespective of the particular content of their creed?

Anthropologists have stressed how difficult it is to give a general definition of the term 'religion', an equivalent of which is lacking in many languages. Religions have several aspects, such as rituals, social organizations, casts of priests, architecture, moral prescriptions, art, and a set of beliefs, the relative importance of which may be different in different religions. There are even religions in which some of these aspects are lacking altogether, so that the philosopher will say that the notion of religion is a family resemblance concept. However, since natural theology is concerned with religious creeds only, the following general definition of the term 'religion' seems to be appropriate for our purposes.

As far as their doctrinal contents are concerned, I define religions as systems of beliefs based upon the factual claim or presupposition that there are hidden powers, such as spirits, demons, and gods (or that there is one hidden power, called Allah, God, Yahweh, etc.), which cannot be discovered by biological research although they resemble human persons to some extent, and which may influence natural events, the course of human life, and our life after death, if there is such a thing.[4] Accordingly, most methods of religious research at the lowest level of generality (a) aim at finding out which hidden powers exist, what these powers require of us, and what they intend to do. Domain-specific methods of research in religion are intimately related to methods of influencing the intentions of these powers, such as rituals, prayer, and sacrifice.

---

[4] In this restricted sense, austere forms of classical Buddhism are not religions, although they are religions in the wider family resemblance sense of the term.

Since the powers postulated by religions are assumed to resemble human persons to some extent, specific methods of religious research at level (a) are all modelled upon modes of communication between humans, such as listening to voices, receiving revelations from someone, or reading off intentions from external signs. Sometimes these methods are based upon a more specific assumption that some type of extraordinary human behaviour in fact contains a message communicated by a power or god, which has to be interpreted by a religious expert or priest. For instance, what we now diagnose as the foaming of epileptics was interpreted until recently, and is still interpreted in large parts of the world, as an intervention by a demon or a god.[5] To mention yet another example, recent geological research of the oracle-site in Delphi has shown that probably the Pythia was intoxicated by gases such as methane, ethane, and ethylene, low concentrations of which can induce states of trance in human subjects.[6] Priests of the oracle interpreted the confused utterances of an intoxicated Pythia as communications by Apollo.

In modern Western culture, religions are confronted with an embarrassing problem concerning such methods of religious research at the lowest level of generality (a). We may safely say that nowadays all domain-specific methods of religious research have lost their intellectual respectability in the light of scientific and scholarly advances, both because they fail tests of empirical validation and because the background theories which they presuppose have been superseded by better theories. For example, predictions of future events based upon revelations, clairvoyance, reading divine signs such as the flight of birds, or oracles have turned out to be unreliable, and in the developed countries we now, on good grounds, consider epileptics as people who suffer from an illness instead of regarding them as subject to influences of demons or as mediums for divine messages.[7] The same fate has struck all methods of influencing the course of events by communicating with postulated hidden powers, such as ritual healings, sacrifice, voodoo, or prayer. Attempts at empirical validation have not been able to show that, apart from incidental placebo effects, such methods have any discernable result.

Let me illustrate this claim by the example of prayer. Both historians of religion and anthropologists will confirm that in pre-scientific cultures, prayer to gods is generally considered an effective means of influencing the future course of events, although the perceived effectiveness of prayer depends upon the assumed reliability and moral qualities of the god concerned. But prayer is not only an instrument of influencing the future; it is also a method of religious discovery. This dual function of prayer resembles the dual function of methods of physical manipulation used at level (a) in the sciences. Prayer may even be used to stage a crucial experiment, which aims at showing which of two rival religious creeds is true. In the tradition of Jewish, Christian, and Islamic monotheism, the best example of such a use of prayer for an

---

[5] Cf. Carrazana et al. (1999); Dewhurst and Beard (1970); Jilek-Aall (1999); and Wohlers (1999).
[6] Cf. J. R. Hale et al. (2003). Of these gases, ethylene is most hallucinogenic.
[7] This is still different in developing countries. Cf. Jilek-Aall (1999). Of course, one would commit a genetic fallacy if one deduced from the fact that the confused utterances of the Pythia or the foaming of the epileptic have a purely secular explanation, that no god or demon is communicating with us through such utterances. Yet on any acceptable probability distribution, discovering the natural causes of these phenomena lowers considerably the probability that a god or demon is at work.

experimentum crucis is provided by the story of the prophet Elijah at Mount Carmel, as related in the Old Testament, 1 *Kings* 18.20–46.

According to this text, Prophet Elijah once asked King Ahab, son of Omri, to gather all people of Israel together at Mount Carmel, including the four hundred and fifty prophets of Baal. There, Elijah challenged these rival prophets and proposed the following test. Both the prophets of Baal and Elijah would prepare a bull on a pyre as a sacrifice to their god, without putting fire to the wood. Then they would pray and ask their god to ignite the pyre. The god who would answer by fire would be the real god. When Elijah put forward this proposal, all the people accepted the test – 'It is well spoken,' they said – and Elijah let the prophets of Baal proceed first. However, after they had built their altar, prepared their bull, and had prayed for a long time, there was no fire or voice, and no one answered. When Elijah then mocked the prophets of Baal, saying that perhaps their god was on a journey or asleep, they 'cried aloud, and cut themselves after their custom with swords and lances, until the blood gushed out upon them'.

After this exposure of Baal as a false or non-existing god, Elijah prepared his own altar, asking people to pour water over the wood three times in order to make the test even more decisive. He alone prayed to God, and, as the reader will expect, 'the fire of the Lord fell, and consumed the burnt offering'. When the people witnessed this miraculous intervention, they were convinced immediately, fell on their faces, and said repeatedly: 'The Lord, he is God'. Then Elijah asked the people to seize the prophets of Baal and he himself killed them all near the brook Kishon. In this story, prayer is used as a method of discovering which god is real, and a god is real only if (s)he unambiguously answers the prayer of a devoted believer by doing what the believer asks for. Can this method of prayer be validated by modern means as a reliable method of religious research? There are two recent double blind tests that provide an answer to this question.

One is the Mantra II research into the effectiveness of petitionary prayer for patients in hospital, as reported in *The Lancet* of 16 July 2005. The religious prediction that prayer has a wholesome effect on patients was tested by a 2x2-randomized trial concerning 748 patients undergoing percutaneous coronary intervention or elective catheterisation in nine centres in the United States during the years 1999–2003. Prayer groups of believing Christians, Muslims, Jews, and Buddhists said prayers for half of the patients during and up until four weeks after an operation, and the patients did not know whether such groups prayed for them or not. Researchers could not detect any influence of the prayers on the subsequent medical condition of the patients.[8]

Even more impressive in terms of size and sophistication is the three-year Study of the Therapeutic Effects of Intercessory Prayer (STEP), sponsored largely by the John Templeton Foundation, the results of which were published in the *American Heart Journal* of 4 April 2006. It examined 1,802 patients undergoing heart-bypass surgery. A team of on average 33 Protestants and Roman Catholics said petitionary prayers during the two weeks of hospitalization for the health of 604 patients who did not know that prayers were said, and also for the health of 601 different patients who were

---

[8] Krucoff et al. (2005). Cf. also Galton (1872).

told that people prayed for them. No prayers were said for a third group of 597 patients, who did not know whether prayers were said for them or not. Whereas there turned out to be no statistically significant differences in medical condition between the first and the third group, surprisingly the second group performed significantly worse than the other two.[9]

Monotheists might object to these investigations that they cannot be single- or double-blind if one of the test-subjects is omniscient. Indeed, why would an omniscient God answer prayers said in order to test him?[10] Furthermore, the marginal effect of the prayers in the experiments might be undetectably small in comparison to the effects of all prayers said in the world at the same time for all the sick. And which hypothesis did the investigators want to test, given the fact that they let prayer groups of different religions say prayers at the same time for the same patients? Do religions not typically claim that only their own prayers are effective? Although these and other objections cannot be refuted easily, they do not show that believers may go on saying petitionary prayers with a good intellectual conscience. For the fact is that up to the present it has not been possible to validate the method of petitionary prayer, since empirical investigations suggest that prayers have no effect at all on those for whom they are said.

If in our times neither the domain-specific methods of religious research, nor the related technologies of influencing the course of events that are inspired by religious hypotheses or creeds, can be validated by empirical tests, natural theology faces a serious methodological impasse. There is no doubt that, originally, religions such as Christianity, Hinduism, Islam, or Judaism pretended to possess a great amount of factual religious knowledge, which allegedly had been verified by domain-specific methods of religious research. Since serious attempts to validate these methods of religious research and communicative technology have failed, whereas the background beliefs which supported them have been superseded by scientific and scholarly developments, one might characterize religions somewhat metaphorically as 'degenerating research programmes' in the sense defined by Imre Lakatos.

Of course it will be possible to save the 'hard core' of religious research programmes from empirical disconfirmation, as one might do with any outdated theory, by adding a protective belt of auxiliary theories. For example, the hard core of theism, consisting of the thesis that as a matter of fact there is an almighty, omniscient, and infinitely good god who created the universe, and who is like a good father to mankind, may be rescued from empirical refutation by the auxiliary theory that this god has good reasons for hiding himself from most humans and for not answering their prayers, even though, due to the limitations of our mental powers, we will never be able to grasp these good reasons. As we saw in Chapter 3, Alvin Plantinga's extended A/C model, and indeed all versions of theism that are not dogmatically creationist, imply that God was hiding himself *completely* from humanity for a very long time, that is,

---

[9] *Scientific American*, June 2006, p. 12; Benson et al. (2006). Various hypotheses are proposed in order to explain this surprising result, such as the hypothesis that patients who know that prayers are said for them suffer from 'performance anxiety'. Cf. Casatelli (2006).
[10] Cf. Swinburne (2006), p. 13. This objection is avoided by *ex post* investigations, such as reported in Galton (1872), who came to a negative conclusion as well.

from the supposed Fall or the advent of the human race onwards until the emergence of monotheism. If we assume that the alleged fall must have occurred very shortly after the evolutionary origin of the first modern humans, say between 400,000 and 250,000 years ago, and that the first (short-lived) monotheistic experiment in human history took place under Pharaoh Akhenaton in the fourteenth century BCE, God's self-revelation to humanity is a late one indeed. In order to reconcile this first auxiliary theory with the theistic idea that God is a good father, the theist will need a second auxiliary theory, such as the hypothesis that God wants us to choose freely whether we want to have faith, and so on.[11] But a serious investigator will not invest in such theories any more, because they are interlarded with ad hoc hypotheses and fail to have what Lakatos calls a 'positive heuristic', that is, the resources to suggest novel auxiliary hypotheses, which are fruitful in that they imply the possibility of new methods of research and interference.[12]

One may describe the predicament of religions in modern Western culture yet in another way, inspired by Emile Durkheim's sociology of religion. Like many empirical students of religions, Durkheim stressed that apart from social structures, rituals, cults, and festivals, religions also comprise a system of ideas, a view of the world. In this respect, however, 'scientific thought is [...] a more perfect form of religious thought', Durkheim wrote in 1912, so that it is only to be expected that 'the second will gradually give way to the first'.[13] This has happened in Western culture during the last four centuries. Because of scientific progress and the increasing institutional differentiation of Western cultures, ever more areas of human life, such as medicine or mental health, have been transferred from the institutional grip of religions to the sphere of scientific and scholarly research. Without any doubt, religions still have some functions that science and scholarship will never be able to fulfil, such as providing rituals for marriage or funerals and enforcing social bonding.[14] Religious beliefs may not disappear, because they have a number of psychological and social functions. But it seems that religions can no longer have a legitimate role in the cognitive and technological domains.

This modern predicament of religions in Western cultures raises the methodological dilemma for natural theology that I announced above. If natural or rational theology aims at rationality$_5$ with regard to methods of investigation, that is, at applying correctly adequate methods of research in the pursuit of truth, objectively speaking, it seems that its methods must be able to pass the same stringent tests of validation as scientific and scholarly methods. In order to be intellectually respectable, the natural theologian should not only practise adequate methods developed by

---

[11] Cf. Chapter 14.11–12, below, for discussion of the 'argument from hiddenness'. To the hypothesis that God wants us to choose freely whether we believe in him (and that he exists) or not, it might be objected that believing is not an action, which we can choose to perform (doxastic involuntarism), so that this hypothesis is incoherent.

[12] Cf. Lakatos (1978a), Chapter 1: 'Falsification and the Methodology of Scientific Research Programmes'.

[13] Durkheim (1912), p. 613: '... la pensée scientifique n'est qu'une forme plus parfaite de la pensée religieuse. Il semble donc naturel que la seconde s'efface progressivement devant la première'.

[14] Of course, private citizens may invent rituals of their own, independently from any religion.

philosophers at the highest level of generality (c), but also apply methods at the intermediate level (b), such as statistics when investigating the effectiveness of prayer, for example, and develop proper domain-specific methods at level (a), which will yield detailed religious knowledge and which can be validated. However, given the degenerating track record of religions at these lower levels of method, this first option for natural theology does not seem to promise much success.

The alternative option is to argue that the area of the divine is so totally different (*totaliter aliter*) from all areas of scientific and scholarly research, that the methods of natural theology must be very different across the board, without any possibility of validating them by the usual types of tests. This option is preferred by many continental religious philosophers, such as Heidegger, Levinas, Derrida, or their followers like Jean-Luc Marion, and by scientists such as Stephen Jay Gould, who claimed that science and religion are *Non-Overlapping MagisteriA* (the NOMA principle).[15] For example, Heidegger's domain-specific method of doing research about 'Being' in his later works is 'listening to the voice of Being', and Derrida urges us to track 'traces'. The problem for this second option is, of course, that these methods seem to have zero reliability in the pursuit of truth, if we can make sense of them at all in spite of their elusive vagueness.[16] Since religions postulate the existence of specific spiritual powers, or of one infinite spiritual power, they make a factual claim. Why should this factual claim be exempted from the necessity of being assessed by applying the usual type of empirical methods?

In short, the methodological dilemma for natural theologians in contemporary Western culture is that they either have to opt for methods of factual research that are intellectually respectable in the light of the present state of the sciences and scholarship, or for alternative methods, which are practised in religious investigations only, and which cannot be validated. In the first case the probability of success is low, whereas in the second case natural theologians will forego intellectual respectability, because they cannot meet the requirements of rationality$_5$. They may claim that the contents of a specific book, such as the Bible or the Koran, are really revealed to humanity by a god, whereas all similar claims with regard to other books, such as the Book of Mormon, are false. Accordingly, they may tell us that their specific method for religious research consists in reading this holy book with a humble spirit of obedience, in following the commandments revealed in it, in practising the liturgy of their religion, in opening their heart to God, and in never expecting that God will answer their prayers, because His mind is unfathomable and His freedom is absolute. But the claim that one book is revealed by a god whereas the other books are not, is a factual claim. How can this claim be justified, except by methods of empirical confirmation?

---

[15] For a critical analysis of the religious theme in Heidegger's later philosophy of Being, see my (1998), section 11. For the NOMA principle, see Gould (1999).
[16] According to Wolterstorff (2005), Heidegger's conception of religious epistemology converges with reformed epistemology. But unfortunately, the 'Being' of Heidegger's later works is very different from the god of reformed epistemologists, so that reformed epistemologists are confronted by the problem of religious diversity again.

6.4 STRATEGIC OPTIONS

Landed in this methodological dilemma, the religious believer has a number of options, apart from accepting one of its horns. The most radical option is to reject the presupposition of the dilemma, to wit, that religions make factual existence claims about spiritual powers such as demons, gods, or God. According to the resulting non-cognitivist interpretation of religious language, saying things such as 'God exists' or 'Allah is the only god and Mohammed is his prophet' is not making a factual statement, which is true or false, in spite of the fact that surface grammar suggests this. Rather, it is expressing a deep commitment to a certain way of life, or expressing the decision to assess the major events of life within the framework of a religious terminology, or conveying a deep longing of the human heart. The non-cognitivist interpretation has the great advantage that one's creed is immunized from rational criticisms to the effect that it is false or unjustified, and, probably, this was the motive for proposing it in the first place.

Influenced by his readings of Kierkegaard, Ludwig Wittgenstein seems to have developed such a non-cognitive interpretation of religion, as is documented by his scattered remarks on the subject. I do not think that Wittgenstein intended this non-cognitivist conception as a historical interpretation of the existing religions, such as traditional varieties of Christianity. Rather, he presented it as *the only defensible* interpretation of religion in our age of science, which would protect faith from devastating criticisms. This is clear from the wording of fragments such as the following:

> It appears to me as though a religious belief could only be something like a passionate commitment to a system of reference [das leidenschaftliche Sich-entscheiden für ein Bezugssystem]. Hence although it is *belief*, it is really a way of living, or a way of assessing life. It is a passionately seizing hold of *this* interpretation.[17]

But followers of Wittgenstein such as D. Z. Phillips have defended the view that the non-cognitivist interpretation of religion is the correct analysis of the deep grammar of religious language *as such*, as it is used in all historical periods. According to these authors, both rational critics of religious creeds and natural theologians who support their religious beliefs by argument and evidence misinterpret the religious language game and are guilty of superstition.

As I have said in the preface to this book and in Chapter 2.4, however, non-cognitivist views of religious language are implausible as a historical interpretation of the vast majority of religious creeds, such as most traditional versions of theism. In the seventeenth century, for example, there was a large industry of scholarly attempts to calculate the age of the universe, the Earth, and man on the basis of the genealogies of the Old Testament. Hundreds of different estimates of the age of the universe since creation were published, varying between about 6,500 and 3,600 years BCE. The best-known instance of this industry is Bishop Ussher's calculation, according to which

---

[17] Wittgenstein (1980), pp. 64 and 64e (MS 136 16b: 21.12.1947); I modified the translation somewhat.

creation started on the evening of Saturday, 22nd October of the year 4004 BCE.[18] The point of this wave of publications concerning biblical chronology was that the authors wanted to predict when the Earth would end, and when mankind would be subjected to the Last Judgement. Clearly, then, the Last Judgement was interpreted as a real event in the future, concerning which it makes sense to calculate the time of its arrival with some precision. A Wittgensteinean non-cognitivist interpretation of belief in the Last Judgement clearly is mistaken as an account of these seventeenth-century texts, because according to this interpretation, the religious idea of the Last Judgement must be radically unlike any factual hypothesis.[19] Indeed, Wittgenstein would have characterized as superstition the idea that the Last Judgement is a future event.

If a non-cognitive view of religious language is inadequate as an interpretation of most religious creeds as they were conceived of in the past, is such a view at least a fruitful reinterpretation of religions for present-day believers, who live in the age of science? I do not think so, mainly because the non-cognitive interpretation of religious language implies a view of religion that is too reductive. There is no doubt that saying things such as 'God exists' usually involves a commitment to a certain way of life, and expresses the decision to adopt a specific system of reference in order to describe and assess one's actions. But claiming, like Wittgenstein, that this is the *only* thing we are doing when we say that God exists, is counterintuitive and unsatisfactory. For we typically adopt a system of reference, such as Western chronology or the language of theism, because, among other things, we want to make specific statements by using it, which are true or false. Furthermore, the very forms of life of Christianity or Islam appear pointless unless one presupposes that in fact God or Allah exists, so that we have certain moral obligations to him, such as worship. Why would one pray to a personal god, for example, if one does not believe that in fact this god exists? Indeed, what can it *mean* to 'pray to God' if one does not presuppose that God exists?

Defenders of Wittgenstein's view of religion tend to draw an analogy between existence statements in mathematics and in theology. If the mathematician says that there are infinitely many prime numbers, he is not making a factual statement, the truth of which should be investigated by empirical methods. Rather, he is asserting that in the conceptual system of natural numbers, infinitely many concepts of a certain kind can be generated. Mathematicians with a penchant for philosophy might misinterpret mathematical existence statements, being misled by their surface grammar. They might think that such statements posit the existence of peculiar entities, which have unusual properties. On the one hand, Platonist mathematicians say, numbers are real entities like chairs or stars. But on the other hand, numbers are very unlike chairs and stars, since it does not make sense to say that the number seven exists at a specific time or place, for example. They conclude that there must be a realm 'outside' of space-time in which entities such as numbers exist timelessly, and to which the mind of the mathematician has access in some mysterious way.

God as defined by theists also has unusual properties. Although he is said to be omnipresent, he is not in space because he is bodiless, and according to many theists he is not in time either. Should we not conclude, then, that a statement such as 'God

---

[18] Ussher (1650); Cf. Jackson (2006), Chapter 2, for an overview of this literature.
[19] Cf. Wittgenstein (1966), pp. 53ff., and (1980), p. 32.

exists' is like statements of the form 'the number N exists'? If so, may the follower of Wittgenstein not plausibly argue that religious believers misinterpret the statement that God exists if they conceive of it as a factual hypothesis of existence, just as mathematicians misinterpret existence statements about numbers when they conceive of them as factual statements about entities of some kind? I do not think so, since the two types of statements differ at a crucial point. Whereas no mathematician would attribute causal powers to numbers, God as defined by theism essentially has causal powers. He is said to have created and to sustain the universe, and to interact with humans in many different ways. Hence, claiming that God exists cannot be identical merely with a passionate commitment to a conceptual system of reference, as Wittgenstein says.

We have to conclude that the non-cognitive option is both incorrect as a historical interpretation of the existing religious creeds and problematic if it is intended merely as a reinterpretation of religion in our age of science. Let me mention two other types of reinterpretation that are even less promising, because they lack the philosophical sophistication of the Wittgensteinean view. One is redefining the term 'God' in such a way that the statement 'God exists' is known to be true. This result is achieved by identifying God with something that obviously exists, such as the universe, the phenomenon of love, or something else for which the believer has a predilection. Another type of reinterpretation reduces the content of a traditional religious creed such as theism to its vanishing point, as when the believer says: 'there must be something, I don't know what'. All these strategical options are in fact varieties of weak crypto-atheism, in that one refuses to endorse any traditional claim that a specific god exists, although one conceals this by continuing to use religious language, reinterpreted in such a way that one does not run any cognitive risk of believing something that is false.

It follows that all substantial religious views may be interpreted as positing the factual existence of a god or more than one god. For those who endorse this interpretation, there remains only one option that avoids the methodological dilemma sketched above. This is the option of Alvin Plantinga's reformed objection to natural theology, which I discussed and rejected in Chapters 3 and 4. It seems, then, that all religious believers who refuse to water down their religion to crypto-atheism are landed in the methodological dilemma. For a critical philosopher of religion, analysing the works of theologians who opt for the second horn of this dilemma is without much interest. When they argue that religious methods of research are completely different from all other methods of factual discovery and cannot be validated by the usual type of tests, their flight from intellectual responsibility and rational methodology is seen through too easily. Also, theologians who choose this horn never develop their methodological views in any detail. This is why I prefer to analyse in this book the grand strategy of a natural theologian such as Richard Swinburne, who radically opts for the first horn of the methodological dilemma. Or, at least, this is what he appears to do at first sight.

6.5 THE TENSION

In his introductory book *Is There a God?*, published in 1996, Richard Swinburne claims that '[t]he very same criteria which scientists use to reach their own theories lead us to move beyond those theories to a creator God who sustains everything in existence'.[20] It seems, then, that Swinburne is opting squarely for the first horn of the methodological dilemma that I sketched above, and this is what I call the first strand in Swinburne's grand strategy. Accordingly, he interprets the traditional doctrine of monotheism as a factual hypothesis, which ultimately explains the existence of everything else there is. He says in *The Existence of God*, the second and substantially revised edition of which was published in 2004, that theism is a 'large-scale theory of the universe'. He also stresses in this book 'the close similarities that exist between religious theories and large-scale scientific theories'.[21] And in *The Coherence of Theism*, the revised edition of which was published in 1993, Swinburne defends a largely literalist interpretation of theism in order to show that theism really can be very much like a scientific theory.[22]

However, there is a second strand in Swinburne's grand strategy, which consists in his solution to a delicate problem that is raised by choosing the first horn of the methodological dilemma. On the one hand, natural theologians who aspire to rationality$_5$ have to claim that their method is very much like scientific or scholarly methods, and that their theistic theory closely resembles large-scale scientific theories, or factual hypotheses in history. But on the other hand, it is clear that if their method and theory resemble scientific methods and scientific theories too closely, their chances of success are negligible, and they put religion at great risk. With regard to method, both scientists and historical scholars will hold that one cannot practise scientific or scholarly method at the abstract level (c) without also inventing and validating domain-specific methods at level (a), and practising methods at level (b). But we have seen that at these lower levels, the track record of religious methodology is a dismal one.

Furthermore, scientists typically hold that no theory should be seriously considered if, given background knowledge, it does not contain the prospect of implying new predictions. As Lakatos argued,

a theory is "acceptable" or "scientific" only if it has corroborated excess empirical content over its predecessor (or rival), that is, only if it leads to the discovery of novel facts. This condition can be analysed into two clauses: that the new theory has excess empirical content (*"acceptability$_1$"*) and that some of this excess content is verified (*"acceptability$_2$"*).[23]

Similar requirements of testability hold for explanatory hypotheses in historical research. But the track record of religions, to the extent that they have attempted to explain specific facts and predicted future events, is discouraging as well. Although in

---

[20] Swinburne (ITG), p. 2.   [21] Swinburne (EG), p. 3.
[22] Cf. Swinburne (CT), p. 72: 'If theology uses too many words in analogical senses it will convey virtually nothing by what it says'.
[23] Lakatos (1978a), pp. 31–2.

the past, Christian theism has been used to explain natural facts such as the stability of the solar system or the adaptive complexity of biological organisms, and to make predictions such as the second coming of Christ shortly after his crucifixion, these explanations have been superseded by better, scientific explanations, and the predictions have not come true. It seems, then, that natural theologians who claim that their creed is a theory of roughly the same kind as scientific theories run the risk that they will have to give up their religious doctrine. Very few theologians are really prepared to run this risk.

Let me call this problem, raised by choosing the first horn of the methodological dilemma, *The Tension*, because it is the result of two requirements that pull in opposite directions. According to the first requirement, the rationality of rational or natural theology should resemble scientific and scholarly rationality in order to be respectable in the age of science. According to the second requirement, it cannot resemble scientific and scholarly rationality too much, because in that case the natural theologian will be doomed to failure.

Concerning this second strand in Swinburne's grand strategy, then, I propose the following interpretative hypothesis. His strategy is designed, I suggest, in order to resolve The Tension by balancing on a very narrow ridge. On the one hand, the method and theory of the natural theologian should be *sufficiently similar* to scientific method and theories in order to be intellectually respectable. On the other hand, his method and theory should be *sufficiently dissimilar* from scientific or scholarly method and theories in order to exclude a fate which many scientific theories and historical hypotheses have suffered in the past: that they were eliminated by superior theories in the course of scientific or scholarly progress. According to my interpretative hypothesis, this is a fate that natural theologians such as Swinburne refuse to risk.

If this is Swinburne's grand strategy, it requires, for example, that the very possibility of competition between theological explanations and properly scientific or scholarly explanations should be excluded a priori, because in the past such competitions have always been won by properly scientific or scholarly explanations. Furthermore, the strategy should exclude that the natural theologian be under the obligation to develop domain-specific methods of theological research at level (a), for all such methods have failed in the past. If one holds, however, that choosing the first horn of the methodological dilemma implies that the natural theologian include such things, instead of excluding them, one will come to the conclusion that Swinburne does not really opt for the first horn at all. Instead, he attempts to go between the horns of the dilemma.

One may wonder whether a grand strategy that resolves The Tension is possible at all. Is it not contradictory to claim that the doctrine of theism is like a scientific theory while refusing to predict new facts, or to develop domain-specific methods for testing it? The answer to this critical question will depend, in part, on one's views about methods at level (c). For example, if one endorses the methodology of scientific research programmes as developed by Imre Lakatos, one will accept his criterion of demarcation for scientific theories, which I quoted above. A theory then counts as 'scientific' at a given time only if it has some excess empirical content over its rivals, that is, if it predicts some novel, hitherto unexpected fact, and if some of its excess empirical content has been corroborated by experiments or observations. Given this time-relative criterion of demarcation, accepting theism in the beginning of the twenty-first century cannot be rational₅ in the sense of applying scientific method. Theism as it is understood nowadays by most Christians, including Swinburne, does not predict any novel and hitherto unexpected facts, and, indeed, the empirical content of ramified Christian theism, though a considerable one in earlier times, has been drastically reduced since the scientific revolution in the sixteenth and seventeenth centuries. In this case, the probability that a natural theology of theism will succeed is zero, because its very project is contradictory.

If, however, one holds that scientific and scholarly method at level (c) merely consists in rules of inference to the best explanation, which enable us to assess how probable a hypothesis is in the light of an evidence-set, one will reject the requirement of novel predictions. Given a purely probabilistic account of scientific method at level (c), it does not make a difference whether elements of the evidence-set are discovered in the past or in the future. As Keynes argued in 1921, the probability of a hypothesis given the evidence cannot possibly be influenced by the moment at which the evidence is produced, before or after the invention of the theory, since this probability depends only on the logical relations between theory and evidence.[24] The reader will not be surprised to hear that Swinburne accepts a purely probabilistic account of scientific method at level (c) of generality, and that he rejects the requirements for a theory being scientific proposed by Popperians such as Imre Lakatos.[25]

We cannot conclude at the very start, then, that the project of natural theology necessarily involves contradictions. Whether the grand strategy that I attribute to Swinburne is successful or not depends on whether he is able to resolve The Tension. As I said, the apologetic philosopher of religion has to argue on the one hand that natural theology resembles scientific or scholarly disciplines sufficiently for being intellectually respectable, and on the other hand that natural theology does not resemble scientific or scholarly disciplines too closely, so that risks of refutation are minimized or even excluded altogether. Formulated in negative terms, religious apologists have to avoid both the Scylla of their natural theology not resembling real science and scholarship enough, and the Charybdis of their natural theology resembling real science and scholarship too much.

In this book, I focus on the grand strategy in natural theology developed by Richard Swinburne, because I think that from an apologetic point of view, this is by far the most promising strategy, at least at first sight. In other words, Swinburne's strategy is the 'toughest case' for the critical philosopher of religion, as I hope to have shown. In order to see whether this grand strategy can succeed, we have to find out whether Swinburne is able to resolve The Tension, taking the traditional doctrine of theism as our example. And because Swinburne's grand strategy pretends that religious reason is of the same general type (at level c) as scientific reason, what we are aiming at is a *Critique of Religious Reason*.

Part II of this book investigates to what extent the doctrine of theism resembles large-scale scientific theories. Can we decide unambiguously what the theory means, or does it consist of irreducible metaphors? To what extent does the theory of theism have predictive power? Are theists able to exclude all risks of refutation without forsaking

---

[24] Cf. Keynes (1921), p. 305, and Lakatos (1978a), p. 39.
[25] Cf. Swinburne (EG), p. 69.

the intellectual respectability of their theory? In Part III we shall investigate to what extent theism is rendered probable by the existing evidence. Is the logic of confirmation used by Swinburne exactly the same as the one used by scientists? In other words, does he correctly claim that '[t]he very same criteria which scientists use to reach their own theories lead us to move beyond those theories to a creator God'? Or is it rather that he uses this logic or these criteria in an idiosyncratic manner, which would be unacceptable to scientific investigators?

PART II
THEISM AS A THEORY

7

Analogy, Metaphor, and Coherence

In Part I of this book it has been argued that religious beliefs cannot be justified, and that endorsing them cannot be rational or reasonable, unless they are rendered plausible by a natural theology which aims at rationality$_5$ (objective diachronical rationality, cf. Chapter 5.4–5). Since the view that there are gods, or that there is a god, is best interpreted as a factual contention, the most promising strategy for the natural theologian in the age of science turned out to be the one advocated by Richard Swinburne and kindred thinkers. They conceive of the doctrinal content of monotheism at least in part as an explanatory theory or an existential hypothesis, which can be confirmed by empirical evidence. But as we shall see in the remainder of this book, even the most promising strategy for the natural theologian will be confronted with many serious problems.

Pivotal to Part II is the question whether theism can be an explanatory theory or hypothesis at all, which, in principle, can be confirmed by empirical evidence. The conception of theism as a theory may be challenged for various reasons, some of which will be spelled out in the four chapters of Part II. A first challenge is whether the words used by theists in order to describe their god are not employed metaphorically or analogically through and through. Many theologians have argued that terms such as 'wise', 'powerful', 'good', and 'free' get their ordinary meaning when used for human characteristics, and that they can be applied to the god of theism by analogy or metaphor only. As a consequence, all descriptions of God would be irreducibly metaphorical or analogical. If that is the case, can we still consider theism as a theory, which has clearly defined empirical implications? In the present chapter I shall discuss one version of this objection.

Some philosophers of language have argued that all language is metaphorical or analogical in a sense. Others deny this, whereas theologians and philosophers who investigate religious language employ many different conceptions of metaphor and analogy.[1] In order to render my discussion reasonably precise, I shall adopt the concept of analogy developed by Richard Swinburne in his book *The Coherence of Theism*, which allows us to make a gradual distinction between literal and analogical uses of words. Furthermore, I endorse Swinburne's claim that if theists cannot articulate their religious view except by using the key terms in an irreducibly analogical

---

[1] Cf. Soskice (1984) for a prominent book in this field.

manner, theism cannot be a theory or hypothesis, which is confirmable by empirical evidence. As he rightly stresses, '[i]f theology uses too many words in analogical senses it will convey virtually nothing by what it says'.[2] Accordingly, Swinburne argues that in the standard articulations of theism nearly all the defining terms are used with their literal meanings. Does he succeed in substantiating this tenet?

7.1 IRREDUCIBLE METAPHORS

Metaphors and analogies are not only used in natural theology; they also play an indispensable role in scientific research. For example, conceiving of human organisms as machines was central to the view that Descartes developed of our nervous system. It was also crucial for the evolution of physiology in the nineteenth century, and today the brain is often conceived of as a very complex information-processing device.[3] To mention another example, in 1913 Niels Bohr used the visual metaphor of a miniscule solar system when he wrote his first paper on atomic theory in order to explore a complex physical system not yet well-understood: the atom.[4] In all such cases, conceiving of the thing to be investigated as if it were something better known is a powerful heuristic device.

Yet the scientific use of analogy or metaphor has clear limits. If one conceives of one's primary object of research O as if it were another entity of type E, which is better understood, the scientist knows very well that O is not an E, and that the analogy holds to a certain extent only. In other words, O will have many properties that are not also properties of an E, and the scientist aims ultimately at a precise literal characterization of O. For instance, when in *The Origin of Species* Darwin used the model of artificial selection by breeders in order to develop his theory of natural selection, he also discussed the many differences between these two processes, and gave a literal description of the latter. In the traditions of theistic theology, however, many authors have claimed that no literal description of God is possible. Since God, being infinite in some sense and radically transcendent, is said to be infinitely different from all finite creatures, whereas our language has been developed for describing the created world, our words can apply to God at best in an analogical sense. As a consequence, we will never be able to substitute a literal description of God for the analogical one, so that the description is irreducibly analogical or metaphorical.

If this is true, however, it seems that theism cannot be a theory containing a testable existential hypothesis, the hypothesis that God exists. In principle, there are two methods only for providing a proper name such as 'God' with a (putative) referent. One is the deictic procedure consisting in a kind of baptism. We point to something that is perceptually present to several speakers of the language and decide that it will be referred to by the proper name N. But in the case of the name 'God' this procedure is impracticable. Although some believers claim to have perceived God, the speakers of the language cannot be perceptually aware of God in the same unambiguous and public manner in which human beings, cities, dogs, or mountains can be perceived.[5]

---

[2] Swinburne (CT), p. 72.   [3] Cf. Paton (1992).   [4] Cf. Miller (2000).
[5] Cf. Chapter 15.4–11, below, for a discussion of the question whether a perception of God is possible.

Hence, no intersubjective referential use of 'God' can be established by such deictic methods.

As a consequence, one is only able to provide the proper name 'God' with a possible referent by giving a description of what the name is supposed to refer to. If no literal description is possible of an entity to which the word 'God' allegedly refers, however, since that entity, if it exists, can only be hinted at by irreducible metaphors and analogies, one should conclude that we could never succeed in providing the word 'God' with a referent. Indeed, we have no clear idea what kind of entity we are hinting at by using these irreducible metaphors. And if the word 'God' lacks a clearly defined referential use, the sentence 'God exists' cannot express a meaningful existential hypothesis.

In the next section, I shall present an argument to the effect that the standard theistic descriptions of an entity called 'God' are irreducibly metaphorical or analogical. Let me quote as an example of such a description the explanation of the proposition that God exists endorsed by Richard Swinburne:

> I take the proposition "God exists" (and the equivalent proposition "There is a God") to be logically equivalent to "there exists necessarily a person without a body (i.e. a spirit) who necessarily is eternal, perfectly free, omnipotent, omniscient, perfectly good, and the creator of all things". I use "God" as the name of the person picked out by this description.[6]

Most of the descriptive terms in this quotation, such as 'person', 'free', 'knowledgeable' (implicit in 'omniscient'), and 'good', have their literal use or meaning when applied to human beings. Why would one think that they cannot be used literally when applied to a putative entity called God? The argument that I shall summarize in the next section is derived from the semantics or logical grammar of our everyday psychological terms, which has been a hot topic of philosophical debate during the last century.

7.2 THREE SEMANTICAL VIEWS

I shall now summarize three different views on the semantics of everyday psychological terms, two of which are rejected as being clearly inadequate by most experts. From the third view, which I consider as by far the most plausible one, it will follow that all uses of such terms in order to characterize a putative entity called 'God' are irreducibly analogical or metaphorical. Of course, one might reject the third view for this or for some other reason, and hold that terms such as 'person', 'free', or 'wise' can be used literally in order to describe God. But in that case, one has to argue that this can be done on the basis of yet another semantical doctrine concerning psychological or personal terms, such as functionalism, for example, and one has to show that this semantical doctrine is superior to the third view discussed below.

---

[6] Swinburne (EG), p. 7; cf. (CT), p. 2. In writing 'God' with an upper-case initial, Swinburne intends to use the word as a proper name, although it is assumed that this name applies to one entity only, so that it is a proper name of a rather special sort. Cf. Swinburne (CT), pp. 234–8. Hence, it is grammatically incorrect to speak of 'a God', as Swinburne does in this quotation, and as is customary in the literature. I write 'god' with a lower-case initial whenever I use the word as a common noun for postulated spiritual powers.

According to the first view, that of Cartesian substance dualism, (pure) everyday psychological terms are used to refer to states, actions, or dispositions of an immaterial entity called our 'soul', which is linked to the human body only contingently.[7] Cartesian dualism has been abandoned by most scientific experts for many kinds of reasons. For example, even if substance dualism were a coherent conception, it would be implausible from a scientific point of view, because it does not fit in well with the gradualism implied by the theory of evolution. Furthermore, substance dualism is difficult to square with the thorough dependence of our mental life on processes in our body, which is revealed in ever more detail by modern brain research.

Philosophers have rejected Cartesian dualism for multiple reasons as well.[8] From an ontological point of view, it seems to be inconceivable that a putative non-physical entity can have causal interactions with a physical substance such as the human body.[9] Furthermore, there can be no conceivable entity without criteria of identity. But no criterion of identity has been, and, indeed, can be given for immaterial substances such as souls, so that it is unclear what makes the difference between one soul and another soul. This difficulty is conceptual rather than epistemic: no rule is given for determining what would count as one soul as opposed to one million souls, all having exactly the same thoughts and feelings.[10]

From an epistemological point of view, Cartesian dualism seems to imply that we cannot know anything at all about the mental life of other persons, because the alleged souls are not publicly observable things.[11] According to traditional dualists, such knowledge can only be based on arguments from analogy. One would be able to discover regularities between one's own mental life and one's bodily behaviour. Then one might conclude by an argument from analogy that if others display certain types of bodily behaviour, their souls are in similar mental states, etc. as one's own soul, when one displays similar types of behaviour.

But this view on the knowledge of other persons' mental life seems to be inadequate for at least six reasons. First, we are usually not aware of drawing such inferences by

---

[7] According to substance dualism, we have to distinguish between pure psychological terms, which are used to refer to states, etc. of a soul, and mixed psychological terms, which refer both to states, etc. of a soul and to bodily states or behaviours. 'Saying hello' is an example of the latter. Cf. Swinburne, (ES), p. 7.

[8] Cf., for example, Hacker (2007), pp. 289–94.

[9] As Jaegwon Kim (1993) says, '[e]ver since [that is, after 1643] there has been a near consensus that the problem of mental causation cannot be solved within the terms set by Descartes' substantival dualism' (pp. 158–9). Like Ernest Sosa, Kim thinks that the 'pairing problem' cannot be solved by Cartesian dualism: since according to substantival dualism there cannot be a spatial relation between a mental cause and its alleged physical effect (or vice versa), there seems to be no way in which the right cause can be linked to the right effect (ibidem, pp. 160ff., and Sosa (1984); cf. also Kim (1998)).

[10] Cf. Strawson (1966), p. 37: 'If we are to make any legitimate employment of the crucial concepts of unity or numerical identity through time, we must apply them, in the light of empirical criteria, to objects encountered in experience. But if we abstract entirely from the body and consider simply our experiences or states of consciousness as such [. . .], it is evident not only that we do not, but that we could not, encounter within this field anything which we could identify as the permanent subject of states of consciousness'; and Hacker (2007), pp. 291–2. Cf. for an attempt to answer such objections by introducing the notion of 'thisness' (or *haecceitas*), Swinburne (ES), New Appendix D: two individuals have 'thisness' if they are distinct and yet (can) have all the same (monadic and relational) properties. But since we cannot point to different souls, saying '*this* one is not that one', the criterion of *this*ness does not make sense with regard to souls.

[11] Cf. Swinburne (ES), p. 155: 'souls [. . .] are not publicly observable things'.

analogy. Very often, we just see that someone is in pain, feels elated, or is angry. Second, these inferences by analogy would commit a fallacy of hasty generalization, since the correlations would be established on the basis of evidence concerning our own case only. Third, the strength of our conviction that, say, someone else is in pain or angry, typically exceeds by far the strength of such a putative argument by analogy. Fourth, we are usually not aware of the relevant bodily behaviour in our own case, such as our facial expressions as perceived by someone else. As a consequence, we cannot even establish many of the relevant correlations in our own case, at least not at the early age at which we start to ascribe confidently mental properties and occurrences to others.[12] Fifth, philosophers of language have argued that if primitive (undefined) psychological terms referred to states or actions of a mental substance, which would be related to bodily behaviour only contingently, one would never be able to learn their uses from others, since the soul of other human beings is not a publicly observable entity. But we clearly learn the uses of psychological words, such as 'pain' or 'pleasure', from other human beings by some kind of public training. And finally, Cartesian dualism would imply that our personal pronouns are strangely ambiguous. When I say, 'I sat down', the pronoun 'I' would refer to my body, whereas it would refer to a different entity, my soul, when I say things such as 'I feel sad'.[13]

Under the influence of Logical Positivism, a very different view of everyday psychological terminology was developed in the 1930s. According to this view, called Logical or Analytical Behaviourism, the meaning of psychological words can be spelled out exhaustively in terms of behavioural repertoires and dispositions to behave. To ascribe pain or intelligence to someone would simply mean that this person behaves in specific ways or has dispositions to do so. But in many cases this view seems to be inadequate as well. Although experiences of pain are often accompanied by typical behaviour and facial expressions, we all think that the word 'pain' signifies bodily feelings of a specific kind, and not these types of behaviour or facial expressions. Moreover, it turned out to be impossible to spell out the meaning of, say, the sentence 'John dreamt that he was a genius', in terms of a specific set of behavioural dispositions. All such attempted specifications had to make substantial assumptions about the remainder of the subject's *mental* make-up (desires, beliefs, etc.). For example, suppose that John remembers with delight a love affair of the past. Can the meaning of this psychological attribution be explicated in terms of inclinations to behave in certain ways, if he has very good reasons never to tell anyone about it?[14]

Given the shortcomings of Cartesian dualism and Logical Behaviourism, a third view of the semantics of everyday psychological terminology seems to be the most plausible one, which was developed by Ludwig Wittgenstein in his later writings.[15]

---

[12] Cf. Plantinga (1993b), pp. 65–70.
[13] As Swinburne admits, 'few philosophical positions are as unfashionable as is substance dualism' (ES, p. ix). I shall discuss Swinburne's own argument for substance dualism below in this chapter. Cf. for a sustained criticism of substance dualism as displayed in the writings of neuroscientists: Bennett and Hacker (2003).
[14] Cf. Block (1981).
[15] There are some other views on everyday psychological language that I regard as implausible. One is that everyday psychological terminology is theoretical through and through, part of the theory of 'folk psychology'. I have criticized this view in Philipse (1990). Cf. also Bennett and Hacker (2003), pp. 366–77. Yet another view is semantical functionalism. In his (1987), William Alston

Wittgenstein rejected both the Cartesian myth of the mental as a private world completely hidden to others and the logical reduction of psychological statements to descriptions of behaviour. What is more, Wittgenstein also repudiated the very conception of human behaviour that Descartes and the Logical Positivists took for granted, that is, the misleading notion of human behaviour as a set of meaningless physical movements of human bodies. This physicalist conception of human behaviour is nothing but a corollary of Cartesian dualism, which the Logical Behaviourists inherited from the philosophical tradition. Like Merleau-Ponty and other phenomenologists, Wittgenstein argued that, initially and typically, human behaviour, including our facial expressions, *displays* our mental states, instead of screening them off, although we may learn later to hide our feelings and thoughts. We simply *see* that someone else is cheerful or sad, for example. We do not have to conclude this from inductive arguments by analogy, or from inferences to the best explanation on the basis of observing purely physical movements or states of this person's body. Accordingly, we often cannot describe human behaviour without using psychological terms, as when we say that a person gestures angrily or jumps up joyfully.[16]

Wittgenstein held that in using psychological language, we ascribe specific types of capacities, inclinations, states, or occurrences to *human beings*. We can learn psychological language from others because it is a feature of psychological concepts that behavioural expressions of these capacities and occurrences function as *criteria* for attributing them to people.[17] For example, that someone uses a word correctly, and gives an adequate explanation of its meaning, are criteria justifying the statement that this person *knows* what the word means. One may wonder what explains our ability to use behavioural expressions as criteria for attributing mental states, or occurrences, etc. to other people. Perhaps the celebrated discovery of the mirror neuron system will provide part of a proximate explanation of the most elementary instances of this ability.[18]

Admittedly, adult human beings may feel a deep love for another person, for example, without expressing it in behaviour of any kind, if they have learnt to conceal their feelings and if there are compelling reasons to hide this love. Yet there is a broad spectrum of natural expressions of love in human behaviour, without which we could never learn to apply the word 'love' in the first place. That there is this range of natural behavioural expression is essential to the meaning of the word 'love'. As a consequence, we cannot meaningfully attribute love to a thing that is unable in principle to display at least some of these natural expressions in behaviour. Moreover, many psychological descriptions presuppose that the subject is capable of using a language. For example, to

---

argued on the basis of functionalism that psychological terms can be applied to God literally. But functionalism is faced with difficulties similar to those I mentioned with regard to Logical Behaviourism. See for an instructive discussion of the resemblances and differences between Logical Behaviourism and Wittgenstein's later conception of psychological terminology: Hacker (1990), pp. 239–53.

[16] Cf. Wittgenstein (1953), I, §§243–693, and Hacker (1990).

[17] Cf. Wittgenstein (1953), I, § 580: 'Ein 'innerer Vorgang' bedarf äußerer Kriterien' ('An "inner process" stands in need of outward criteria').

[18] Cf. Preston and De Waal (2002); Decety and Jackson (2004); Gallese (2001); and Keysers and Gazzola (2007).

say that John thinks that the train to London departs at 4 p.m. would not make sense if John were a baby of two months old.[19]

It follows that the proper home of psychological expressions is neither a putative soul nor bodily behaviour conceived of as a series of physical movements, but the human being. As Wittgenstein wrote:

only of a living human being and what resembles (behaves like) a living human being can one say: it has sensations; it sees; is blind; hears; is deaf; is conscious or unconscious.[20]

Of course we can apply psychological expressions meaningfully to other animals to a certain extent, and we do so by using their behaviour as a criterion. For example, we can say of a dog that it expects to get its food when it wags its tail upon seeing that the master lifts its bowl. But we cannot meaningfully say that the dog expects to get a good meal next year at its anniversary, since there is nothing in its behavioural repertoire that would or *could* count as exhibiting such an expectation.[21]

One can also apply some, but not all, psychological expressions to human artefacts, such as texts or computers, and to human institutions, such as a parliament or a board of directors, even though texts, computers, or institutions neither have animal bodies nor the criterial behavioural capacities. We may say, for example, that the intention of a passage in a text is to say that *p*, and that parliament decided that *q*. Clearly, we can do so meaningfully because of the relations these artefacts or institutions have to their producers, users, or to the individuals who constitute them. But when people state that the book they read is awake, that their computer is in pain, or that at a given moment parliament is conscious as opposed to unconscious, we do not have a clue as to what they intend to say, and they have to explain these metaphors to us in other terms. The reason is that books, computers, or parliaments do not have the proper behavioural repertoire which can function as a criterion for saying that they are awake, in pain, or conscious.

We may now state as follows the first challenge to the claim that theism is a theory or existential hypothesis. How can one meaningfully say that God knows everything,

---

[19] Cf. for an elaborate defence of this Wittgensteinean view of psychological language: Hacker (2007).
[20] Wittgenstein (1953), I, § 281. Cf. Kenny, 'God and Mind', in Kenny (2004), p. 48.
[21] The criterial relation between behavioural repertoires and (third-person ascriptions of) mental states or occurrences is stronger than that of a merely inductive correlation. Indeed, in order to establish inductive correlations, we have to identify mental states or occurrences in the first place, which we cannot do concerning others without using behavioural criteria. But clearly, the relation is weaker than that of entailment, since, for example, we may feel pain and hide it, or may simulate pain when we don't feel it. A behavioural repertoire *B* is a criterion for attributing a mental state or occurrence *M* to a subject if it is a conceptual truth that a display of *B* is (defeasible) evidence for *M*. One cannot object to this Wittgensteinean notion of criteria that we might imagine that human beings would behave very differently when they feel pain from the way they in fact behave, so that the criterial connection cannot be a conceptual one (Plantinga (1993b), p. 74). The reason is that many conceptual connections depend for their usefulness upon the factual constitution of our human nature, so that they would not have been established if human nature had been quite different. In other words, if humans had been different, the criterial conceptual relations would have been different as well. As Hacker writes: 'Our concepts here are erected upon the normality of a web of connections, namely, between injury, reactive behaviour, avowal, and subsequent behaviour'. Hacker (1990), pp. 556–7.

listens to our prayers, loves us, speaks to us, answers (or does not answer) our supplications, etcetera, if God is also assumed to be an incorporeal being? For the stipulation that God is an incorporeal being *annuls the very conditions for meaningfully applying* psychological expressions to another entity, to wit, that this entity is able in principle to display forms of bodily behaviour which resemble patterns of human behaviour. In other words, the very attempt to give a meaning and a possible referent to the word 'God' as used in theism must fail, because this attempt is incoherent. On the one hand, the theist ascribes psychological predicates to this putative entity. On the other hand, the theistic assertion that God is bodiless implies that a necessary condition or presupposition for ascribing such predicates meaningfully does not obtain in this case.[22]

It follows that all uses of psychological terms in an alleged description of a putative entity called 'God' must be irreducibly metaphorical or analogical. What is meant by 'irreducibly' here is that one cannot explain the metaphors in terms used literally, so that, strictly speaking, we do not know what the metaphors mean. The psychological terms are used 'in the absence of the criteria which give them their meanings in the language games in which they have their home', whereas they have not been provided with a new meaning.[23] The upshot of these considerations has been expressed as follows by Anthony Kenny: 'Philosophy in this area leads to the same conclusion as that of those theologians who have said that when we speak of God we do not know what we are talking about'.[24] Even this is too generous for the theist, however, since we cannot speak *of* God if we have no intelligible idea of what we are speaking of. If the words one wants to use in order to describe the being to which the word 'God' refers are used in an irreducibly metaphorical manner, no clear referential use of the word 'God' has been specified, so that the word cannot be used meaningfully to refer at all.[25]

If this is so, one might object, how are we to explain the fact that the word 'God', and sentences such as 'God loves me', *appear* to be used meaningfully in monotheistic language? But explaining this is not difficult. The religious uses of the putative proper name 'God' are parasitic upon, and resemble to a large extent, the ordinary uses of proper names and psychological expressions for human beings. What religious believers fail to notice is that by substituting 'God' for an ordinary proper name in sentences such as 'John loves me', or 'Paul will condemn him', they cancel the conditions for using meaningfully the words 'loves' and 'condemns'.

Monotheistic believers often are vaguely aware that the meaning of words eludes them when they utter sentences containing the word 'God'. But they misinterpret this fact as symptomatic of the spiritual depth of religious discourse. They think that the profoundly mysterious nature of monotheistic language points to a transcendent reality, which cannot be grasped by us, limited human beings.[26] In this case, however, the impression of profoundness is caused by a mere misuse of language.

---

[22] Cf. for a defence of this objection: Kenny (2004), Chapters 1, 4, and 5.
[23] See for this definition: Kenny (2004), pp. 16–17.
[24] Kenny (2004), p. 80.
[25] The view that all predicates of God are used metaphorically has been defended by many authors. Cf., for example, McFague (1982).
[26] Even Kenny claims that it 'may well be that the use of such metaphors is essential if we are to have a proper understanding of the world in which we live', and that 'the metaphorical nature of religious language does mean that it is profoundly mysterious [. . .]. For it means that when we talk in

7.3 LITERAL AND ANALOGICAL USES OF WORDS

A natural theologian who holds that theism is a theory has to deny that descriptions of the putative entity called 'God' are irreducibly metaphorical or interlarded with indispensable analogies. I already quoted Richard Swinburne, who correctly stresses that '[i]f theology uses too many words in analogical senses it will convey virtually nothing by what it says'.[28] But how can the theist refute the argument to the effect that religious language is irreducibly metaphorical?

In his book *The Coherence of Theism*, Swinburne uses two different strategies to this effect, a negative and a positive one. Negatively, he attempts to refute the Wittgensteinean conception of psychological concepts, according to which these concepts stand in need of behavioural criteria. According to Swinburne, the argument that we can only apply psychological predicates to other beings that are able to give public behavioural expression to their mental life is 'far too quick' for two reasons. One is that 'quite clearly a person may on occasion give no public expression to his mental states'. Apparently, Swinburne thinks that concerning this issue it is obvious that what is sometimes possible might always be possible: 'Why should it not always be the case for some person that [...] he does not ever give expression to his mental states?' The other reason is that, according to Swinburne, a bodiless spirit such as God can very well give expression to his wants or fears, etc.: 'If he fears that a certain man will make a wrong choice, he may make marks on sand conveying a message to him', for example.[29]

These objections to the Wittgensteinean conception of psychological concepts are not very convincing, however. The first objection fails because here, as in many other cases, it is not true that what is sometimes possible is always possible. For example, people belonging to a linguistic community speak a language and lie sometimes, but it is not logically possible that they speak a language and always lie. The reason is that the ability to lie is a secondary capacity, which is parasitic on the mastery of a language. If they were always to lie, the words they use would lack meaning.

Similarly, we may meaningfully attribute feelings, intentions, and other psychological characteristics to a person on an occasion, even though this person does not express these feelings or intentions on that occasion. But we can do so meaningfully only because the person has the behavioural *repertoire* of natural expressions of these feelings or intentions. Indeed, hiding one's feelings, such as pain, fear, or joy, by suppressing the concomitant natural behavioural expressions, is a secondary capacity, which has to be learned during education. Consequently, if we attribute fear to a thing of a kind that completely lacks the behavioural repertoire of expressing fear, such as a chair or a mountain, what we say has no sense.

---

the language of the divine metaphor, we do not really know what the metaphors are about': Kenny (2004), p. 45.

[27] Wittgenstein (1953), I, §111 (Wittgenstein's italics).
[28] Swinburne (CT), p. 72. [29] Swinburne (CT), pp. 108–9.

Swinburne's second objection is equally unconvincing. Can we say without metaphor that a bodiless person may very well express its mental states by making marks on sand, for example? It does not seem so, because in the ordinary sense of the words, 'making marks on sand' means that someone produces these marks by bodily behaviour. We may conclude that Swinburne's negative strategy fails, since it is an unsuccessful attempt to refute the Wittgensteinean account of psychological terminology. Let us turn, then, to his positive strategy for refuting the objection that religious language is irreducibly metaphorical. This positive strategy presupposes Swinburne's own conception of religious language, which is developed in Part I of *The Coherence of Theism*.

The sentences used by believers in order to articulate their religious creed are called 'credal sentences'. Swinburne defends the view that the words which occur in the credal sentences he uses in stating theism as a theory, mostly are ordinary words employed in their normal senses, or technical terms such as 'omnipotent' defined in terms of such words.[30] Sentences can be used to make statements, which are true or false, and a statement is coherent if it makes sense to suppose that it is true. As its title indicates, it is the objective of *The Coherence of Theism* to show that the statements by which theism as a theory is spelled out are coherent, both individually and taken together.

Coherent statements are called *synthetic* or factual if their negation is coherent, and they are called *analytic* if their negation is incoherent, so that the analytic/synthetic distinction exhausts the set of coherent statements or propositions.[31] Swinburne holds that the statement that God as characterized by theism exists, is factual or synthetic and not analytic.[32] This implies that all purely conceptual arguments for the existence of this god, such as the many varieties of the so-called ontological argument, are ruled out in advance. In other words, it is coherent to suppose that God does not exist (cf. Chapter 8, below).

Although Swinburne rejects all versions of the principle of verification, according to which a sentence is meaningful only if a statement made by using it can be tested empirically, he holds that the *words* in terms of which theism is defined as a theory must be 'empirically grounded'. This means that we must be able to learn the uses of these words, such as 'wise' or 'person', either by learning on the basis of experience of public phenomena to what type of things or properties these words refer, or by being given a description of the intended referent(s) in terms of words that we have learned directly on the basis of experience. Without such an 'empirical cashability' of the words, the credal sentences by which theism is defined would not be meaningful.[33]

---

[30] Swinburne (CT), pp. 2–3.

[31] Cf. Swinburne (CT), pp. 14–22. It is important to stress that Swinburne applies this analytic/synthetic distinction to statements and not to sentences, as Quine did in his (unconvincing) critique of the distinction in 'Two Dogma's of Empiricism' (Quine, 1953, Essay II). Cf. for a critique of Quine's critique: Hacker (1996), pp. 193–227.

[32] Swinburne (CT), p. 22; cf. pp. 274–5; (EG), p. 148, and Chapter 8, below.

[33] Swinburne (CT), pp. 36–7. Swinburne is right, I think, in not endorsing a holistic semantics à la Quine for the terms used in his definition of 'God'. Quine embraced such a holistic semantics even for terms of ordinary perceivable objects, such as trees or human beings, because he claimed that our only empirical evidence consists of stimuli or surface irritations of our body, and that even ordinary

Some theologians have argued that, God being entirely different (*totaliter aliter*, in Karl Barth's phrase) from anything we know of, the words of credal sentences must be used in completely new senses. But Swinburne argues correctly that no intelligible procedure can be described by which we might learn these new meanings, and by which the new theological uses of words might be established in a community of believers.[34] In particular, believers cannot give new theological meanings to ordinary words by pointing to new standard examples of their correct application provided by so-called religious experiences, since typically such experiences do not arrive simultaneously and in the same situation to different believers, whereas descriptions of their contents in ordinary words diverge widely. As a consequence, what is given in a religious experience is not a sufficiently public and objective phenomenon to be used as a sample in order to confer a new meaning to a word.

Swinburne distinguishes two methods by which we can characterize things that are different from all phenomena we are acquainted with by using ordinary words in their ordinary meanings, or in their 'mundane' meanings, as he calls them. As Hume already argued, we may *combine* words to describe imagined things that are unknown to us, such as unicorns, and we may ascribe properties to imagined things in *degrees* that do not exist in our world, such as omniscience in the sense of 'knowing everything that can be known'.[35] Yet Swinburne agrees with most theologians that theism cannot be formulated coherently without a third method, the method of analogy. If theists use language analogically, they do not give an entirely new sense to words such as 'wise' or 'loving' or 'person', which have their mundane meanings when applied to human beings. Rather, the meanings of these words are extended or loosened up. How can this be done, if no new standard examples or samples for these terms can be introduced by religious experiences?

Swinburne distinguishes between two types of rules by which the meanings of descriptive terms are fixed, and which we learn (often implicitly) when we learn a language. He calls these rules *semantic* and *syntactic* rules, using the term 'syntactic' in a somewhat special sense, because his syntactic rules are not rules of ordinary grammar. Semantic rules for the use of a word '*W*' indicate or describe paradigmatic examples of things correctly called '*W*' and of paradigmatic contrasting things that are not-*W*, containing a similarity rider in the case of general names. Syntactic rules state in words how '*W*' is used correctly by giving a verbal definition, for example, or by pointing out other conceptual connections between '*W*' and related words, such as listing synonyms. We have succeeded in giving meaning to a descriptive term by semantic and syntactic rules if users generally agree whether the word applies in standard circumstances.[36] In many cases it is not necessary for learning the uses of words that these rules are formulated explicitly.

---

macroscopic objects are theoretical 'posits'. I think that this behaviourist conception of stimuli as evidence is incoherent, and that we could never learn a language if it were correct. In fact, we learn to perceive medium-sized objects before learning our first language, which is excluded by Quine's theory.

[34] Swinburne (CT), pp. 56–8.   [35] Swinburne (CT), pp. 52–5.
[36] Swinburne (CT), pp. 31–6.

Swinburne holds that, without entirely changing our semantic rules by giving new standard examples, we may extend or loosen up the meaning of a word '*W*' by analogy in two interconnected ways. One way is 'to modify the role of the standard examples in the semantic rules' and the other way is 'to abandon some of the syntactic rules'.[37] In the case of general names, the first way is in fact a modification of the similarity rider. Usually, a semantic rule for a general name '*W*' provides standard examples, and a similarity rider which says that any other object is also correctly called '*W*' if it resembles the standard examples in the respect in which they resemble each other to the extent to which they resemble each other. The modification that enables us to use the word analogically then consists in saying that an object *O* may also be called '*W*' correctly even though it does not resemble the standard examples to the extent that they resemble each other in the relevant aspect, provided that *O* resembles the standard examples in the relevant respect *more than it resembles standard examples of objects which are not-W*. This modification of the similarity rider in a semantic rule has the effect that the word will now apply to a wider class of objects than it did before, whether actual or merely conceivable, provided that some of the syntactic rules for the word are modified as well or dropped altogether.[38]

This conception of analogy, which Swinburne defends, is a gradual conception, so that there are degrees of analogy.[39] That it can be quite permissive is shown by the following example. Suppose that we spell out the semantic rule for the word 'person' by pointing to adult human beings as paradigmatic examples and to stones and lakes as paradigmatic counter-examples. We may then apply the word 'person' analogically to porcupines, because they resemble humans more in the relevant respect of behaviour than they resemble stones or lakes, provided that we drop a great number of syntactic rules, such as the rule that adult persons have mastery of a language, or the rule that adult persons have reflexive self-knowledge. It clearly follows that, as Swinburne stresses, 'the more words are used in analogical senses, and the more stretched those senses (i.e. the more the rules for the use of the words are altered), the less information does a man who uses the words convey'.[40] As a consequence, theism can be a theory only if this analogical use of words is restricted severely, and Swinburne's positive strategy consists in an attempt to show that this can be done. There is yet another reason for avoiding analogy if theism is to be a theory. If words are used analogically, it is difficult to show that a theory is coherent.

---

[37] Swinburne (CT), p. 59. [38] Swinburne (CT), pp. 59–60.
[39] Swinburne's conception of analogy is different from the two traditional conceptions of analogy, which were inspired by Aristotle's *Categories*. Traditionally, to say that eating fruits or doing exercise is 'healthy' is called an analogy of attribution, because these things causally contribute to our health in the non-analogical sense of bodily well-being. More generally, something is called *F* by analogy of attribution (*pros hen* in Aristotle) if it has some specific relation to the paradigmatic thing called *F* non-analogically. The second traditional notion of analogy, called 'analogy of proportionality', does not depend on relationships to one paradigmatic type of thing but on a similarity of relationships. For example, we may call old age the 'autumn of life' because old age relates to the end of life as the autumn relates to the end of the year.
[40] Swinburne (CT), p. 72.

7.4 COHERENCE AND THE DILEMMA OF ANALOGY

In a minimalist sense, which Swinburne adopts, a statement or a theory is coherent if it does not contain concealed contradictions and it makes sense to suppose that it is true.[41] Accordingly, we can demonstrate the incoherence of a theory or a statement by deriving an explicit contradiction from it. Demonstrating coherence is somewhat more complicated. We cannot show that no contradictions can be derived even from one predicative statement, because there is no method to make sure that we have listed all entailments.[42] What we can do, however, is to show the coherence of the statement that p by deducing it from another statement, r, if we have good reasons for assuming that r is coherent, since if r is coherent, no contradictions can follow from it.[43] Swinburne calls such a demonstration of coherence of p, which presupposes that another statement r is coherent, a *direct* proof of coherence. Direct proofs of coherence can settle disputes about the coherence of the statement that p if there is enough initial agreement about the coherence of the statement that r, which often is the case.

However, if words in p are used analogically, such a direct demonstration of coherence is quite problematic, if not excluded altogether. The reasons are, first, that the more syntactical rules for the meaning of a word 'W' in its analogical use we drop, the more difficult it becomes to derive statements containing the word from other statements, for these derivations depend crucially on the syntactical rules. Second, if we introduce the new similarity rider defining analogy in Swinburne's sense, a new object can be called 'W' provided that it resembles standard examples of 'W' more than standard examples of not-W in the relevant respect. However, in that case we are 'a great deal less clear about what kind of objects there could be which we have not experienced which would be correctly called "W"', as Swinburne observes. He concludes that 'once we give analogical senses to words, proofs of coherence or incoherence become very difficult'.[44]

In such cases, the best thing we can do according to Swinburne is to give what he calls an *indirect* argument for coherence, that is, to give arguments to the effect that p is true, because if p is true, p must be coherent. If p is a factual statement, such as the statement that God as defined by theism exists, an indirect proof of coherence has the logical form of an inductive argument for the truth of p, which uses factual statements as premises. Swinburne stresses that an indirect argument from factual premises for *in*coherence is not possible, because evidence that a statement is false is not evidence for its incoherence.[45] However, Swinburne's idea of an indirect proof of coherence is not unproblematic. Does it really *make sense* to aim at providing arguments for the

---

[41] Cf. Swinburne (CT), p. 38: 'It follows that if a statement does not entail any contradiction, then it expresses a coherent supposition'. A theory can be conceived of as one statement, consisting of a conjunction of many statements, so that this definition also applies to theories. Of course one might give more demanding definitions of coherence for theories.

[42] Cf. Swinburne (CT), p. 38, who claims that '[o]ne cannot list all the entailments of a statement, for any statement entails an infinite number of statements'. This is true only in the trivial sense that p entails $p \vee q$, $p \vee q \vee r$, etc., but these entailments are irrelevant for demonstrating the coherence of p.

[43] Swinburne (CT), p. 39.   [44] Swinburne (CT), pp. 63, 62.

[45] Swinburne (CT), pp. 45–50, 62–3.

truth of a theory if it does not make sense to suppose that this theory is true, that is, if the theory is incoherent? In Chapter 8.6, below, I shall investigate whether an indirect proof of the coherence of theism is possible at all.

Although introducing analogical uses of words imperils the possibility of showing that a statement or theory is coherent, sometimes the introduction of analogy may offer the only way out. This is the case whenever a statement is incoherent if the words are taken non-analogically. For example, the statement that the number five is red is incoherent if interpreted literally. Assuming that one can speak meaningfully of numbers as objects at all, surely such objects are not spatial objects. But saying that something is red entails that it is spatially extended. Hence, if interpreted non-analogically, saying that the number five is red would be incoherent, because it implies a contradiction. The contradiction may be avoided by saying that the expression 'is red' is used in some analogical sense.

On the basis of these considerations about analogy, we may now formulate the *dilemma of analogy* for a natural theologian who wants to spell out theism as a theory. On the one hand, it seems that theism can be coherent only if quite a lot of analogy is introduced. For example, how can theists claim without contradiction that God is a person and yet bodiless? In the ordinary non-analogical sense of 'person', persons are human beings, and human beings are bodily entities. When we say that we cannot attend the meeting 'in person', what we mean is that we cannot be physically present. So it seems that to define God as a bodiless person is contradictory, unless one uses the term 'person' analogically in Swinburne's sense. There are a great many other apparent contradictions in the traditional theistic conception of God. For example, it seems that an incorporeal spirit cannot be 'present' at some place in the normal mundane sense of 'present', let alone that it can be omnipresent. If these and other apparent contradictions in the theistic conception of God can be avoided only by having recourse to analogy, the theory of theism cannot be coherent unless one introduces abundant irreducible analogy.

On the other hand, however, the theory of theism can be meaningful and informative only if the use of irreducible analogy is restricted drastically. We saw that this is also Swinburne's position, since he stresses that '[i]f theology uses too many words in analogical senses it will convey virtually nothing by what it says'. Although Swinburne thinks that '[t]he "analogical sense" card is a legitimate one', he holds that 'it must not be played too often – for the more it is played, the less information will be conveyed by what is said'.[46] We saw as well that introducing much analogy has the disadvantage that direct proofs of coherence become impossible. And an attempt to argue for the coherence of theism indirectly, by adducing factual premises that confirm the theory and make its truth probable, will be excluded if theism does not have a clear informational content. It seems, then, that it will not be possible to show the coherence of theism as a theory, because there are two necessary conditions for this possibility that are incompatible: the condition that theism involves abundant analogy and that it involves very little analogy.

It is Swinburne's official position in *The Coherence of Theism* that the card of analogy is 'a joker', which is played only once, because 'it would be self-defeating to

---

[46] Swinburne (CT), p. 72.

play [it] more than two or three times in a game'.[47] As we shall see in the next chapter, Swinburne explicitly plays the joker of analogy in order to resolve a contradiction with regard to the necessity of God's properties. In the remainder of this chapter, I shall investigate another issue: is analogy in theism as formulated by Swinburne restricted to the issue of necessary properties? Can he really resolve all other apparent contradictions in the theistic concept of God without having recourse to analogy? In an earlier section of this chapter (7.2), it has been argued that psychological predicates must be used in an irreducibly metaphorical way if applied to an allegedly bodiless entity. I shall now focus on the apparent contradiction in the claim that God is a bodiless *person*.

7.5 GOD AS A BODILESS PERSON: A POSITIVE PROOF OF COHERENCE

In Chapter 7 of *The Coherence of Theism*, Swinburne argues for the coherence of the claim that God is a *bodiless person* by using two methods. In order to show that no analogy is needed in this case, he attempts to construct a 'direct' or 'positive proof' of its coherence, in the sense defined above (section 7.4), and he attempts to refute arguments to the effect that the claim is incoherent. I shall attempt to show that the direct proof fails, because Swinburne surreptitiously introduces analogy, and, in the next section, that his rebuttals of arguments for incoherence are not sound, since they rely on a false assumption.

'That God is a person, yet one without a body, seems the most elementary claim of theism', Swinburne insists. Monotheists introduce young children to the concept of God by saying many things that we can only say of persons, such as 'that God always listens to and sometimes grants us our prayers, he has plans for us, he forgives our sins', and by then adding that God does not have a body.[48] Yet there is a strong presumption that this most elementary claim of theism is incoherent unless it is interpreted analogically, because according to the dictionary a person is a *human being*, considered as having a *character* of his or her own.[49] Since human beings are bodily beings, persons in the non-analogical sense are bodily beings, so that the statement that God is a person without a body entails a contradiction if interpreted non-analogically. Yet Swinburne wants to show in Chapter 7 that there is 'no need for the theist to plead "analogical senses" for his words at this point'. His reason for having this objective follows from our considerations above: 'If the words are to be understood in analogical senses, [one] can give no straightforward proof of the coherence or incoherence of the

---

[47] Swinburne (CT), p. 282. Showing this is Swinburne's positive strategy to refute the thesis that religious language is irreducibly metaphorical, to which I referred in section 7.3.
[48] Swinburne (CT), p. 101.
[49] Cf., for example, the *Longman Dictionary of Contemporary English* (1978). The word 'person' is also a status term, and the status of being a person, that is, a subject of rights and obligation in a moral community and a legal system, can be given only to beings that have specific capacities, such as being able to use a language and to be conscious of oneself. I shall not investigate these ramifications of the notion of a person here. Cf. Hacker (2007), Chapter 10.

claim', and theism will not be a theory with sufficient informational content for being susceptible of empirical confirmation.[50]

What is a person, to start with? Swinburne correctly stresses that we (begin to) learn to use the word 'person' by having paradigmatic examples pointed out to us, that is, by learning what he calls the semantic rule for the word 'person': 'we ourselves are said to be persons, and so are our parents, our brothers and sisters, other children and their fathers and mothers'. Following Peter Strawson's classic *Individuals*, Swinburne then specifies what he calls the syntactic rules for the concept of a person by pointing out that we may ascribe to persons both 'M-predicates' (material or corporeal predicates), such as 'weighs eighty pounds' or 'is six foot tall', and 'P-predicates' (psychological predicates), such as 'is smiling', 'is going for a walk', 'is in pain', 'thinks hard', etc. On Strawson's account, persons are distinguished from inanimate things because they are also subjects of P-predicates.[51] However, since many P-predicates may be ascribed to animals as well, whereas we do not call animals 'persons', we should list a subset of P-predicates that are ascribable meaningfully to humans or persons only, negatively or positively, such as 'speaks Chinese', 'understands relativity theory', 'is opposed to abortion', 'feels remorse', and so on. Indeed, many predicates we use for persons are neither M-predicates nor P-predicates in Strawson's sense, such as 'has been elected as a member of Parliament'. And, as Swinburne points out, persons typically have many capacities that animals lack, such as the ability to use a complex language and to form moral judgements.[52]

Now it seems obvious that God cannot be said to be a person non-analogically if he is denied a body, granting that we endorse Swinburne's own definition of analogy and Swinburne's own account of the literal meaning of the word 'person'. The reasons are that, first, the paradigmatic examples of persons specified by the *semantic* rule are all human beings, that is, bodily beings, whereas, second, many of the predicates specified by the *syntactic* rules cannot be ascribed meaningfully, either positively or negatively, to an allegedly bodiless god. This holds not only for M-predicates such as 'weighs eighty pounds' but also for a great many P-predicates, such as 'masters Chinese', 'intends to go for a walk', 'smiles buoyantly', or 'is in pain'. Clearly, a bodiless being cannot resemble the standard examples of persons in the respects in which and to the degrees to which they resemble each other, so that the similarity rider for literal uses of the word 'person' excludes God. Equally clearly, many of the syntactic rules have to be dropped if we want to apply the term 'person' to a bodiless being. How can Swinburne maintain, then, that we do not need analogy in order to avoid the incoherence of saying that God is a person without a body? Let us turn to his direct proof of coherence first.

Swinburne says that '[i]t is easy to spell out in more and more detail' what the supposition of there being a person without a body amounts to.[53] He prepares this

---

[50] Swinburne (CT), p. 100.

[51] As Hacker (2007), p. 312, points out, this Strawsonian account of the concept of a person is still 'overly Cartesian', because of the 'dichotomous division of predicates into P- and M-predicates'. Indeed, many predicates of persons do not fit into one of these two classes, such as 'is in debt' or 'has been appointed as president'.

[52] Cf. Swinburne (CT), pp. 102–3; cf. Strawson (1959), Chapter 3.

[53] Swinburne (CT), p. 106; cf. (ES), p. 152.

direct proof of coherence by asking 'what it is for a person to have a body', and answers this first question by answering a second one: 'what is it that I am saying when I say that this body, the body behind my desk, is *my* body?'[54] Following Jonathan Harrison, Swinburne holds that I claim five things when I am saying this: (1) that the disturbances in this body cause me pains or aches, etc.; (2) that I feel the inside of this body, such as the emptiness of this stomach; (3) that I can move directly many parts of this body, such as this arm; (4) that I look out on the world from where this body is, and (5) that my thoughts and feelings are affected non-rationally by goings-on in this body, such as: getting alcohol into this body makes me see double. Swinburne then concludes as follows:

(Conclusion A)
Now clearly a person has a body if there is a material object to which he is related in all of the above five ways. And clearly a person does not have a body if there is no material object to which he is related in any of the above five ways.[55]

He then argues that it is easy to spell out coherently the supposition that you do not have a body by imagining that (1*) you stop having pains or aches when this body of yours is disturbed; that (2*) you stop feeling this body from the inside; that (3*) it turns out that you can move directly by your will many things that are not parts of this body at all, while 'you also find yourself able to utter words which can be heard anywhere, without moving any material objects'; that (4*) it turns out that you 'come to see things from any point of view which you choose, possibly simultaneously'; and that (5*) pouring alcohol into this body no longer affects your thoughts. According to Swinburne, this amounts to a direct proof of coherence concerning the notion of a bodiless person, so that he concludes:

(Conclusion B)
Surely anyone can thus conceive of himself becoming an omnipresent spirit. So it seems logically possible that there be such a being. If an opponent still cannot make sense of this description, it should be clear to many a proponent how it could be spelt out more fully.[56]

However, this is not clear at all, and there are quite some confusions and fallacies in Swinburne's direct proof of coherence. Let me point out the most important ones only.

First, Swinburne interprets our use of sentences such as 'John has a muscular body' or 'Jane has a beautiful body' by modelling them on the use of sentences such as 'John has a powerful car' or 'Jane has a beautiful dog'. In other words, he interprets them as sentences that are used to describe a (five-fold) *relationship* between John or Jane and some physical object, five ways of being related which in fact obtain but might just as

---

[54] Swinburne (CT), p. 104 (Swinburne's italics).
[55] Swinburne (CT), p. 105. My summary of the 'five ways' partly consists of quotes from Swinburne (CT), pp. 104–5. Cf. Harrison (1973–4), and Swinburne (ES), p. 146: 'A person has a body if there is a chunk of matter through which he makes a difference to the material world, and through which he acquires true beliefs about that world'. According to Harrison and Swinburne, embodiment is a matter of degree, because one might fulfil some of these criteria and not others. Cf. (CT), p. 105.
[56] Swinburne (CT), p. 107. Cf. for his definition of 'omnipresence': (CT), p. 106.

well not obtain, as Swinburne states in conclusion (A). But this not only begs the question; it is also a misinterpretation of the usage of these sentences. When we say of John that he has a muscular body, or that he has a powerful mind, we are not describing a relationship between John and another entity, his body, or between John and a third entity, his mind. In other words, we are not assuming that there are three entities involved here – John, his body, and his mind – which are somehow related to each other. Rather, we are ascribing properties to John as a human being, while at the same time indicating that these properties are *bodily* properties or *mental* properties *of this human being*, respectively.

It follows that it does not make sense to speak of John's *relation* to his body or John's *relation* to his mind in the way Swinburne does, as if there were two or three entities involved, which are related to each other in many different ways. The main reason is that from an ontological point of view it does not make sense to speak of the 'relation between a property and its bearer'. It is nonsensical to wonder, for example, what the relation is between this paper and the property of its being white. Admittedly, we do sometimes speak of John's relation to his body. But what we mean is not that there is some relation between two entities, John and his body. Rather, we are stating how John feels about his bodily properties. For example, we may mean that John is frustrated by his lack of muscular strength or is ashamed of the fact that he is obese. Similarly, John might lament that he has such a miserable mind or such a bad memory. What he means is that he would like to be more intelligent or remember more things. In spite of the similarity of surface grammar, the logical grammar of sentences such as 'John has such and such a body', 'John has such and such a mind', or 'John has such and such a memory', is very different from 'John has such and such a house'. Misinterpreting the logical grammar of 'having such and such a body' on the model of ownership or possession, which would imply the existence of distinct entities that are related to each other, clearly leads to an absurd multiplication of entities. Suppose, for example, that John has a muscular body, a bad memory, a feeble intelligence, mediocre eyesight, vast geographical knowledge, a great gift for football, etc., how many interrelated entities would be involved?

Although it makes perfect sense to suppose that John does not have this car but has that car, or that he has no car at all, it does not make sense to suppose that John is bodiless, because human beings *are* bodies in the sense of being biological organisms, subject to physical forces such as gravity, and located in time and space.[57] It follows that in our language *we simply have no use* for sentences such as 'this body, the body behind my desk, is my body', except perhaps when pointing to a picture. But even in this latter case, we would say, when pointing to a figure on the picture: 'That is me', and not: 'That is my body'. As a consequence, the second question that Swinburne is asking, the question as to 'what is it that I am saying when I say that this body, the body behind my desk, is *my* body', is a nonsensical question. Statements (1–5) are

---

[57] Note that the logical grammar of '*having* such-and-such a body' differs from 'human beings *are* bodies' (in the sense of spatio-temporal material continuants). As Hacker stresses, '[t]he animate body that a human being is, is the human organism. Clearly, a human being does not *have* the organism that he *is*. So the body – the organism – that he is, is not the body that he has' (2007, p. 271; Hacker's italics).

true, of course, but they do not spell out five ways of being related to a body, which determine that, contingently, *this* body is *my* body.

It follows that imagining the falsity of (1–5), as in (1*–5*), is not conceiving of oneself as a bodiless person or spirit. Of course you may very well imagine that (1*) a disturbance of your body, such as cutting off your legs, does not cause you pain; that (2*) you no longer feel the inside of your body, such as the emptiness of your stomach; or that (5*) you can drink a lot of alcohol without becoming drunk. But this amounts to conceiving of specific kinds of insensitivity, which may be caused by an anaesthetic, rather than conceiving of oneself as a bodiless person.

Let me discuss (3*) and (4*) in somewhat greater detail. You can indeed imagine that 'you find yourself able to move directly anything which you choose'. But this is conceiving of the possibility that you have powers of telekinesis rather than beginning to conceive of yourself as a bodiless person. And you can imagine that you 'find yourself able to utter words which can be heard anywhere', but again, this is not imagining that you are a bodiless person. Rather, it is imagining that acoustics works differently from the way it in fact works. When Swinburne adds that you might utter words 'without moving any material objects', things become somewhat more dubious. Does he mean: without moving your lips? If so, what can he mean by 'uttering words'? For surely, what we mean by 'uttering words' is precisely that we produce them orally.

Later in his book *The Coherence of Theism*, Swinburne assumes without further argument that it makes sense to ascribe a voice to spirits or bodiless persons. In an argument to the effect that we may check the memory of a spirit in order to establish its identity, he says that '[i]f a voice of a spirit made claims about goings-on [. . .] and acknowledged that previous claims made in the same voice were his claims . . . ', etc.[58] But surely we cannot meaningfully ascribe a voice to a spirit in the non-analogical sense, for someone's voice is the voice that we hear when someone is opening his or her mouth, and is articulating words or singing a song, or when we hear a recording thereof.[59]

It is even more difficult to conceive consistently (4*) that 'you also come to see things from any point of view which you choose, possibly simultaneously'. A point of view in the non-analogical sense logically presupposes a body with eyes, or a camera, and it is difficult or impossible to suppose consistently that a body with eyes occupies simultaneously different points of view. Perhaps we may imagine this by imagining that our body has many octopus-like tentacles with eyes at their extremities, which can stretch into other galaxies. You will object that this is stretching our imagination too far. However that may be, imagining all these things is *not even beginning* to conceive of a person without a body. We cannot but conclude that Swinburne's direct proof of coherence with regard to the supposition that there can be a bodiless person or spirit, fails altogether.

---

[58] Swinburne (CT), p. 112; cf. (ES), p. 173.
[59] Cf. Alston (2005), p. 233: 'To say that I *spoke* to you has part of its meaning that I made sounds by the use of my vocal organs. But because God has no vocal organs, that cannot be part of what it means for God to speak to someone' (Alston's italics).

7.6 PERSONAL IDENTITY: THE ARGUMENT FOR CARTESIAN DUALISM

One will come to the same conclusion with regard to Swinburne's attempt to refute philosophers who argue that the notion of a bodiless person is incoherent. These philosophers endorse what Swinburne calls *an empiricist theory of personal identity*. According to such a view, there are various criteria of personal identity, which are somehow balanced against each other, in particular the criteria of bodily continuity, continuity of character, and continuity of (personal) memory. If one holds that the criterion of bodily continuity as a human being is crucial to the concept of personal identity, as is clearly the case in legal and other everyday contexts, the notion of a bodiless person is ruled out as incoherent from the very start. But even if one were to hold that the criterion of bodily continuity could be overruled in principle by the other two criteria, the notion of a bodiless person is incoherent. The reason is that the criteria of continuity of memory or character can be applied only to creatures exhibiting bodily behaviour, as I argued above (section 7.2).

In a digression into the nature of personal identity, Swinburne aims at refuting simultaneously all such arguments to the effect that the notion of a bodiless person is incoherent. He attempts to do so by arguing that what he calls empiricist theories of personal identity are 'not at all on the right lines'.[60] According to Swinburne, the main flaw of these theories is that they fail to keep distinct two very different questions about personal identity, namely the question (A) what it *means to say* that a person P_2 at t_2 is the same person as a person P_1 at an earlier time t_1, and the question (B) what kinds of *evidence* we might have for such a claim.[61] Swinburne holds that empiricist theories of personal identity give a correct answer to question (B), but mistakenly assume that this answer also applies to question (A).

The logical structure of Swinburne's refutation of empiricist theories of personal identity is simple, and it may be summarized by the following three propositions:

(a) Empiricist theories of personal identity imply that sometimes there will be no right answer to the question whether P_2 is the same person as P_1.
(b) But there always is, or must be, such a right answer, even though it may be that we do not have any evidence for deciding what the answer is.
(c) Hence, empiricist theories of personal identity are mistaken.

Swinburne concludes as a reply to question (A) that 'personal identity is something ultimate', and that it is 'unanalysable into conjunctions or disjunctions of other observable properties'. This would be the 'only alternative' to empiricist theories of personal identity.[62] What, then, does our personal identity consist in, according to Swinburne's alternative theory? He answers by endorsing Cartesian dualism, in spite of the fact that very few philosophers of mind still think that this is a defensible

---

[60] Swinburne (CT), p. 115.
[61] Swinburne (CT), p. 112. Swinburne spells out these types of evidence in (ES), Chapter 9.
[62] Swinburne (CT), p. 122.

option: 'it is not that we are non-spatial things, but that we have as our essential part a non-spatial thing, a soul', while 'we may also have a body as a non-necessary part'.[63] Of course, if even persons in the ordinary sense of human beings essentially are spirits, and have their bodies as non-necessary parts only, and if indeed it makes sense to say this, it must be coherently conceivable that there is a god who is a bodiless person as well.

What should we think of this refutation of 'empiricist theories' of personal identity? We should accept Swinburne's distinction between (A) the notion of personal identity on the one hand, and (B) the evidence that we might have for a specific statement of personal identity on the other hand. In spite of what Swinburne claims, however, there is no good reason for thinking that empiricist accounts of personal identity necessarily conflate these two issues. According to a plausible empiricist account, it is part and parcel of our ordinary notion of personal identity that global bodily continuity of a human being is a fundamental and necessary condition for personal identity, which cannot be overruled by other criteria. A person, call her Jane, may lose her memory in old age, and her character may be changed quite tragically by a debilitating mental illness. But we would all say that it is still Jane who lost her memory and is mentally ill, because there is global bodily continuity of Jane as a human being, although we may add that she is 'not her old self any more'.

Yet someone who holds such an 'empiricist' view of personal identity will admit that in many cases we might have no empirical evidence whatsoever for substantiating a claim of personal identity. For example, it may be that police inspectors cannot identify a suspect because no fingerprints or genetic materials are available, because the person has changed his or her looks by plastic surgery, and refuses to say anything when caught. Moreover, in most cases we do not keep track of someone else's bodily continuity, so that we use other evidence or grounds for re-identifying human persons, such as their name, physical appearance, voice, character, function, typical behaviour, or recollections.

We should also accept Swinburne's first premise (a). Following contemporary philosophical fashions, Swinburne tells a science fiction story involving a mad surgeon, which is such that the empiricist criteria for personal identity allegedly yield no answer to the question of whether a person P_2 is identical with a person P_1.[64] Imagine, for example, that a mad surgeon removes P_1's brain and divides it in half. He then transplants each of the resulting hemispheres into one of two bodies of recently deceased people, from which half of the existing brain has been removed. Suppose that the transplants succeed, the two bodies are reanimated, and that as a result two persons P_2 and P_3 come into existence, who both behave to the same extent like P_1, have to the same extent the same memories as P_1, and so on. Can we now say that P_1 is identical with P_2, or with P_3? I agree with Swinburne that on the basis of an empiricist

---

[63] Swinburne (CT), pp. 125–6. See for Swinburne's defence of Cartesian dualism also his book *The Evolution of the Soul* (ES), Part II, and (EG), pp. 197–200. The arguments for Cartesian dualism in ES and CT are essentially the same. For further discussion, cf. the last section of this chapter and Chapter 13, below. Of course I cannot examine here all recent defences of Cartesian dualism, such as Zimmerman (2010).

[64] Swinburne (CT), pp. 118–19.

theory of personal identity, we sometimes simply do not know what to say concerning such science fiction cases.[65]

However, this should not be considered a drawback of empiricist accounts. Indeed, it is exactly what we should expect. An account of a concept should give us an illuminating overview of the various uses of the relevant term in the language. However, our language is not made for a fanciful science fiction world, but for the real world in which we are living. The concept of a person is tailored to the human condition. It is a useful concept because the rules that govern the word 'person' and the notion of personal identity are such that in all common situations of real life, we know whether the concept applies or not. Such ordinary concepts are typically somewhat vague – does the word 'person' apply to a newborn baby? – and they may break down or lose their grip if we want to apply them to far-fetched science fiction cases that never occur in real life, or to exceptional cases, such as when we wonder whether people suffering from multiple personality disorder are one or more persons.

We may draw two lessons from these considerations. First, it is a methodological mistake to test an analysis of an ordinary concept such as the concept of a person or the concept of knowledge by attempting to apply it to science fiction examples. Although in neuroscientific research, uncommon brain lesions that cause serious impairments of our psychological faculties may yield new knowledge about the normal workings of the brain, conceptual investigations are different. One cannot elucidate the normal use of a word, and analyse the concept it expresses, by wondering how we should apply it to imagined situations that never or very rarely occur, and in which the concept loses its grip. Second, it is misleading to call such an analysis a 'theory', which should cover all kinds of imaginary situations. What Swinburne calls the 'empiricist theory of personal identity' simply is an attempt to spell out explicitly the rules we all use when we apply the concept of a person. This attempt may be defective in some respects, as in fact it is, but the remedy is not to introduce a philosophical theory that is even further removed from the concept we use.[66]

As a consequence, the second premise (b) of Swinburne's refutation of empiricist theories of personal identity, that there must always be an answer to questions as to whether P_i is identical with P_j, is false. Swinburne does not state this premise explicitly, but he needs it in order to stage his *reductio ad absurdum* of empiricist theories, which proceeds as follows.

If there must always be an answer to such questions, at least one absurdity arises. It may be that the empiricist theory, as applied to the science fiction case described above, allows for duplication, so that P_1 is both identical with P_2 and with P_3. But then the absurdity follows that one and the same person P_1 can be at two places simultaneously, which is contradictory according to our normal concept of a person. Hence, theories that allow for duplication must be rejected (the 'duplication difficulty'). However, if empiricist theories rule out duplication, other absurdities result. If one rules out duplication by saying that a subsequent person at a later time cannot be identical with the earlier person if another person at that later time satisfies the criteria

---

[65] Although in this science fiction case, most of us would probably say that P_1 is dead, as is evident when one examines P_1's corpse on the operation table. Cf. Hacker (2007), p. 307, and the main text below.
[66] Cf. Hacker (2007), pp. 308–9.

for personal identity equally well, it follows, quite absurdly, that who I am may depend on whether or not you exist, and that a man could ensure his own survival by ensuring the non-existence of future persons too similar to himself. However, if one rules out duplication by putting a threshold value in the criteria for personal identity, requiring for example that P_2 can only be identical at t_2 with P_1 at t_1 if P_2's body contains 90% of the matter of P_1's body, this involves an 'essential arbitrariness', which Swinburne spells out at length (the 'arbitrariness difficulty').[67] Also, there will always be borderline cases, which should be excluded if we want a clear-cut answer to questions of personal identity in all cases, which is required by Swinburne's second premise (b).[68] As I said, however, (b) is false, so that Swinburne's *reductio ad absurdum* of what he calls empiricist theories of personal identity does not succeed. And since his rebuttals of arguments to the effect that the notion of a bodiless person is incoherent, are based on this *reductio*, they also fail, since they rely on a false assumption.

Swinburne's argument against empiricist views on personal identity and in support of Cartesian dualism in Part II of his book *The Evolution of the Soul* is essentially the same as the argument in *The Coherence of Theism*. Again, Swinburne imagines a mad surgeon story, inspired by Bernard Williams, according to which the surgeon is going to transplant your left cerebral hemisphere into one human body, and your right one into another.[69] He again assumes that two living persons will result from this operation, and that there will be exactly the same amount of bodily continuity between you, P_1, as you were before the operation, and each of the resulting persons P_2 and P_3. Also, they will perform equally well on the other empiricist criteria for personal identity with P_1. The mad surgeon then tells you that he is going to torture one of the resulting persons and to give a great gratification to the other, so that it is very important for you whether after the operation you will be P_2 or P_3.

Like in the argument discussed above, Swinburne assumes in *The Evolution of the Soul* that (b) there always is, or must be, a right answer to such questions of personal identity, even in science fiction cases. It is, he suggests, 'a factual matter whether a person survives an operation or not'. As he stresses, '[t]here is a truth here that some later person is or is not the same as some pre-operation person'.[70] But since in the science fiction case of the mad surgeon, P_2 has exactly the same amount of bodily continuity with P_1 as P_3 has, and exactly the same continuity of mental life and character, whereas P_2 cannot be identical with P_3, we allegedly have to infer that there is something else to the continuity of a person than any continuity of (parts of) the brain or the body or mental life or character. Excluding the idea of a *partial* survival of P_1, Swinburne concludes that personal identity and continuity can best be conceived

---

[67] Cf. for these labels Swinburne (ES), New Appendix C, p. 328.
[68] My summary of Swinburne's *reductio* is partly quoted from CT, pp. 115–22; cf. also ES, pp. 328ff.
[69] Swinburne (ES), p. 149, jo. Williams (1970).
[70] Swinburne (ES), p. 147; cf. pp. 182–3: 'If you divide my cat's brain and transplant the two halves into empty cat skulls and the transplants take, there is a truth about which subsequent cat is my cat which is not necessarily revealed by knowledge of what has happened to the parts of my cat's body'. Cf. also p. 329: 'only someone already in the grip of a strong philosophical dogma could deny that there is a truth about whether or not I have survived an operation...'

of as consisting in the identity and continuity of an immaterial soul as an independent substance, which in principle can exist without the body and can survive death.[71]

In the 'Prolegomenon to the Revised Edition' of *The Evolution of the Soul*, Swinburne claims that his arguments for the existence of the soul are 'of immense strength'. Indeed, he seeks again 'to persuade [his] philosophical colleagues and more generally the scientific world and the wider educated public, of the immense strength of those arguments'.[72] Furthermore, he suggests in the introduction to this book that taking seriously the facts of personal continuity, and of our mental life and its causal efficacy, might trigger a scientific revolution, because '[s]cientific revolutions occur when data and coincidences previously regarded as unimportant are taken seriously and made the focal point of understanding'.[73] But as we have seen, Swinburne's arguments for the existence of the soul as an independent spiritual substance, which might exist without the body, are inconclusive because they rely on the untenable premise (b).[74] Let me add two other criticisms of Swinburne's argument for substance dualism, both of which are decisive as well.

First, Swinburne assumes without further argument that P_1 did not simply die on the operation table. But undoubtedly, this is what we should say, because the criterion of global bodily continuity is essential to our ordinary concept of personal identity. In the case of the imagined brain operation, none of the resulting persons (P_2, P_3) meets the criterion of global bodily continuity for being identical with me (P_1), because each of them merely incorporates one hemisphere of my brain. Hence, we should conclude that the operation would cause my death, as all observers of my corpse after the operation would confirm.

If this is the correct conclusion, there is no need to introduce a substantial soul in order to understand what personal identity over time of human beings consists in. Swinburne would reject this conclusion, however, because he avers that 'I go where my brain goes'.[75] But by saying this, he introduces a new criterion of personal identity, which is not part of our ordinary concept of a person.

Second, it should be noted that Swinburne does not provide the alleged souls as separable substances with any clear criteria of (diachronical) identity, although

---

[71] Swinburne (ES), Chapter 8, pp. 147–55. For a definition of substance dualism, cf. (ES), p. 146: 'The body is separable from the person and the person can continue even if the body is destroyed. [...] The soul, by contrast, is the necessary core which must continue if I am to continue; it is the part of the person which is necessary for his continuing existence'.

[72] Swinburne (ES), p. ix.

[73] Swinburne (ES), p. 3.

[74] Swinburne also offers a 'modal argument', which runs as follows. (1) It is logically possible that a human person continues to exist although his body is destroyed entirely. But this entails that (2) this person actually has an essential part other than a bodily part, which can continue, and which might be called his soul. In other words, '[f]rom the mere logical possibility of my continued existence [when my body is destroyed] there follows the actual fact that there is now more to me than my body; and that more is the essential part of myself', that is, my soul (Swinburne (ES), p. 154, and New Appendix C, pp. 322–32). In order to substantiate the premise of this argument, Swinburne has to tell a story similar to the one I discussed in the previous section concerning the coherence of the idea that a person can exist without a body. My criticisms are similar as well, so that I reject the premise (1) of the modal argument. Cf. for a more formal critique of the argument, Kretzmann and Stump (1996).

[75] Swinburne (ES), p. 147.

positing an entity without specifying criteria of identity does not make sense.[76] Even worse, he *cannot* provide souls with such identity criteria when he uses the argument from brain transplants and similar arguments. Suppose, for example, that one adopts sameness of personal memories as a criterion of identity over time for souls. Then either the criterion will enable us in principle to decide who is me after the operation, because there is more sameness of personal memories with my old 'me' in the one person than in the other, or it will not. But in the first case the criterion may be used to solve the identity problem created by the imagined brain transplants without postulating substantial souls – if we assume that memories supervene on brain states, for example. In the second case, postulating souls will not help us to resolve the identity problem, since in the imagined case it is in principle impossible to decide to which new human being *my* soul migrated, if at all. If there really were a fact of the matter who of the two new human beings is me, we would not be able *in principle* to discover what that fact is. This argument against the necessity of introducing a soul holds with regard to any diachronic identity criterion one might propose for souls.[77]

For all these reasons, Swinburne has not shown that the notion of God as a bodiless person is coherent by establishing that even in the case of us humans our personal identity consists in the identity of a soul, which might exist without the body. We have to conclude, then, that the description of God as a bodiless person implies a contradiction, unless both the predicate 'person' and, indeed, all P-predicates are applied in an irreducibly analogical manner. If that is the case, however, it seems that theism cannot be a theory, which is confirmable by empirical evidence, as Swinburne admits. In other words, theism is either an incoherent theory or not a theory at all.

I started this chapter by saying that Richard Swinburne's grand strategy is the most promising one for an apologetic philosopher of religion: the strategy according to which theism is a theory or existential hypothesis, which can be confirmed by empirical evidence. Readers who are convinced by the arguments of this chapter will conclude, however, that the strategy miscarries even before he attempts to implement it. But perhaps not all readers are convinced by my arguments. Some of them will reject the Wittgensteinean account of our psychological vocabulary, and they will hold, with Swinburne, that (b) there must be a fact of the matter who is I even after the brain-transplant operation by the mad surgeon. Let us assume for the sake of argument, then, that theism can be defined by using nearly all of the defining terms in their literal senses, and also that theism is not obviously incoherent. Can we safely conclude on these assumptions that theism is a theory, which can be confirmed by empirical evidence?

---

[76] As Quine (1969) said, 'no entity without identity' (p. 23). In the case of souls, it has to be specified when a soul at a later time t_1 is identical with a soul at an earlier time t_0.

[77] The (degenerate) identity criterion of 'thisness' cannot function as a criterion for *diachronic* identity, and it is inapplicable to the postulated souls because one cannot identify a soul by *pointing* to it and saying 'this soul'. On 'thisness' or *heacceitas*, cf. Swinburne, (ES), New Appendix D on 'The Nature of Souls; Their Thisness', pp. 333–44.

8

God's Necessity

On several grounds, theologians have claimed that God exists necessarily. Supposing that the existence of all contingent things stands in need of a causal explanation, they feel that God must be the ultimate explanation of everything, so that his existence needs no explanation. If God exists necessarily, no further questions concerning his existence can be asked, and the regression of causes comes to a halt. Moreover, mathematics has often been regarded as the paradigm of true knowledge. Should natural theology not aspire to that status? Hence, the existence of God is sometimes compared to the existence of numbers, which, if they exist, exist necessarily. Inspired by these and other considerations, natural theologians from Anselm to Hartshorne and Gödel have attempted to construct purely conceptual proofs of the existence of God, such as the ontological arguments.

However, the idea that God exists necessarily seems to challenge the view that theism is a theory or existential hypothesis, which should be confirmed by empirical evidence. We might express this second challenge for theism as a theory in the form of a dilemma, which may be called the *dilemma of God's necessary existence*, or *the dilemma of necessity*, for short. Either God's existence is necessary or it is not. In the first case, it seems that his existence, like the existence of numbers, cannot and need not be confirmed by empirical evidence. In the second case, it seems that postulating God's existence in order to explain empirical data raises the question as to how his existence itself is to be explained. But that is the very question which theologians wanted to avoid.

The best tactic for theists consists in going between the horns of this dilemma. On the one hand, they should deny that God's existence is a purely conceptual matter, like the existence of numbers. In other words, God's existence cannot be logically necessary, but must be a factual issue. On the other hand, theists should affirm that God's existence is necessary in a different sense, in order to exclude that we can sensibly ask what explains the existence of God. In this chapter, I discuss Richard Swinburne's views on God's necessity, as developed in Part III of *The Coherence of Theism*. These views constitute a paradigmatic example of escaping between the horns of the dilemma. I first investigate the necessity of God's existence, and then turn to the necessity of God's properties.

8.1 THE NECESSITY OF GOD'S EXISTENCE

Philosophers from Parmenides to Kripke have proposed and discussed a large number of concepts of necessity. A proposition may be necessarily true; it may be instrumentally necessary to do something in order to obtain a specific result; an event may be called necessary given certain sufficient causal conditions or if it is fated; it may be necessary to do something because one is forced to, or because morality prescribes it; and something may be necessarily identical with something else. Also, philosophers have often criticized and rejected specific notions of necessity. For these reasons, theists are obliged to explain precisely what they mean when they say that God's existence is necessary, or that God necessarily has certain properties.

In Chapter 13 of *The Coherence of Theism*, Richard Swinburne defines and discusses six concepts of necessity, labelled A–F. In *The Christian God*, he introduces a seventh notion. He then investigates in Chapter 14 of the former book in which sense God's existence can, and cannot be called necessary, and in what sense God may be said to possess certain properties necessarily. The reader of these texts will notice that even though some of the defined notions of necessity are commonly used in philosophy and ordinary language, Swinburne stipulatively introduces new notions, which may be quite complex. For example, the seventh notion, called *metaphysical necessity*, is defined as follows:

A statement is metaphysically necessary if it is true and it reports the occurrence of some everlasting event E, and there is no cause (active or permissive) of E, apart from any cause whose backwardly everlasting existence with certain properties has no cause and whose properties are such as to entail it actively or permissively causing E (either directly or through a chain of causes).[1]

Readers of *The Christian God* will not understand the point of this surprising stipulative definition of metaphysical necessity, with its cumbersome apparatus of epicycles, until they grasp that it is needed for attributing 'necessary existence' to all three persons of the Christian Trinity. Since in this book I am concerned merely with apologetic strategies in defence of bare theism, I shall not discuss Swinburne's concept of metaphysical necessity: it figures only in his rational reconstruction of the ramified theism of Trinitarian Christianity.[2] Here, I shall focus on the most salient senses in which God's existence can, and cannot be called necessary, assuming for the sake of argument that the theistic notion of God as a bodiless person makes sense (cf. Chapter 7, above). Is Swinburne able to resolve the dilemma of God's necessary existence, and, if so, at what price?

The latter issue is not unimportant, because it may be that the price for resolving the dilemma is too high for a believing theist. Let me start by quoting a passage in which Swinburne explains the main motive why he and other theists claim that God exists necessarily:

Basically because they have seen that God cannot be the cause of his own existence or of his being the kind of being which he is. Since no other agent can be the cause of these things, it follows if it is not to be a matter of chance that God happens to exist and be the kind of being which he is, then it must be somehow in the nature of things, unavoidable. For a being whose existence and nature was a chance matter would hardly deserve worship.[3]

---

[1] Swinburne (ChrG), p. 118.
[2] Cf. Alston (1997) for a critical analysis of *The Christian God*.
[3] Swinburne (CT), p. 264.

In other words, if Swinburne cannot but conclude from his analysis of God's necessary existence that God exists as a matter of chance, if God exists at all, he pays too high a price for resolving the dilemma, at least according to his own standards. We might call this the *non-contingency criterion* of success for his solution.

It would be cumbersome to discuss each of the six remaining concepts of necessity defined by Swinburne. Therefore, I shall now mention the two main senses of necessity in which according to him God's existence is not necessary. Then I shall specify in the next section the central sense in which he claims that God's existence *is* indeed necessary, if God exists. After that, the reader will be able to assess the price one has to pay for a solution to the dilemma of God's necessary existence.

According to Swinburne, God's existence is not necessary in the purely logical or conceptual sense of the word. He uses a common definition of logical necessity, labelled necessity A, according to which a proposition is logically necessary if and only if it is analytic, that is, 'if and only if it is coherent and its negation is incoherent'.[4] We find in Swinburne's works at least eight arguments to the effect that the core statement of theism 'God exists' does not express a logically necessary proposition, and these arguments are of unequal strength. For example, it is not a good argument to say that God's existence cannot be logically necessary because 'a world without a particular substance or a particular kind of substance seems always to be a coherent supposition [. . .]; no set of propositions which describe such a world seem to entail a self-contradiction'.[5] Clearly, this argument risks being an *ignoratio elenchi*. The question is not whether we can describe a world without God coherently, since we might do so without using the theistic concept of God. Rather, the question is whether, having mastered the theistic concept of God, we may consistently suppose that he does not exist.[6]

Another somewhat unfortunate argument adduced by Swinburne starts from the following premise. If the proposition that God exists were logically necessary, it would be possible to prove his existence on the basis of some logically necessary principle, as ontological arguments purport to do. Swinburne avers that this would make God dependent upon something external to him, which should be unacceptable to the theist:

But God would seem less than totally supreme if he depended for his existence on something quite other than God – for instance, on such a general logically necessary principle.[7]

This argument seems to contain a fallacy of ambiguity, because it confuses the causal or ontological dependency of objects with the logical dependency of propositions. That the conclusion 'God exists' depends logically upon premises holds for all valid or correct arguments for the existence of God. If Swinburne thinks that this logical

---

[4] Swinburne (CT), pp. 243–4. In other words, a proposition is logically necessary if and only if it is coherent and its negation entails a self-contradiction; cf. Swinburne (ChrG), p. 144.

[5] Swinburne (ChrG), p. 144.

[6] Given the context, it is clear that this is what Swinburne means in (ChrG). In footnote 23, he endorses Hume's argument that 'no negation of a fact can involve a contradiction' (p. 144). However, whereas he accepts this argument in (ChrG), provided that Hume's maxim applies to substances and not to abstract things such as numbers, he rejects it in (CT), pp. 273–4.

[7] Swinburne (ChrG), p. 145; cf. (CT), p. 301.

dependence of the conclusion 'God exists' on premises makes God himself less ultimate or less worthy of worship, he should not offer any arguments for the existence of God and abandon his project of a natural theology.[8]

A third unconvincing argument, put forward in *The Existence of God*, runs as follows. Assuming that God's act of creation fully explains the existence of the universe, and that the existence of the universe is logically contingent, Swinburne argues that the proposition 'God exists' cannot be logically necessary. The reason is that in a full explanation, a statement describing the *explanandum* is deducible from the *explanans*, whereas 'you cannot deduce anything logically contingent from anything logically necessary'.[9] If this argument is meant to be complete as it stands, it clearly is not sound. According to the standard theistic explanation of the universe, which is also Swinburne's explanation, the existence of the universe is not deduced from the existence of God as the only premise. Another premise is needed, namely that God choose to create a universe like ours, and, assuming with Swinburne that God's choice is absolutely free, this second premise is logically contingent.[10]

Fortunately, however, Swinburne also provides convincing arguments for his claim that the proposition 'God exists' is not logically necessary. He says, for example, that if this proposition were logically necessary, all propositions entailed by it would also be logically necessary. But a proposition such as 'it is not the case that no one knows everything about the past', though entailed by 'God exists', clearly is not logically necessary. For we may safely assume that its negation, 'nobody knows everything about the past', does not contain any concealed contradiction.[11] Another convincing argument to the effect that 'God exists' is not logically necessary relies on the fact that theists conceive of God as a being with causal powers, whereas '[a]ll other things which exist of logical necessity, such as numbers, concepts, and logical truths are pale lifeless things which cannot exercise causal influence'.[12] The fact that all known versions of the ontological argument are shown to be invalid will further support inductively the conclusion that 'God exists' is not logically necessary.[13] And the believing theist may be reconciled to this conclusion by Swinburne's final comment on the issue: '[f]or the theist the existence of God is a tremendous thing, the most fundamental truth about the universe. It seems to trivialize it to say that it holds for the same reason as does the truth that all bachelors are unmarried'.[14]

---

[8] According to Swinburne (personal correspondence), I misunderstood this argument. Its real point is that ontological arguments imply that there be a *reason which explains why God exists*, such as that the existent is more perfect than the non-existent. According to Swinburne, theists should not accept that one can *explain* God. But this is compatible with propounding empirical arguments for the existence of God on the basis of things *explained by* God's activities, as Swinburne does. However, the text of ChrG, p. 145, does not state this point very clearly.

[9] Swinburne (EG), p. 79; cf. p. 148.

[10] Cf. Swinburne (CT), p. 232: 'Note that God being creator of the universe in my sense does not entail the existence of our universe or any other material universe – only that if any universe does exist God created it or permitted some other being to do so. Theists have always held that God did not have to create the universe'.

[11] Swinburne (CT), pp. 274–5.

[12] Swinburne (CT), p. 301.

[13] Cf. Sobel (2004), pp. 29–167, and Swinburne (ChrG), p. 145: 'All ontological arguments known to me that purport to show the logical necessity of God's existence seem to me unsound'.

[14] Swinburne (CT), p. 275.

Let me now turn to a second main sense in which according to Swinburne the existence of God is not necessary. This is what he calls 'physical necessity' or criterion F, which might also be labelled 'causal necessity' or 'causal inevitability'. Something is causally necessary if and only if there exists a full explanation of its occurrence. There is a full explanation of an event E or of the existence of an entity E if, given the explanatory factors or causes mentioned in the explanation, E could not but occur.[15] In that case, the set of explanatory factors is a sufficient condition for the occurrence of E. Swinburne argues convincingly that God cannot be the sufficient causal condition for his own existence (*causa sui*), since the notion of a *causa sui* is incoherent.[16] Even if one could say that earlier temporal stages of God's existence are a full cause of the later temporal stages of his existence, this does not imply that God can be the cause of all (infinitely many) stages of his existence. Swinburne further argues not only that there cannot be a sufficient cause of God's existence, but also that God's existence cannot have any cause at all.

A first argument that he adduces in support of this latter thesis runs as follows. 'Being eternal' is one of God's necessary properties. Hence, if God exists, he exists at all moments of time. But since a cause of God's existence must precede it, 'no moment of time would be early enough for an agent to bring about God's existence'. It follows that 'it is not logically possible that any agent could bring about God's existence'.[17] A second argument to the same effect starts by stating that among the defining properties of God is 'being the creator of all things', that is, of 'all logically contingent states of affairs, apart from his own existence and anything entailed thereby'. Now '[t]here cannot be causation in a circle'. It follows that, '[s]ince all logically contingent things depend on God for their existence, he cannot depend on them'.[18] Indeed, if one of these contingent things did not depend on God for its existence, 'then God would not have been able to make it exist or not exist, and so would not be omnipotent'.[19] Hence, if God exists, nothing can be a cause of God's existence.[20]

---

[15] Swinburne formulates all his definitions of 'necessity' in the formal mode. Criterion F reads: 'A proposition p is necessary if and only if it is true and there exists a full explanation of what it states to be the case' (CT, p. 260). Because causal necessity primarily applies to events or the existence of entities, I have stated this condition in the material mode. Swinburne defines the notion of a full explanation as follows: 'we have a full explanation of an event E if the explanation cites explanatory factors which brought about E, such that a statement of their occurrence entails the occurrence of E' (CT, p. 137). This formulation is unfortunate, however, since we cannot say of laws of nature that they 'bring about' an effect, whereas without laws of nature (or universal premises) there cannot be an entailment. A somewhat better definition is provided in (EG), pp. 25, 76; cf. also (CT), p. 269.

[16] Swinburne (CT), pp. 268ff.

[17] Swinburne (CT), p. 265.

[18] Swinburne (CT), p. 276.

[19] Swinburne (EG), p. 99. This latter argument is not conclusive if one defines 'omnipotent' in Swinburne's own way (Swinburne (CT), p. 165): 'a person P is omnipotent at a time t if and only if he is able to bring about the existence of any logically contingent state of affairs x after t, the description of the occurrence of which does not entail that P did not bring it about at t, given that he does not believe that he has overriding reason for refraining from bringing about x'. The point of this cumbersome definition is to avoid well-known inconsistencies concerning simpler notions of omnipotence.

[20] Swinburne (EG), pp. 98–9.

Clearly, if God's existence cannot have any cause, it a fortiori is not necessary in sense F of causal inevitability. Given the fact that God's existence is not logically necessary either, in what sense can we still say that if God exists, his existence *is* necessary? And will this sense of necessary existence satisfy the theists, who claim that if it were a matter of chance that God happens to exist, he would hardly deserve worship? As Swinburne says elsewhere, '[i]t seems to me that a theist, if he is to worship a God worthy of worship, must hold that God's necessity is necessity of the strongest kind that the being described so far could possess'.[21] What is this strongest kind of necessity, which pertains to God's existence, if at least he exists?

8.2 ONTOLOGICAL NECESSITY AS RADICAL CONTINGENCY

In order to prepare his answer to this question, Swinburne introduces a notion of necessity that was suggested by the writings of apologetic philosophers of religion such as John Hick.[22] He stipulatively defines as follows this concept of necessity D, which he calls 'ontological necessity':

[D] A proposition *p* is necessary if and only if it is true, but the truth of what it states is not (was not, or will not be) dependent on anything, the description of which is not entailed by *p*.

The word 'dependent' in this definition is explained in terms of causality or physical probability, and not as logically or conceptually dependent. Accordingly,

A state of affairs S_1 is dependent on another state S_2, if the existence of S_2 makes a difference to whether or not S_1 exists. This will be so if S_2 brings about S_1 (S_1 would not have existed but for S_2) *or* S_2 makes it more probable that S_1 will exist.[23]

Swinburne points out that if there is no creator god, 'the universe exists' is a necessary proposition in sense D. As he says, 'in that case the universe would not depend for its existence on anything else [...;] it would be an ultimate brute fact'.[24]

This latter observation shows clearly that [D] is a stipulative definition, which provides the word 'necessary' with a somewhat surprising new meaning. If there has been a first state of the universe, such as a hot Big Bang some 13.7 billion years ago, and if this first state cannot be said to have been caused by anything else, nobody would say that the universe exists necessarily, or that the Big Bang was 'ontologically necessary'. Agreeing with Swinburne that in this case the existence of the universe would be an 'ultimate brute fact', we would conclude that its existence is not necessary in any sense, but radically contingent. Indeed, the notion of necessity D, which Swinburne calls 'ontological necessity', is not opposed to any plausible sense of 'contingency'. What is necessary in sense D is both causally and logically contingent. Should we not conclude that something of which the existence is an ultimate brute fact, hence radically contingent, exists merely 'by chance'?

---

[21] Swinburne (EG), p. 95.
[22] Cf. Hick (1961) and Franklin (1964), referred to by Swinburne (CT), p. 258, note 13.
[23] Swinburne (CT), pp. 258–9 (Swinburne's italics).
[24] Swinburne (CT), p. 259.

The reader of *The Coherence of Theism* will be quite surprised, then, by the fact that Swinburne answers as follows the question in which sense God's existence should be called 'necessary' if God exists:

> Necessity on criterion [D] seems to me to be the kind of necessity which the theist wishes to attribute to God's existence. To say that "God exists" is necessary on this criterion is to say that God does not depend for his existence on himself or on anything else. No other agent or natural law or principle of necessity is responsible for the existence of God. His existence is an ultimate brute fact. Yet being the sort of being which he essentially is, everything else in the universe depends on him, and must do so – for he is by his nature the ultimate source of things. Hence his existence is not merely *an* ultimate brute fact, but *the* ultimate brute fact. All other logically contingent facts depend on this one.[25]

What else is Swinburne achieving here than on the one hand admitting that if God exists, his existence is a brute fact, or radically contingent, and on the other hand rescuing the traditional idea of God's necessary existence by a purely verbal move, using the stipulatively defined sense D of ontological necessity?

But perhaps we should prefer a more charitable reading of *The Coherence of Theism* on this point. By spelling out precisely which senses might be given to the imprecise statement that God exists necessarily, Swinburne has discovered an important truth that escapes most theists, and which is as follows. The only important sense in which God's existence may be called 'necessary' is the stipulatively defined sense D. It follows that the necessity of God's existence is not opposed to its contingency. On the contrary, it just means that God's existence, if God exists, is a brute fact and radically contingent. This crucial discovery enables Swinburne to escape from the dilemma of God's necessary existence by going between its horns. On the one hand, the necessary existence of God can be confirmed in principle by empirical evidence, because it is a purely factual matter. On the other hand, this does not imply that we should raise the question as to what caused God's existence itself, since this existence, if it obtains, is an ultimate brute fact.

Readers will now be able to decide for themselves whether the price Swinburne pays for resolving the dilemma of necessity is too high for a theist. Is a God whose existence is an ultimate brute fact still worthy of worship? It may seem that the answer must be a negative one on Swinburne's own criterion. For should we not say that an entity the existence of which is a brute fact, exists as a matter of chance?[26] And did Swinburne not claim that 'a being whose existence [...] was a chance matter would hardly deserve worship'? Was this not the reason why the theist wanted to claim that the

---

[25] Swinburne (CT), p. 277 (Swinburne's italics). Cf. (EG), p. 96: 'To say that "God exists" is necessary is, I believe, to say that the existence of God is a brute fact that is inexplicable – not in the sense that we do not know its explanation, but in the sense that it does not have one'.

[26] In EG Swinburne explicitly denies that God's existence is a matter of chance (p. 95). Cf. also (CT), p. 277: 'There is no chance here, because the nature of things is that what is and what is not depends on the will of God, and that can only be if he exists'. But of course, from the premise that the existence of created things is not a matter of chance, because their existence depends upon God's will, it does not follow that God's existence is not a matter of chance.

8.3 NECESSARY PROPERTIES

According to theism, God not only exists necessarily, but he also necessarily has a number of his properties. So the theist should specify which properties are of God necessarily, and in which sense of 'necessarily'. We saw above, for example, that according to Swinburne, God necessarily possesses the property of being 'creator of all things except Himself'. This premise was needed in an argument to the effect that God's existence cannot have a cause. But since Swinburne also holds that God freely decided to create the universe, it seems that God could have decided to create nothing whatsoever. Can we then say that God *necessarily* possesses the property of being 'creator of all things except Himself'?[29] If God freely decided to create, this property is not even necessary in sense D. For the fact that God has the property, assuming that he has it, then depends upon something the description of which is not entailed by the proposition that God created all things except himself, to wit, God's free decision.

What is more, can we not consistently suppose that the universe (not including God) has always existed, so that its existence is backwardly eternal? This is what many physicists assumed in the nineteenth century. As we saw above, Swinburne holds that if God's existence is backwardly eternal, it cannot be caused, since 'no moment of time would be early enough for an agent to bring about God's existence'. Similarly, one might conclude from the premise that the existence of the universe is backwardly eternal, that the universe cannot be caused. Yet the theist will want to say that even in that case, one can consistently assume that God exists. If so, however, God cannot necessarily have the property of being the creator of the universe, because on these assumptions he cannot have that property at all.[30] Let us ask again, then, which properties theists should attribute to God necessarily, and in which sense of 'necessarily'.

At first sight, the best answer to these questions seems to be that the necessity involved is merely conceptual or analytic in sense A, a consequence of an optional and stipulative definition of 'God', and that it is up to religious believers to decide which properties pertain necessarily to the god or gods they believe in. In other words, the necessity of God's properties would be that of God's nominal essence only, as spelled

---

[27] See the quote above from Swinburne (CT), p. 264.

[28] Perhaps we can console the theist by stressing that God's existence is also necessary in Swinburne's sense E: 'A proposition is necessary if and only if it is a true proposition which states how things are at some time and if things always will be and always have been as it states that they are at that time' (Swinburne (CT), p. 260, jo. p. 276). Since God exists eternally by definition, the proposition 'God exists', if true, would be necessary in this sense. But as Swinburne says, this use of 'necessary' 'may seem to us an unnatural one' (ibidem, p. 259). If the universe exists eternally, its existence would have to be called 'necessary' in this sense as well.

[29] Swinburne's solution to this problem (CT, p. 232) is to define the property of being the creator of the universe conditionally as follows: 'if any universe does exist God created it or permitted some other being to do so'.

[30] In that case, he cannot even have the property as defined conditionally (CT, p. 232; cf. the previous note), if at least we are speaking of diachronic causation.

out by an individual theist. But this conception has disadvantages that most theists want to avoid.

One may have different views on the relation between the nominal essence of a god, as specified by a proposed definition or description, and the truth conditions of an existence claim concerning this god. On the one hand, one might hold that the existence claim concerning a god G is true if and only if an entity exists that has all the properties included in G's nominal essence.[31] But this option has the following drawbacks. If many properties are included, the risk that the god does not exist is considerable. Furthermore, if adherents to a religion propose slightly diverging defining descriptions of G, as they usually do, they would not refer to one and the same god by using the proper name 'G'. As a consequence, communication between believers would break down easily, albeit perhaps imperceptibly.

On the other hand, one might for these reasons prefer a cluster theory of proper names for descriptive definitions of gods. According to such a cluster theory, the claim that G exists is true if and only if there is in fact an entity that has most or (almost) all properties included in the nominal essence of G. The statement 'G exists' will then be false if and only if there is no individual that satisfies more than a few of these properties, whereas it will be a borderline case for being true if there is in fact an individual that has quite a few but not (almost) all properties mentioned in the defining description of G.[32] However, this view on the truth conditions of 'G exists' has another drawback, to wit, that the individual G does not have any specific property necessarily. What is necessary to be G is merely that an individual has most or (almost) all properties mentioned in the descriptive definition or nominal essence of G. The drawback becomes even more serious if one argues that the description specifying what kind of entity G is does not function as a definition of the proper name 'G' at all, but merely as an identifying description that provides 'G' with a referent.[33] For sometimes one might succeed in identifying an individual by using some description, even though this description is largely false of that individual.

For these reasons, theists who claim that God has some *specific* properties necessarily may want to argue that they are using another notion of necessity than the merely conceptual necessity due to a stipulative definition of 'God'. Yet another reason may motivate theists to argue for this conclusion. The conceptual necessity due to a stipulative definition of a god is purely *de dicto*. If an existing entity at first has enough properties mentioned in the god-definition for being properly picked out by 'G', so that we may say that G really exists, nothing in our concept of G seems to exclude that this individual at some point of time loses a number of these properties. However, traditional theists want to rule out that the entity called 'God' can lose his defining properties. They will hold, for example, that God in spite of his omnipotence can neither forfeit his property of being forwardly eternal by committing suicide, nor lose his omniscience by growing demented.[34]

---

[31] This is Swinburne's view: (CT), pp. 237–8.
[32] Cf. Swinburne (CT), p. 237.
[33] This is also Swinburne's view: (CT), p. 236.
[34] Cf. Swinburne (CT), pp. 266ff.

This view makes sense only if the necessity pertaining to God having these properties is not *de dicto* but *de re*. In other words, it must be due to the very nature or real essence of God that he cannot lose or forfeit certain properties, and not merely be a matter of a verbal definition. But is there a legitimate notion of *de re* necessity? The philosophical fortunes of the idea of *de re* necessity have varied in intellectual history. Whereas Aristotle held that each natural kind has a distinct real essence, this notion became suspect during the scientific revolution. Descartes argued, for example, that there is only one real essence for all material entities, and Locke held that the real essences of things cannot be known, if they exist at all. Phenomenologists such as Edmund Husserl claimed that there are real essences and that they can be known a priori by a procedure called eidetic intuition or *Wesensschau*, but this doctrine has been rejected on good grounds by most analytic philosophers. In the 1970s, the notion of *de re* necessity was revived by Saul Kripke and some others, such as Hilary Putnam. Using a possible world semantics for modal logic, Kripke and Putnam argued that both identity statements involving proper names such as 'Phosphorus is Hesperus', and statements about microstructures involving names of natural kinds such as 'water is H_2O', have *de re* necessity, although their truth has to be discovered a posteriori by empirical methods.[35]

To put it briefly, their basic idea is that proper names such as 'Phosphorus', natural kind names such as 'water', and chemical formulae such as 'H_2O' are *rigid designators* in the sense that they designate the same entity or have the same extension in all possible worlds – that is, in all conceivable situations – in which these entities exist (and do not designate anything else in possible words in which these entities do not exist). If we then discover empirically that Phosphorus is Hesperus, or that water is H_2O, these identity statements are a posteriori but also *necessarily true*, since in each of these statements the two denoting terms (e.g. 'Phosphorus', 'Hesperus') pick out the same entity or stuff *in all possible worlds*. Asked how names can be rigid designators, Kripke answered by proposing a causal theory of names, according to which a name gets attached to its bearer by an original baptizing act, and remains attached to it if the uses of the name later in history are causally connected in a correct manner to this inaugurating use. In the case of general names for natural kinds, such a simple causal theory will not do, however, since the initial baptism by which some stuff was called 'water' for the first time cannot link the name to all instances of water. Therefore, the initial baptism for natural kind terms has to be accompanied by an implicit similarity rider such as 'this stuff, and all the stuff that is essentially the same, is water'. It would then be the task of scientific research to discover what is essentially the same stuff as the original and paradigmatic instances of water by investigating their microstructure. Accordingly, Kripke's and Putnam's modal view of natural kinds is often called *scientific essentialism*.

Swinburne uses this Kripkean notion of *de re* necessity in order to specify in which sense of 'necessity' God has certain properties necessarily. The notion is labelled 'necessity B' and defined as follows:

---

[35] Cf. Saul Kripke: 'Identity and Necessity' (1971) and 'Naming and Necessity' (1972). Hilary Putnam and Keith Donnellan developed similar views around the same time.

130   *Theism as a Theory*

[B] A proposition is necessary if and only if it is incoherent to suppose that the individuals in fact picked out by the referring expressions in the sentence which expresses it do not have the properties and/or relations claimed by the proposition.[36]

And Swinburne claims that calling the statements in which these properties and/or relations are attributed to God necessary on criterion B 'gives the theist all he could possibly want in the way of necessity'. For it would link his defining properties to God by 'a very strong bond indeed; by the bond of the incoherence of supposing that God could lack them'.[37]

8.4 GOD'S MINIESSENTIAL KIND

It is important to notice, however, that when he applies Kripke's notion of *de re* necessity to God, Swinburne uses it selectively at least in three respects. First of all, he rejects the causal theory of proper names for the name 'God'. Second, whereas for Kripke a statement containing a proper name as a referring expression can be necessarily true in his new sense only if it is an identity statement such as 'Hesperus is Phosphorus', Swinburne wants to attribute Kripkean necessity to propositions ascribing properties to God. And third, attributing these properties to God is not the result of an empirical and scientific investigation into the hidden essential microstructure of an entity called 'God', as is the case with 'water is H_2O' according to Kripke or Putnam. Rather, Swinburne attributes necessity B to propositions about God that are identical with, or entailed by, his lexical or stipulative definition of the proper name 'God'. In this section, I shall first elucidate these three points and then raise a number of critical questions.

Swinburne is right, it seems, in repudiating the causal theory of proper names for 'God'. Although he considers the word 'God' as it is used by monotheists to be a proper name and not a mere shorthand for a definite description, Swinburne argues that in the case of 'God' a descriptive account of this proper name is clearly superior to Kripke's causal account:

The history of the word 'God' down the ages is quite irrelevant. Men could have put forward in our time the theory that there is an omnipresent, etc. spirit, suggested 'God' as the name of the postulated spirit, and then discussed whether God exists. Their question would be the same question as our question whether God exists.[38]

Elsewhere, he even says that he takes the proposition 'God exists' to be logically equivalent to 'there exists necessarily a person without a body (i.e. a spirit) who necessarily is eternal, perfectly free, omnipotent, omniscient, perfectly good, and the creator of all things'.[39]

The second point needs a somewhat longer elucidation. One might think initially that with regard to statements containing proper names, Kripkean necessity holds for true identity statements only, such as 'Tullius is Cicero'. Statements attributing certain

[36] Swinburne (CT), p. 244.
[37] Swinburne (CT), pp. 278–9.
[38] Swinburne (CT), p. 237; cf. p. 235.
[39] Swinburne (EG), p. 7.

properties to an individual picked out by a proper name would be Kripke-contingent. Indeed, the claim that we can always conceive consistently the falsehood of such a statement was Kripke's main argument for concluding that ordinary proper names function as rigid designators and that all descriptive theories of proper names are mistaken. But if this is correct, and if the word 'God' is a proper name, we cannot say that statements such as 'God is omniscient' are necessarily true in sense B of Kripke-necessity.

In order to solve this problem within a Kripkean framework, one has to forge a link between proper names and natural kinds. This can be done by stressing two points. First, if we say that a proper name rigidly designates a specific individual, we are implicitly relying on identity criteria for that individual.[40] Second, one may argue plausibly that, given our linguistic practices for proper names, one of these identity criteria is that the individual belongs to a certain natural (or cultural) kind. For example, it is part of our common practice of using proper names such as 'John' and 'Saul' that they are used to refer to male human beings, and it is part of our naming practice that 'Amsterdam', 'Geneva' or 'London' are used to refer to cities.

Accordingly, if in fact the proper name 'Saul' rigidly designates a specific individual human being, we might claim that a statement such as 'Saul is a human being' is necessarily true in the sense of Kripke-necessity. Or, as Swinburne prefers to say, this proposition states an 'essential kind' to which Saul belongs:

A proposition states an essential kind to which an individual belongs if the proposition states what sort of a thing the individual is (e.g. "is a person", "is a car", "is a number") and if it is not coherent to suppose that that individual no longer be that sort of thing yet continue to exist by being something else instead.[41]

Since there is a taxonomic hierarchy of kinds, if 'a is ϕ' states an essential kind of an individual a, and if 'a is ϕ' entails 'a is ψ', the proposition that a is ψ will also state an essential kind of a. For example, if Saul essentially is a human being, Saul will also be essentially a mammal and a vertebrate, which are kinds higher up in the hierarchy.

Furthermore, the lower an essential kind is in the taxonomic hierarchy, the more properties an individual belonging to that kind can be inferred to possess necessarily. Given that Saul essentially is a human being, we can infer that necessarily he has all the properties belonging to human nature. If we can say only that Saul essentially is a vertebrate, we can infer fewer essential properties. Hence, it is important to specify what is the lowest essential kind to which an individual belongs. Swinburne calls such a lowest essential kind a 'minimum essential kind'. This label may be misleading, however, because the lowest essential kind to which an individual belongs is minimal only from the point of view of its extension. It is maximal with regard to the properties it implies.

Swinburne defines what he calls a *minimum essential kind* of a as follows:

"a is ϕ" states a minimum essential kind to which a belongs if there is no proposition "a is ψ" which states an essential kind to which a belongs and which entails "a is ϕ" but is not entailed by "a is ϕ".

---

[40] Swinburne (CT), p. 245.
[41] Swinburne (CT), p. 249.

If this is the case, Swinburne also says that ϕ is the *miniessential* kind of a, or that a is *miniessentially* ϕ.[42] Although there is some arbitrariness involved in this notion of a miniessential kind of a, since we may disagree about what is a's miniessential kind, its logic is unproblematic. For example, if Nixon is miniessentially human, it is essentially impossible that he turn into a chimpanzee, and the statement that Nixon is a human being is necessarily true in the sense of Kripke-necessity. But if Nixon is miniessentially a mammal, we may suppose that he turn into a chimpanzee, but not that he metamorphose into an alligator.

Swinburne applies this technical terminology to the case of the proper name 'God' as follows. If we assume that the proper name 'God' in fact picks out an existing individual who is omniscient, omnipotent, etc., and we suppose that God is miniessentially *a person*, omniscience, omnipotence, etc. will be *accidental* properties of God. In that case, God could deprive himself of his omnipotence or omniscience without ceasing to be God. Or he could commit suicide, thereby annulling his accidental property of being eternal. However, if God's lowest essential kind is that of a *personal ground of being*, that is, a person having all the defining properties of God, each of these properties will be an essential property of God, so that it is incoherent to suppose that the individual picked out by the name 'God' might lack one or more of the defining properties traditionally attributed to him by theists. According to Swinburne, the theist should suppose that God is miniessentially a personal ground of being, since God would be 'more obviously worthy of worship if he is miniessentially a personal ground of being than if he is miniessentially divine', for example.[43]

This application to God of Kripkean necessity B shows clearly that the specification of God's essence or miniessential kind by Swinburne is not the result of an empirical investigation. Admittedly, Swinburne attempts to adduce empirical evidence for the *existence* of God in his book *The Existence of God*. But he does not develop specific methods for investigating empirically the very *nature* of God, methods that would be the theological analogies of, say, methods of chemical analysis of liquids such as water. Indeed, it is difficult to imagine comparable and valid methods of empirical theology at level (a), as we saw in Chapter 6.3, above. As a result, Swinburne's method of determining the miniessential kind to which God belongs – supposedly a (super) natural kind with one instance only – is a far cry from the type of investigation Kripke and Putnam prescribed for determining the essence or microstructure of natural kinds. This is the third respect in which Swinburne's use of Kripkean necessity is selective or eclectic.

With regard to Swinburne's selective usage of Kripkean necessity to the effect that God has all his defining properties necessarily *de re*, we may raise four critical questions. First of all, is the very notion of Kripke-necessity a valid notion? Have Kripke and Putnam really discovered a new type of necessity, which is 'metaphysical' in some unobjectionable sense? The notion may be invalid in two respects. It may be that certain claims made by Kripke or Putnam are simply incoherent. If one asserts, for example, that the extension of the term 'water' was 'really' determined by the chemical composition of water even before it was discovered that its composition is H_2O, one

---

[42] Swinburne (CT), p. 251.
[43] Swinburne (CT), p. 292 and pp. 299–300.

implies that the meaning or the rules for the use of a word can be unknown to all competent users of the language at a certain time, say, in the year 1700. But this is incoherent, since the rules for the use of a word at a time are given by what is accepted at that time as the correct explanation of the meaning of that word. Accordingly, when it was discovered that most of what was called 'water' has the chemical composition H_2O, scientists did not reveal some metaphysical essence *à la* Kripke. Rather, they decided to adapt their use of the word 'water' to this empirical discovery after the event by defining different kinds of water. Apart from pure water in the more precise sense of H_2O we now have other types of water, such as heavy water (D_2O).[44]

The other respect in which Kripke's account of necessity is invalid is that it is inadequate with respect to its paradigmatic instances. For example, the proper names 'Hesperus' and 'Phosphorus' do not function as rigid designators, because they connote a certain way of appearing (in the evening, or very clearly in the early morning) of a planet, as Frege already said.[45] And if one assumes that the microstructure which allegedly is the essence of a natural kind has to be one and the same in all instances of that kind, Kripke's essentialism cannot apply to biological species and other taxa, for example, since typically there is genetic variation within a species. If one accepts a cladistic classification of taxa in biology, a biological species is a historical entity and not a natural kind in the essentialist sense of Kripke and Putnam.

A second critical question is as follows. Supposing for the sake of argument that Kripke's account of *de re* necessity is a tenable one, we may wonder whether Swinburne's selective use of it is coherent. For example, can one reject the idea that the name 'God' functions as a rigid designator and yet claim that the statement 'God is omniscient', if true, has Kripkean *de re* necessity? And can one coherently claim that if one's determination of God's miniessential kind does not result from an empirical investigation into His hidden microstructure, but rather from a somewhat arbitrary decision concerning which god would be more worthy of worship, one has discovered God's metaphysical essence rather than merely stated again one's preferred nominal essence? Clearly, the answer to these two questions must be negative.

It follows that we have to give a negative answer as well to the third critical question one might raise. As we saw, Swinburne claims that calling the statements in which defining properties and/or relations are attributed to God 'necessary' on criterion B, 'gives the theist all he could possibly want in the way of necessity'. For it would link his defining properties to God by 'a very strong bond indeed; by the bond of the incoherence of supposing that God could lack them'.[46] But does Swinburne's selective application of Kripkean necessity to God's defining properties really give theists all they could possibly want in the way of necessity? Even if Kripke's own notion of metaphysical necessity were coherent and adequate, Swinburne's selective use of it

---

[44] Cf. Hacker (1996), pp. 250–3.
[45] G. Frege, 'Über Sinn und Bedeutung' (1892), p. 26: 'den Sinn des Zeichens [. . .], worin die Art des Gegebenseins enthalten ist'.
[46] Swinburne (CT), pp. 278–9.

cannot deliver what theists want. After all, this use does not amount to more than simply restating God's nominal essence and claiming *de re* necessity for it without using the Kripkean apparatus of rigid designators and empirical research that would justify this claim. However, even if we should answer each of these three critical questions in the affirmative, a fourth issue arises, with which I shall deal in the next section.

8.5 ANALOGY AND COHERENCE AGAIN

From the foregoing considerations, we cannot but conclude that calling God's properties 'necessary' is as much of a purely verbal manoeuvre as calling God's existence 'necessary'. Whereas the necessity of God's existence was produced by a stipulative redefinition of the term 'necessity' as radical contingency (D), the necessity of God's properties results from Swinburne's stipulative rule for using the word 'God' as a proper name for an individual that miniessentially is a personal ground of being. However this may be, the thesis that God miniessentially is a personal ground of being yields a serious problem of consistency, which Swinburne discusses at length.[47]

God is said to be a person, and, as Swinburne stresses, the semantic and syntactic rules for the word 'person' limit the kind of persons that there can be. In the usual sense of 'person', people may remain one and the same person while losing many of their mental capacities, because they have Alzheimer's, for example. But if God is miniessentially a personal ground of being, a person who is God cannot lose his defining properties, such as omniscience. In other words, there is a contradiction between on the one hand the thesis that God is a person and on the other hand the thesis that God miniessentially is a personal ground of being. With regard to this contradiction, the two usual methods of resolving it without having recourse to analogy will not work. We can neither introduce restrictions into the definition of God's properties, nor prefer weaker versions of these properties to stronger versions, because in this case the contradiction is not generated by the nature of God's properties at all. It arises because certain properties are said to be of God necessarily, whereas persons in the usual sense of the term might lose such properties while remaining one and the same person. This is why Swinburne is compelled to play, 'at long last, the analogical card', as he says.[48]

Swinburne uses this 'joker' of analogy as follows. There is no doubt that we learn the notion of a person partly by examples of adult human beings, which specify the *semantic* rule for the word 'person' (cf. Chapter 7.3–5, above). The use of the term is further fixed by what Swinburne called *syntactic* rules, which link the word 'person' to other words such as 'thought', 'emotion', 'action', 'intention', and so on, since a person is (analytically) an animate being who can have thoughts and intentions, and who is able to act in a great number of ways. All these other terms are also introduced partly by examples of thoughts, emotions, actions, etc., which are typical of adult

---

[47] Swinburne (CT), pp. 279–87.
[48] Swinburne (CT), p. 281.

human beings. It follows that if we are to use the term 'person' analogously, the analogy will spread out over these other terms. Let me call the entire cluster of such terms the network of *person-terminology*. Swinburne now introduces analogy into this network by the method described in Chapter 7.3, above. Since one cannot point to God, or to his thoughts and intentions as new paradigmatic examples, which redefine the person terminology, Swinburne has to retain the old standard examples by which the semantic rules for 'person' are fixed. What he does, instead, is to change the similarity rider.

In our non-analogous uses of a term, all entities belonging to its extension should resemble the standard examples in the respect in which they resemble each other to the extent to which they resemble each other. Swinburne introduces analogy into his use of the word 'person' and the related network by changing this similarity rider into another: an entity in the new, analogously enlarged extension only needs to resemble the standard examples more than it resembles things which are clear cases of non-persons, non-thoughts, etc., and it does not need to resemble the standard examples as much as they resemble each other.[49] Since Swinburne takes as standard examples of non-persons 'houses or trees or tables', it follows that God can be called a person analogously if he resembles human persons more than he resembles houses or trees or tables.[50] One might think that this analogous extension of the word 'person' is very generous, and that we might now apply the word 'person' also to worms, or to clouds that resemble human faces. But this does not follow, for Swinburne wants to retain the entire system of syntactical rules for 'person', with one major exception. We should allow for a person such as God, whose defining epistemic, moral, and potentiary properties are undetachable from him. In other words, God is a person, analogically speaking, who cannot remain individually the same person if he loses one of these properties.[51]

There is no doubt that God, so described, is 'a person of a very different kind from ordinary persons', as Swinburne avers.[52] The reason he adduces is that God has properties necessarily which ordinary persons would not have necessarily, if they had them. In sections 7.2 and 7.5, above, I have argued that theists already introduce a large amount of analogy into the concept of a person in another respect, by saying that God is a bodiless person. However, even if the analogy introduced into the concept of a person were limited to one respect only, that of necessary properties such as omniscience or omnipotence, Swinburne acknowledges that this limited amount of analogy has two serious implications for the prospects of theism as a theory.

First, and in line with the theological tradition, he has to endorse 'the inability of man in any way fully or adequately to understand what is being said when it is claimed that there is a God'. Indeed, in the *Conclusion* to *The Coherence of Theism*, Swinburne stresses that his account of God, if true, would vindicate Aquinas's claim that 'no created intellect can see God's essence' by 'its own natural powers'. As Swinburne writes:

---

[49] Swinburne (CT), p. 283. [50] Swinburne (CT), p. 283.
[51] Swinburne (CT), pp. 283–4. [52] Swinburne (CT), p. 287.

Man does not have the concepts in terms of which to think about God adequately. He must stretch words in order to talk about God and does not fully understand what is being said when the words are stretched.[53]

But can theism be a theory or existential hypothesis, which is open to empirical confirmation, if so much is admitted? Imagine that cosmologists would say a similar thing of super-massive black holes. That is, they would not merely say that we do not yet understand black holes adequately, or that we have to develop our concepts and theories further in the light of empirical research, but rather that man is *unable in principle* to understand adequately what they are, since they cannot be described without using irreducible analogy. I do not think that any scientist would then take seriously the hypothesis that there are black holes.

The second implication is equally worrying for the theoretical theist. As we have seen in Chapter 7.4, a direct proof of coherence becomes difficult if one introduces analogy. For this reason, Swinburne raises the question whether we can now prove the coherence of the statement 'God is miniessentially a personal ground of being and a person'.[54] His answer is as follows:

I cannot now myself prove either that the quoted statement is coherent or that it is incoherent. The stretch of meaning of the words involved has left me without arguments of the normal kind for or against coherence.[55]

If analogy is introduced, one cannot, by using syntactical rules, derive an explicit contradiction from the statement that God is both a person and miniessentially a personal ground of being, thereby proving its incoherence. This was the point of introducing analogy in the first place. Can one derive a contradiction by using the semantic rules for 'person'? Given the new similarity rider, this will be very difficult, as Swinburne stresses. One would have to show that there cannot be an entity which necessarily is a personal ground of being and yet resembles ordinary persons more than it resembles houses, tables, and trees. But 'who knows what kind of beings there can be wildly dissimilar from those known to us?' We would have to list all the kinds of being of which it is logically possible that they exist, and there is not the slightest reason to suppose that we can do so.[56]

However, if a direct proof of incoherence is now impossible or very difficult, the same holds for a direct proof of coherence. As we saw in Chapter 7.4, a direct proof of the coherence of a statement *p* has the form '*o* entails *p*', where the coherence of *o* is either obvious or not in dispute. Given the coherence of *o*, deriving *p* from *o* shows that *p* is coherent, because no contradiction can follow from a coherent statement. But since such derivations have to rely on syntactical rules, it will be more difficult to prove the coherence of *p* directly to the extent that we introduce analogy into *p*, thereby relinquishing syntactical rules. Swinburne holds that even by playing the joker of analogy only once, he has introduced too much analogy into the conception of God for a direct proof of coherence to be feasible. Moreover, we probably lack sufficient agreement about the coherence of instances of *o* in this case. For these reasons,

---

[53] Swinburne (CT), p. 307.
[54] Swinburne (CT), p. 288. For reasons of ease of exposition, I use the word 'person' in my summaries instead of Swinburne's criterion for personal identity over time. But this simplification in my exposition has no influence on the validity of the arguments involved.
[55] Swinburne (CT), p. 288.   [56] Swinburne (CT), p. 288.

Swinburne says in the *Conclusion* to his book with regard to his theory of theism that '[t]his doctrine I have been unable either to prove coherent or to prove incoherent by normal direct means'.[57] Surely this is a remarkable conclusion for a book called *The Coherence of Theism*.

8.6 AN INDIRECT PROOF OF COHERENCE?

Swinburne does not conclude that theism should not be considered a theory at all, which is open to empirical (dis)confirmation. He claims that, even though a direct proof of coherence is not feasible, we may have a convincing indirect argument for coherence. In the case of theism, such an argument should be an inductive argument from factual premises to the effect that God as defined by theism exists. Clearly, if theism is true, it must be coherent, so that according to Swinburne each direct empirical argument for its truth is an indirect argument for its coherence. In *The Coherence of Theism*, Swinburne does not give such indirect arguments for coherence. He refers the reader to *The Existence of God*, the book in which he attempts to show that the existence of the god of theism is more probable than not.[58] However, apart from all empirical evidence we may wonder whether in the case of theism there can be an indirect argument for its coherence at all. Since I do not think so, I shall suggest that theism is not eligible as a theory or existential hypothesis open to empirical confirmation.

In order to substantiate this sceptical conclusion, I discuss two examples of indirect arguments for coherence, which Swinburne provides. The first example is concerned with the statement 'there is a strip of paper with only one surface'.[59] We may think that this statement involves a contradiction, since all pieces of paper have two surfaces. Then someone shows us a Möbius strip and demonstrates that one can draw a line in the middle of the strip and parallel to its boundaries, which starts at a point on the surface and comes back to this point, covering all surface of the strip there is. If this is accepted as a sufficient criterion for 'having one surface only', it has been shown indirectly that the statement is coherent. Similarly, it may be shown that 'there is a strip of paper with only one boundary' is coherent. This first example is less relevant to our concerns, however, for it does not involve analogy. What makes the Möbius strip surprising is, that whereas usually the different criteria for counting surfaces yield the same result, in this case they yield different results.

Swinburne's other example of an indirect proof of coherence is the phenomenon of light.[60] Traditionally, physicists thought that light either is a stream of particles, as Newton had argued, or consists of waves, originally conceived of by Christian Huygens as waves in a material medium. Although these conceptions may seem to exclude each other, some phenomena fit in well with the wave-theory, such as diffraction and interference, whereas other phenomena rather seem to confirm a particle-theory, such as the photoelectric effect or the Compton effect. As Swinburne says, all these

---

[57] Swinburne (CT), p. 305.
[58] Cf. Swinburne (CT), pp. 73, 289, 306.
[59] Swinburne (CT), p. 49.
[60] Swinburne (CT), pp. 67–71.

phenomena can now be predicted with great precision by using quantum theory. He then introduces the notions of analogy and of an indirect proof of coherence by the following argument.

Of course we may consider the quantum theory merely as a mathematical predicting device, and this is what the standard Copenhagen interpretation prescribes. But Swinburne argues that in doing so, we would abandon another function of science, the function of telling us what the world really consists of. What, according to Swinburne, should physicists say if they want to tell us what light really is? They cannot say that (a) light is sometimes particles and sometimes a wave, because one and the same beam of light may show both wave-like and particle-like effects. They cannot say either that (b) light is always particles and always a wave, since this is self-contradictory. What physicists should say, Swinburne claims, is 'that light is both "particles" and "wave", only in extended senses of the terms which do not exclude each other'.[61] To put this differently, Swinburne claims that the physicist should use the terms 'particle' and 'wave' analogously, using the similarity rider which defines analogy. By claiming that light is both particle-like and wave-like, the physicist then would claim merely that light resembles standard examples of particles more than it resembles standard examples of non-particles, and that it resembles standard examples of waves more than it resembles standard examples of non-waves. Swinburne further claims that 'this second, realist alternative is that adopted by the majority of scientists'.[62]

In other words, the realist physicist as conceived of by Swinburne would remove the contradiction between the wave- and the particle-conception of light in the same way as Swinburne removes the contradiction between 'God is a person' and 'God is miniessentially a personal ground of being', to wit, by introducing analogy. And theists would be able to argue indirectly for the coherence of this conception of God by adducing empirical evidence for its truth, exactly like physicists are able to argue for the coherence of their realist conception of light by pointing to wave-like phenomena such as interference and particle-like phenomena such as the photoelectric effect, which both belong to one and the same beam of light. Unfortunately, however, I do not think that many physicists would endorse Swinburne's account of what they should do or are doing.

One reason is that, when physicists speak of the particle-like nature of light and the wave-like nature of light, they are not at all relying on some vague analogy. Rather, they are giving a well-defined meaning to these terms by the wave function and by Max Planck's discovery that electromagnetic energy is emitted in discrete packets, which was verified by Einstein in 1905 when he explained the photoelectric effect. In other words, the 'analogy' which the physicist is using, if any, is supported by a very precise scientific theory, and such a theory is lacking in the case of theism. Another reason is that most physicists are not realists with regard to quantum mechanics. And if they are, they will not be satisfied with the view that light really is both particle-like and wave-like in an analogous sense. They will rather say that we do not yet know precisely what light is, and that, perhaps, we have to develop a new theory.

We should also reject for two reasons Swinburne's idea that the coherence of theism may be shown indirectly by adducing empirical arguments for its truth. First,

---

[61] Swinburne (CT), p. 68.
[62] Swinburne (CT), p. 69.

Swinburne's comparison of theism with the Möbius strip or the dual nature of light is misleading, since light and the Möbius strip are things we can see and study empirically, whereas God as conceived of by theism is merely a postulated entity. If theists adduce empirical facts in order to make probable the existence of God by an argument to the best explanation, they will be confronted by what philosophers of science call the problem of underdetermination. In principle, it will be possible to explain the very same set of facts equally well by postulating entities different from God. In such a case, we need criteria for theory-selection apart from the criterion of empirical adequacy, such as simplicity, consilience with background knowledge, and additional predictive power. One of these criteria is consistency. I do not think that from the large set of alternative theories we will ever select a theory the coherence of which cannot be demonstrated independently, apart from the empirical evidence. But this is precisely the case of theism as defined by Swinburne, so that we will never select theism in order to account for a set of facts. Hence, an indirect proof of the coherence of theism as proposed by Swinburne is not possible.

The second reason is a subsidiary one. Even if an indirect empirical proof of the coherence of theism were possible in principle, it is not feasible in the case for which Swinburne requires it. This case is not, like the examples of the Möbius strip or the dual nature of light, a situation in which we wonder whether an entity E can possess two properties M and N, which seem to exclude each other. Rather, Swinburne needs an indirect proof of coherence for the claim that God as a person possesses certain properties M and N *necessarily*. Even if we were able to show by an argument to the best explanation that an inferred entity has properties M and N, no empirical argument can show that this entity has these properties *necessarily*, so that an indirect proof of coherence is impossible in the very case for which Swinburne needs it.

I conclude from these arguments that theism as defined by Swinburne is not eligible as a theory which is open to empirical confirmation. Readers who are not yet convinced of this conclusion by the present chapter and Chapter 7 will find more arguments to the same effect in Chapter 9.

9

The Predictive Power of Theism

Suppose a monotheist interprets theism as an explanatory theory, which posits God in order to account for the existence of the universe and some of its features. Then we should wonder to what extent theism possesses the theoretical virtues that we usually require of explanatory hypotheses. For example, has theism a broad explanatory scope and does it account for all phenomena within its domain? Is theism empirically adequate in that testable consequences derivable from it are in agreement with the results of experiments or observations? Can theism be used to predict phenomena that are novel in the sense of not having been known or taken into account when the theory was formulated? Is theism falsifiable in some strong or weak sense? And so on.[1]

Clearly, theism interpreted as a theory or existential hypothesis will be claimed to possess some of these virtues to a greater extent than others. For example, it seems that theism has a broad explanatory scope. It has been affirmed by theists that 'theism purports to explain everything logically contingent (apart from itself)', that is, 'the universe and all its characteristics'.[2] It follows that '[t]he theist argues from all the phenomena of experience, not from a small range of them'.[3] But one cannot reasonably expect of theism that it yield precise quantitative predictions, for example. In contradistinction to scientific theories such as quantum electrodynamics, theism offers a *personal* or *intentional* explanation of phenomena. In this respect, theism resembles explanatory hypotheses that we find in the writings of historians, rather than 'large-scale scientific theories'.[4]

In this chapter I focus on one theoretical virtue only, which I shall call theism's predictive power. First, a minimalist notion of predictive power will be defined, which should be acceptable to theists. It will then be argued that there are good grounds for supposing that theism barely has predictive power even in this minimalist sense. As a consequence, it will not be easy to confirm or disconfirm theism by empirical evidence, so that many empirical arguments for or against theism come to nothing. Some theists have attempted to solve this problem regarding the predictive power of theism, and I shall discuss the solution proposed by Richard Swinburne in (the revised editions of) *The Existence of God* (2004) and *The Coherence of Theism* (1993).

---

[1] Cf. Hempel (1983), pp. 87–8.
[2] Swinburne (EG), pp. 66 and 108.
[3] Swinburne (EG), p. 71; cf. pp. 93, 99.
[4] Cf. Swinburne (EG), pp. 2–3: 'My use of confirmation theory [...] enables me to bring out the close similarities that exist between religious theories and large-scale scientific theories'.

9.1 PREDICTIVE POWER

If one takes the everyday meaning of the verb 'to predict' as a guide, one will define the predictive power of a theory or hypothesis as its ability to yield forecasts of future events, which can be derived from the theory together with auxiliary assumptions and specifications of initial conditions. Without such predictive power, technological and other practical applications of theories would be impossible. Yet this initial attempt at a definition is too narrow. If we aim at maximizing the growth of knowledge, we are not only interested in forecasting future events, but also in discovering items in the present or past that are new in the sense of not yet being known. Accordingly, the predictive power of a theory or hypothesis should be defined as its capacity to yield specifications of data that are not currently available.

Most scientists consider predictive power in this first sense as an indispensable virtue of hypotheses and theories. It is often added that predictions must be testable, and that the ability to make testable predictions of new data marks a decisive border between science and pseudoscience.[5] However, natural theologians who stress the 'close similarities' that exist between the theory of (bare) theism and large-scale scientific theories are landed in a dilemma at this point.[6] In order to be credible they claim that the theory of theism has predictive power. But will they also accept the implication of the first definition (if applied to bare theism), namely that bare theism should yield testable specifications of new data?

Either the defenders of theism endorse the requirement of predictive power as defined above, or they opt for a different definition, which does not demand that a theory predicts new data. In the first case, they run the risk that theism will be disconfirmed by future tests. For example, if a believer predicts on the basis of theism that specific prayers to God will make a statistically significant difference to the health of seriously ill people for whom they have been said, wide-ranging statistical research might disconfirm theism (cf. Chapter 6.3, above). Sophisticated contemporary theists may not be prepared to run this risk, so that they will opt for the second horn of the dilemma. Although theists who interpret bare theism as a theory cannot reject the requirement of predictive power, they will prefer a definition which does not imply that theories should entail testable predictions of new phenomena. They will stress that theism does not imply such predictions, and that many perfectly decent historical explanations do not do so either.[7]

Choosing this second horn of the dilemma is not without risks for the theist either. Practising scientists will now deride theists when they claim that the theory of bare theism closely resembles large-scale scientific theories. But, fortunately for theists, we find definitions of predictive power in the literature on the philosophy of science that meet their demands. These are definitions in terms of logical relations only, since for the assessment of the logical relations between a theory and the descriptions of the

---

[5] Cf. Timmer (2006) and Chapter 6, above.
[6] Cf. Swinburne (EG), p. 3.
[7] Cf. Swinburne (EG), p. 70: 'Hence it is in itself no objection to the hypothesis that there is a God, that it does not yield predictions such that we can know only tomorrow, and not today, whether they succeed'.

phenomena that it predicts, it is not relevant whether these phenomena were known when the theory was proposed, or whether they were used in constructing it. If they are prudent, theists will settle for such a logical definition of predictive power. Here they have much elbow room, for there are many logical definitions.

According to the classical deductive-nomological model of scientific explanation, for example, the predictive power of a theory is measured by specifying the set of descriptions of phenomena that can be deduced from the theory together with statements about initial conditions, auxiliary theories, and other background information. From this logical point of view, there is no difference between derivations of known phenomena and predictions of new phenomena. Accordingly, the predictive power of a theory is now defined as the total set of descriptions of phenomena deducible from the theory, irrespective of whether these phenomena were known when the theory was proposed.

Predictive power in this deductive sense has a number of dimensions. One is the *scope* of the theory (or the size of the field it is concerned with), which is broader to the extent that descriptions of more diverse types of phenomena are derivable from it. Another dimension is the degree of *completeness* of the theory within its domain or scope. How great is the theory's ability to account for the phenomena falling within its domain? Does it enable us to derive all of them or only some? Furthermore, one may consider the *precision* with which the derived descriptions specify phenomena as a third dimension of a theory's predictive power. The greater the predictive power of a theory is in these dimensions, the more favourably the theory will be regarded prior to empirical testing.[8]

However, this initial attempt at a logical definition of the predictive power of a hypothesis is too narrow for the following reason. In many cases, such as statistical theories or hypotheses of personal explanations, we cannot deduce the description of phenomena from the theory. What we can deduce is, typically, that there is a certain probability that a phenomenon exists. Or, to put this differently, we may say that, given the theory and the relevant background information, there is a certain probability that a specific phenomenon occurs. In the symbolism of probability theory, the predictive power (or *likelihood*) of a theory or hypothesis h with respect to an event or phenomenon e may now be defined as $P(e|h\&k)$, that is, the probability P that e will occur, given the truth of h and the background knowledge k. If the description of e can be deduced from h and k, this probability is 1, so that the deductive-nomological definition of predictive power is a limiting case of the probabilistic definition.

If we use the definition of predictive power in terms of logical probability, we have to say that the predictive power of a hypothesis h has a fourth dimension apart from the three dimensions mentioned above: its *strength*.[9] For each (description of a)

---

[8] In EJ, Swinburne defines the scope of a theory in such a way that it includes completeness and precision (p. 82). But it is useful to distinguish these different dimensions of scope in Swinburne's global sense.

[9] Like Swinburne, I use the expression 'logical probability' in order to stress that what is at issue here is how probable it is that a proposition p is true given that another proposition q is true, irrespective of who is doing the calculation or the assessment. In other words, $P(p | q)$ is an objective relation between propositions, and does not measure a subject's degree of confidence, for example. Cf. Swinburne (EG), pp. 15–16, and (EJ), Chapter 3. Inductive probability has the same objective logical nature, and cannot be reduced to a subjective degree of belief as expressed in betting behaviour, for

phenomenon e, this strength just is the probability $P(e|h\&k)$, which lies between 1 and 0. In some cases we will be able to determine this strength by assigning a precise number to P, when h is a statistical hypothesis, for example, but often we can rely only on intuitive estimates.

We may now wonder whether there is some threshold value of minimum probability, below which we should not consider a theory or hypothesis h as admissible at all. Should we not require, for example, that $P(e|h\&k) > ½$ for each e within the scope of h? One might think that if this probability is less than ½, h is irrelevant for the occurrence of the phenomenon e, so that e, if it occurs, will not be good evidence for h. In the past, some philosophers of science have proposed a substantial threshold as a condition of relevance, and they have denied that a hypothesis has predictive power if the threshold condition is not met.

Yet one cannot require a substantial and uniform threshold value in all cases, as is clear from a lottery example. Suppose that h is the hypothesis that Deborah bought a ticket in a fair lottery in which only one in a million tickets will win. If e_1 is the event that Deborah loses, we have a decent explanation of e_1 if it occurs, since $P(e_1|h\&k)$ is 999,999/1,000,000. Indeed, it is difficult to imagine a better explanation of e_1, since the probability that it occurs given that Deborah bought a ticket is very high. Suppose, now, that e_2 is the event that Deborah wins. The probability of e_2 given $h\&k$ is very low, namely 1/1,000,000. But what if e_2 occurs? How should we explain it? The answer is, of course, that Deborah bought a ticket in a fair lottery. This explanation h is an acceptable explanation of e_2 in the sense that it leaves nothing to be desired in our everyday explanatory context, assuming that the lottery is indeed fair, even though the probability that e_2 occurs given $h\&k$ is very low.

What lessons should we draw from this example for our probabilistic notion of predictive power? What makes h a good explanation of e_2 is that although winning in a fair lottery is very improbable if one buys a ticket, it is even more improbable if one does not buy a ticket. So what one should usually require of a good explanation or prediction is that $P(e|h\&k) > P(e|k)$, or in words, that the probability of e on the supposition that the hypothesis is true, given our background knowledge, is greater than the probability of e, given our background knowledge only.[10] While the *predictive* power of h with regard to e is defined as $P(e|h\&k)$, one sometimes defines the *explanatory* power of a hypothesis h with regard to e as

$$\frac{P(e|h\&k)}{P(e|k)},$$

and it is said that if (and only if) this factor is greater than 1, the occurrence of e will provide us with a correct confirming argument or C-inductive argument for h,

---

example (Cf. Mahler (2006)). Usually, inductive probability is defined as being concerned with the probability of hypotheses on evidence. But according to a more general definition, which I prefer, inductive probability is concerned with the probability of one proposition on another (Cf. in the same vein Swinburne (EJ), p. 62). Accordingly, the labels 'logical probability' and 'inductive probability' will be used as equivalents in this book.

[10] Or: than the probability of e, given some rival hypothesis. Cf. Chapter 11.4 for further discussion of various views on inductive support, such as Bayesianism and likelihoodism.

that is, an argument that raises to some extent the initially assumed probability that h is true.[11] The condition that $P(e|h\&k) > P(e|k)$, or, alternatively, that $P(e|h\&k) > P(e|\sim h\&k)$, may now be called the 'relevance condition' or the 'relevance criterion'.[12]

What are the implications of these definitions for our assessment of the predictive power of theism? It has often been objected to traditional empirical arguments for the existence of God that theism does not make very probable the evidence which is adduced in its favour. It is said, for example, that the cosmological argument for the existence of God is not a good argument since the existence of our universe cannot be deduced from the existence of God, and may not even be quite probable on the assumption that God exists and is free to create or not to create. We now see that although this objection is correct if one uses a deductive-nomological conception of predictive power, it is mistaken if one endorses the probabilistic definition. When putting forward the cosmological argument for the existence of God, theists do not need to claim that the hypothesis of theism, if true, would make the existence of our universe very probable. What they need to argue is merely that the existence of the universe is more probable on the assumption that God exists than on the assumption that God does not exist.[13] If that is so, the existence of the universe yields a correct C-inductive argument for the existence of God.

Theologians who interpret bare theism as a theory, or as an existential hypothesis, should agree that it must have a predictive power of this minimal strength with regard to each e_i that is adduced as evidence for it. Let me call this for short the *minimal predictive power* of a hypothesis with regard to e_i. If we translate our result into the terminology of threshold values, we may say that the threshold value for predictive power will vary from case to case, since in each case $P(e_i|h\&k)$ must be greater than $P(e_i|k)$, so that 'being greater than $P(e_i|k)$' functions as the local threshold value for each e_i. The central question of this chapter is to what extent the traditional theory of bare theism can have such a minimal predictive power with regard to established facts e. I shall argue that this issue is problematic indeed.

Admittedly, one correct C-inductive or confirming argument may not help theists very much. What they should do is to invent a cumulative case strategy in which many correct C-inductive arguments will be added up. Considering the total evidence for and against theism by taking into account the relevant probabilities, the theist will then have to argue that this cumulative case yields a correct *P-inductive argument* for

---

[11] Cf. for the notion of C-inductive arguments and P-inductive arguments: Swinburne (EG), p. 6; Cf. also Carnap's distinction between incremental and absolute confirmation in Carnap (1950), preface to the second edition (1962), p. xviii.

[12] Cf. Mackie (1969), Howson and Urbach (1993), Chapter 7, Swinburne (EG), Chapters 1, 3, 6. One might give examples in which we speak of an *explanation*, although the relevance condition has not been met: cf. Van Fraassen (1980), p. 108. I shall not dwell here on the many notions of explanation that philosophers of science have developed. In the sense of 'explanatory power' defined, a theory cannot have explanatory power without also having predictive power. This is why I am focusing on this latter, quite technical, notion.

[13] As the reader will have noticed, the technical definition of predictive power is now removed quite far from the ordinary meaning of 'prediction'. But clearly, if theism has no predictive power in this minimalist sense, it will have no predictive power in more substantial senses either.

the existence of God, that is, an argument which raises the probability that God exists to a value above ½, or to an even higher value if that is required for legitimate religious belief.[14] In this chapter, I focus on the issue of the minimal predictive power of theism only, without going into any of the further problems which such a cumulative case strategy has to face.

9.2 DOES BARE THEISM HAVE PREDICTIVE POWER?

What does the theory of bare theism consist in? According to its standard version, which is endorsed by many Christian, Jewish, and Muslim authorities, bare theism is the claim that there is only one god, a bodiless person (a spirit), who is eternal, perfectly free, omniscient, omnipotent, perfectly good, and who in fact created and sustains all things apart from himself.[15]

To this core thesis of bare theism each of the monotheistic religious traditions adds other claims, such as the Christian tenet that God incarnated himself in Christ, or the Muslim doctrine that Mohammed is God's messenger. Accordingly, it is customary to distinguish between *bare* theism and the more developed doctrines of the various *ramified* theisms. In this chapter, I shall be concerned with the predictive power of bare theism only, since natural theologians have to establish this core thesis before they can begin to argue that one specific revelation is reliable in contradistinction to others, so that one ramified theism is true and the others are false to some extent. In other words, since the details of ramified theisms are derived from alleged revelations, whereas we cannot endorse a revelation without relying on the arguments of natural theology (Chapter 1), theists cannot use these details for providing theism with predictive power within the context of natural theology.

Theists typically contend that their theory explains a great many things, such as the existence of the universe and of man. As we saw, theists even claim that the scope of theism is universal, since whatever happens, happens because God directly or indirectly makes it or permits it to happen.[16] However, the degree of completeness of theism within this universal scope has varied during its long career. For example, whereas Paley could still claim in his *Natural Theology* of 1802 that the functional anatomical complexity of the first specimina of each biological species should be explained directly by postulating special creative acts of God, sophisticated theists nowadays prefer explanations within the framework of the modern synthesis of neo-Darwinism, since such accounts of the facts are much more detailed and adequate. Presently they will confine themselves to some global explanation of the evolutionary process, and leave the task of local explanations to the scientist.

---

[14] The arguments should not only be correct C- and P-inductions, but also 'good' inductive arguments in the sense that their premises are known to be true by those who dispute about the conclusion. Cf. Swinburne (EG), pp. 6–7.
[15] Cf., e.g., Swinburne (EG), p. 7.
[16] Cf. Swinburne (EG), pp. 71, 93, 99. Of course, saying that God permits something is not a full explanation of that thing. It is only a necessary condition for its existence.

Champions of theism rarely dwell on the incompleteness of theism within its domain. Yet this may be a problem for theism, since many empirical data, which theism never predicted, may perhaps be interpreted as disconfirmations. Assuming with many theists that God's ultimate aim in creating the universe was to engender human persons on Earth or elsewhere, we should wonder, for example, why God created such a vast universe with hundreds of billions of galaxies, which is largely inhospitable to life, or why humans appeared only very recently in the evolution of life on Earth, which now lasts some 3.5 billion years, and will probably exist during a minute fragment of cosmic time only.[17] Such scales to the universe do not seem to make sense on the theistic hypothesis, if at least creating humankind was God's ultimate aim. Monotheists often claim that theism as a global theory of the universe 'makes more sense' of everything there is than its secular rivals.[18] But the more knowledge we acquire, the more theism may seem to be disconfirmed, since ever more data appear to be anomalies for theism.

Yet this kind of criticism shares an important assumption – if only for the sake of argument – with those who propose bare theism as a theory: the assumption that bare theism has some predictive power. In order to investigate to what extent this fundamental assumption is justified, we first have to specify the type of explanation that theism aims to provide. In scientific explanations, we typically derive a description of the explanandum from a set of laws of nature in conjunction with initial conditions and background assumptions. Theistic explanations are of a different type, which is called a personal or intentional explanation. When we give a personal explanation, we explain a phenomenon or event e by saying that it is brought about by a rational agent or person P doing some action A intentionally, that is, voluntarily, deliberately, on purpose, for a reason, or out of a motive.[19]

How can personal explanations satisfy the requirement of a minimal predictive power? Or, more specifically, how can the hypothesis h that God exists function as a personal explanation-with-predictive-power of the universe and some of its features? If we do not yet know who or what caused some e, and put forward the hypothesis h that a person P was present when and where e occurred, this hypothesis has predictive power as a personal or intentional explanation of e only if the following condition is satisfied. There must be background knowledge k about P, which can be justified independently of the fact that e exists, and which makes it likely that P had the intention to do A in order to bring about e. Of course $h \& k$ should also imply that the person P had the power to do so. Since Swinburne defines the hypothesis h of theism as the thesis that God exists, and because God is conceived of as omnipotent, in the case of theism the value of $P(e|h \& k)$ will depend only on how likely it is that God had the intention to bring about e.[20]

---

[17] Cf. for this 'argument from scale': Everitt (2004), Chapter 11, and Chapter 15.2, below.
[18] Cf. for example, Pannenberg (1973).
[19] In an extended sense, one may also speak of a personal explanation if e is an unintended effect of P's intentional action; cf. Swinburne (EG), p. 38. But I shall use the term in the narrower sense, as equivalent with 'intentional explanation'.
[20] Cf. Swinburne (EG), p. 72: 'Note that I take h simply as "there is a God". By itself it provides merely a partial explanation of e. It needs to be conjoined with an intention to bring about e in order to provide a full explanation of e. The value of $P(e|h \& k)$ will, for the various e, depend on how probable it is that God will have that intention'.

Suppose, to take a mundane example, that Margaret dies in the mountains by falling from a rock (= e), and that there are no physical indications of what caused her fall. Now we propose the hypothesis h that her husband Henry was present at the accident. This hypothesis has predictive power as a *personal* or *intentional* explanation of e only if we have independently attested background knowledge k, which makes it likely, for example, that Henry had the intention to kill his wife and attempted to do so by pushing her off the rock. The example resembles the theistic hypothesis with regard to the universe and its characteristics, although there is one minor difference. In Henry's case we already know that he exists, whereas we merely postulate the existence of God.

We may say, then, that apart from physical traces, a personal explanation of e can have a convincing predictive power only if there are good reasons to attribute to the person P specific intentions to bring about e. Attributing a specific intention to P should not be completely arbitrary. Let me call this the *condition of non-arbitrariness* for adequate personal explanations. Moreover, if we want to avoid circular explanations or explanations that are purely ad hoc, these reasons should be justified independently from our knowledge that e in fact exists or occurred.[21] Let me call this requirement the *condition of independence* for good personal explanations.[22]

Such independent reasons may be derived from many types of background knowledge about the person P. It may be that knowledge about the situation P is in, and about P's character, profession, or ambitions, provides us with good reasons for attributing a specific intention. For example, if in the case of Margaret's fatal fall Henry's firm is nearly bankrupt and Margaret leaves him a fortune when she dies, and he is known to be a ruthless businessman who started hating his wife a long time ago, the hypothesis that her death was the result of his presence, intention, and action has some predictive power. When we also know that Margaret was in high spirits and a good mountaineer, whereas she and Henry walked on an easy route over the rock, the hypothesis that her death was brought about by Henry instead of being an accident clearly meets the requirement of minimal predictive power. That is, the hypothesis satisfies the relevance condition $P(e|h\&k) > P(e|k)$, or, alternatively, $P(e|h\&k) > P(e|\sim h\&k)$.[23]

Apart from independent knowledge about P's situation and character, other types of background knowledge may enable us to attribute specific intentions to P non-arbitrarily and independently, thereby providing a personal explanation with predictive power. Having specific intentions in situations of a certain type may be characteristic for persons belonging to a specific culture. Our biological nature provides us with ample background knowledge for attributing intentions to people, such as wanting to have a meal, a rest, or a companion. These four types of background knowledge about human actors, that is, concerning their situation, character, culture,

---

[21] Of course, these reasons are often not independent from knowledge of e within the *context of discovery*. In many cases, we only start our search for intentions or motives of possible actors after the event. But they should be independent within the *context of justification*.
[22] Cf. on this condition of independence: Sober (1999) and Sober (2008), section 2.12. Sober also requires independently attested auxiliary hypotheses about the Designer's intentions.
[23] Cf. Chapter 11.4 for a likelihood reconstruction of such arguments.

and biology, may be ranked in terms of increasing generality. And finally, P may simply avow that he or she had specific intentions.

We are now sufficiently prepared to answer the main question of this chapter: to what extent can one attribute to the god of bare theism specific intentions to act or create, in such a way that the attribution satisfies the conditions of non-arbitrariness and independence? It seems that this is the only way of providing bare theism with predictive power. In other cases of personal explanations, the hypothesis that a certain individual brought about *e* may have predictive power because the person is a human being, whose body is involved in his intentional actions. For example, if we suppose that Bill committed a burglary, this hypothesis makes it probable to some extent that we will find Bill's fingerprints or DNA on the open safe. But since God is defined as a bodiless person, who is supposed to exercise his spiritual powers without using any material mechanism by means of which he can be identified, the issue of whether bare theism can have predictive power with regard to a specific *e* depends entirely upon the possibility of attributing intentions to God in a non-arbitrary and independent way.[24]

It seems clear, however, that there is no background knowledge whatsoever about God, which would enable us to attribute intentions to him in a non-arbitrary and independent manner.[25] Since God is bodiless, there can be no biological background knowledge. Because God does not participate in a culture, no cultural background knowledge about God is available. Furthermore, in spite of anthropomorphic statements in religious sources such as the Old Testament, according to which God is jealous, for example, the theory of bare theism in itself does not imply that God has specific character traits.[26] We cannot say either that before he created, he was in a specific situation, for, *ex hypothesi*, he then was the only thing existing. The same argument from absence holds for all other types of background knowledge that enable us to attribute intentions to persons non-arbitrarily and independently in the case of human actors. Furthermore, because God is supposed to be unique in his kind, we cannot have statistical knowledge that permits us to predict something about God's intentions. Consequently, we must be very cautious in attributing specific predictive powers to the theistic hypothesis concerning any known facts.[27]

---

[24] I shall not discuss here another vital issue for theism: whether the notion of agency or personal causation can have a meaning at all with regard to a god who is claimed to be bodiless. Cf. Rundle (2004), Chapter 4, for a negative answer to this question.

[25] Believers in revelations will deny this. But we are here concerned with the first stage of the global strategy in defence of theism, in which data derived from alleged revelations are not (yet) available.

[26] Cf. *Exodus* 20.4: 'for I the LORD your God am a jealous God'. Some theists will consider God's 'goodness' as a character trait. To what extent God's goodness enables us to attribute to him intentions non-arbitrarily and independently, will be discussed in the next two sections of this chapter.

[27] However, one might easily *imagine* things, the occurrence of which would be much more probable if theism were true than it would be if no supernatural spirit existed. For example, if suddenly from our point of view on Earth all the stars in the sky formed a sentence in Hebrew saying 'I, Yahweh, am the only true god and atheists will be punished for their unbelief', this *e* would be vastly more probable given theism than given atheism. For only persons can use language, and God, being omnipotent, has the power to rearrange stars and galaxies. If we also assume that he created man, directly or indirectly, it is plausible to attribute to God the intention to communicate with us in a spectacular and unambiguous manner, so that the independence requirement is satisfied. And because such an event is vastly more probable given theism than on our natural background knowledge, its occurrence might raise the probability that God exists to above ½ even if the prior probability of theism is very low.

One might object that the God-hypothesis has enormous predictive powers by definition, since God is *defined* as an omnipotent bodiless person *who in fact created and sustains all things apart from him*. But this objection fails for two reasons, an external and an internal one. First, the objection would trivialize the theological explanation of the universe, since any arbitrary explanation of an e by postulating the existence of an entity x can be provided with predictive power by adding 'and x caused e' in the definition of that entity. The second, internal reason is that according to the standard versions of bare theism, God was free to create or not to create. Indeed, as against varieties of Neoplatonism, most Christian, Jewish, and Muslim versions of monotheism stress God's absolute freedom.[28] If the creation is supposed to have been such an act of freedom pure and unrestrained, we should wonder what can have motivated God to create at all, and what can have motivated him to create this rather than that. If we have no good independently attested reasons for attributing to God the intention to create, and to create specific things rather than others, the hypothesis of bare theism has no predictive power concerning these things.

Theists who believe in a revealed religion, such as Judaism, Islam, or Mormonism, will reply that we do have independent information about God's intentions. Did God not reveal his creative projects to his prophets some thousands of years ago, or to Mohammed, or to Joseph Smith Jr. in 1827? Do such revelations not contain reliable and independent information about God's plan?[29] But in order to show that a specific revelation is reliable, one first has to argue within the framework of natural theology that bare theism is probably true, and, on this basis, that a specific revelation probably is a reliable one (cf. Chapter 1, above). And one cannot show this if bare theism has no predictive power concerning established phenomena.

The problem of predictive power is all the more pressing for theists because it seems that, being perfect, their postulated god is without needs, since needs are traditionally seen as imperfections. Why would a perfect being without needs intend to create something else apart from itself? One might answer: 'because it would be good that these other things exist'. But one might just as well answer: 'from God's point of view, it would be better if nothing existed apart from him'. Theism assumes that if things are created, they will be less perfect than the Creator. By creating less perfect beings apart from him, God would mar the absolute perfection (or diminish the average perfection) of what exists, since without a creation, everything there is would be absolutely perfect. Hence, it may seem to be excluded a priori that a perfect god creates anything whatsoever. Aristotle, who thought deeply about a monotheistic god, was convinced on good grounds that such a god would never create, and that his eternal activity would consist merely in thinking about his own thinking.[30]

If it is as easy to invent reasons why the hypothetical god would not create as it is to invent reasons why he would create, the condition of non-arbitrariness is not met. I conclude that one must be very cautious in attributing specific predictive powers to

---

[28] Cf. Swinburne (CT), p. 145: 'All that the omnipresent spirit who is God does, he does because he chooses to do. Nothing makes him do what he does. He did not, for example, have to create the world'. Cf. (EG), p. 49: 'God is supposed to be perfectly free'.

[29] The problem with this answer is, of course, that in the course of scientific progress, too many predictions based upon revelations have been disconfirmed by the facts (cf. Chapter 1.1, above).

[30] Aristotle, *Metaphysics* XII, 1072b–1074b.

bare theism as a theory. And to the extent that bare theism has no predictive power, all empirical arguments for and against theism fail, since they assume incorrectly that for some *e* the hypothesis of theism raises or diminishes the probability that *e* occurs.[31] Are theists able to counter this objection?[32] I discuss the most prominent attempt to do so in the next section.

9.3 THE MORAL ACCESS CLAIM

In Chapter 11 of *The Coherence of Theism* and Chapter 5 of *The Existence of God*, Richard Swinburne offers a solution to the problem of the predictive power of theism, which may be summarized by the following seven theses:

(1) Moral judgements to the effect that this action is morally good or that one is morally bad are true or false.[33] Some of these judgements are indeed true.
(2) Since God is omniscient, God knows all moral truths.
(3) Since God is perfectly free in the sense of not being subject to temptations, God always chooses to do what he believes to be the morally best action, or, if there is no morally best action, one of the morally good actions.[34]
(4) The actions that God intends to do are in fact the morally best actions, or, if there is no morally best action, they belong to the set of good actions (this follows from (3) and God's omniscience (2)).
(5) God always performs the morally best actions or one of the set of good actions, so that God is perfectly good (this follows from (4) and God's omnipotence).
(6) We humans have a reasonably good grasp of moral truths.
(7) We have a real though limited access to God's intentions and motives, independently of our empirical knowledge of the world, so that we are able to attribute intentions to God in an independent and non-arbitrary manner (this follows from (4) and (6)).[35]

---

[31] Of course, if theists assume both that theism is true *and* that God had specific creative intentions, *they* may be refuted empirically. This is what Darwin did at more than 22 passages in his (1859), for example by arguments from bad design and arguments from biogeography.
[32] An anonymous referee raised the following objection to my argument. If it is as easy to imagine reasons why God would not want to create as it is to imagine reasons why God would want to create (our universe, for example), might one not say that we should assign equal probability intervals to each of these divine intentions? So the theist might estimate the probability that God did not want to create somewhere between 1/5 and 1/2, and he might estimate the probability that God wanted to create a universe also somewhere between 1/5 and 1/2. Then he might argue that theism has sufficient (that is, minimal) predictive power concerning the existence of the universe if this fact is less probable than 1/5 if God does not exist. What I argued, however, is that all these probability estimates are arbitrary. One might just as well locate the probability that God does not want to create between 1/2 and 1, and the probability that God wanted to create our universe between one billionth and two billionth.
[33] Swinburne (EG), p. 99.
[34] One cannot suppose that God will always do the best action there is. For there may be an infinite number of incompatible actions, each of which God could do but each not as good as some other action. In that case, God will inevitably intend to do an action less good than some other action that he could do. Cf. Swinburne (ChrG), pp. 70–1.
[35] For this argument, see Swinburne (EG), pp. 99–109 and 112 ff. and (CT), pp. 184–209.

The core idea of Swinburne's solution to the problem of predictive power is the thesis that to some extent we have access to God's intentions, because the truths of (human) morality are also guiding God. Or, to formulate this more carefully, it is the thesis that we can attribute intentions to God non-arbitrarily because some (sub)set of moral truths that guides human actions also guides God's actions. Hence, our moral judgements about what it is good to do *for us* also provide some insight into what is good *for God* to do. This is a very strong thesis, which I call *the moral access claim*. Apart from defending the moral access claim, the theist has to apply it and to argue on the basis of our moral insights that in fact it was good for God to create, and that it was good for God to create this existing world rather than something else. I discuss Swinburne's defence of the moral access claim here and examine his main applications of it in section 9.4.

Michael Dummett once observed that '[a]t one time it was usual to say that we do not call ethical statements "true" or "false", and from this many consequences for ethics were meant to flow'.[36] Similarly but inversely, when Swinburne claims that (1) moral judgements *are* true or false, he intends to derive from this principle together with the theistic definition of 'God' and premise (6) the moral access claim. But does the moral access claim really follow from thesis (1) etc.? The answer to this question depends on how one interprets this premise.

Even a prescriptivist moral philosopher such as Richard Hare held that we call moral utterances or propositions of the form 'this *a* is a good *a*' true or false. He argued that we do so because apart from their prescriptive meaning such statements also have a descriptive meaning, which is due to the *standards* we use for assessing *a*'s goodness. Such standards typically have the logical form 'if an *a* has properties M_{1-n}, it is a good *a*', where the properties M_{1-n} are non-moral properties.[37] However, if one holds that the moral standards for assessing human actions are not necessarily the same as the standards for assessing God's actions, the moral access claim does not follow from (1). Admittedly, (2) would still be true. Since God knows everything, he would know all moral standards for assessing human actions. But this knowledge would be like the knowledge of eminent anthropologists, who can tell you everything about the morality of a primitive tribe without being obliged or even feeling inclined to apply its ethical standards to themselves.

Clearly, then, Swinburne needs a strong interpretation of (1) for the moral access claim to flow from it. This interpretation should eliminate the dependence of the truth or falsity of moral judgements to the effect that a specific action is morally good or bad, on standards that hold for humans (or other creatures) only, at least concerning the moral judgements that theists apply to God. In Chapter 11 of *The Coherence of Theism*, Swinburne aims at providing such a strong interpretation, which he calls the thesis of 'moral objectivism'.[38] Three subsections of Chapter 11 are devoted to this

---

[36] Dummett (1959), p. 3.
[37] Hare (1989), p. 26.
[38] On (CT) p. 188, Swinburne seems to define moral objectivism *merely* as the view that moral statements are true or false. But of course, the main issue is how one should understand this claim. For example, moral *descriptivism* holds that statements such as 'abortion is wrong' describe (truly or falsely) states of affairs, as empirical statements do. This position is also called moral *realism*. A moral *cognitivist* might hold, however, that such universal moral statements are not descriptive but resemble mathematical statements in being necessarily true, so that not all moral cognitivists are moral

topic.[39] Let me now answer the following three questions. How, exactly, does Swinburne define the thesis of moral objectivism? Is his argument to the effect that moral objectivism is true, convincing? And does Swinburne's moral objectivism entail or make probable the moral access claim, given his other premises?

Initially, Swinburne defines the thesis of moral objectivism as the view that universal moral judgements such as 'truth-telling is always right' or 'euthanasia is always wrong' are 'true or false *in the way in which* factual statements are true or false, such as 'we are now living in England' or 'grass is green in summer'.[40] In other words, the moral objectivist seems to hold that such judgements ascribe a contingent moral property to all actions of a certain type, whereas subjectivists claim that these judgements merely express (dis)approval, or prescribe or recommend actions of this type. Having distinguished some varieties of moral objectivism, such as identity naturalism (moral properties just are natural properties) and intuitionism (moral properties are non-natural properties), Swinburne then argues from the premise that moral properties supervene on natural properties for a version of naturalism according to which the fact that an individual action has certain natural properties *entails* that it has a specific moral property.[41] Let me call this version 'entailment supervenience naturalism'. From this Swinburne concludes that universal moral judgements such as 'euthanasia is always wrong' or 'truth telling is always right', if true, are *logically necessary*, since they warrant such entailments.[42]

Something must have gone wrong in Swinburne's first section on moral objectivism, however, since universal moral judgements such as 'truth-telling is always right' cannot be *both* true in the way in which factual statements are true *and* logically necessary. In other words, Swinburne's two definitions of moral objectivism seem to contradict each other. What has gone wrong? I suggest that in arguing for his entailment-naturalist version of objectivism *via* an argument from supervenience, Swinburne commits an error that I dub the Inverted Lewis Carroll Fallacy.[43] He interprets as a rule of logic or a logically necessary truth what is in fact a normative premise of a moral syllogism. If one concludes (c) 'saying that *p* was right' from the minor (b) 'saying that *p* was telling the truth', one relies on the implicit major (a) that 'truth-telling (in such-and-such situations) is always right'. Clearly, (a) is not a rule of logic, or a necessary or analytic truth, because its negation is not self-contradictory or incoherent.[44] A logically necessary or analytic truth cannot express a substantial moral

---

descriptivists or realists. A Kantian moral philosopher such as Christine Korsgaard will be a cognitivist and an objectivist without being a moral realist or descriptivist. Cf. Korsgaard (1996), pp. 35–7 and 44–8.

[39] Swinburne (CT), pp. 188–209. Cf. also Swinburne (1976).
[40] Swinburne (CT), pp. 188–9 (my italics).
[41] Swinburne (CT), p. 192: 'The naturalist claims that when an object *a* has a certain moral property, say *M*, its possession of it is entailed by it possessing certain natural properties, say *A*, *B*, and *C*. Then it is a necessary truth that anything which is *A*, *B*, and *C* is *M*'.
[42] Swinburne (CT), p. 192. In fact, Swinburne holds that moral judgements of this necessary kind are more complex, since they are true only if the actions are described in much more detail.
[43] Cf. Carroll (1895).
[44] This is Swinburne's own definition of 'logically necessary'. Cf. CT, p. 15: 'I shall call a coherent statement whose negation is incoherent, an analytic or logically necessary proposition'. And he claims *expressis verbis* that true universal or 'pure' moral judgements such as 'truth-telling is always right' are 'analytic truths' in the sense of 'logically necessary propositions': (CT), pp. 203–5, 207. Swinburne

rule of conduct, which is what (a) is supposed to do.⁴⁵ It is not a contingent factual truth either. Rather, it expresses the *norm* 'one should always tell the truth (in such-and-such situations)' in the logical form of a statement. Only if one assumes like Swinburne that all statements must be either synthetic in the sense of factual or analytic in the sense of logically necessary, will one misinterpret moral norm-expressions (and value judgements) either as factual statements or as analytic statements.⁴⁶ In his first section on moral objectivism, Swinburne manoeuvres from the first mistake to the second.

Having answered various objections to moral objectivism in his second section on *The Objectivity of Moral Judgements*, Swinburne proposes a positive argument for it in the third section, which runs as follows:

(x) There are established procedures for reaching agreement in moral matters. Starting from a common basis of moral agreement, we can try to reach consensus on disputed moral statements by arguments of various kinds. Consensus on these procedures is not more difficult to get than it is to reach consensus on procedures in philosophy, for example.

(y) The existence of such procedures for reaching agreement is a sufficient condition for the disputed moral statements to be either true or false.

(z) Moral statements are true or false.⁴⁷

From this conclusion (z), Swinburne directly infers that 'an omniscient person [. . .] will know of any action, the characteristics of which are fully set out [. . .], whether or not that action is morally good or bad. While we have rather cloudy feelings that abortion and euthanasia are evils, he will know the truth about these matters (whatever it is) with crystal clarity'.⁴⁸

But why should we endorse (y)? In many areas, such as democratic policymaking, we have procedures for reaching agreement, such as argumentation, discussing, and voting, even though there is no reason to call the results of these procedures true or false. When an American president declares, 'We are going to invade Iraq', he is not necessarily stating a prediction that is true or false. Rather, he is expressing a decision, on which the members of the United States' government have reached agreement in

---

will have to say that the negation of a true moral judgement, such as, for example, 'telling the truth is always good (in such-and-such situations)' is contradictory or incoherent. But this is an extremely implausible view, which cannot be argued for merely by drawing an analogy between mathematics and ethics, for example.

⁴⁵ Cf. Mawson (2002), pp. 3–7 for further arguments to this effect.
⁴⁶ Cf. Swinburne (CT), pp. 14–22. Swinburne defines synthetic or factual statements as statements whose negation is coherent, and true analytic or logically necessary statements as statements whose negation is incoherent (pp. 14–15). But norm-expressions such as 'speaking the truth is always good' are neither logically necessary/analytic, since their negation is not contradictory, nor factual in the sense in which 'we are now living in England' is factual (which is the sense of 'factual' Swinburne uses). Incidentally, one should distinguish between norm-expressions and the descriptive or factual statement that a norm *N* holds in a specific social group or country.
⁴⁷ Swinburne (CT), pp. 205–6: 'If we can find that agreement on procedures and results is as easy or hard to get in one of these clearly objective disciplines as in morals, that will indicate that there is enough agreement on procedures and results in morals for us to term it an objective discipline', etc.
⁴⁸ Swinburne (CT), p. 208.

conformity with established procedures. In other words, procedural objectivism is not a sufficient condition for alethic objectivism.[49]

Moreover, even if our agreement and the existence of procedures for reaching it would warrant calling certain statements true or false, this does not imply that the truth or falsity of these statements is independent of human evaluative attitudes. For example, a statement such as 'wearing pumps is now fashionable' may be true, but the truth of the statement depends upon what a majority of a population likes at a certain time, or on what the fashion industry produces. This example shows that alethic objectivism does not entail the type of objectivism which Swinburne needs, to wit, that the truth of moral statements is independent of what human beings happen to think or like, or what their evolved and culturally informed moral sense urges them to do.

What is more, even if the truth of moral statements were independent of human evaluative attitudes, this would not yet yield the strong type of moral objectivism that Swinburne needs for deriving his moral access claim, as I shall argue now. Let us accept for the sake of argument that both general moral statements of the form 'doing A is always right' and particular moral statements of the form 'action a is morally good' may be called true or false. Let us also accept for the sake of argument that the truth or falsity of these statements is independent of what human beings happen to think and of how their moral sense has happened to evolve in the past. We now have to enquire whether moral objectivism in this strong sense entails the moral access claim, given the other premises of Swinburne's argument.

I do not think it does, for the following reason. Moral statements of the form 'doing A is always right' are equivalent to norm-statements of the form 'one should always do A (in the relevant situations)', such as 'one should always speak the truth'. By saying that moral statements of the former form are true, we mean that the corresponding moral norm is valid.[50] However, moral norms typically apply to a specific set of norm-subjects, who are the referents of the word 'one' (or 'you' or 'thou') in the norm-statement. For example, the set of norm-subjects to which the commandment 'You shall not commit adultery' applies, is the set of human beings who are married or eligible for marriage. This commandment cannot apply to the god of bare theism as a norm-subject, since God is not a conceivable candidate for marriage or adultery, neither being a bodily person nor participating in a culture that counts marriage among its institutions.[51]

---

[49] A common argument for alethic moral objectivism in the sense of moral realism is the so-called Argument from Moral Experience, which runs as follows. All of us 'take moral value to be part of the fabric of the world'. From this premise, it is concluded that probably, moral value properties are indeed part of the fabric of the world. According to Dancy (1986), pp. 172 and 175, this is 'perhaps the only argument for moral realism'. However, as Loeb (2007) has argued, the premise of this Argument from Moral Experience is dubious indeed, whereas its support for the conclusion is very problematic. Cf. also Mackie (1977), p. 35: 'The claim to objectivity, however ingrained in our language and thought, is not self-validating'.

[50] Let me stress again that one should distinguish between the normative statement that norm N is valid and the descriptive statement that norm N holds in or is accepted by some specific community.

[51] Let me not attempt to fathom the mysteries of the Christian Trinity. But I suppose that even if God incarnated himself in Jesus, and Jesus is eligible for marriage, this does not entail that God is eligible for marriage (at least not in a non-metaphorical sense).

We see now that in deriving the moral access claim from his doctrine of moral objectivism, Swinburne implicitly makes a substantial assumption: that all or most or at least some moral norms which hold for humans as norm-subjects also hold for God as a norm-subject. But this assumption is not self-evident, as may be illustrated by yet another example. We will agree that it is good to be courageous, whether this is obligatory (in some situations) or supererogatory (in other situations). But clearly, God cannot be a norm-subject for the norm 'one should act courageously'. The reason is that the very idea of courage makes sense only with regard to norm-subjects who are vulnerable and mortal, so that they are inclined to flee in the face of danger. Since God is bodiless, eternal, and omnipotent, he is neither vulnerable nor mortal, and the idea of courage does not make *any* sense with regard to God.

In general, it seems plausible to suppose that all human virtues, and the contents of all substantial moral judgements and evaluations that we deem to be true, are intrinsically related to the nature of human beings and to human flourishing.[52] This holds both for norms of the form 'one ought to V', which state moral obligations, and for supererogatory norms of the form 'it is good to V'. Indeed, it has been argued by many authors on the basis of comparative empirical research that the normative content of human morality depends on the kind of animals that we are and, to some extent, on the specific culture that particular communities have. As primatologist Frans de Waal writes:

> Morality is a group-oriented phenomenon born from the fact that we rely on a support system for survival. A solitary person would have no need for morality [. . .].[53]

Darwin made a similar point perceptively in *The Descent of Man*, when he argued 'that any animal whatever, endowed with well-marked social instincts, [. . .] would inevitably acquire a moral sense or conscience, as soon as its intellectual powers had become as well, or nearly as well developed, as in man'.[54] But since the social structures of different social animals diverge widely, different emotional impulses (or other proximate mechanisms that trigger moral behaviour) are adaptive for different species of animals and, as a result, will spread in their populations by the mechanism of natural selection. Since human morality is firmly anchored in such emotional impulses, Darwin concluded that (many of) the norms produced by our moral sense are contingent products of evolution. If other social animals would develop a moral sense or conscience, it would prescribe norms to them that might be very different from the ones we endorse. Darwin illustrated this important conclusion by the following thought experiment:

---

[52] Cf. Mawson (2002), pp. 16–17: 'It is most plausible to think of contingent moral truths as similarly species variable, i.e. to endorse the weaker reading of the strong version of Swinburne's meta-ethical assumption of objectivism'. But this 'weaker reading' is insufficient for substantiating the moral access claim, as I argued.

[53] De Waal et al. (2006), pp. 161–2. Cf. also Joyce (2006); Katz (2000); Kitcher (1998); Ruse (2006); Sober and Wilson (1998). Of course, solitary persons would have a need for morality in the restricted sense that they would have to exercise self-control, and display courage, prudence, and industry. But even these virtues cannot apply to God.

[54] Darwin (1871), pp. 120–1.

If, for instance, to take an extreme case, men were reared under precisely the same conditions as hive-bees, there can hardly be a doubt that our unmarried females would, like the worker-bees, think it a sacred duty to kill their brothers, and mothers would strive to kill their fertile daughters; and no one would think of interfering.[55]

If this is correct, our human moral capacity will not give us access to moral norms or moral values that are 'objective' in the extremely strong sense needed by Swinburne's moral access claim: that they hold or are true independently of our human species. Informed by evolutionary theory, philosophers such as Michael Ruse and Richard Joyce have argued that moral norms are intrinsically species-relative, and that our tendency to regard them as 'objectively' valid should be explained as an adaptive illusion, which might strengthen the human motivation to behave morally.[56]

We may conclude not only that Swinburne has failed to establish his moral access claim, but also that according to many authors this contention would be quite implausible if evaluated from an evolutionary perspective.[57] Many theists reject the moral access claim as well, while Aristotle argued that '[i]f we go through the list we shall find that all forms of virtuous conduct seem trifling and unworthy of the gods'.[58] Accordingly, the moral access claim cannot figure as an assumption underlying the arguments of natural theologians for the existence of God. As Swinburne stresses, such arguments can be good ones only if their premises are 'known to be true by those who dispute about the conclusion'.[59]

9.4 ANTHROPOMORPHIC PROJECTIONS

If we endorse the hypothesis, defended by Darwin, Ruse, Joyce, Street, and others, that the ethical insights provided by our moral sense are a function of human nature and culture, so that the moral access claim is implausible, we will expect that Swinburne's attributions of intentions to God on the basis of our moral judgements are anthropomorphic projections. In order to test this prediction and, if the test turns out to be positive, to confirm the hypothesis, I single out for discussion the two most important

---

[55] Darwin (1871), p. 122.
[56] Ruse (1986), p. 103; Joyce (2006), Conclusion (pp. 221–30); Street (2006). Cf. for an early criticism of moral objectivism on five grounds: Mackie (1977), pp. 30–49.
[57] If one thinks that the arguments by Ruse and others are convincing, one will have to conclude that theists never thought sufficiently precisely about the issue in which sense God can be called 'good'. For if the existence of God implies moral objectivism (as opposed to species-relativity of ethics), and moral objectivism is an implausible doctrine, the existence of God is implausible as well. But if the existence of God does not imply moral objectivism, whereas our morality is relative to the human species and does not hold for God, God will do what he knows is best according to his own norms (if there are any norms holding for God). Since in that case we do not have a clue as to what these norms are, we know nothing about the extension of 'good' as applied to God. In this case the term 'good' can only mean that God always does what his own norms prescribe, irrespective of whether the result is good according to human morality. However, most theists will not be satisfied with this conclusion.
[58] Aristotle, *Nicomachean Ethics*, X, 1178b, 17–20. Cf., for example, Van Inwagen (1991), p. 150: 'My position is [...] that for all we know our inclinations to make value-judgments are not veridical when they are applied to cosmic matters unrelated to the concerns of everyday life'.
[59] Swinburne (EG), pp. 6–7.

intentions that Swinburne attributes to God: the intention to create at all, and the intention to create what he calls *humanly free agents*.

Human beings have numerous good reasons to create, for they have many needs and aspirations, such as the need for shelter and food or the aspiration to transcend death by creating works of art. But God as defined by theism lacks such needs and aspirations, because he is perfect and eternal. So why would he create at all? As I have argued, a perfect and perfectionist god may not deem it good to create, since by creating things less perfect than he is, he would diminish the average perfection of all that exists. Would it not be much better for a perfect god, even from a moral point of view, to engage eternally in the sublime activity of contemplating his own contemplation, as Aristotle thought? It is startling that in *The Existence of God*, a book of more than 350 pages, Swinburne dedicates only one paragraph to this crucial issue. He says:

Plausibly, it is better for God to bring about the existence of something beyond himself rather than [. . .] to refrain from bringing about anything [. . .]. God must bring about the existence of other things. This is affirmed by a principle that Aquinas often invokes, and sometimes attributes to Dionysius, that "Goodness is by its very nature diffusive of itself and (thereby) of being". Norman Kretzmann has spelled out and justified this principle, understanding it as the principle that a good being will inevitably try to make other good things; and so a good God to whose power there is no limit will inevitably go on making more good things.[60]

Swinburne's reference at this point to Kretzmann's book *The Metaphysics of Theism* is somewhat surprising. Kretzmann argues that the principle according to which Goodness is by its very nature diffusive of itself and (thereby) of being, does not apply to the god of theism if he is supposed to be perfectly free.[61] Rather, it applies to a monotheistic god as conceived of by Neoplatonism, who can be compared to a horn of plenty that overflows and creates necessarily.[62]

It may be that Swinburne took notice of this point, for he continues as follows:

Hence God must inevitably bring about the existence of things apart from himself, a consequence from which Aquinas backs away when it becomes explicit, in view of his wish to defend the normal Christian view, which is that God did not have to create anything apart from himself.[63]

Since Swinburne also defends the normal Christian view, according to which in principle God (even the god of bare theism) is perfectly free to create or not to create, it remains unclear in *The Existence of God* why he does not back away from Dionysius' principle as well, as Aquinas did.[64] We may conclude that Swinburne fails to solve

---

[60] Swinburne (EG), p. 117.
[61] Kretzmann (1997), pp. 223–5.
[62] Kretzmann tries to combine these two notions of God in his interpretation of Aquinas, to the effect that Aquinas' god creates necessarily but is free in choosing what he creates: 'As I see it, then, God's will is necessitated as regards whether to create, but fully free as regards what to create' (1997, p. 225). For Swinburne, however, God is also perfectly free to create or not to create. Cf., for example (CT), p. 145: 'All that the omnipresent spirit who is God does, he does because he chooses to do. Nothing makes him do what he does. He did not, for example, have to create the world'. Furthermore, there is the traditional worry whether this 'ontological' goodness of a Neoplatonist God can be equated with 'moral' goodness.
[63] Swinburne (EG), p. 117.
[64] The solution to this problem of exegesis will become clear only when one reads Swinburne's ChrG, p. 178, note 5, according to which the Dionysian principle is 'adequately expressed in the

explicitly the crucial problem of why an absolutely free and perfect god would intend to create anything at all. Relying on human moral notions, he simply projects onto God our own urge and obligation to create. I now turn to the second point: to what extent can we non-arbitrarily attribute to God intentions to create this rather than that?

According to Swinburne, God created the material world mainly in order to enable finite spirits of a certain kind, called *humanly free agents*, to have significant free choices. As he says:'[i]f God is to create creatures with limited free choice to make deeply significant differences to themselves, each other, and the physical world for good or ill, he must make them embodied. Humanly free agents need bodies, and thus to be placed in a physical universe that God has made'.[65] But why would God want to create such humanly free agents in the first place? There are two sides to this issue, which should be dealt with separately. One is why God would want to create other spirits, and another is why he would want to create the specific type of spirits that Swinburne calls humanly free agents.

To the question of why God would want to create other spirits in general, such as other divine persons, Swinburne answers that '[a] solitary God would be a bad state of affairs. God needs to share, to interact, to love, and he can do so most fully with equals'.[66] What should we think of this answer? Of course we cannot exclude that our grasp of human morality permits us to see what would be morally good to do for beings that differ from us vastly. As Swinburne stresses, we 'are able to judge to some extent whether [specific] actions would be good or wrong for us to do, as the case may be, if we were beings of different kinds – if we were very powerful or had created the people who are now starving'.[67]

In this case, however, Swinburne's attribution to God of the intention to create other spirits, because allegedly God 'needs to share, to interact, to love', clearly is an anthropomorphic projection. It is typical for social animals such as humans that they need to love and to interact. The presence of this need in people is explained by the theory of evolution, since without our innate tendency to love and interact, humans would not survive and procreate. We humans cannot interact and love without entering into stable relationships with individuals of the same species. But God as conceived of by bare theism cannot have such needs or tendencies, since he is perfect and unique, the only specimen of his kind. The more we reflect on the way God is conceived of by bare theism, that is, as a unique and omnipotent bodiless spirit, the less we shall understand what can be meant by attributing a need to love to God, for example. So it seems to be arbitrary to attribute to God the intention to create other spirits.

---

continual mutual sustenance of the Trinity'. I am grateful to Swinburne (personal correspondence) for drawing my attention to this passage.

[65] Swinburne (EG), p. 130.
[66] Swinburne (EG), p. 119. Cf. ChrG, p. 177.
[67] Swinburne (EG), p. 113. I am not concerned here with the question as to whether God would have specific moral obligations to living beings that he created, such as the obligation not to kill them gratuitously. Even this might be debated, however, since is the obligation not to kill not species-relative? Most of us do not think there is a moral obligation not to kill mosquitoes or flies. And would it not be God's right to take away a gift he gave freely? The issue here is rather whether it would be good from God's perspective to create specific entities in the first place.

Theists will object that there are no anthropomorphic projections involved here, because the existence of spirits such as deities other than God clearly is a good thing in itself. Since God is perfectly free, he will do what is good objectively, and because the existence of other deities is a good thing, he will intend to create them, and will do so in fact. But arguments of this kind are problematic for two reasons.

First, if we try to avoid anthropomorphic projections and abstract from our own interests, we will end up by confessing that we have no moral views whatsoever concerning the issue as to whether the existence of spirits or deities apart from God would be a good thing. The existence of many types of entities is morally indifferent even to us humans, such as that of other galaxies or dark matter or black holes. Is it a good thing that they exist or not? We do not have a clue, unless their existence would be relevant to our life on Earth. If there were such a thing as objective goodness or badness in Swinburne's sense, there also must be something like objective moral indifference. Why would it not be morally indifferent whether other spirits exist apart from God? Second, and less importantly, the argument that the existence of other deities than God is objectively good in itself, so that God will create them, will be problematic for genuine monotheists. As a consequence, they will have to reject its premise and claim that the existence of such deities is not objectively good at all.

Let us now turn to the question as to why God would create humanly free agents. According to Swinburne, humanly free agents are especially valuable in themselves and hence to God because they are able to make *significant free choices*, that is, choices between good and evil. God lacks this type of freedom, since he is perfectly good. Why would it be valuable to create other spirits that have such a risky type of freedom? Swinburne answers this question as follows: '[t]he goodness of significant free choice is, I hope, evident. We think it a good gift to our own children that they choose their own path in life for good or ill...'.[68] But this argument from analogy is fallacious. Given the fact that a human couple has children, the parents may have good reasons to enable their children to choose freely their own path in life. This is the case in our modern culture, which stresses the value of individual freedom. However, good reasons for giving freedom to children who already exist in a specific culture are not identical to good reasons for having children in the first place. A fortiori, they are not good reasons for God to create humanly free agents, so that Swinburne fails to solve this third problem as well.

Again, theists might object that objectively speaking, the existence of humanly free agents who can make significantly free choices is a great good. But is this really the case? If we seriously try to avoid anthropomorphic projections and abstract from the human perspective, we might just as well think that the existence of humans would be as morally indifferent as is the existence of spirits or dark matter. Given the existence of humans, it surely is a great good (for us) that there are children, even though demographic experts may hold that presently there are far too many children on Earth. And given the existence of children, it is a great good that these children are educated in order to be free, and to take responsible moral decisions. But if we try to abstract from the human perspective, it is unclear what would be the 'objective' moral value of the human race, or of children, or of human freedom, and as unclear as it is

---

[68] Swinburne (EG), p. 119.

what would be the objective value of dinosaurs, for example. In short, I do not think that one can determine non-arbitrarily how probable it is that God had the intention to create other spirits or humanly free agents.

According to Swinburne, the predictive power of theism in these cases is huge. He claims that if God exists, the probability that 'there will be rational beings other than a single divine person is 1', and that '[t]he probability that there will exist humanly free agents (and so a physical universe) is ½'.[69] I have argued, however, that these attributions of creative intentions to God are arbitrary anthropomorphic projections, and that bare theism does not have any predictive power concerning these issues. Consequently, the very idea that bare theism can be a theory with significant predictive power runs the risk of being nothing but an all-too-human illusion.

---

[69] Swinburne (EG), p. 123. Swinburne's extremely high estimate of the first probability is motivated, I suppose, by his wish to argue for the existence of the Christian Trinity. Cf. Swinburne (ChrG), Chapter 8. But Unitarian Christians and Muslims might prefer a much lower estimate, somewhere near to zero.

10

The Immunization of Theism

In Part I of this book it was concluded that theism cannot be credible in our science-informed age, unless it is supported by arguments of the same type as used by scientists and scholars in favour of other factual claims of existence. Accordingly, theism must be interpreted as an existential hypothesis or theory, which can be confirmed by empirical evidence. Although some problems for this conception of theism were raised in Chapters 7–9, I now assume counterfactually and for the sake of argument that theists are able to solve these problems.

Let us suppose, then, that theism is a meaningful and coherent hypothesis, and that it really has the substantial predictive power which a natural theologian such as Richard Swinburne claims for it. As he says, 'theism purports to explain everything logically contingent (apart from itself)'.[1] This means that '[t]he theist argues from all the phenomena of experience, not from a small range of them', so that 'all our empirical data are among the things to be explained'.[2] In other words, the predictive power of theism must be such that the occurrence of each phenomenon which in fact occurs, or at least the occurrence of all these phenomena taken collectively, is more likely on the hypothesis that God exists than it would be otherwise.[3] The reason for this is that, according to theism, God is omnipotent, so that 'whatever happens happens because [God] makes it or permits it to happen'.[4] Theism, then, is a theory of the widest possible scope, since it purports 'to explain the universe and all its characteristics'.[5]

However, if the predictive power of theism has such a wide scope, and if it has a reasonable degree of completeness and precision within this scope, another serious problem arises for the believing theist. Suppose that theism makes some specific conceivable phenomenon more probable than it would be otherwise. Then theism runs two risks. One is that the phenomenon in fact does not occur. The other is that the phenomenon shows up, but that a rival theory makes it more probable than theism does. In both cases, theism is disconfirmed to some extent. An example of the first risk was briefly discussed in Chapters 6 and 9, above. If God is like a good father, as theists often stress, one might expect that petitionary prayers to him for the sick will make a

---

[1] Swinburne (EG), p. 66.
[2] Swinburne (EG), pp. 71 and 93.
[3] Swinburne (EG), pp. 71–2.
[4] Swinburne (EG), p. 99.
[5] Swinburne (EG), p. 108.

statistically significant difference to their health.[6] But it turns out that no effect can be detected. Should we not conclude that such cases disconfirm theism?

As we saw in Chapter 9.1, theists will attempt to exclude this first risk by redefining the expression 'predictive power' in such a way that they do not need to derive from theism predictions of unknown events or phenomena. As Swinburne avers, on the basis of his purely probabilistic notion of predictive power 'it is in itself no objection to the hypothesis that there is a God, that it does not yield predictions such that we can know only tomorrow, and not today, whether they succeed'.[7] Of course this tactic raises the critical question as to whether theists should be taken seriously with regard to their claim that there are 'close similarities [...] between religious theories and large-scale scientific theories'.[8] Do theists not introduce striking dissimilarities as well, in order to immunize theism against empirical disconfirmations? How can Swinburne reconcile his tenet that theism does not imply predictions concerning unknown data with his claim that theism is a theory of the widest possible scope, because it purports to explain the universe and all its characteristics? But let us suppose for the sake of argument that this first tactic for solving what I called *The Tension* (Chapter 6.5) is legitimate. Then the theist still has to deal with the second risk: that the theory of theism enters into competition with scientific or scholarly theories, which purport to explain the very same phenomena.

This second peril for theism is the topic of the present chapter. It may be called the *Dilemma of God-of-the-Gaps*. As we shall see in the first section, theists of the past often did not want to avoid this risk. On the contrary, they argued that many specific natural phenomena yield a strong confirmation of their theory, since theism allegedly provides the best, or even the only possible, explanation of these phenomena. But the history of science taught many contemporary theists that it is too risky to appeal to particular empirical phenomena in support of theism. In countless cases, scientists or scholars came up with more precise and detailed explanations of the phenomena, so that religious explanations were massively superseded. Should we not conclude by a pessimistic induction that this is always likely to happen, or that it is at least a real possibility? Consequently, a sophisticated theist such as Richard Swinburne employs stratagems designed to prevent contests between theism and properly scientific or scholarly explanations. In the later sections of this chapter these stratagems are reviewed in order to answer two questions. To what extent do they succeed? And can theism still be sufficiently like a scientific theory if the stratagems are successful?

10.1 THE DILEMMA OF GOD-OF-THE-GAPS

In the General Scholium to the third book of his *Principia*, Isaac Newton declared that the solar system 'could only proceed from the counsel and dominion of an intelligent

---

[6] Cf. Swinburne (EG), p. 285: 'God has the reason of friendship to seek living interaction with people whom he has made [...]. Hence one would expect him to intervene in the natural order occasionally in response to the human situation, especially in answer to request (that is, petitionary prayer) for good things'.
[7] Swinburne (EG), p. 70.
[8] Swinburne (EG), p. 3.

and powerful Being'.⁹ By this remark, Newton revived one of the most prominent ancient arguments from special design, and he did so on the following grounds. Whereas in the classical tradition of mathematical astronomy from Ptolemy to Copernicus it had been assumed that circular orbits are the natural movements of heavenly bodies, which did not require any special explanation, this dogma had been overhauled by observations of comets and the introduction of Kepler's laws. However, if the motions of the planets and other heavenly bodies are to be explained on the basis of Newton's three axioms and the law of gravity, the peculiar and relatively stable arrangement of the planets and their satellites may seem to be miraculous. How can it be that the masses and speeds of these bodies are so fine-tuned that the system does not either collapse by gravitation or disintegrate by centrifugal motion?

Newton had refuted Descartes' vortex theory as a physical explanation of the solar system. This theory could neither easily accommodate Kepler's laws, nor could it explain adequately the movements of comets or the orbits of planetary satellites. Since it seemed to Newton that no other physical mechanism could be invented that accounts for these phenomena, whereas the likelihood that random change was responsible for the arrangement of planets and their satellites seemed to be extremely small, he had recourse to the refuge of a divine explanation. But in 1796 Newton's design argument from the solar system to God turned out to be a mere *argumentum ad ignorantiam*. In his *Exposition du système du monde*, Pierre Simon Laplace endorsed Newton's premise of the miniscule likelihood that random change can generate the unidirectional movement and the stable rotation of the planets and their satellites, or the small eccentricity of planetary orbits. But he proposed his nebular hypothesis as an explanation, sophisticated versions of which are still endorsed by astronomers today, and argued that this physical theory could explain the phenomena adequately.¹⁰

Since Laplace had solved in 1786 with great ingenuity another problem with regard to which Newtonians had invoked a hypothesis of divine intervention, the so-called Jupiter/Saturn problem of the stability of the solar system, many philosophers concluded at the end of the eighteenth century that one should never resort to a theological hypothesis in order to account for astronomical phenomena.¹¹ Perhaps it was partly for this reason that in his *Natural Theology* of 1802, William Paley observed that astronomy 'is *not* the best medium through which to prove the agency of an intelligent Creator'.¹² Instead, Paley inferred the existence of God from the assumption that the adaptive complexity of a great number of biological phenomena cannot be explained otherwise than by recourse to God's wisdom and omnipotence. But fifty-seven years later, Darwin refuted this assumption. Indeed, *The Origin of Species* of 1859 can be read as one continuous argument to the effect that Darwin's theory of descent with modification is vastly superior to the traditional theological hypothesis of special creation.

What lesson can theologians learn from these and innumerable other examples (e.g. in medicine or concerning natural catastrophes) of refuted arguments from

---

⁹ Newton (1729), Vol. 2, p. 544.
¹⁰ Laplace (1796), vol. 2, pp. 301–4. Immanuel Kant proposed the hypothesis as well, so that it is called the Kant–Laplace nebular hypothesis.
¹¹ Cf. Hahn (2005), Chapter 5.
¹² Paley (1802), p. 199 (Paley's italics).

(or rather to) special design? The logical form of these arguments may be construed as follows. First it is claimed of a particular empirical phenomenon E that its occurrence is very improbable and cannot be the product of random processes. Second, it is assumed implicitly or argued explicitly, that a scientific or natural explanation of E is impossible. From these two premises it is concluded that one should adopt a supernatural explanation of E, to wit, that E is designed or caused by a Supreme Being. The central weakness of all such arguments to design is obvious and it resides in the second premise.[13] As the numerous examples of refuted arguments to special design show, people who endorsed this premise were committing the informal fallacy of an argument from ignorance. They inferred from the fact that up to their time no natural explanation of E had been given, or that they did not see how this could be done, that such an explanation could not be given in principle.

In his Lowell Lectures given at Boston in 1893, the evangelical lecturer Henry Drummond already criticized those 'reverent minds' that argue for the existence of God from the things science cannot yet explain, that is, from 'gaps which they will fill up with God'. As Drummond wrote, their 'interest in Science is not in what it can explain but in what it cannot', and their 'quest is ignorance not knowledge', so that their 'daily dread is that the cloud may lift'.[14] Many theologians have taken his criticism to heart, and argue that believers should not be committed to arguments for 'God-of-the-gaps'. As a consequence, they should avoid special intelligent design arguments, for example. But how easy it is for natural theologians to avoid God-of-the-gaps depends upon the apologetic strategy they prefer.

Philosophers of religion who hold that the reasons religious believers should advance for the truth of their creed are of a totally different kind from the reasons scholars and scientists adduce in support of factual existence claims, will not be tempted by God-of-the-gaps arguments. I have suggested in Chapter 6.3–4, however, that in our age of science this apologetic strategy is not very convincing. Why should the arguments for a factual existence claim in theology not need to be supported by empirical evidence obtained by validated methods of investigation? For this reason the bulk of this book is devoted to analysing another strategic option for a natural theologian, according to which arguments for the existence of God must be of the same logical type as those for other factual existence claims.

Is it not unavoidable that this empiricist strategy in apologetic philosophy of religion is confronted by the problem of God-of-the-gaps? If the predictive power of theism really has the universal scope Richard Swinburne claims for it, theism seems to be landed in the following dilemma, which I call the *Dilemma of God-of-the-gaps*. Either theism enters into competition with (possible) scientific or scholarly explanations of the phenomena it claims to explain, or it does not. In the first case, theism will risk being disconfirmed as soon as a good scientific or scholarly explanation has been found. As the numerous historical examples show, such explanations are more empirically adequate than theological explanations, so that by now theistic explanations of specific phenomena have been abandoned massively. In the second case, the theist has to argue that although theism indeed explains all the phenomena of experience, it

---

[13] Cf. for other weaknesses: Chapter 9 on the predictive power of theism and Chapter 11.5–10 on the prior probability of theism. Cf. also Manson (2003), pp. 5–8.
[14] Drummond (1896), Chapter 10, p. 426.

explains them in a sense different from that in which empirical science or scholarship explains phenomena, so that competition between scientific and theistic explanations is a priori excluded. But this move threatens to destroy the initial credibility of the empiricist strategy in apologetic philosophy of religion.

Let me illustrate this latter claim by a historical example. There is a long tradition in philosophy, which runs from Plato to Leibniz and Berkeley, according to which there are two different types of causality. Science would only investigate so-called secondary causes, whereas God would be the first cause of everything. Because they are concerned with different types of causality, theological explanations allegedly are of a type different from scientific explanations, and although both of them are concerned with the same phenomena, they could never be in competition or conflict. In other words, even if all explanatory questions of science and scholarship were answered, there would still be room for a theological explanation, and the god of monotheism would never be God-of-the-gaps.

Perhaps we may locate Newton's view of the origin of the solar system in this tradition. According to Newton, all relative movements of the Sun, the planets, and the planetary satellites in the solar system can be explained if we apply the laws of classical mechanics to specific initial conditions. But Newton held that by doing so, we can never explain the stable orbits of planets and satellites in the first place. They have to be accounted for by postulating God as a first cause. In other words, God must have established the *original* initial conditions of the distribution of matter in the universe. It is interesting to note that when Laplace arrived in Paris and took up his teaching post at the École militaire in 1769, one of the textbooks used was *Histoire des causes premières*, in which abbé Charles Batteux gave a historical review of all attempts to distinguish primary from secondary causes.[15] Laplace then explicitly proposed his nebular hypothesis as an explanation of the first cause of the stable orbits in the solar system, which would make God superfluous as a first cause.[16]

The lesson we may learn from this example is the following. If apologetic philosophers of religion opt for the second horn of the dilemma of God-of-the-gaps by endorsing some version of the doctrine of double causality, they are confronted with a new dilemma. Either the two senses of 'causality' or 'explanation' distinguished by the theist are both substantial in the sense that they fit in with the empiricist apologetic strategy, or one of them is not. In the first case, it seems that one can never exclude that a scientific explanation will compete with the theological 'first cause' explanation. In the second case, the tactic of distinguishing between two senses of explanation and causality is a merely verbal move of immunization.

Suppose that the theologian says, for example, that God's causality differs from scientific causality because it is a causality of *sustaining* and not of *producing*, one should ask what would happen if God withdrew his sustenance of a particular phenomenon. If, in that case, the phenomenon would change, or disappear completely, there is a conceivable event, for which, if it would happen, scientists might attempt to find a competing explanation. But if theologians answer that *whatever* happens, God sustains it, their move is merely verbal, and they have abandoned the empiricist

---

[15] Batteux (1769).
[16] Cf. Roger Hahn, 'Laplace and the Mechanistic Universe', in Lindberg and Numbers (1986), esp. pp. 270–1.

apologetic strategy.[17] As Karl Popper stressed on good grounds, a theory that can explain all conceivable phenomena, so that it does not rule out or make improbable any logically possible course of events, in fact explains nothing and is a pseudo-theory.

10.2 TOO ODD: MIRACLES

Richard Swinburne uses two different stratagems in order to deal with the dilemma of God-of-the-gaps. One tactic is to restrict the scope of theism's predictive power to empirical phenomena of which he claims that a scientific explanation is impossible in principle. Another tactic consists in applying a version of the doctrine of double causality, at least implicitly. In both cases, the aim is to preclude that theism as a theory enters into competition with rival scientific or scholarly hypotheses, that is, to immunize theism against elimination by superior explanations of the same phenomena. In this section, I review one example of the first tactic.

Swinburne distinguishes between two types of phenomena that cannot be explained in principle by science and secular scholarship. One type consists of phenomena that are 'too *odd* to be fitted into the established pattern of scientific explanation', and the other type consists of 'phenomena that are too *big* to be fitted into any pattern of scientific explanation'.[18] If they exist, miracles are examples of the first category (too *odd*), and the existence of the universe allegedly is an instance of the second (too *big*). Swinburne also argues that the phenomena of mental–physical interaction, and, indeed, the existence of souls, instantiate the first category, whereas there are three main instantiations of the second. In sections 10.2–5 of this chapter, I discuss the first example of 'too odd'. I shall do so rather extensively in order to introduce some empirical content into my philosophical arguments. Our question is: can one succeed in excluding that theological explanations enter into risky competition with scientific or other secular explanations in the case of miracles?

Let me begin by attempting to account for a difference between Swinburne's formulations of the categories 'too odd' and 'too big'. In the latter case, he requires that a phenomenon is too big to be fitted into *any* pattern of scientific explanation, whereas in the former case he merely requires that a phenomenon is too odd to be fitted into *the established* pattern of scientific explanation. Why is he less demanding in the former case than in the latter? The answer may be related to problems concerning the traditional definition of a miracle, which Swinburne endorses.

If one wants to argue from the occurrence of a miracle to the existence of God, one has to start by establishing (a) that a specific event E occurred, and (b) that this event E is a miracle. From (a) and (b) one first concludes (c) that there cannot be a natural explanation for E, so that its cause, if any, must lie outside of the system for which the laws of nature hold. Excluding as extremely improbable that E is an uncaused event, one further argues (d) that E is likely to be caused by God, and (e) that E is more likely

---

[17] As Rundle argues (2004; §4.3), an appeal to a sustaining cause of some phenomenon is needed only under very specific conditions, such as, for example, when there is a disintegrating factor to be countered or inhibited (p. 88). Cf. Chapter 12.5, below, for further discussion.

[18] Swinburne (EG), p. 74 (Swinburne's italics).

to be caused by God than by other non-natural causes, so that (f) the occurrence of E yields a good confirming argument for God's existence.

How should one define the notion of a miracle in order to argue in this manner for the existence of God? If one defines a miracle merely as 'an event caused by God', the argument will run the risk of being viciously circular. This is why in traditional modern apologetics a miracle is often defined as *a violation of laws of nature*.[19] If one does not endorse this definition, one will not be able to argue for lemma (c). But the definition is ambiguous, and spelling out the ambiguity we shall see that theists are landed in a dilemma. By the expression 'laws of nature' one might mean either law statements that scientists now endorse, or true law statements, which scientists will perhaps discover in the future.

If the expression 'violation of laws of nature' is interpreted in the first sense, the notion of a miracle is merely an epistemological category. In this case, a miracle is an event E that we cannot explain in terms of the *established* patterns of scientific explanation. But since we all admit that the law statements in our established patterns of explanation may not be strictly true, and that future scientists may reject the theories and law statements we now accept because they have formulated better ones, we cannot exclude that an event E, which violates at least one law statement we now endorse, might be explained by future science. What is more, should we not consider such an event E, if its occurrence has been established beyond reasonable doubt, as a disconfirmation of the relevant law statement(s) we now accept? Hence, the epistemological category of a miracle will not suffice for an argument from miracles to God, and it risks to involve a God-of-the-gaps fallacy.

If, however, we take the expression 'violation of laws of nature' in the second sense, so that the notion of a miracle is an ontological category, two other problems seem to arise for theists who want to argue from miracles to God. First, the notion of a violation-miracle in this latter sense is self-contradictory. The occurrence of an event E cannot *violate* a law unless the description of E *contradicts* the law statement. If, however, this law statement is true, as it is according to the second interpretation, the description of E must be false, so that it is logically excluded that E as described occurs. Clearly, this problem of contradiction arises irrespective of which conception of laws of nature one holds.[20]

In order to avoid the conclusion that the notion of a violation-miracle is self-contradictory, theists will have to reinterpret what it means 'to violate' the true laws of nature, and, indeed, what true laws of nature are. They might argue, for example, that all true laws hold for nature or the universe as a closed system apart from God. In this

---

[19] Minimally, being a violation of *at least one* law of nature has to be considered a *necessary* criterion for E being a miracle. There may be other necessary criteria, such as religious significance, but they do not have a role in my argument. Cf. Everitt (2004), Chapter 6 for an introductory discussion. As Mackie said, '[i]f miracles are to serve their traditional function of giving spectacular support to religious claims [...], [w]e must keep in the definition the notion of a violation of natural law' ((1982), p. 19).

[20] Cf. Earman (2000), pp. 9–10. If one holds, alternatively, that true laws of nature are such that they admit of exceptions, an exception will not *violate* these laws, so that violation-miracles are logically impossible too. Cf. Everitt (2004), pp. 118–20, for further discussion of this point.

case, God might cause an exception to a true law of nature by intruding into the universe from without.[21] The resulting event E, which is an exception to a true natural law, would not *violate* it in the strict sense of being a counter-instance to the law, because there would be an (often implicit) *ceteris paribus* condition in all true natural law formulations that they hold for nature as a closed system only. So one might solve this first problem by redefining a miracle as an event E, the occurrence of which cannot be accounted for in terms of the true laws of nature, which hold for the universe as a closed system only.[22] A miracle then is a *prima facie* violation of at least one true law, or a *counterfactual-violation* in the sense that an event E is a miracle if and only if it *would* violate a true law when nothing interfered from without with the workings of the closed system of nature, for which the true natural laws hold.[23]

If theists accept this interpretation of the second horn, however, they will seem to be faced, secondly, with an insurmountable difficulty of demonstration. In order to establish that an event E is a miracle, they would have to show that no future scientific theory or explanation, superior to the ones we now endorse, will be able to account for E. But it is impossible in principle to predict which scientific theories humanity will develop, since in order to formulate such a prediction in detail, one must already possess the future theory. How, then, should theists define the notion of a miracle, if they want to argue from miracles to God?

Swinburne's solution to this dilemma consists in endorsing its second horn of the ontological definition, and in attempting to resolve the difficulty of demonstration. On the one hand, he defines a miracle as 'a "violation" of laws of nature'. And he defines a 'violation' of a law of nature as 'the occurrence of an event that is impossible, given the operation of the *actual* laws of nature'.[24] It seems that by 'actual' laws of nature, he means the laws that in fact hold and not the law statements that we now deem to be true. Accordingly, he should redefine what it means to 'violate' the true laws of nature along the lines indicated above, in order to avoid the conclusion that his

---

[21] Perhaps David Hume meant something like this when he wrote in his essay 'On Miracles': 'A miracle may be accurately defined, *a transgression of a law of nature by a particular volition of the Deity, or by the interposition of some invisible agent*' (1748, Sect. X, Part I, §90; Hume's italics). Cf. also Mackie (1982), pp. 19–22.

[22] Another solution might be to endorse David Lewis' account of natural laws, according to which a (true) law of nature is a contingent generalization that 'appears as a theorem (or axiom) in each of the true deductive systems that achieves a best combination of simplicity and strength' (Lewis (1973), p. 73). According to this conception, there may be exceptions to (or 'violations' of) true deterministic laws of nature. Cf. Swinburne (2003). The problem is, however, that on this conception of laws of nature, we cannot conclude from the fact that an event E is a violation of a true law of nature, that its cause (if any) must be external to nature as a closed system. In other words, we cannot argue for thesis (c). Cf. for discussion Luck (2005).

[23] This is the view of Mackie (1982) and it seems also to be the view of Rowe (1993), Chapter 9. Cf. Curd (1996) for further discussion. Mumford (2001) has argued that in order to do justice to our '*modal*' intuition that miracles are logically possible' (p. 191, his italics), we should adopt a deontic interpretation of laws of nature, according to which they *prescribe* how natural things should behave, so that laws of nature have some '*regulative* force' (cf. pp. 194, 195, 198). But I regard this as a desperate proposal. There is a clear logical distinction between legal and moral laws on the one hand, which are prescriptive, and laws of nature on the other hand, which are descriptive.

[24] Swinburne (EG), p. 277; my italics. This is his definition for deterministic laws. For the case of probabilistic laws, Swinburne defines the notion of a 'quasi-violation': (EG), pp. 280–1. I shall not discuss here Swinburne's early views on miracles in his 1970 book (CM).

notion of a miracle is self-contradictory. And he seems to do so implicitly where he says that any cause of such an event E 'would lie outside the system of natural laws'.[25]

On the other hand, Swinburne stresses that sometimes we can have sufficient evidence for E being a miracle in the second sense if E is an exception to a law statement which we *now* accept, that is, to the *established* pattern of explanation. This can be so if we have good grounds for thinking that the laws we now deem to be true are indeed true. That is, these laws must have great explanatory power and a high prior probability, whereas increasing their explanatory power by amending the laws, for example in order to explain the event E, would lead to a considerable loss of prior probability. Furthermore, we must have good evidence that E is not a repeatable exception to a law statement, since in that case the violation would be evidence that probably the law statement is not true or at least not fundamental.[26]

As Swinburne admits, the judgement that E is a non-repeatable exception to an established pattern of explanation, and that the law statements used in this pattern are true, is corrigible.[27] In other words, our judgement (b) that E, if it occurs, is a miracle, is open to revision in the light of future scientific research. Hence, by saying that an event E is 'too odd to be fitted into *the established* pattern of scientific explanation', Swinburne has not really avoided the risk of a God-of-the-gaps fallacy. But this risk may be negligible with regard to specific miracles proposed by particular religions or sects. For instance, if a religious believer claims that a holy damsel was decapitated, that her head was lying separated from her body for two weeks, and that after these weeks the decapitated and decomposed corpse stood up, put the head on its neck, and that the resurrected lady lived on happily until her natural death, we all tend to agree that future science will never succeed in explaining such an event, if at least it really happened. According to Swinburne, the bodily resurrection of Christ, if it occurred, is an equally convincing example of an undisputable miracle. We must conclude that with regard to unproblematic examples of miracles, the risk of God-of-the-gaps fallacies has been eliminated effectively. *If* such events occur, they probably are 'too odd' to be fitted into *any* scientific pattern of explanation.

But of course this does not imply that arguments from miracles to God are immunized against refutations by science or historical scholarship. Apart from the

---

[25] Swinburne (EG), p. 279. Cf. also p. 278, where he says that there can be a violation 'if some cause from outside the system of law-governed objects intervenes to bring about an event not permitted by laws'. Given Swinburne's preferred view of natural laws, the so-called substances-powers-and-liabilities account, according to which laws of nature are nothing but 'summaries of the powers and liabilities of substances that have the same powers and liabilities as other substances of the same kind' (p. 277), it may seem natural to conceive of laws of nature as holding for a closed system of substances only. Yet the difficulty of a contradiction arises here too, since the hypothesis of a miracle would contradict the true account of the powers and liabilities of the substances involved.

[26] Cf. Swinburne (EG), pp. 278–80, for a development of these ideas. A believer in the miracle of transubstantiation will either have to contest Swinburne's non-repeatability criterion for E being a miracle, since the transubstantiation allegedly occurs millions of times a day, or to reinterpret transubstantiation as merely symbolical, so that it is not a violation-miracle. The same holds for those who believe that petitionary prayers are really answered by God more than once. And indeed, the non-repeatability condition for miracles seems odd for God: if he can perform a miracle once, he will be able to perform the same miracle many times, and of course he can, since he is omnipotent. Cf. Everitt (2004), p. 113.

[27] Swinburne (EG), p. 279: 'Claims of this sort are, of course, corrigible – we could be wrong; what seemed inexplicable by natural causes might be explicable after all'.

170   *Theism as a Theory*

premise (b) that an event *E* is a miracle in the sense of a (quasi-)violation of at least one true law of nature, these arguments require the equally fundamental premise (a) that *E* in fact occurred. Typically, this premise (a) is based mainly upon existing testimony by alleged eyewitnesses.[28] The religious apologist explains the complex historical facts *T* of the existence of such testimony by saying that *E* really occurred, was perceived by eyewitnesses, and that, consequently, these eyewitnesses told the truth. However, historical scholars without apologetic intentions will usually prefer another explanation of *T*. According to them, the very fact that the occurrence of *E* would be a miracle makes it excessively unlikely that *E* occurred. As many authors have stressed, the evidence that supports premise (b) will undermine premise (a), so that it will be very hard to sustain the double burden of arguing for both (a) and (b).[29] Hence, scholars without apologetic intentions will deny that we have to postulate the occurrence of *E* in order to explain *T*. In the next sections, I shall explore the logic of this contest and illustrate it amply by the example of Christ's resurrection.

10.3 THE RESURRECTION OF JESUS: THE TESTIMONY

In section X, part I of his *Enquiry Concerning Human Understanding*, David Hume proposed as 'a general maxim worthy of our attention' that 'no testimony is sufficient to establish a miracle, unless the testimony be of such a kind, that its falsehood would be more miraculous, than the fact, which it endeavours to establish'.[30] As we have seen, in order to establish the existence of a miracle, we have to show (a) that an event *E* occurred, and (b) that this event *E* is extremely unlikely and very surprising for two reasons: (1) it is a *prima facie* violation of at least one law statement which we presently endorse; (2) it is very probable in the light of abundant evidence that this law statement is true, and that no future scientific explanation of *E* is possible. Clearly, evidence for (a) will risk counting as evidence against (b-2), and all evidence for (b-2) will count as evidence against (a), so that establishing the occurrence is quite a demanding endeavour.[31]

What Hume's maxim implies, then, is that if alleged eyewitnesses tell us that *E* occurred, this makes credible the occurrence of a miracle only if it is even more

---

[28] There might be other types of evidence as well, such as physical traces.
[29] Cf. Mackie (1982), p. 26: 'those who accept this as a miracle have the double burden of showing both that the event took place and that it violated the laws of nature. But it will be very hard to sustain this double burden'; cf. also Everitt (2004), p. 115; Swinburne (EG), p. 283: 'For the very fact that, if *E* occurred, it would have been violation of a natural law is in itself of course evidence against its occurrence, as Hume classically argued'; Reppert (1989), p. 36: 'Since experience strongly supports the stability of the laws of nature and only weakly supports human veracity, there is a very strong presumption against miracles, and it is for all practical purposes impossible for testimonial evidence to defeat this presumption'.
[30] Hume (1748), Sect. X, Part I, §91. Cf. Everitt (2004), Chapter 6, for a more ample discussion of Hume's argument.
[31] One cannot explicate the unnaturalness of a miracle merely in terms of the *improbability* of an event *E* given the laws of nature, since many events are improbable given these laws, although we would not consider them as miracles. Take, for example, winning a lottery in which only one lot in a billion can win. What makes a miracle surprising is that the event *E* would be *impossible*, unless the natural world were interfered with from without, so that the occurrence of *E* would force naturalists to reject their world-view. Cf. Reppert (1989), pp. 43–6.

unlikely that the testimony is false, given the number, mutual (in)dependence, and reliability of the witnesses, than it is unlikely according to the reasons for (b) that E occurred.[32] Hume argues in part II of section X that *in fact* this condition is never met. All historical testimony of alleged miracles is insufficient for many reasons. There never are enough witnesses. They are not independent of each other. They usually are not disinterested. They are rarely well educated and often come from backward cultures or classes. Their testimony is cancelled out by the testimony for miracles of rival religions. They are subject to temptations of vanity to appear as an 'ambassador from heaven'. Most human beings love to believe extraordinary stories. The witnesses often contradict each other, and so on.

Hume overstated his case by concluding: 'no human testimony *can* have such force as to prove a miracle'.[33] However unlikely we esteem the occurrence of E because of our reasons for premise (b), it is always logically possible, assuming that the probability of E is not zero, that there are enough independent and reliable witnesses who tell us that E occurred to make it even more unlikely that they were mistaken.[34] As Charles Babbage argued in his *Ninth Bridgewater Treatise* of 1838, 'if independent witnesses can be found, who speak the truth more frequently than falsehood, *it is ALWAYS possible to assign a number of independent witnesses, the improbability of the falsehood of whose concurring testimonies shall be greater than that of the improbability of the miracle itself*'.[35] In other words, we can develop no a priori argument against miracles in general, as Hume sometimes pretended to have done, if at least the assumed definition of a miracle in terms of a violation of a law of nature is not self-contradictory.

What we can and should do, however, is investigate each actually recorded case of a miracle, and, using Hume's maxim, argue that in each particular case the testimonial evidence is not sufficiently strong to establish that E really occurred.[36] The best way of doing so is to show that apart from the religious explanation of the existing testimony T by positing E as a cause for T and God's action as a cause of E, there is an alternative explanation of the existence of T, which is purely secular and superior to the religious explanation in terms of the usual criteria for theory evaluation. I shall now discuss the alleged miracle of Christ's bodily resurrection in order to illustrate this procedure. It goes without saying that the resurrection miracle is of overwhelming importance for traditionally minded Christians. As Swinburne argues, if God wanted to proclaim by a miracle that suffering and death have been overcome, he had to resurrect Jesus with his

---

[32] Cf. Earman (2000), p. 38ff., for alternative translations of Hume's maxim in terms of conditional probability. Cf. also Mackie (1982), pp. 23–4 for some qualifications of Hume's maxim, and Holder (1998) for a Bayesian interpretation and critique.

[33] Hume (1748), Sect. X, Part II, §98 (my italics).

[34] If a miracle is defined as a quasi-violation of at least one law of nature, the probability of a miracle will not be zero unless it is somehow a priori excluded that something interferes from without with the closed system of nature for which the true laws of nature hold. As Reppert (1989) has argued, if the probability of miracles given theism is not zero, the probability of miracles can be zero only if the probability of theism is zero (p. 43). And this is only the case if theism is self-contradictory or meaningless (cf. Chapter 7 for an argument to this effect). However, in this chapter I have assumed for the sake of argument that theism is not self-contradictory or meaningless. I shall discuss below the issue of what will happen to our probability assessment of a miracle if, apart from the laws of nature, we also take theism into account.

[35] Quoted by Earman (2000), p. 54 (italics in the original). Cf. also Holder (1998), p. 53.

[36] Cf. Earman (2000), §§ 16–20.

previously damaged body. Merely resurrecting Jesus in an embodied state with a totally new body would not have been enough.[37]

In order to examine the case of Christ's alleged bodily resurrection, we first have to investigate conscientiously the nature of the facts T. How many witnesses tell us that Christ was resurrected and in which sense of 'resurrected'? Are our sources really written by eyewitnesses of the resurrection? To what extent were these witnesses mutually independent? Should we assess the witnesses as unprejudiced or rather as partisans? To what extent do the contents of the testimonies concur or diverge?[38] Then the two rival explanations of T, a miracle explanation and an alternative secular explanation, are to be compared. How likely are the facts T given each of these explanations? And how well confirmed by background evidence is each of the rival explanations?

1. Since we do not have writings in Aramaic, which Jesus and his disciples probably spoke, produced by eyewitnesses of Jesus' bodily resurrection, we have to rely on the earliest secondary sources that tell us about this alleged event. It is a first striking fact about these sources that they all have been written by members of (pre-) Christian communities on the basis of oral traditions. Consequently, the sources cannot be considered mutually independent or impartial. As far as we know, neither the Jewish historian Justus of Tiberias nor the Jewish scholar Philo of Alexandria, who were contemporaries of Jesus, mentioned him or his bodily resurrection in their works. Although the Jewish writer Flavius Josephus tells us about a Jesus in his *Jewish Antiquities* published in 93 CE, this so-called *Testimonium Flavianum* probably was rewritten in part by later Christian copyists.[39] When Roman writers such as Pliny the Younger, Tacitus, and Suetonius started to speak of 'Christians' for the first time in the years 110–120 CE, they portrayed them as superstitious troublemakers, and did not mention a resurrection of Jesus.[40] Perhaps there is no reason to expect that either Justus or Philo would have referred to the leader of a minor Jewish sect. But in order to evaluate the historical significance and reliability of the extant testimony, it is crucial to stress that followers of Jesus gave or produced all of it.

2. A second important fact concerning the source material is that with one possible exception (Paul), the extant testimony is not authored by eyewitnesses of the bodily resurrected Christ. For example, the synoptic Gospels were written between thirty-five and seventy years after the crucifixion by later Christians who never knew Jesus personally.[41] Whereas we learn from these texts that the twelve disciples of Jesus were, like him, lower-class men from rural Galilee, who spoke Aramaic and probably were illiterate, the authors of the canonized Gospels were well-educated, Greek-

---

[37] Swinburne (RGI), p. 1.

[38] Cf. for an extensive and erudite analysis of the existing testimony: Wright (2003), Chapters 5–17. N. T. Wright is the Bishop of Durham and also a prominent New Testament scholar.

[39] Josephus (1966), pp.18, 63–4, and 20, 200. Cf. Theissen and Merz (1996), §3.1.

[40] Cf. Tacitus, *Annales* XV, 44; Pliny the Younger, *Epistulae*, X, 97–8; Suetonius *De Vita Caesarum*, Nero, 16.

[41] Cf. Ehrman (2009), p. 143: 'the Gospels are full of discrepancies and were written decades after Jesus' ministry and death by authors who had not themselves witnessed any of the events of Jesus' life'. Cf. for a thorough examination of the sources concerning Jesus: Theissen and Merz (1996), §§ 2–4.

speaking Christians. From their relative ignorance of Palestinian geography one cannot but infer that they lived elsewhere in the Roman Empire and probably had not visited Jerusalem.[42] Moreover, many of the extant sources are not acknowledged as legitimate by the established Christian churches, since they excluded them as heretical in order to authorize their own unified account. This happened during a process of canonization, which took several centuries. For example, whereas the *New Testament* contains four gospels only, the gospels attributed to Matthew, Mark, Luke, and John, other gospels are preserved at least in part, such as those of Thomas, Peter, Philip, Mary, the Ebionites and/or the Hebrews, Judas, and the Egyptians. Some of these gospels do not even mention a resurrection of Christ, and their content shows that there were widely divergent oral traditions within the various Christian communities.[43]

For a conscientious reconstruction of historical events it is not decisive whether a source has been canonized or not, so we should study all extant sources, using reliable historical methods.[44] Scholars tend to agree that the gospels of Matthew and Luke are in part based on Mark, directly or indirectly, and on a lost source, called Q.[45] If this is correct, the authentic letters of Paul (50–61 CE) are the oldest extant sources concerning a resurrection, while the first version of Mark is dated by many scholars around 70 CE or later. The fact that the synoptic gospels containing tales about Jesus' life were all written in Greek a long time after Jesus' death by authors who had not known Jesus, may be explained by many factors. Illiteracy was widespread in the Roman Empire.[46] For the first followers of Jesus, the many divergent oral traditions will have been sufficient, whereas the disciples probably had the eschatological expectation that the Kingdom of God would arrive during their lifetime.[47] It was only when it became clear to his followers that the Kingdom of God would not arrive soon, and the oral traditions were weakening, that the need for writing down the stories passed on among Jesus' followers began to be felt.

3. With regard to the earliest extant sources, we have to face an important problem of content, which I mentioned in Chapter 1.1. Whereas the author of Mark tells us the story of the empty tomb, suggesting that Jesus was resurrected with his earthly or terrestrial body, in the earlier source of Paul's first letter to the Corinthians (15.40–50), which was probably written within the range of 53–57 CE, we read that Jesus was resurrected with a celestial body. Indeed, Paul argues explicitly that the Christ was raised not with his terrestrial or physical body but with a new,

---

[42] Cf. Ehrman (2009), pp. 104–7, 143–4.
[43] Cf. Wright (2003), p. 403, note 8: 'It is noticeable that the non-canonical gospel traditions contain very little parallel material to the canonical gospels on this topic' [of Jesus' resurrection]. For example, in the worldview of the Gospel of Thomas, resurrection 'is simply ruled out' as a matter of principle, as Wright stresses (p. 537).
[44] Theissen and Merz (1996), §2.1. I am assuming here that criteria of historical reliability were not *very* prominent in the procedures of canonization.
[45] Cf. Gowler (2007) for a brief review of recent historical scholarship. Cf. also Theissen and Merz (1996), §2.2.
[46] Cf. Harris (1989) and Hezser (2001).
[47] Cf. *Mark* 9.1: 'Truly, I [=Jesus] say to you, there are some standing here who will not taste death before they see that the kingdom of God has come with power'; cf. also *Mark* 13.24–30.

spiritual body (*sooma pneumatikon*), and he does not mention an empty tomb in any of his letters.[48] Accordingly, he could claim that the risen Jesus also had appeared to him, Paul, and he lists himself as a witness of an appearance of the resurrected Christ on equal footing with the other witnesses (1 Corr. 15.8). This may suggest, thirdly, that the story of the empty grave and the resurrection of Jesus with his physical body is a somewhat later invention or development, which was ever more embellished in subsequent gospels.[49] It is striking that according to the author of Mark (16.8), the three women who discovered the empty tomb 'said nothing to anyone' about it. In the context of the story, this is unlikely, since by keeping silent the women would have disobeyed the order of the angel to tell the disciples about the resurrection (16.7). Should we interpret this incongruous text as an attempt to explain why the story of the empty tomb was unknown in early Christian communities?[50]

4. Although the present text of Paul's first letter to the Corinthians mentions a large number of witnesses to appearances of the resurrected Christ with a celestial or spiritual body, to wit, Cephas or Peter, more than five hundred brethren at one time, James, the twelve apostles, and Paul himself, the canonical gospels tell us that Jesus as resurrected with his earthly body appeared to a few only. According to Mark 16.9–14, the risen Jesus showed himself only to Mary Magdalene and to the eleven apostles. But since these verses are lacking in the oldest manuscripts of Mark, and were perhaps added in the second century or later, their historical value is very dubious.[51] Furthermore, there are a great number of differences between the canonical gospels with regard to the questions by whom, when, and where the resurrected Jesus was seen, and what the appearance was like. For example, whereas according to Matthew and Mark the disciples saw Jesus in Galilee, Luke says that they did not leave the environs of Jerusalem.

Clearly, then, the nature of the testimony *T* is not as unambiguous and convincing as may be wished by those traditional Christians who defend the view that Jesus really was resurrected with his earthly body. According to the extant sources, there were no eyewitnesses of the resurrection-event itself, but only of an empty tomb and of Jesus-appearances. We do not have writings of these alleged eyewitnesses, however, since

---

[48] Although Thiselton (2000) denies that the difference between the terrestrial body of Jesus and the celestial or spiritual body is a matter of composition (pp. 1276–81), he stresses also that 'the **body** that rots in the grave is emphatically *not* the **body** of the ... resurrection' (p. 1267; Thiselton's italics and bold).

[49] The story of *Matthew* 28.11–15, according to which the priests and elders invented the lie that Jesus' disciples had stolen his body from the grave, may have been such an embellishment as well.

[50] Cf. Deschner (1996), p. 116. According to the Jesus Seminar (a group of about 150 biblical scholars and laymen founded in 1985 by Robert Funk and John Dominic Crossan), Jesus did not rise bodily from the dead, and there was no empty tomb. Cf. Gowler (2007), p. 38, and Funk (1998), p. 533. It is perhaps most likely that Jesus' body was thrown into a mass grave after the crucifixion. Cf. Hick (1978), p. 25; Kent (1999), p. 88; and Spong (1994), pp. 239–41. Of course, as soon as the story of the empty tomb became popular, it will not have been difficult to find one. Cf. for a historical approach to the resurrection also: Lüdemann and Ozen (1995). As against a near consensus of historical Bible scholars, the reality of the empty tomb is defended by Swinburne (RGI), pp. 160–3 and 174ff, and by Wright (2003), p. 321.

[51] Cf. *The New Oxford Annotated Bible*, ed. by H. G. May and B. M. Metzger (Oxford: Oxford University Press, 1962), pp. 1238–9, note.

probably they were not able to write and they were relatively uneducated. All the alleged eyewitnesses and authors of the early sources that mention them were followers of Jesus or of Paul, so that they were neither disinterested nor mutually independent. If Yahweh really wanted to stage the 'super-miracle' of a bodily resurrection, would he not have done so before the eyes of more independent witnesses, such as Pilate and the Roman soldiers, in order to convince humanity at large?[52] Furthermore, the tale of the empty grave and of Jesus' apparitions with his mutilated earthly body are absent from the oldest sources, whereas the number of alleged eyewitnesses of these things is very small.

Moreover, if one wants to believe the second-century or later additions to Mark, one should conclude that the earliest (and perhaps only real) witness of Jesus as resurrected with his earthly body is completely unreliable.[53] For the author of Mark 16.9 says of Mary Magdalene that Jesus had 'cast out seven demons', so that we may assume that she was either hysterical or suffered from a more serious mental disorder.[54] There are diverging descriptions of what exactly the eyewitnesses experienced, and, as I said, there are many contradictions between the sources concerning the number and identity of the eyewitnesses and the places where they saw apparitions of Jesus. Finally, the genre to which the gospels and Paul's letters belong is not that of scholarly tracts written by professional historians. These texts were written primarily to preach, to proselytize, to moralize, and to convince the readers of an eschatological view.[55]

10.4 EXPLAINING THE EXISTING TESTIMONY

If this is the nature of the testimonies T about the resurrection of Jesus, how are they to be explained? According to the traditional Christian explanation, there really was an empty grave and a bodily resurrection E, although Jesus' earthly body may have become somewhat transformed in the process. There really were eyewitnesses of E, or, more precisely, of the empty grave and the appearances of Jesus resurrected. The diverging stories about who they were, what they saw, and where they saw Jesus as

---

[52] This problem was discussed already by Celsus and embarrassed Origen. Cf. Deschner (1996), p. 120. Swinburne solves it as follows: 'God [...] must not make his presence and his intentions for us too obvious', since otherwise man would lose the freedom to choose between good and evil (RGI, p. 172; cf. EG, pp. 267–72, 285, 290, and *passim*). However, if in God's presence man were always to obey him, it becomes a mystery how according to *Genesis* Eve could eat the apple in Paradise.

[53] I add 'and perhaps only real' because according to Mark 16.12, the resurrected Jesus appeared to (two of) the disciples 'in another form'.

[54] In his celebrated *Vie de Jesus*, Ernest Renan points out in a footnote that Mary Magdalene was 'le seul témoin primitif de la résurrection': Renan (1864), pp. 119–20. He then argues that the love and enthusiasm for Jesus of this only real witness of the resurrection caused her to *hallucinate* a resurrected Jesus (p. 356). The hallucination thesis has been defended by many modern scholars. Cf. Habermas (2001) for an overview of recent naturalistic accounts of the resurrection story.

[55] Swinburne judges otherwise. Of Paul he says: 'when he gives us narrative, he means it to be taken as a literal historical truth. If we question the genre of those letters, we would have to question the genre of every historical document ever written' (RGI, p. 70). And concerning the synoptic gospels, Swinburne writes: 'I conclude that the three synoptic Gospels purport to be history (history of cosmic significance, but I repeat, history all the same). That is their genre' (RGI, p. 74). Historical Bible scholars will strongly disagree with Swinburne at this point.

resurrected must be explained by the vacillations of the oral traditions preceding the writings of Paul and the authors of the gospels. But how can this explanation compensate for the extreme improbability of E, given the arguments for premise (b) that E 'violates' at least one law of nature? What the sources say about the number and quality of the alleged eyewitnesses clearly cannot outweigh this extreme improbability of E, so that, using Hume's maxim, we should conclude that the bodily resurrection E did not occur. This conclusion is endorsed by most historical Bible scholars.[56]

Can we counterbalance the extreme improbability of E given premise (b) by arguing that the occurrence of E was quite likely given T and the theistic hypothesis, that is, the hypothesis of bare theism? This is the argument Swinburne develops in *The Resurrection of God Incarnate*.[57] In order to do so it has to be argued, first, that the truth of the theistic hypothesis is rather likely given the arguments of natural theology supporting it, apart from the argument from resurrection. Second, theists have to contend contra Spinoza and many others that if he exists, God will occasionally 'violate' laws of nature or at least inhibit their operation. Third, it has to be demonstrated that God's self-incarnation in Jesus, Jesus' crucifixion, and his bodily resurrection are all quite likely given the truth of the hypothesis of bare theism.

It might be argued, for example, that in view of the sorry moral state of humanity, the human race 'ought by sacrificial action to show its contrition to its creator'. But since an adequate atonement 'was not within the capacity of a fallen race', God 'could insist on the sacrifice of none other but himself'.[58] The apologist will then profess that God's *unique* incarnation in Jesus is quite likely, given intentions which we may plausibly attribute to God, and given Jesus' character and teachings; that Jesus' crucifixion is quite likely, given the need for an atonement and God's obligation to show solidarity with human suffering; and that Jesus' bodily resurrection is also quite likely, given God's intention to provide a moral guide to humanity and to put a 'divine signature' on the life of his incarnation.[59] Having concluded on the basis of these arguments and of the testimony T that Christ's bodily resurrection probably occurred, an apologetic philosopher of religion such as Richard Swinburne will then use this event E as a further confirming argument for the theistic hypothesis that God exists.[60]

---

[56] Cf. Funk (1998), p. 533. According to Wright (2003), p. 7, this is part of the 'dominant paradigm' concerning Jesus' resurrection, a paradigm that Wright wants to challenge. But unfortunately, Wright does not challenge the paradigm in his Chapter 18 within a neo-Humean framework, even though he admits that historical explanations should be assessed in terms of probability. In other words, he does not take into account the intrinsic excessive improbability of Jesus' bodily resurrection, when he argues that '[t]he proposal that Jesus was bodily raised from the dead possesses unrivalled power to explain the historical data at the heart of early Christianity' (p. 718). For this reason, philosophers will not be convinced by his arguments, unless they already believe that God exists and that He probably was incarnated in Jesus, etc.

[57] Swinburne (RGI), pp. 25, 30; cf. (EG), p. 284: '... Hume's main mistake was his assumption that in such cases our knowledge of what are the laws of nature is our main relevant background evidence. Yet all other evidence (...) about whether there is or is not a God is also relevant, since, if there is a God, there exists a being with the power to set aside the laws of nature that he normally sustains'.

[58] Swinburne (EG), p. 289. One problem with this argument is, of course, that it does not respect the maxim that 'ought implies can'.

[59] Swinburne (RGI), Chapters 2–8, 12–13.

[60] Cf. Swinburne (EG), pp. 284 and 287. Cf. for a formalization of the argument in terms of Bayesian probability, Swinburne (RGI), pp. 204–16 and Earman (2000), pp. 65–7. The argument

However, such a theistic explanation of *T* by postulating *E* may be contested on four types of grounds. First, many Muslims will argue that God's incarnation in Jesus, an atonement by his crucifixion, and the bodily resurrection are extremely unlikely given theism, since the unity and simplicity of God exclude an incarnation. Secularists may concur because, for example, the above argument for the likelihood of atonement by a crucifixion of God's incarnation in Christ presupposes that, quite absurdly, God would accept a (fake) death penalty for himself because of the sins freely committed by humans. In general, the secularist will object that too many far-fetched premises are used in apologetic arguments to the effect that given theism, the occurrence of Jesus' resurrection is likely.

Second, Swinburne endorses the idea that if 'the production of the Koran really was a miracle, the Resurrection could not have happened'.[61] This is an illustration of Hume's thesis that '[e]very miracle [...], as its direct scope is to establish the particular system to which it is attributed; [...] has [...] the same force [...] to overthrow every other system', which contradicts the first.[62] Hence, apologetic Christian theists should balance their theological arguments in favour of the bodily resurrection *E* against the theological arguments in favour of the Koran as a miracle.

This second line of attack may not seem to be very promising, however, because it can plausibly be argued that writing the Koran was not a miracle in the sense defined. As Swinburne says: 'however great a work that is, there is little reason to suppose that for an uneducated prophet to write it constitutes violating a law of nature'.[63] Yet the burden of proof for Christian theists is somewhat more substantial than this. Even if the Koran is not a miracle in the technical sense defined, it might contain God's word or many true statements about God. If so, a divine incarnation in Christ might be excluded, since according to the Koran, Jesus is neither God's son nor his incarnation (19.30–36). It follows that in order to succeed, Swinburne should have completed *The Resurrection of God Incarnate* by an argument to the effect that the Koran is not God's word and that the relevant passages are false.

The traditional theistic explanation of *T* will be contested, thirdly, because arguably the improbability of *E* given the reasons for premise (b) is not outweighed at all by the quality of the existing testimony *T* for *E*, even if taken together with the probability of the theistic hypothesis and the likelihood of *E* given that hypothesis.[64] Let us not forget that in order to be a miracle, *E* has to be excessively unlikely, because physically

---

from miracles to God may seem circular to those who overlook its Bayesian structure. But it is not circular, if at least the argument from miracles is not used in calculating the specific prior probability of theism that functions in calculating the posterior probability of theism given the miracle(s).

[61] Swinburne (EG), p. 288.
[62] Swinburne (EG), p. 288; jo. Hume (1748), Sect. X, Part II, §95.
[63] Swinburne (RGI), p. 62.
[64] At least this will be true for someone who is in the process of deciding whether or not to accept theism partly on the evidence of a miracle. From the perspective of those who already firmly believe that theism is true, and that given theism God's incarnation in Christ, the crucifixion as atonement for human sins, and Jesus' resurrection are very probable, the evidence will be weighed differently. Indeed, theists might even claim that all natural evidence against the occurrence of a miracle is simply irrelevant, since 'our evidence for how the world behaves when God does not intervene is irrelevant to how the world behaves when God does intervene', as Otte (1996) argues (p. 153). But clearly, an event *E* can only be used as evidence for the existence of God if it has been established independently of theism that *E* is impossible given the natural course of events. This independence

178   *Theism as a Theory*

impossible, given all that we know about the universe. Furthermore, the facts *T* as summarized above will give us no great confidence in the reliability of the testimony. Hence, Hume would have been right to conclude that with regard to the resurrection of Jesus with his terrestrial body, the falsehood of the testimony is much less miraculous or improbable than the event which it relates, even if one takes the theistic hypothesis into account as background knowledge.[65] This argument will be reinforced by a fourth type of criticism of the theistic explanation of *T*, to wit, that there is another and purely secular explanatory hypothesis, which is vastly superior to the theological explanation.

10.5 COGNITIVE DISSONANCE AND COLLABORATIVE STORYTELLING

Apart from a wealth of data obtained by historical and archaeological scholarship concerning Jewish, Greek, Hellenistic, and Roman cultures in the first and second centuries, in which miracle stories abound, a secular explanation of the testimony *T* must use concepts and results of psychological and sociological research concerning phenomena such as testimony, collaborative storytelling, cognitive dissonance, and source amnesia.

As empirical investigations concerning groups of interdependent witnesses have shown, people are inclined to base their subjective certainty about what happened on the opinions of others if they are confronted with information that they cannot verify by direct observation. In such situations, we may speak of 'collaborative storytelling' or 'social contagion', that is, a cumulative process of mutual reinforcement of ideas, which occurs among people who have to interpret ambivalent information.[66] Since human memory functions as an updating machine, which often retains information without also retaining knowledge about its source, people may think that what they remember stems from their own experience, whereas in fact they rely on communication by others. These mechanisms of collaborative storytelling and source amnesia explain the astounding fact, for example, that in some cases the suspect of a murder may honestly confess under the influence of protracted and suggestive interrogations by the police, although in fact someone else committed the crime.[67]

Another concept that is pertinent to an explanation of the testimony about Jesus' resurrection is the notion of cognitive dissonance, which was originally developed by Leon Festinger. Cognitive dissonance may be defined as an unpleasant mental tension arising from a perceived incompatibility within one's set of opinions (including known opinions of others one has no reason to reject), attitudes, feelings, behaviours,

---

condition is stressed by Curd (1996), who argues that the standard attempts to avoid a contradiction in the very notion of a miracle will result in a violation of the independence condition.

[65] Of course this should be argued in detail as against Swinburne's (RGI).

[66] Cf. Wagenaar and Crombag (2005), p. 166. They also refer to: Asch (1956), Bavelas et al. (2000), Edwards and Middleton (1986), Gould and Dixon (1993), Hardin and Higgins (1996), Kotre (1995), Kraus (1987), Loftus (1979), Loftus (2003), McGregor and Holmes (1999), Novick (1997), Pasupathi (2001), Tversky and Marsh (2000), Victor (1993), and Weldon and Bellinger (1997).

[67] Cf. for the analysis of such a case, Wagenaar and Crombag (2005), Chapter 9.

and values. A habitual smoker will experience cognitive dissonance when hearing about the severe health risks of this habit, and the open expression of disagreement in a group will lead to cognitive dissonance in its members. Festinger and others performed many experiments with subjects experiencing cognitive dissonance. They found that people often attempt to reduce this unpleasant tension by using or inventing stories that mask the incompatibility, by adapting their behaviour, or by attempting to convince other people of their beliefs.[68]

Festinger and his colleagues also did participating field research on a religious sect, the members of which believed that just before a cataclysmic flood were to engulf most of the American continent, the sect would be saved by flying saucers, which would take its members to another planet. Whereas the sectarians kept away from the press before the time at which the predicted events were to take place – presumably the space in the flying saucers would be very limited – they accommodated their theology as soon as it had become obvious that neither flood nor flying saucers had arrived. They also invited public and press in order to convince outsiders of this modified theology, according to which God had saved the world because of the enlightening behaviour of the group.[69]

How can one explain that after the predictions of the sect had been refuted by the facts, its members attempted to convince others of their somewhat transformed beliefs? The theory of cognitive dissonance accounts for this surprising behaviour as follows. Since the core members of the sect had quit their jobs and abandoned possessions because of the predicted flood, they were so deeply committed to their beliefs that they could not discard them quickly. When neither flood nor flying saucers arrived at the predicted dates, they experienced an agonizing cognitive dissonance, which they had to reduce. The core members assembled in the home of the leading woman, a Mrs Keech, who allegedly received messages from Guardians in outer space. During this gathering they mutually supported each other in constructing an accommodated theology, which partially reduced cognitive dissonance. Further reduction was achieved by attempting to convince the outside world of their glad tidings.[70]

An analogous explanation can be given for the accommodated theology and proselytizing zeal of the members of the Jesus sect, if at least we suppose plausibly that Jesus really has existed. We may assume that before Jesus' crucifixion, the members of this group had utopian or apocalyptic beliefs somewhat similar to those of some other contemporary Jewish sects mentioned by Flavius Josephus or Philo of Alexandria. With God's support their leader would bring an end to the Roman reign of Palestine, transform human society, and restore the twelve tribes of Israel, perhaps by becoming their king, so that God's kingdom was imminent. Many apostles must have been deeply committed to these beliefs, since presumably they had left everything in order to follow Jesus. According to *Matthew* 19.27–9, Jesus promised them as a compensation that they would 'sit on twelve thrones, judging the twelve tribes of Israel', and that they would 'receive a hundredfold' reward.[71] When Jesus went to

---

[68] Cf. Festinger (1962).
[69] Festinger et al. (1956).
[70] Festinger (1962), pp. 252–9.
[71] Cf. Sanders (1993), Chapters 11–12. As Sanders notes on p. 190, this 'hundredfold' reward appears to refer to an expected earthly kingdom.

Jerusalem, he and his followers will have expected that this new utopian reign be near. Indeed, if Jesus really entered Jerusalem on a colt or ass (*Matt.* 21.7–9), he must have believed that he was fulfilling a prophesy in Zechariah (*Zech.* 9.9), and by so doing implicitly declared himself a king. But then, Jesus was convicted as a blasphemous troublemaker and crucified.

This unexpected and horrifying event will have caused a devastating cognitive dissonance within the flock of his followers. Since they had invested so much in Jesus' eschatological promises, they could not abandon their beliefs easily. As the theory of cognitive dissonance predicts, they engaged in collaborative storytelling instead, slightly accommodating and mutually reinforcing their eschatological ideas, using existing cultural resources. During this process, they created a somewhat modified theology, according to which the fact of Jesus' crucifixion acquitted all dedicated followers from their sins, and announced the imminent heavenly kingdom. Somewhat later, members of the sect started to proselytize feverishly, as the theory of cognitive dissonance also predicts.[72]

In accordance with a contemporary Jewish tradition concerning prominent righteous people and martyrs, the followers will have believed initially that Jesus had been resurrected and had ascended into heaven on the day of his death, or on the third day after his death, with a new celestial and luminous body, like Henoch, Abraham, Isaac, Jacob, Moses, Ezra, Jeremiah, and Baruch.[73] His old, earthly body remained in the mass grave into which it was probably thrown. Interpreting their dreams and vivid memories of Jesus as spiritual experiences of their resurrected leader, they construed Jesus' resurrection as a promise that they too would overcome death very soon, as is testified by Paul's authentic letters. Stories about appearances of deceased people were common in the ancient world as indications that they had entered into heaven.[74] Also, Jesus' followers will have interpreted these experiences in accordance with their cultural traditions as justifying proselytizing activities.[75]

Only somewhat later, when younger members of Jesus sects evangelized more amply in the Roman Empire, they may have started to use the story of the empty tomb as a proof of Jesus' resurrection.[76] Since this story was inspired by ascension myths concerning Greek or Roman heroes such as Adonis, Aeneas, Hercules, Theseus, and Romulus, in which the absence of earthly bones or bodies was often considered a

---

[72] Cf. Strauß (1835/36) for an early version of this view.

[73] De Jonge (1989), pp. 34, 36, 38. See for Henoch: *Sirach* 44.16; for Abraham, Isaac, and Jacob: 4 *Macc.* 16.25; for Moses: Philo, *Vita Mosis* II, 291, *Ass. Mos.*, and *Mark* 9.2–8; for Ezra: 4 Ezra; for Jeremiah 2 *Macc.* 15.14, and for Baruch: Syr. Apoc. Bar. Cf. also 2 *Macc.* 7, and Kleinknecht (1988). The notion that the resurrection took place on the third day after death was taken from Hosea 6.2: 'After two days he will revive us; on the third day he will raise us up', and from other sources, such as *Jona* 1.17. Whereas in Hebrew, 'some' or 'a pair' could mean 'three', this is not possible in Greek, so that Paul had to write that the resurrection took place after three days.

[74] De Jonge (1989), p. 42. Such stories also abounded in the pagan world: Livius I:16, and Vergilius, *Aeneïs* II, pp. 771–89.

[75] Cf. *Exodus* 3.10; *Judges* 6.14, and De Jonge (2002).

[76] Cf. Lowder and Price (2005). It is interesting to note that Mark uses the story of the empty tomb as a 'proof' of the resurrection but leaves out (in the original version; *Mark* 16.9–20 cannot have been part of the original text) the appearance stories. Since both types of stories served the same objective of making probable the reality of the resurrection, the author(s) of *Mark* may have thought that the story of an empty tomb would suffice.

compelling proof of ascension, it was perhaps more convincing to their new audiences, who probably were longing for a bodily resurrection.[77] We may conjecture that the story of the empty tomb in *Mark* and the idea of Jesus' resurrection with his earthly body are symptoms of the Hellenization or Romanization of Christianity in the late first and early second centuries.[78]

If we now compare the traditional Christian explanation with this rival secular explanation of the existing testimony T, we must come to the conclusion that the latter is vastly superior to the former.[79] First, the secular explanation does not postulate a miraculous event E, the occurrence of which is excessively improbable or impossible given our scientific and everyday background knowledge. Second, the psychological theories and historical assumptions used in the secular explanation are confirmed by extensive empirical research and historical scholarship, whereas the theistic hypothesis is merely supported by abstract philosophical arguments, the force of which is called into question even by quite a number of expert apologetic philosophers of religion.

---

[77] De Jonge (1989), pp. 43–5. Cf. for Heracles: Diodorus Siculus IV, 38; for Romulus: Livius I, 16, 2; and for Aeneas: Dionysius Halicarnasseus, *Antiquitates Romanae* 1, 64, 4. Cf. also Carrier (2006).

[78] De Jonge (1989), p. 45. This is also the conclusion of the Jesus Seminar. Cf. Gowler (2007), p. 38, and Funk and the Jesus Seminar (1998), p. 533. Some archaeologists are convinced that Jesus' tomb and bones have been discovered. Cf. Tabor (2006).

[79] Wright (2003) argues for an opposite conclusion concerning an explanation *à la* Festinger on pp. 697–701. But his arguments are unconvincing for the following reasons. First, he does not give any weight in his probability assessment to the fact that the bodily resurrection would be a miracle, that is, an excessively improbable event in itself given our natural background knowledge, although he stresses (p. 706) that '[w]hat we are after is high probability'. Second, he avers that most experiments Festinger et al. did on cognitive dissonance are 'hardly relevant to the historical study of a group of first-century Jews' (p. 697), so that only the case study of the flying-saucer cult would be relevant. But this claim violates the principle of the consilience of inductions, which is fundamental to scientific and scholarly research. Wright should have taken into account a mass of later socio-psychological research on topics such as collaborative storytelling, as I suggest in the main text. Third, Wright rejects the study of the flying-saucer cult on the ground that the sociologists who infiltrated the rather small group may have influenced and even virtually directed the proceedings. Although this risk is clearly present, it should be established that it occurred in this case before one dismisses Festinger's results. Fourth, Wright stresses that whereas the flying-saucer group did not survive for a long time after the refutation of its predictions, the Jesus-sect became very successful. But this can be explained by taking into account the cultural context and the facts of Jewish diaspora within the Roman Empire. Fifth, Wright argues that the followers of Jesus, in contradistinction to the members of the flying-saucer cult, 'were not refusing to come to terms with the fact that they had been wrong all along', but 'were indeed coming to terms with, and reordering their lives around, dramatic and irrefutable evidence that they had been wrong' (p. 700). But of course, this is true for the members of the flying-saucer cult as well. They came to terms with the fact that neither flood nor flying-saucer had arrived by transforming their eschatological doctrines, and the followers of Jesus, who may have expected him to be king, came to terms with the crucifixion by transforming their original utopian and eschatological doctrines into a theology according to which by being crucified and resurrected, Jesus inaugurated the kingdom of God, which would arrive within the time of their life. Finally, Wright stresses that 'many of the messianic movements between roughly 150 BC and AD 150 ended with the violent death of the founder', whereas this did not happen to the Jesus movement (p. 700). But is the hypothesis that Jesus was bodily resurrected really the most plausible explanation of this fact? However this may be, Wright finally concludes that '[t]he proposal that Jesus was bodily raised from the dead possesses unrivalled power to explain the historical data at the heart of early Christianity' (p. 718), and that 'the historian, of whatever persuasion, has no option but to affirm both the empty tomb and the "meetings" with Jesus as "historical" events [...]: they took place as real events' (p. 709).

Third, in order to argue that Jesus' bodily resurrection was quite likely given the truth of theism, theists have to use implausible premises, such as the tenet that God would accept a (fake) death penalty for himself because of the sins freely committed by humans.

Fourth, many aspects of the existing testimony T are much more likely on the secular explanation than on the theistic hypothesis. For example, if there really had been an empty tomb of Jesus, Paul would have mentioned it. Since in 1 Corinthians 15 he wanted to prove that Jesus had been resurrected, he would not have left out the best possible evidence, an empty grave, had he known of it.[80] Also, according to Paul there is no temporal space between Jesus' resurrection with a spiritual body and his reception in heaven. For him, these two events are one, which excludes a bodily resurrection on Earth. Indeed, the resurrected Jesus simply is the heavenly Jesus.[81] And if Paul had assumed a bodily resurrection on Earth, he would also have spoken about a separate and later ascension of Jesus into Heaven. But Paul never mentions such an event, in contradistinction to Matthew and Luke.[82]

However this may be, it is obvious that adducing miracles as evidence for theism does not exclude a competition between theism and rival secular views. Admittedly, genuine miracles are without any doubt 'too odd' to be explained by science or secular scholarship. But the real evidence to be explained does not consist in the miracle itself, but rather in the existing testimony T, or in alleged physical traces of a miracle. If religious apologists postulate a miracle in order to explain this evidence, its occurrence will be contested, and alternative secular explanations will be put forward.[83] As we shall also see in later chapters, by focusing on postulated events or on evidence that seems to be 'too odd' to be explained by secular science and scholarship, one will not safely avoid the cul-de-sac of God-of-the-gaps. In the next section, I shall investigate provisionally to what extent limiting the alleged evidence for theism to facts that are 'too big' to be explained by science or secular scholarship is a more promising strategy.

---

[80] I do not agree with Wright's claim in his (2003) that '[t]he fact that the empty tomb itself, so prominent in the gospel accounts, does not appear to be specifically mentioned [in 1 *Corinthians* 15], is not significant' (p. 321). Wright's argument that Paul mentions Jesus' burial in verse 4a in order 'to indicate that when Paul speaks of resurrection in the next phrase it is to be assumed [...] that this referred to the body being raised to new life, leaving an empty tomb behind it' is very weak given what Paul says about the resurrected body in verses 35–50. Paul does not mention an empty tomb in any of his letters.

[81] Cf. 1 *Corinthians* 15.46–7: 'But it is not the spiritual [body] which is first but the physical, and then the spiritual. The first man was from the earth, a man of dust; the second man is from heaven'. Cf. also 2 *Corinthians* 5.1: 'For we know that if the earthly tent we live in is destroyed, we have a building from God, a house not made with hands, eternal in the heavens', and vs. 8: 'We are of good courage, and we would rather be away from the body and at home with the Lord'. Clearly it was Paul's view that our earthly body would rot away in the grave, and that some time after our death God would give us a new, spiritual body in heaven.

[82] Cf. De Jonge (1989), p. 34, and 1 *Cor.* 15.48–9; 1 *Thess.* 1.10; 2 *Cor.* 5.1–8.

[83] Cf. Swinburne (EG), p. 277: miracles are certain events 'the occurrence of which is normally disputed'. Swinburne also argues that the connections between our mental life and brain processes are too odd for science to explain. Cf. (EG), Chapter 9, and (ES), pp. 186–93. Since there clearly are such connections, the critic of Swinburne will have to argue in this case that a scientific explanation cannot be excluded a priori: cf. Chapter 14.1–4, below.

10.6 TOO BIG FOR SCIENCE

There are three main large-scale phenomena that allegedly are 'too big' for science to explain: the existence of the universe as a temporal whole, the fact that the most fundamental laws of nature are what they are, and the fact that the universe is 'fine-tuned' for certain things to develop.[84] Can theists avoid the dilemma of God-of-the-gaps by limiting the explanatory tasks of theism to these large-scale phenomena? Let me discuss them briefly in this order.

According to a simplified covering law model of scientific explanation, a state of affairs is fully explained scientifically by showing that according to laws of nature, it had to occur given a preceding state of affairs and specific boundary conditions. Let us suppose that, ideally, one might collect all states of affairs that make up a time-slice of the universe to which the present moment belongs, and let us call this collection the present state of the universe *a*. Let us further suppose that this present state is preceded by a series of earlier states of equal finite duration. Then we may say that, in principle, if the laws are deterministic there can be a full scientific explanation of *a*, which shows that according to laws of nature *a* had to occur given the preceding state of the universe *b*. Equally, *b* may be explained scientifically given its preceding state of the universe *c*, and so on. If we call *b* the full cause of *a*, and *c* the full cause of *b*, etcetera, we may say that *c* is the cause of the collection of temporal states of the universe {*b*+*a*}, and that *d* is the cause of the collection {*c*+*b*+*a*}.[85] Accordingly, Swinburne proposes the principle that 'in so far as a finite collection of states [backwardly in time up to the present] has a cause, it has its cause outside the collection'.[86] Let us call this maxim the *principle of collective causality*.

Most cosmologists in the nineteenth century assumed that the series of states of the universe backwards in time is infinite. If this is the case, science can in principle attempt to trace the causes of each finite collection of states of the universe. But when cosmologists such as Einstein and De Sitter started to apply the theory of general relativity to the universe in 1917, the hypothesis of an expanding universe became conceivable. This hypothesis was confirmed by Hubble's discovery in 1929 of a redshift of all galaxies, the magnitude of which is proportional to their distance from Earth. The theory of the expanding universe suggested in its turn the hypothesis of a hot Big Bang, some sort of colossal explosion that initiated the expansion of the universe between 12 and 15 billion years ago. Hot Big Bang theory was first confirmed in 1963–5, when Penzias and Wilson discovered by accident the cosmic background radiation.

If Big Bang theory is correct, there may seem to be clear limits to possible scientific explanations. Indeed, before so-called Planck time, shortly after the postulated Big

---

[84] Swinburne (EG), pp. 74–5, 142–3, 160, and 235. On pp. 277 and 326, Swinburne mentions two other phenomena that might be 'too big' for science to explain, but these can perhaps be subsumed under the phenomena for which laws of nature allegedly are 'fine-tuned'. Cf. for discussion, Chapters 12 and 13, below.

[85] This notion of 'cause' is not the everyday one, according to which an actor causes a result by acting upon something else. Cf. Rundle (2004) for an illuminating analysis of our ordinary notion of causality. Cf. also Chapter 12.3–5, below.

[86] Swinburne (EG), p. 142.

Bang, all known science fails us, and it seems that we cannot use scientific methods of investigation to discover what existed 'before' the Big Bang.[87] If the question as to what caused the Big Bang is meaningful at all, it will seem 'too big' for science to answer. Applying the principle of collective causality to the finite set of all temporal states of the universe from the Big Bang to the present, we may now say that if this collection of states has a cause, its cause must be external to the universe from the Big Bang onwards, and cannot be discovered by science. For this reason, it seemed to Pope Pius XII that Big Bang theory offered a new prospect for natural theology. In his lecture *Una Ora* of 22 November 1951, he argued that Big Bang theory supports the *Genesis* doctrine of a creation *ex nihilo* of the universe by God, so that this spectacular scientific finding confirms the existence of a transcendent Creator.[88]

We may wonder, however, whether Pope Pius XII was able to avoid the risk of a God-of-the-gaps fallacy. In 1951, long before the discovery of background radiation, there still was a rival cosmological theory in the race, the steady-state theory developed by Hermann Bondi, Thomas Gold, and Fred Hoyle. According to this theory, the universe is infinitely old, so that, in principle, there is a scientific explanation of each temporal state of the universe in terms of scientific laws and an earlier temporal state. If the steady-state theory had prevailed over the theory of a hot Big Bang, the cosmological argument from the Big Bang to God would have been refuted by scientific results. Hence, a natural theologian who wants to avoid the risk of a God-of-the-gaps fallacy should not rely on scientific theories, which, in principle, may be refuted by future research.

Swinburne argues that even if the universe stretches backwards in time infinitely, an application of the principle of collective causality leads to the result that if the collection of all its temporal states has a cause, this cause must lie outside the collection. He writes: 'The whole infinite series will have no full explanation at all, for there will be no causes of members of the series lying outside the series'. Hence, even the existence of a complex physical universe over infinite time backwards allegedly is something 'too big' for science to explain.[89] But clearly, this application of the principle of collective causality is fallacious. The principle only applies to finite collections of temporal states up to the present, since only in these finite temporal series is there a first state, which cannot be explained with reference to an earlier state belonging to the series. If the universe stretches backwards in time infinitely, however, there is no first state. In that case, each temporal state of the universe can be explained scientifically with reference to an earlier state, which belongs to the infinite series, so that there is no causal question concerning any state of the universe that is 'too big' for science.[90]

It is at this point that Swinburne introduces a doctrine of double causality. May there not be a 'deeper explanation' than that of scientific causality, he wonders, 'one in which the explanatory factors themselves are explained by a personal cause'? Might

---

[87] Cf. Chapter 12.6–9, below, for further discussion.
[88] Pius XII (1951). Cf. Krach (1996), pp. 256–9. Incidentally, the book *Genesis* does not contain a doctrine of *creatio ex nihilo*.
[89] Swinburne (EG), p. 142.
[90] Cf. Parsons (1989), pp. 74–6 and 88–91, and Chapter 12.4, below, for other criticisms of Swinburne on this point.

God not 'bring it about by a basic action at each period of time that the laws of nature *L* operate', and by so doing, '[keep] the universe in being'?[91] But as I argued above and shall further argue in Chapter 12.3–5, below, by referring to such a 'deeper explanation', which would apply whatever happens in the universe, natural theologians abandon their empiricist strategy in defence of theism.

Moreover, as we saw in Chapter 8.1, Swinburne argued with regard to God that if His existence is backwardly eternal, 'it is not logically possible that any agent could bring about God's existence'. The reason he adduced is that since a cause of God's existence must precede it, 'no moment of time would be early enough for an agent to bring about God's existence'.[92] But if this holds for God, it also holds for a universe that stretches backwards in time infinitely. Hence, Swinburne cannot argue without inconsistency that God must be *un*caused, whereas it makes sense to say that an infinitely old universe *is* caused by some agent. We may conclude that it is not easy for theists who advocate an empiricist strategy to persevere in it, and to avoid the peril of God-of-the-gaps. If a Big Bang theory is correct, what precedes Planck time may be 'too big' for science to explain. But in this case, the fate of theism will depend on the course of scientific progress. If, however, the universe stretches backwards in time infinitely, there is in principle no state of the universe that is 'too big' for science to explain.[93]

There is a similar problem concerning the second large-scale phenomenon that according to Swinburne is too big for science. Quite often, it is possible to explain the operation of one law of nature by the operation of other laws, which are more fundamental. Swinburne mentions the example of Galileo's law of fall, the operation of which is predicted to a high degree of accuracy by the laws of Newton's classical mechanics.[94] Now it may be that there is a set of ultimately fundamental laws, so that the explanation of laws by more fundamental laws has a terminus. In this case, the question as to why the ultimately fundamental laws hold is too big for science to explain. But can theists avoid the peril of God-of-the-gaps if they claim that the operation of the ultimately fundamental laws of nature has to be explained by God's action?

At a determinate moment in history, scientists may be justified in regarding a specific set of natural laws as ultimately fundamental, given the collection of empirical data known at that time. Swinburne argues, for example, that in the later eighteenth century natural scientists were justified in considering the laws of classical mechanics as the most fundamental laws of nature. Does it follow that the question as to why Newton's laws operate is 'too big' for science to explain? As the later history of science has shown, this does not follow at all. Many phenomena discovered by scientists in the nineteenth and twentieth centuries could not be explained on the basis of Newton's laws, and we now account for the fact that classical mechanics yields quite good predictions in many limited domains by relying on more fundamental laws and

---

[91] Swinburne (EG), pp. 142–3.
[92] Swinburne (CT), p. 265.
[93] The reader might object: surely science cannot explain why there is something rather than nothing at all! Cf. Rundle (2004), and Chapter 12.3–4 for further discussion.
[94] Swinburne (EG), p. 81.

theories, such as quantum mechanics and the theories of relativity.[95] Since the same fate may happen to these latter laws and theories, when scientists succeed in formulating tenable versions of quantum gravity or string theories, theists run the risk of God-of-the-gaps if they pretend that the question as to why a *specific* set of natural laws holds, is 'too big' for science to explain.

Can theists avoid the risk of God-of-the-gaps by saying that *whatever* laws are ultimately fundamental, the question as to why these fundamental laws operate is 'too big' for science to answer?[96] Although by this move the explanandum becomes completely unspecified, the stratagem may not be sufficient either. Possibly, the elements of each specific level of physical things are constituted by even more elementary elements. Suppose that the interactions of the elements at each level can be described by laws which hold for that level only, and that the operation of these laws at each level can be explained by the laws of the next lower level, together with the way in which the elements of the higher level are constituted out of those of the lower level. In this case, there is an infinite regress of ever more fundamental laws, and there is no ultimate level, the explanation of which is in principle 'too big' for science to explain.

With regard to the first two large-scale phenomena that allegedly are 'too big' for science to explain, we reach the following conclusion. If in fact there has been a first time-slice of the universe, and if in fact there is a specific set of most fundamental scientific laws, the questions as to what caused this first time-slice, or what explains the fact that these specific fundamental scientific laws hold, are 'too big' for science to answer (if they are meaningful; cf. Chapter 12). However, the assumption that there has been such a first time-slice, or that a specific set of laws is the most fundamental one, may be refuted by scientific progress, so that the risk of a God-of-the-gaps fallacy has not been avoided.

Perhaps theists will hope to avert the risk of God-of-the-gaps by giving a very vague or general formulation of the explananda. They might say that what science cannot explain in principle is 'the existence of our universe as a whole' or 'why there are any states of affairs at all', and why 'the whole series' of ever more fundamental laws is operative.[97] Or perhaps they will say that what science cannot explain is the fact that relatively simple regularities and laws hold for many phenomena.[98] But if the explanatory power of theism is so unspecified, it is misleading to say that there are 'close similarities [...] between religious theories and large-scale scientific theories'.[99] For no scientist would take seriously a theory with such an unspecified predictive power. In

---

[95] Cf. Swinburne (EG), pp. 82–3.
[96] Cf. Swinburne (EG), p. 160: 'while the operation of non-fundamental laws may be explained by the operation of fundamental laws, that these are fundamental laws of nature [...] is something "too big" for science itself to explain'.
[97] Cf. Swinburne (EG), p. 75 and p. 158, note 3. Of course, if the existence of God were a brute fact, as Swinburne claims (cf. Chapter 8.1–2, above), God's existence would be a contingent state of affairs as well. In that case, theism cannot explain either 'why there are any states of affairs at all'.
[98] Cf. Swinburne (EG), p. 166. It might be argued by theists, for example, that many alternative sets of laws might have served God's ultimate aims equally well, so that it is arbitrary which specific set God created. Hence, although theism has no predictive power concerning a *specific* set of laws, it has predictive power for a relatively simple *generic* nomic structure of the universe. Cf. Swinburne (2005), pp. 923–4.
[99] Swinburne (EG), p. 3.

other words, it is very questionable whether theists are able to resolve what I called *The Tension* (Chapter 6.5).

At first sight, it may seem that the well-known teleological arguments from fine-tuning to the existence of God offer a solution to this problem, because these arguments rely on the existence of specific empirical phenomena, such as mankind. The structure of design arguments from fine-tuning may be sketched as follows. Using our human moral intuitions as a guide for attributing intentions to God, theists claim that quite probably, God may have wanted to create human free agents, that is, animate substances having a significant free choice between good and evil.[100] Since such agents need bodies and a physical environment, which enables them to learn from experience in order to make significant choices, God will also have wanted to create a physical universe, in which evolution results in producing human beings.[101] But as physicists have discovered, the evolution of life resulting in humans would have been impossible if the values of the constants in many laws of nature had not been situated in specific intervals, which are very narrow compared to the ranges of all logically possible values.[102] These findings are often expressed by saying that the universe is 'fine-tuned' for human life. Then it is argued that if God does not exist, it is very improbable that these values all lie in these narrow intervals. If God exists, however, this is quite probable. Hence, the facts of fine-tuning are good confirming arguments for the existence of God, if at least fine-tuning cannot be explained equally well or even better by another religious or a scientific hypothesis.

If one assumes that fine-tuning for human life of constants in more superficial laws of nature cannot be explained in terms of more fundamental laws, unless these also contain constants fine-tuned for human life, the (alleged) facts of fine-tuning are 'too big' for science to explain. Does this mean that arguments from fine-tuning both refer to specific explananda, such as human life, and avoid the peril of God-of-the-gaps? However one assesses the force and soundness of these arguments (cf. Chapter 13.4–5, below), it is clear on two grounds that they cannot completely avoid this latter peril.

First, the assumption may turn out to be false. Second, science has revealed the vastness of the universe, both in time and in space. Compared to the stretch of time between the Big Bang and the present moment, human life on Earth emerged only in the last microsecond, so to say. Compared to the infinity of future time during which the expanding universe will exist, the human race will occur at a negligibly short moment only. Furthermore, most planets are inhospitable to human life, whereas the number of stars in the universe is estimated in the order of a magnitude of 10^{21}. Even if there were humans on other planets than Earth, the number of planets on which human life exists would be negligible compared to the total number of planets in the universe. Hence, saying that the universe is fine-tuned for human life in particular seems to be as absurd, as would be the claim put forward by a little lonely fly in the palace of Versailles that this palace was constructed especially for it.[103]

---

[100] Cf. Swinburne (EG), pp. 118–19.
[101] Swinburne (EG), pp. 123–31.
[102] Cf. Swinburne (EG), pp. 172–88.
[103] Cf. Bradley (2001, 2002) for criticisms of the fine-tuning argument, and Chapter 13.4–5, below.

We come to the following conclusion. The attempt to immunize theism and to avoid God-of-the-gaps arguments by restricting its predictive power to phenomena that are 'too odd' or 'too big' for science to explain, either fails altogether or risks to reduce the explananda of theism to unspecified general facts, such as the fact that the universe exists, or the fact that there are (relatively simple) laws of nature. In Part III of this book we shall investigate whether such unspecified facts can yield good confirming arguments that make theism more probable than it would be otherwise, and, indeed, whether a cumulative case of all these arguments makes it more probable than not that God exists.

PART III
THE PROBABILITY OF THEISM

11

Ultimate Explanation and Prior Probability

The belief that God exists is justified, I argued in Part I, to the extent that its truth can be established by appealing to evidence which is public, at least in part. Adopting a religious belief on the basis of a revelation, of divine grace, or at the internal instigation of a holy spirit, is not reasonable for conscientious believers unless it is supported by natural theology. Furthermore, in our scientific age natural theology has to apply a methodology sufficiently similar to that of the empirical sciences in order to be credible, and to meet the standards of rationality$_5$ as defined in Chapter 5. These results imply that bare theism (the thesis that God in fact exists) has to be interpreted as an existential hypothesis or explanatory theory, which should be confirmable by empirical evidence. Whereas this interpretation of theism may represent a minority position among contemporary religious believers, prominent Christian philosophers of religion such as Richard Swinburne rightly defend it.

In Part II it became clear that the interpretation of theism as a theory is confronted by a number of serious problems. Let us suppose for the sake of argument, however, that theists are able to solve these problems, at least to the extent that bare theism turns out to be a meaningful and coherent theory with some predictive power. Let us also require that theists succeed in avoiding the risk of God-of-the-gaps. Then we should wonder how probable it is that bare theism is true, given the total permitted evidence for and against it. The objective of Part III is to find the correct answer to this hypothetical question. In the present chapter, I raise a number of preliminary conceptual issues.

Theism is often presented as a particularly satisfying explanation of the universe as a whole and of man, because it is an ultimate one. But to what extent is this really the case in the sense of 'ultimate' as defined by theists? Furthermore, how should we analyse the logic of confirmation and disconfirmation of explanatory hypotheses? Many philosophers of science use the theorem of Bayes in order to formalize this logic, but it will become clear that the argumentative strategy adopted by theists will work only if they apply the theorem in a disputable manner.

Finally, confronted with the problem of under-determination, theists may want to argue that irrespective of empirical (dis)confirmations, theism is more likely to be true than infinitely many conceivable rivals, which explain equally well the same set of empirical data. Theists such as Richard Swinburne hold that the criterion of simplicity is decisive in determining this prior probability of theism, because *simplex veri sigillum* (simplicity is the seal of truth), and they rule out other criteria, such as fit with empirical background knowledge. But can the criterion of fit with background knowledge be eliminated? How does the criterion of simplicity relate to other criteria, such as the scope of a theory? And is the criterion of simplicity, as used by theists, really truth-conducive?

11.1 FULL, COMPLETE, AND ULTIMATE EXPLANATIONS

Theists who interpret theism as a theory are tempted to argue 'that the existence of God forms a more natural stopping place for explanation than, say, the existence of the universe'.[1] The explanation by theism is considered a final or an ultimate one, the proper terminus of explanation, and this is often presented as an advantage of theism over scientific explanations. The idea seems to be that only God can satisfy our longings for an intellectually fulfilling explanation of the universe and some of its features, such as the existence of man. Theism allegedly achieves this by putting an end to the regress of why-questions, so that all explanatory chains would ultimately end in this single explanation.[2] But how can the theory of bare theism succeed in this respect? In what sense and to what extent is theism an ultimate or a final explanation?

In this first section I introduce some definitions in order to formulate these questions more precisely. A strong notion of ultimate explanation has been defined by Richard Swinburne in his book *The Existence of God*. Let me develop his definition here and wonder in the next section whether, in principle, theism can be an ultimate explanation in this robust sense, as Swinburne suggests.[3] My conclusion will be that at best, bare theism can be an ultimate explanation in a weaker sense only, if at least we do not want to introduce questionable metaphysical assumptions. Whether theism is indeed a good ultimate explanation in this weak sense will be investigated in Chapter 12 on cosmological arguments.

Swinburne defines the notion of an *ultimate* explanation in terms of *full* and *complete* explanations. An event or state of affairs E is *fully explained causally* if and only if the explanation mentions both causes C and reasons R which, taken together, necessitate the occurrence of E. On the traditional deductive-nomological account of a causal scientific explanation, the causes are initial conditions, whereas the 'reasons' are causal laws of nature, and the explanation is a full one iff the propositions stating C and R, taken together, entail a description of E.[4] In the case of personal or intentional explanations, the cause is (the action of) a person with certain basic powers, whereas the reasons are this person's intentions and beliefs, hopes, etc.[5] If explanations are not full, they are partial or probabilistic.

Personal explanations that merely mention a human being and this person's powers, intentions, and beliefs will rarely be full explanations. This is because humans typically have intentions upon which they do not act, although their beliefs imply that they

---

[1] Swinburne (EG), pp. 73–4; cf. p. 108: 'if theism is true, then, of logical necessity, God's action provides a complete and ultimate explanation of what it explains. For [...] it follows from God's omnipotence and perfect freedom that all things depend on him whereas he depends on nothing.'
[2] Cf. Gale (1991), p. 280.
[3] Cf. for what follows Swinburne (EG), Chapter 4.
[4] As is usual, I shall sometimes use 'iff' as an abbreviation of 'if and only if'.
[5] I slightly modify the definition Swinburne provides on (EG), p. 76; cf. p. 25. Often, causal laws will have a role in personal explanations as well.

should do so and they do not lack the relevant capacity.[6] But in God's case, to say that God exists and has the considered intention to cause an event E at time t, will always amount to a full explanation of E at t, or so it seems. For by definition God is not subject to temptations that divert him from realizing his intentions, and, being omnipotent, he will be able to execute what he intends to do in all cases. In other words, God can always do what he intends, and never suffers from *akrasia* or weakness of will.

Full explanations may be synchronic or diachronic, in the sense that they refer either to causes that are operative at the time the resulting event occurs, or to causes that precede this event, respectively. An example of the former would be to say that a vase does not fall because the table on which it is placed sustains it. Clearly, there may be further synchronic causes that explain this same effect indirectly, since the table is sustained by the floor of the house, the floor is sustained by the walls, the walls are sustained by the foundations, and the foundations are sustained by the rock, sand, or mud on which the house has been built. Assuming that the series of contemporary sustaining causes is finite, one might define a *complete* synchronic explanation of E as a full explanation of E which mentions all direct or indirect contemporary causes of E, and all reasons for which these causes are operative, so that there is no further synchronic explanation (either full or partial) of their existence or operation in terms of factors operative at the time of their existence or operation.[7] Analogously, one might spell out the notion of a diachronically complete explanation.

An *ultimate* explanation may now be defined as an explanation that is both synchronically and diachronically complete.[8] In the case of such an ultimate explanation, at least one diachronic cause or set of causes must be referred to that fully explains all subsequent causes in the series, either directly or indirectly, whereas there is no further explanation of *its* existence or operation. In other words, it is a necessary condition for an explanation being an ultimate one that it refers to first diachronic causes and ultimate reasons, for which there is no further explanation. That these first causes exist and these ultimate reasons are operative is in that case an ultimate brute fact. As I argued in Chapter 10.6, we can have an ultimate causal explanation of an event E only if the series of diachronic causes backwards in time is finite. If the causal sequence is infinite, there clearly can be no ultimate explanation, since in that case each cause must be explained by earlier causes.

---

[6] In other words, it is not true that in the case of basic actions, 'E is fully explained when we have cited the agent P, his intention J that E occur, and his basic powers X, which include the power to bring about E; for given all three, E cannot but occur' (Swinburne (EG), p. 36).

[7] The latter part of this definition is quoted from Swinburne (EG), p. 78. By modifying Swinburne's definition, I am making explicit what he assumes, namely that the set of synchronic causes mentioned in a complete explanation is finite.

[8] On (EG), pp. 78–9, Swinburne fails to make explicit the distinction between synchronically and diachronically complete explanations (cf., however, p. 77 jo. note 3 on that page). In fact, he defines a complete explanation in the way I have defined a synchronically complete explanation, and then claims that an ultimate explanation is 'a special kind of complete explanation' (pp. 78, 79). But he intends an ultimate explanation to be diachronically complete as well, for he says that we have an ultimate explanation if it is synchronically complete and we can 'also state the factors that originally brought C and R about, and which factors originally brought those factors about, and so on until we reach factors for the existence and operation of which there is no explanation' (p. 79).

The notion of an ultimate explanation is mostly used by theists in a derivative sense, as referring to a first cause and ultimate reasons only, instead of as referring to the complete set of synchronic and diachronic causes and reasons. Richard Swinburne says, for example, that if God does not exist and the universe began with a Big Bang in state X at time t_0, and ultimate deterministic laws L governed the further development of the universe, according to which state X produced state Y at t_1, and state Y produced state Z at t_2, which brought about E at t_3, then (X and L) would be an ultimate explanation of E.[9] This is unproblematic if the universe is indeed deterministic, since in that case the description of all later causes can be derived, in principle, from the description of the first causes in conjunction with the laws of nature.

In what follows, I shall use the notion of an ultimate explanation in this derivative sense, in which it refers merely to a first full cause of a series of causes resulting in E, and to ultimate reasons. We may now formulate as follows the theistic contention with which we started this section. According to theists, the existence of God with his creative intentions is a more natural stopping point for the explanation of an E belonging to the universe than a hot Big Bang, for example, because theism is a better candidate for an ultimate explanation than current Big Bang theories. In other words, if we must choose between positing God's creation and positing a hot Big Bang as the ultimate explanation of the presently existing universe, theists claim that we should opt for the former and not for the latter. The proper terminus of explanation is God's free act of creation, since of logical necessity it is an ultimate brute fact (if it is a fact), that is, a fact 'that explains other things, but itself has no explanation'.[10] As Swinburne stresses, 'if theism is true, then, of logical necessity, God's action provides a complete and ultimate explanation of what it explains [. . .]. If God features at all in explanation of the world, then explanation clearly ends with God'.[11]

11.2 THEISM AS AN ULTIMATE EXPLANATION

Let us now wonder whether theism really can be an ultimate explanation in the sense defined by Swinburne of an E belonging to the universe, such as the existence of humanity on Earth. As I said, this sense is a robust one, because Swinburne defines ultimate explanations in terms of complete explanations, whereas he defines complete explanations in terms of full explanations. One might think, however, that theism as such cannot be an ultimate explanation of some E in this robust sense, because it cannot be a full explanation.

One trivial reason for this contention is that, according to Swinburne, the theistic hypothesis merely contains the thesis that God exists.[12] Hence, if one symbolizes theism as h, and supposes that the background knowledge k merely contains the a priori knowledge of logic and mathematics, the probability $P(e|h\&k)$ that the event E described by e occurs given theism will depend on how probable it is that God had the

---

[9] Cf. Swinburne (EG), p. 79.
[10] Cf. Swinburne (EG), p. 111.
[11] Swinburne (EG), p. 108.
[12] Cf. Swinburne (EG), p. 7: 'The claim that there is a God is called theism'; and p. 72: 'Note that I take h [= the hypothesis of theism] simply as "there is a God"'.

intention to cause E.[13] If this probability can be determined non-arbitrarily (cf. Chapter 9), but is not 1, the theistic explanation of E as described by e will be merely probabilistic, so that it is not a full explanation in the sense defined. Hence, it will not be an ultimate explanation either, if at least ultimate explanations form a subset of full explanations. For example, according to Swinburne the probability that God had the intention to create humans can be given the 'artificially precise value of ½', so that the existence of God does not necessitate the existence of humans.[14]

One might remedy this first flaw by conjoining the hypothesis of theism h with the statement i that God had the intention to create humans. It will now seem that, because God is omnipotent and never fails to act on his intentions, $P(e|h\&i\&k) = 1$, so that theism in conjunction with i amounts to a full explanation of E.[15] But is this really the case? That depends, since there is a second reason for denying that a theistic explanation is a full and ultimate one, which derives from the conditions for successfully avoiding the risk of God-of-the-gaps. As I argued in Chapter 10, theists should eschew the ambition to explain directly by a hypothesis of special creation specific phenomena such as the existence of humans as a biological species. By so doing, they would run the risk that a rival and scientific theory provides a much better explanation of these same phenomena. As we saw, sophisticated theists will want to avoid these risks.

Accordingly, Richard Swinburne says that he has 'no wish to challenge Darwin's account' of how the human species evolved.[16] It follows that if theists aim at providing an ultimate explanation of the existence of the human species (= E) by positing both God and a divine intention to create humans, this explanation has to be excessively indirect. The theist would have to say, for example, that God created the first initial conditions of the universe at the moment of the Big Bang, some 13.7 billion years ago, which ultimately produced the hominid lineage between 7 and 5 million years ago, and the first anatomically modern humans about two hundred thousand years ago, in accordance with laws of nature.[17] It follows that, if God created humans, doing so was not a basic action of God, but a mediated action, one that an agent does by doing something else. In this case, the intended result would be an exceedingly distant effect of God's basic action.

With regard to mediated actions, determining the value of $P(e|h\&k)$ is more complicated than in the case of basic actions. We not only have to determine the probability that the agent performed the basic action, but we also must multiply this probability with the probability that the basic action produces the intended result. In order to calculate $P(e|h\&k)$, where h is the hypothesis that God exists, and e asserts the existence of the human species, we have to multiply the stipulated probability of ½ that God had the intention to create humans with countless other probabilities,

---

[13] Swinburne (EG), p. 72.
[14] Swinburne (EG), pp. 123 and 131.
[15] Cf. Swinburne (EG), p. 72: 'Note that I take h simply as "there is a God". By itself it provides merely a partial explanation of e. It needs to be conjoined with an intention to bring about e in order to provide a full explanation of e. The value of $P(e|h\&k)$ will, for the various e, depend on how probable it is that God will have that intention'.
[16] Swinburne (EG), p. 171.
[17] Given the (often) gradual nature of evolution, it will be quite arbitrary which individual is called 'the first anatomically modern human'.

such as the probability that God intended to create humans very indirectly by first creating a Big Bang; the probability that given the postulated hot Big Bang as produced directly by God, there will be stars with planets; the probability that some stars will have planets apt for life; the probability that, given planet Earth (or some other planet) in its state of some 4.5 billion years ago, there will be life on Earth (or on some other planet); and the probability that, given the origin of life on Earth, its evolution will result eventually in the species of homo sapiens.[18]

Nobody has ever reliably estimated all these probabilities, and it is dubious whether this can be done by a finite intellect. But by multiplying them, the probability $P(e|h\&k)$ might become quite small, unless the universe is completely deterministic.[19] In other words, what we might encounter here is a problem of dwindling probabilities, if at least we suppose that there is only one pathway leading from the first initial conditions to us, so that the predictive power of theism with regard to the existence of Homo sapiens might be negligible. It would follow not only that unless the universe is deterministic, the theistic explanation of the existence of man cannot be an ultimate explanation (if ultimate explanations are defined à la Swinburne as a subset of full explanations), but also that its predictive power might be very weak. However this may be, we should conclude that we simply do not know what the predictive power of theism is concerning the existence of human beings, because nobody has calculated all the relevant probabilities on the basis of sufficient empirical evidence.

It is interesting to note that many of the probabilities involved are a topic of animated debates between scientists. For example, the late Stephen Jay Gould invited us to consider the following evolutionary thought experiment. Rewind the tape of life to the moment at which the first animals evolved, some 540 million years ago. Then wonder how the story of life would again unfurl. According to Gould, replaying life's tape starting from the same initial conditions might result in radically different macro-evolutionary outcomes, that is, in forms of life that bear little resemblance to the complex forms that now exist, such as the human species. Gould concluded that 'almost every interesting event of life's history' is a matter of historical contingency.[20] Other biologists have contested Gould's contingency thesis, stressing the importance of convergent evolution. They hold that over immense spans of geological time, evolution will tend to produce similar forms of life.[21]

Of course we cannot decide this scientific debate here. My point is, however, that if Gould is correct, the claim that God had the intention i to create human life (= E) on

---

[18] The first astrophysicist to propose a formula for computing the probability of (extraterrestrial, detectable) technologically advanced life in our galaxy was Frank Drake. For this so-called Drake equation, cf. Drake and Sobel (1992). Here, we would need a different formula, on the basis of which it can be computed what the probability is that a 'human species' develops in the universe at some time given the initial conditions of the Big Bang.

[19] Even if $P(e_{i+1} | e_i)$ is 0.99 for each i, it will still be the case that $P(e_{1,000} | e_1)$ is minuscule, if at least there is only one pathway from e_1 to $e_{1,000}$. Supposedly, there are trillions of pathways, however, because in principle human life might evolve on many of the very huge number of planets in the universe.

[20] Gould (1989), pp. 290–1. Here, Gould clearly assumes that there is only one pathway that leads from specific initial conditions to a specific outcome. Cf. Sober (2008), p. 363.

[21] Cf., for example, Morris (2003). Cf. Sober (2008), p. 363, on the different probability models used by Gould and Morris.

Earth (or elsewhere) indirectly by way of creating the initial conditions of the Big Bang, may seem not to have much predictive power with regard to E.[22] According to that scenario, God's direct action of creating these initial conditions did not make the evolution of the human species very probable. Theists will object that even if the evolution of the universe and the course of macroevolution were radically contingent, so that, given the very same initial conditions, very different forms of life might result after billions of years, God could foresee that the evolution of life on Earth (or on some other planet) would contingently yield the human species, since God is omniscient.

But it seems to me that this objection is an *ignoratio elenchi*. The question is not whether God in his omniscience could foresee the outcomes of a radically contingent macroevolution, but rather whether God's direct action provides a full explanation of a phenomenon E. And I argued that if Gould is correct, this is not the case with regard to the existence of humans on Earth, assuming at least that one has no wish to challenge a Darwinian account of how the human species evolved.[23] Consequently, theism cannot figure as an ultimate explanation of the existence of the human species in the robust sense of 'ultimate explanation', if at least Gould's view of evolution is the right one.

How can theists avoid the conclusion that unless one assumes cosmic determinism, even theism is not an ultimate explanation of the existence of man, for example, because it is not a full explanation? One course might be to say that although diachronically God created man indirectly by the mediated action of causing the Big Bang, he also creates each human being directly and synchronically by a basic action, that is, by 'sustaining' each human being at every moment of his or her existence. In this context, the verb 'to sustain' must have a different meaning from the one spelled out above, according to which I am sustained directly by the chair on which I am sitting and not by God. However, as I argued briefly in Chapter 10.6, and shall further argue in Chapter 12.5, such a doctrine of double sustaining causality, if meaningful at all, should be avoided by theists who intend to interpret theism as a testable empirical theory.

Consequently, theists cannot but retreat to a second line of defence. If they want to persist in claiming that theism provides an ultimate explanation of things, they have to weaken the definition of this notion. They should say that even a probabilistic and partial explanation could be an ultimate one, on condition that it mentions a first diachronic cause and ultimate reasons. Can theism provide an ultimate explanation in this weak sense of an E belonging to the universe?

With regard to this question theists are confronted by a third set of difficulties, which again involve the risk of God-of-the-gaps (cf. Chapter 10.6, above). For

---

[22] Cf. for a recent defence of Gould's view against critics: Powell (2009). Of course the theist does not need to claim that God intended to create humans on Earth, in particular, and Swinburne does not claim this. Any planet hospitable to life might do. This will raise the probability that, given the initial conditions of the universe at the time of the Big Bang, human-like life will evolve *somewhere* in the universe. However, this probability has not been calculated on the basis of sufficient empirical evidence, so that we do not know what the predictive power of theism is concerning this issue.

[23] Cf. my earlier quote from Swinburne (EG), p. 171. Of course, theists might claim that God influenced the contingent course of evolution at crucial points by direct actions. By doing so, however, they would challenge a Darwinian account, and indeed Gould's account, of how the human species evolved.

example, theists have to assume that the universe does not stretch backwards in time indefinitely, since in that case there can be no first state of the universe and, hence, no ultimate explanation by theism of any E, unless E is a miracle.[24] In this respect, Pope Pius XII may have had a point in claiming that Big Bang theory opened up new perspectives for the hypothesis of a divine creation *ex nihilo*.[25] However, alternatives to hot Big Bang cosmology are proposed by present-day cosmologists, such as theories of a bouncing universe. According to these theories, the Big Bang state of the universe was caused by earlier states, so that there is no room for a creative action of God as the diachronic cause of the Big Bang. Consequently, theists who want to avoid the risk of God-of-the-gaps should not claim that theism is an ultimate explanation of anything in the universe, or indeed of the existence of the universe itself.[26]

However, let us suppose that theists are somehow able to solve these difficulties. Why should we then assume that theism is an ultimate explanation in the weak sense that it postulates God's action as a first diachronic cause of a series of causes that ultimately yields an E? Can one not ask what caused God's existence, or which reasons motivated his creative acts, so that even a theistic explanation is not necessarily ultimate? Swinburne claims that theism is an ultimate explanation for both ontological and conceptual reasons. Since a theistic explanation of E posits (a) God's existence and assumes that (b) God probably formed the intention to create E, the explanation can be an ultimate one (in the robust or the weak sense) only if in principle there is no further explanation for each of these two alleged facts (a, b). Can this be argued successfully?

Swinburne advances two strong arguments to the effect that explaining E by positing (a) God's existence necessarily is an ultimate explanation. One of the defining characteristics of God is that, if he exists in fact, he exists eternally.[27] Consequently, the hypothesis that God's existence is itself produced by a diachronic cause C, which is not God, can be refuted by a *reductio ad absurdum*. For suppose that C causes God to exist at time t. Then, by definition, God exists at all times, also at the moments preceding t. That would mean, however, that a cause operative at t could have effects before t. But this is excluded if at least it is conceptually impossible that a cause brings about an effect which precedes it.[28] Hence, God's existence cannot be caused by something that is not God.

Since theists pretend that the notion of a synchronic sustaining cause makes sense in all contexts, one might object (*ad hominem*) that this notion may be applied to God as well. Can we not suppose that if God exists, at each moment of his eternal existence he is sustained by synchronic sustaining causes, so that an explanation by positing God would not be an ultimate one? But this hypothesis is ruled out by another defining

---

[24] Could the theist not require an explanation for the existence of the infinite series of time slices of the universe, that is, for this infinite series 'as a whole'? I shall investigate this question in Chapter 12.4.
[25] Cf. Chapter 10.6, above and Chapter 12, below.
[26] Cf. Chapter 12.6–9.
[27] As Swinburne argues in (CT), Chapter 12, God's 'eternal' existence should be interpreted in a temporal sense in order to avoid contradictions with many other things theists want to say about God.
[28] Swinburne (CT), p. 265, and Chapters 8.1 and 10.6, above. According to a number of philosophers, backward causation is conceptually possible. If so, Swinburne's argument is inconclusive. I am not among these philosophers.

characteristic of God. If God by definition created everything apart from God, if at least there is anything apart from God, he cannot be sustained by something that is not God, assuming that synchronic causation in a circle is conceptually impossible.[29] Alternatively, if a synchronic cause C sustains God at t_1, God must have created C at an earlier moment t_0, so that the explanation by God is ultimate, after all.

These arguments are convincing as far as they go.[30] But what should we say about God having the intention i to produce E at time t? Is this element (b) of the theistic explanation of E also an ultimate one? According to Swinburne, there can be no further explanation of God forming a specific intention, except by God having some wider intention from which it is derived. As he says, 'God's widest intention has no explanation except that he chose this intention'.[31] Furthermore, God's choice of his widest intention 'has no causal explanation at all, since he is perfectly free'.[32] At this point, however, theists are landed in a dilemma, which might be called the dilemma of predictive power.

Either God is perfectly free in a radically voluntarist sense. In other words, God selects his widest intentions without any reasons, as Descartes once held. Then there is no further explanation of the alleged fact that God chooses to have specific intentions, but theism will lack predictive power, as I argued in Chapter 9. Or God forms his intentions on the basis of reasons. According to Swinburne, for example, these reasons are provided by God's omniscience of objective ethical truths. In this latter case, however, the alleged fact (b) that God had an intention to produce E is not an ultimate explanation, since it is the realm of objective ethical truths that explains God's choice. Hence, either theism has no explanatory power, or it cannot provide an ultimate explanation.[33]

11.3 EVIDENTIALISM

Whereas theists may yearn for an ultimate explanation, most scientists and secular scholars will happily welcome the verdict that no such explanation can be had. So let us leave behind the issue of ultimate explanation and turn to some conceptual preliminaries concerning the second, and indeed crucial, question that we have to raise with regard to theism: is theism probably true, given the total available and permissible empirical evidence for and against it? In order to prepare my attempt to answer this question in the next four chapters, I shall introduce briefly some key

---

[29] Swinburne, (EG), pp. 98–9.
[30] Strictly speaking, a theistic explanation of E by a direct action at t merely needs to posit God's existence at t or shortly before t. If one assumes that God's existence at t needs to be explained by God's existence at an earlier moment and his intention to live on, and that this holds for each moment of God's existence, no particular theistic explanation is an ultimate one. Cf. Swinburne, (EG), p. 80. Swinburne attempts to refute such an objection by arguing that God's defining properties, such as his being eternal, have *de re* necessity (CT, Chapter 14; EG, p. 96). I criticized his use of this Kripkean notion in Chapter 8.4, above.
[31] Swinburne (EG), p. 98.
[32] Swinburne (EG), p. 80, cf. p. 49.
[33] Swinburne might reply that an explanation by reasons is not a causal explanation, cf. (EG), pp. 7, 49, 98, 105. But an ultimate explanation should not only mention ultimate causes but also ultimate reasons.

concepts and, in the next section, discuss two proposals for the logical analysis of probabilistic arguments in which evidence is adduced for or against a hypothesis.

As many philosophers have justly (apart from gender-preconceptions) said, 'a wise man proportions his belief to the evidence'.[34] We might call this precept the *evidentialist maxim* and these philosophers may be said to endorse *evidentialism*. As stated, the maxim is in need of some interpretation, however, because the term 'belief' is ambiguous between someone's believ*ing* that *p* and the propositional content or proposition that *p*. Which meaning of the word 'belief' is intended?

Clearly, the maxim applies to beliefs in the first sense, because only they can be 'proportioned', and it says that one should proportion the strength of one's believ*ing* or conviction that *p* to the evidence. But if there is a logic of probability, according to which a specific piece of evidence *e* affects the logical or inductive probability that the belief that *p* is true, the term 'belief' as used in this logic refers to the propositional content that *p*, and the probability of a belief is the logical or inductive probability $P(p | e)$ that this proposition is true given evidence *e*.[35] Let us mark these two senses of 'belief' by labelling the first 'belief$_s$'(for a subject believ*ing* that *p*) and the second 'belief$_p$' (for *what* is believ*ed*, the propositional content of the belief, to wit, that *p*). Using this terminology, we should reformulate the evidentialist maxim as follows. Wise persons proportion the strength of their belief$_s$ to the logical or inductive probability conferred by the evidence upon the belief$_p$.

As probabilities may take all real-number values between 0 and 1 (inclusive), 'probability of belief$_p$' is a gradual notion. The evidentialist maxim, as reformulated, implies that belief$_s$ should be taken as a gradual or analog notion as well: we believe that *p* with a *degree of conviction d*. The maxim then says that if *d* also takes values between 0 and 1, the value of *d* should in each case be identical to the value of $P(p | e)$, where *e* is the available evidence. However, the notion of belief$_s$ is interpreted quite often as a dichotomous one: we either believe$_s$ that *p* or we don't. And if we don't believe$_s$ that *p*, this may be either because we believe$_s$ that not-*p*, or because we suspend judgement as to whether it is true or false that *p*. An evidentialist who wants to use the dichotomous or 'all-or-nothing' notion of belief$_s$ should fix a threshold value of $P(p | e)$ or, equivalently, of *d*, above which the belief$_s$ that *p* is justified by the evidence, so that it is rational to believe that *p*, and typically this threshold will be context dependent.[36]

If we want to act on our beliefs by building a motorway bridge, for example, we will choose a very high threshold value for the belief being justified that a specific construction will not collapse under the weight of the traffic, since we don't want to take any risks in this case. Some defenders of the evidentialist maxim have even

---

[34] Cf. Hume (1748), section X, Part I, §87.

[35] One should clearly distinguish the logical or inductive probability that a belief is true given specific evidence both from physical probability, that is the extent to which a particular event is predetermined by its causes, and from statistical probability, which is a proportion of the members of an actual or hypothetical set that have a certain property. The logical probability that a set of propositions (*q, r, s,*...) makes another proposition (that *p*) true, which does not belong to this set, may be called inductive probability. For a discussion of these concepts, see, for example, Swinburne (EJ), Chapter 3; Mahler (2006); and Howson and Urbach (1993), *passim*.

[36] Determining such threshold values for rational credibility may be complicated, among other reasons because one should avoid Henry Kyburg's (1961) Lottery Paradox and David Makinson's (1965) Preface Paradox. I shall not discuss these complexities here, but one can resolve them by taking the context into account.

claimed, implausibly, that in all cases the dichotomous belief₅ that *p* is justified or rational only if the evidence entails the truth of the proposition that *p*, so that the threshold value should always be $P(p|e) = d = 1$.³⁷ But most philosophers endorse a weaker condition for rational credibility in ordinary contexts. They hold that the dichotomous belief₅ that *p* is justified or rational if and only if given the evidence it is highly probable that *p*.³⁸

Richard Swinburne seems to think, however, that the threshold value for religious belief being justified or reasonable should be low, that is, just above ½. For in *The Existence of God* he defines a 'correct P-inductive argument' as 'an argument in which the premises make the conclusion probable', and stipulates that the premises make the conclusion *h* probable if and only if $P(h|e\&k) > ½$.³⁹ Clearly, this latter use of the term 'probable' is not gradualist. It implies that the threshold for legitimately believing that God exists is merely that it should be more probable (in the usual, gradualist sense) that God exists than that he does not exist. Indeed, this is what Swinburne attempts to show in his book *The Existence of God*. Of course we will be interested to hear why the theist is satisfied by such a low threshold value for a justified dichotomous belief that God exists.⁴⁰

11.4 BAYESIANISM AND LIKELIHOODISM

Let us now enquire whether there is a logic of probability in accordance with which we might confer a specific degree of logical or inductive probability on the hypothesis that God exists, and, if so, how we should formalize this logic. It is instructive to begin with an example of another hypothesis, concerning which the logic of probability is relatively unproblematic. If we want to test the hypothesis *h* that a person, call her Jane, has a specific type of cancer, say breast cancer, and the validating research for the cancer-test procedure *T* has shown that if *h* is true, there is a probability of 0.99 that the test has a positive result *e*, the predictive power $P(e|h\&k)$ has the value of 0.99. Using R. A. Fisher's technical terminology, which is common among statisticians, I shall call this probability $P(e|h\&k)$ of *e* given the hypothesis *h* (and background knowledge *k*, such as that test *T* has been performed) the *likelihood* of *h* with regard

---

³⁷ Adler (1999). As Adler says, he defends 'evidentialism [...] for full – rather than partial or degrees of – belief: one ought to believe that *p* only if one's evidence or reasons adequately support the truth of *p*' (p. 267). He holds that support is only adequate if 'it is not possible for a belief to be justified but false' (p. 275). To the extent that rational credibility can be equated with warranted assertability, Williamson (2000) is arguing for a similar view in his Chapter 11.
³⁸ Cf. Douven (2006), p. 457.
³⁹ Swinburne (EG), pp. 6, 13, 17. Should we really be prepared to act on a hypothesis if its truth merely is more probable than its falsity given our total evidence, as Swinburne says on p. 13? This depends on the context, as I argued above.
⁴⁰ One should distinguish threshold values for dichotomous *justified* belief from threshold values for dichotomous *belief*. In (EJ), Swinburne defends the threshold value of 1/2 for all cases of dichotomous *belief*, arguing that any higher threshold value 'would be extremely arbitrary' (p. 35). However, this would be the case only if one abstracts from the contexts in which the notion of dichotomous belief is used. As I said, the threshold value for dichotomous *justified* belief is context dependent, and this may hold for dichotomous *belief* as well. Theists might justify a low threshold value for religious belief by arguing, for example, that this low threshold leaves room for the free choice to have faith. Cf. Swinburne (FRb), Chapters 4–7.

to e.[41] Suppose that the test yields result e. Can we then calculate the logical or inductive probability that h is true given the evidence e? That is, can we determine the posterior probability $P(e|h\&k)$?

The following considerations show that in order to do so, we need two other probabilities. First, it might be the case that test T always has a chance of 0.99 to yield outcome e, whether the tested person has cancer or not, so that $P(e|k)$ is 0.99 as well. In that case, the test clearly is irrelevant for the hypothesis. As we have seen in Chapter 9.1, the outcome e is relevant as a confirmation of h if and only if $P(e|h\&k) > P(e|k)$. We will also say that the more the first term of this inequality exceeds the second, the more reliable is test T (where k includes the proposition that T has been performed). In other words, if and only if the 'explanatory' power

$$\frac{P(e|h\&k)}{P(e|k)}$$

is greater than 1, the occurrence of e will provide a correct confirming argument or *C-inductive argument* for the hypothesis h, whereas the confirmation for h provided by e will be stronger to the extent that this explanatory power of h with regard to e is greater.[42] Equivalently, the outcome e of test T can only confirm hypothesis h if $P(e|h\&k) > P(e|\sim h\&k)$, that is, if the outcome e of test T is more likely to occur if h is true than if h is false. Hence, we will not be able to compute the posterior probability $P(h|e\&k)$ unless we are able to determine not only the likelihood $P(e|h\&k)$ but also the likelihood $P(e|\sim h\&k)$.

It is clear, however, that by comparing these two likelihoods, the occurrence of e will merely yield a comparative confirmation for the hypothesis h. Given the occurrence of e, it will be more probable that h is true than that $\sim h$ is true to the extent that $P(e|h\&k) > P(e|\sim h\&k)$. We might express this result by the following principle, which Ian Hacking baptized the *law of likelihood*:

Law of likelihood:
The observations e favour hypothesis h_1 over hypothesis h_2 if and only if $P(e|h_1) > P(e|h_2)$, whereas the degree to which e favours h_1 over h_2 is identical to the likelihood ratio $P(e|h_1)/P(e|h_2)$.[43]

Accordingly, if only the two likelihoods $P(e|h_1)$ and $P(e|h_2)$ are available, the evidence e cannot do more than provide a *differential* support to h_1 as compared to h_2. It cannot confer a determinate probability on h_1, since such a differential support of 'favouring' is essentially contrastive.

It follows that, if we want e to confer a determinate degree of probability on h_1, we should put into our calculations yet another probability, the probability $P(h_1|k)$ of h_1 prior to the evidence e, where k stands for background knowledge. If this prior

---

[41] As is stressed by many authors, Fisher's use of the term 'likelihood' is misleading, since in ordinary language 'likelihood' and 'probability' are synonyms, whereas $P(e|h\&k)$, called 'the likelihood of h', expresses a probability of e given h and k, and not a probability of h.
[42] Cf. Chapter 9.1 for this technical notion of explanatory power. Of course there are many different measures of confirmation, which I shall not discuss here.
[43] Cf. Hacking (1965), quoted by Sober (2008), p. 32. Cf. Sober (2008), pp. 32-3, for what follows.

probability is available, we are able to calculate a determinate posterior probability $P(h_1|e \& k)$ from the three probabilities we have now described by using the theorem of Bayes:

$$\text{Bayes' Theorem}: P(h|e \& k) = \frac{P(e|h \& k) \cdot P(h|k)}{P(e|k)}$$

In our medical example, we may determine a prior logical probability $P(h|k)$ of the hypothesis that Jane has breast cancer on the basis of statistical investigations using random trials. Suppose, for example, that Jane is fifty years old, and that 20% of her age group in a specific relevant population has breast cancer. If this statistical probability provides sufficient evidence for fixing the prior logical probability of the hypothesis that Jane has breast cancer at 0.2, we can calculate the probability that she has breast cancer given the outcome of the test by multiplying the prior probability of 0.2 with the likelihood $P(e|h \& k)$ of 0.99 divided by $P(e|k)$.[44]

In doing so, we use Bayes' theorem, which can be proven on the basis of the axioms of probability theory, as a rule for updating probabilities.[45] To the question when we can legitimately use the theorem in this manner, or what such a use shows, two answers can be given. *Subjectivists* argue that Bayes' theorem is nothing but a rule of logic, which shows how various probabilities are related. What our application of the rule demonstrates is merely which value we should assign to the posterior probability $P(h|e \& k)$ *if* we put in specific values for the likelihood $P(e|h \& k)$ and for the prior probabilities $P(h|k)$ and $P(e|k)$. Subjectivists argue in particular that there are no or few constraints on the values we assign to prior probabilities, apart from the axioms of probability theory. Objectivist Bayesians are more ambitious. They want to affirm on good objective grounds that the posterior probability $P(h|e \& k)$ has a specific value, or lies within a specific interval. But this means that they have to provide good objective grounds for their probability assessments of all the inputs of the formula. Whereas the objectivist Bayesian holds that this can always be done, the subjectivist Bayesian denies it for many prior probabilities.

In our medical example, we have assumed that such good objective grounds can be provided. Validation research on a test procedure and statistical research on the occurrence of breast cancer in specific populations allowed us to assign specific values to the prior logical probabilities $P(h|k)$ and $P(e|k)$, and to the likelihood $P(e|h \& k)$, where our background knowledge k in $P(e|h \& k)$ and $P(e|k)$ included the fact that the test has been performed. But what should we say concerning the hypothesis that God exists? Can we adduce good objective grounds for the prior probabilities of this hypothesis and of the evidence apart from the hypothesis, and for the likelihoods involved? In other words, can we determine the inputs of Bayes' theorem with sufficient objectivity for being justified in stating that given the evidence e there is a specific probability that God exists?

In Chapter 9, above, I have argued that we cannot do so with regard to the likelihood or predictive power $P(e|h \& k)$ of theism. But in the present chapter,

---

[44] Here we use as a rule of inference that '[a]ny statistical probability (...), taken by itself, gives a corresponding logical probability to any instance'. Cf. Swinburne (EJ), p. 78–9.
[45] Cf. Howson and Urbach (1993), pp. 28–9, for a proof and various versions of Bayes' theorem.

I have assumed for the sake of argument that theists can determine non-arbitrarily (although perhaps roughly) this likelihood with regard to at least *some* items of evidence they want to adduce. Let us now investigate whether theists are able to provide sufficiently objective grounds for being justified in assigning specific values, however rough or imprecise, to the prior probabilities $P(h|k)$ of theism and $P(e|k)$ of the evidence. With regard to each of these two terms, theists have to face a set of serious problems, which I shall discuss in the next section.

11.5 TWO PRIOR PROBABILITIES

In the medical example of the hypothesis h that Jane has breast cancer, we relied on frequency data in order to determine the prior probability $P(h|k)$. In the case of the theistic hypothesis, however, it seems that no such background evidence is available, or, if it is, the background evidence seems to make the truth of theism extremely improbable (see section 11.6). We might formulate this problem as a dilemma for theists who want to confer non-arbitrarily some prior logical probability $P(h|k)$ on the hypothesis h that God exists. Either there is no relevant evidence for doing so with sufficient objectivity, or there is relevant evidence. In the first case, we will not be able to calculate a determinate probability of $P(h|e\&k)$ with sufficient objectivity either. In the second case we are able to do this, but we would need a very large amount of positive evidence for theism in order to swamp its low prior probability. Let me explain.

The theorem of Bayes is often used for calculating the posterior logical or inductive probability of specific medical hypotheses like the one concerning Jane. But can the theorem also be used to calculate the posterior probability of global scientific theories, such as the theory of general relativity, for example? Many philosophers of science have argued that it cannot, because we have no good objective grounds for fixing a determinate value of their prior probability $P(h|k)$. In the case of general relativity theory, we have no relevant frequency data. Neither do we possess theoretical background knowledge on the basis of which we might be justified in assigning a specific value to the prior probability of general relativity theory. So-called *likelihoodists* in the philosophy of science conclude that concerning general scientific theories, the only possible type of probabilistic justification by evidence is the comparative favouring of one hypothesis over another by applying the law of likelihood.

For example, when Arthur Eddington tested general relativity theory during his expedition to Principe Island in 1919 by measuring the bend in starlight during a solar eclipse, he was in fact comparing the likelihoods $P(e|\text{general relativity}\&k)$ and $P(e|\text{Newtonian mechanics}\&k)$. The result was that the observed bend of light e strongly favoured general relativity theory over classical mechanics, because the first likelihood is much greater than the second.[46] But, as stated in the last section, such a differential confirmation is essentially contrastive, and it cannot yield a determinate

---

[46] In such cases of complex scientific observations, we should probabilify e as well and use the rule proposed by Jeffrey (1965), in which one has a degree of belief in e. For reasons of simplicity of exposition, I am using Bayesianism with strict conditionalization, which can be applied only if we have certainty concerning whether e occurred.

posterior probability for the best theory. It merely says that the three-place relation between e and the theories h_1 and h_2 is such that e favours h_1 over h_2 to the extent that the likelihood $P(e|h_1)$ exceeds the likelihood $P(e|h_2)$.[47]

If there are no objective grounds available for determining the prior probability of theism, it will seem that we must use the law of likelihood and not the theorem of Bayes whenever we want to test theism by adducing empirical evidence. As in the case of general relativity theory, such a likelihood comparison will merely allow us to conclude that given the sum of all empirical evidence e, theism is better (or less well) confirmed than some specified rival theory. But it might also be argued that the case of theism is different from the case of general relativity theory, because concerning theism we do have objective grounds for determining its prior probability. Indeed, at first sight there seems to be a large amount of empirical background knowledge on the basis of which we should estimate the prior probability of theism very near to zero.

Theism is the thesis that God exists, and God is defined as a bodiless spirit. However, both research in animal biology and advances in the neurosciences have shown ever more convincingly that the mental life of animals and of man depends on brain processes. Although perhaps we cannot always claim that our mental life has been *explained* by neural research, all empirical investigations suggest that mental phenomena cannot *exist* without neural substrata. In the light of this growing reservoir of empirical background knowledge, the theistic hypothesis becomes ever more implausible, since it presupposes that mental life can exist without a corresponding neural substratum.[48] This is one reason, I suppose, why Richard Swinburne attempts to argue in *The Existence of God* that no background knowledge whatsoever can be relevant for determining the prior probability of theism. I shall examine his argument to this effect in the next section (11.6). For the moment, it seems safe to conclude with regard to the prior probability of theism $P(h|k)$ that either it cannot be determined non-arbitrarily, or that it is very low. In the former case, we will not be able to apply the theorem of Bayes in order to calculate the posterior probability of theism with sufficient objectivity. In the latter case we can, but the resulting posterior probability risks being very low as well.

Let us now focus on the second set of problems, which arise with regard to the prior probability of the evidence $P(e|k)$. By the law of total probability, $P(e|k)$ can be spelled out as follows:

$$P(e|k) = P(e|h\&k).P(h|k) + P(e|\sim h\&k).P(\sim h|k).$$

So we see that in order to calculate the prior probability $P(e|k)$ of the evidence e, we not only have to know the prior probability $P(h|k)$ of the hypothesis h, but also the prior probability $P(\sim h|k)$ and the likelihood $P(e|\sim h\&k)$ of its negation $\sim h$. But how can we determine these latter probabilities? That depends on what '$\sim h$' stands for.

In the case of our hypothesis that Jane has breast cancer, $\sim h$ is the hypothesis that Jane does not have breast cancer. If in this case the prior probability $P(h|k)$ can be

---

[47] These considerations are inspired by Sober (2008), pp. 24–37.
[48] I criticized Swinburne's arguments for substance dualism in Chapter 7.5–6, above. For further criticisms, cf. Chapter 14.1–4.

estimated on the basis of frequency data concerning the occurrence of breast cancer in specific populations, we can do so as well for the prior probability $P(\sim h|k)$, or simply derive it from the first probability.[49] Furthermore, frequency data obtained by our validation tests will enable us to determine the likelihood $P(e|\sim h \& k)$. But what does '$\sim h$' stand for if we want to apply Bayes' theorem as a logic of confirmation concerning global scientific theories, such as general relativity or the neo-Darwinist synthesis? By negating such a theory, we do not obtain another theory that is equally specific. On the contrary, in such cases $\sim h$ is what is called a 'catch-all hypothesis', since in principle there may be infinitely many theories that are incompatible with general relativity or neo-Darwinism, and which purport to explain the same evidence. The prior probability of e then is a function of the average likelihoods and prior probabilities of all these alternatives to h.[50] But since most of these alternatives to h have not been formulated, whereas it is very difficult to invent such an alternative theory, and also because of the first problem I discussed, nobody can know what their individual likelihoods and priors are. Hence, we cannot compute $P(e|k)$ in such cases.[51]

This conclusion also applies to the hypothesis of theism. Whereas theism is the determinate hypothesis that God exists, adduced in order to explain the facts described in e, its negation says that God as defined by theism does not exist. But this $\sim h$ is not a determinate explanatory hypothesis of e, the prior probability and likelihood of which we might estimate on the basis of frequency data or other background knowledge. Rather, it is a catch-all of infinitely many alternative hypotheses, incompatible with theism, which might be proposed in order to explain the same evidence. Hence, we have to conclude again that we cannot apply the theorem of Bayes in order to compute the posterior probability of theism $P(h|e \& k)$ with sufficient objectivity.

The same conclusion holds for the law of likelihood, if theists want to apply it in order to argue that the evidence e favours theism over all its rivals, so that it is more probable that God exists than that he does not. In applying the law of likelihood, we have to compare the likelihood $P(e|h_1)$ with the likelihood $P(e|h_2)$. But if h_2 is defined as the negation of h_1, and if this negation $\sim h_1$ is a catch-all hypothesis, as it is in the case of theism, we will not be able to determine the value of $P(e|h_2)$. So it seems that there is no logic of probabilistic arguments according to which we may determine the logical or inductive probability of theism *tout court*.

At best, we can compare the likelihood $P(e|h \& k)$ of theism with regard to some specific e with the likelihood of some rival hypothesis, and argue that e favours theism over this specific competitor. For example, if we follow Pope Pius XII in claiming that theism is a plausible explanation of the postulated hot Big Bang, we might argue that the evidence of the Big Bang favours theism over a specific rival explanation, such as a loop quantum gravity theory, which purports to explain the Big Bang state of the universe some 13.7 billion years ago in terms of a bouncing universe. But of course,

---

[49] It follows from the axioms of probability theory that $P(\sim h|k) = 1 - P(h|k)$. Cf. Howson and Urbach (1993), p. 24, for a proof.
[50] As Swinburne says (EJ, p. 222), the prior probability of the evidence $P(e|k)$ arises 'from the probability of e on all the various hypotheses that might hold (whether or not in vogue), weighed by their prior probabilities – whether or not they were recognized at the time of the formulation of h'.
[51] Cf. Sober (2008), pp. 28–30.

theists who want to avoid the risk of God-of-the-gaps should never propose such an argument.

We may conclude that natural theologians who want to use some logic of probability in order to argue that, given the available evidence, the probability that theism is true is above ½, should start by solving two serious problems. First, they should somehow eliminate the background knowledge provided by animal biology and the cognitive sciences, in the light of which the hypothesis of a bodiless spirit is extremely implausible. Second, they should show that even in the absence of such background knowledge, we can determine the prior probability of theism with sufficient objectivity, and that this prior probability is higher than that of all conceivable rival hypotheses, which make the same evidence equally probable. More precisely, theists should attempt to argue either that a higher prior probability of theism outweighs conceivable greater likelihoods and empirical adequacies of rival hypotheses, or, conversely, that a greater likelihood and empirical adequacy of theism outweighs the possible greater prior probabilities of its competitors. In the remaining sections of this chapter, I shall discuss the solutions to these problems proposed by Richard Swinburne in *The Existence of God*.

11.6 THE ELIMINATION OF EMPIRICAL BACKGROUND KNOWLEDGE

The prior probability of a theory is by definition its logical probability before we take into account the evidence that we want to consider. As we saw, a subjectivist Bayesian will allow that each subject assigns some arbitrarily chosen value to the prior probability of a theory or hypothesis h to be tested. The theorem of Bayes is then used merely in order to make explicit how much the resulting posterior probability $P(h|e\&k)$ will differ from this arbitrarily chosen prior probability $P(h|k)$ when we take the evidence e into account. In this case, the output value of an application of the theorem may be as arbitrary as the prior probability of the hypothesis put into the theorem.[52] But theists who want to argue that their belief in the existence of God is justified should not be subjectivist Bayesians concerning this issue. They will hold that there are objective grounds for assigning, however roughly, a specific value or interval of values to theism's prior probability.

According to Richard Swinburne, for example, the prior logical probability of a theory or hypothesis depends on three factors, which he claims can be assessed with sufficient objectivity: its fit with background knowledge, its simplicity, and its scope.[53] If the scope of a theory can be specified without taking empirical background knowledge into account, one might say that both scope and simplicity are *intrinsic* features of a theory as such, which can be assessed entirely a priori. For this reason, Swinburne says that the *intrinsic probability* of a theory is determined by the degree of its simplicity and by its scope, and that this intrinsic probability is 'independent of its

---

[52] Cf. Earman (1992), Chapter 6, on the issue when this is the case.
[53] Cf. Swinburne (EG), p. 53; (EJ), pp. 80ff; (BT), Chapter 1.

relation to any evidence'.[54] Clearly, this is an idealization in many cases. For quite often, we can only determine the scope of a theory if we conjoin it with empirical background knowledge. Even worse, in many cases such relevant background knowledge is not available, so that we cannot determine the scope of the theory at all.[55]

However this may be, it seems that even if the intrinsic probability of theism were considerable, its prior probability would be very low, since theism fits in badly with well-established empirical background knowledge. As I said above, if meaningful at all, the very idea of a bodiless spirit is utterly implausible in view of our increasing knowledge of how mental functions depend on neural processes. Many experts think that mental properties and functions must be seen as emergent in the sense that they supervene on complex bodily properties and functions of the brain.[56] If everything mental is emergent in this sense, or depends for its existence on complex material structures and processes in some other way, a bodiless spirit simply cannot exist. This conclusion is confirmed by the fact that we have never been able to find convincing evidence for the existence of other bodiless spirits, such as poltergeists, incorporeal angels, demons, ghosts, human souls surviving bodily death, and so on, in spite of extensive research.[57] As it seems unlikely that the empirical evidence adduced by believers in support of God's existence will be sufficiently strong to overcome this extreme prior improbability, they seem to be well advised to abandon their case at once.

The only remaining hope for theists at this point is to argue that in the case of the hypothesis that God exists, empirical background knowledge does not count *at all* for determining its prior probability. In *The Existence of God*, Richard Swinburne develops the following argument to this effect. With regard to each hypothesis to be tested, we should decide where to locate the empirical data and other items of empirical knowledge, such as well-confirmed theories, that are relevant to its posterior probability. Do these empirical items belong to the evidence to be explained by the hypothesis h, that is, to e in the formula for the theory's likelihood or predictive power $P(e|h\&k)$? Or do they rather belong to the background knowledge k, which is relevant for evaluating the prior probability $P(h|k)$ of the theory? In the case of Newton's classical mechanics, for example, Kepler's three laws of planetary motion and Galileo's law of fall belong to the knowledge e to be explained, because they can be derived (approximately) from Newton's laws. But if we want to assess the prior probability of a more specific hypothesis, such as the claim that there is life on Mars, there is a large stock of empirical background knowledge k concerning the chemistry of life and of Mars, which is used in assessing its prior probability.

In general we might say, then, that the larger the scope of the theory, the smaller will be the set of empirical background knowledge with regard to which its prior probability should be determined. If more and more of the observational evidence and

---

[54] Swinburne (EG), p. 53.
[55] For example, in the cases of composite hypotheses discussed by Sober (2002a), pp. 26–30.
[56] Cf. Kim (1998) for a discussion of these issues.
[57] The fact that many people hear strange voices very clearly, which in the old days was considered convincing evidence for the existence of spirits and demons, is now explained as a hallucination that frequently occurs in cases of psychosis and schizophrenia. Similarly, many other facts that were once adduced as evidence for the existence of such supernatural persons have now been explained on the basis of empirical and secular science.

well-established theories fall into the category of data *e* that the theory purports to explain, less and less items of empirical knowledge can be relegated to the category *k* of background evidence. It follows that, if there really were a 'Theory of Everything' in the strict sense of this label, there would be no empirical background knowledge whatsoever by which we could determine its prior probability. The prior probability of such a theory would then be identical to its intrinsic probability.[58]

On the basis of such considerations, Richard Swinburne attempts to establish the irrelevance of empirical background knowledge to the prior probability of theism by the following argument.[59] As he says, theism 'is a hypothesis of enormous scope'. If we consider a Theory of Everything as a purely physical theory, the scope of theism allegedly is even greater than that of such a theory, since 'theism purports to explain everything logically contingent (apart from itself)', that is, 'all our empirical data'. It follows that in the case of theism, 'there will be no background knowledge with which it has to fit', since its scope is universal. Hence, 'it will not [. . .] be a disadvantage' to theism 'if it postulates a person in many ways rather unlike the embodied human persons so familiar to us'. Neither will it affect the prior probability of theism negatively that God is supposed to be able to execute his intentions directly without any material mechanism, such as a chain of neural events culminating in bodily movements. Because the hypothesis of God's existence is claimed to explain everything contingent apart from God's existence, no empirical background knowledge whatsoever remains for assessing theism's prior probability. Accordingly, the set *k*, which is relevant for determining this prior probability before all empirical tests of theism, can contain only tautological and other non-empirical knowledge, such as logic and mathematics.[60]

If this argument for the irrelevance of empirical background knowledge were sound, Swinburne would have eliminated one of the most devastating objections against theism. Does the argument succeed? Unfortunately, I do not think it does, because it involves a specific fallacy of ambiguity, which is usually called a *fallacy of division*. Logicians distinguish between two varieties of the fallacy of division. A fallacy of the first variety (also called the *mereological fallacy*) is committed if one argues that what is true of a whole must also be true of its parts, like when one concludes from the fact that Gounod's *Faust* is quite a long opera, that each of its arias lasts long as well. One commits the second type of division fallacy whenever one argues from the properties of a set or a collection of elements to properties of the elements themselves. For example, it is fallacious to conclude from the premise that the average life span of humans in the USA is now 78 years that each individual American citizen will pass away aged 78.

In order to see which variety of the fallacy of division Swinburne is committing when he argues that empirical background knowledge is not relevant to the prior probability of theism, we have to zoom in on his notion of the scope of a theory. In

---

[58] Cf. Swinburne (EG), pp. 59–60: 'as we deal with theories of larger and larger scope, there will be less and less background knowledge with which these theories have to fit, because '[m]ore and more of the observational evidence falls into the category of data that the theory needs to explain' (cf. also p. 146); and (EJ), p. 93: 'When we are considering very large-scale theories, there will be no such background evidence'.
[59] Swinburne (EG), p. 66. Cf. (EJ), p. 104: '*k* then becomes mere tautological knowledge – namely, irrelevant'.
[60] Swinburne (EG), pp. 65–8, 93, 145–6.

Chapter 9.1, I distinguished between four different dimensions of the predictive power or likelihood $P(e|h\&k)$ of a theory, that is, the total set $e_{1\text{-}n}$ of empirical predictions that can be derived from h. A first dimension is its *scope* or domain, which is broader to the extent that descriptions of more diverse types of phenomena are derivable from it. The second dimension is the degree of *completeness* of the theory within this scope, which is greater to the extent that descriptions of more individual items belonging to each type can be derived. Third, there is the degree of *precision* of these descriptions, which is greater when they are quantifiable and to the extent that they are quantified more exactly. Finally, there is the *strength* of h's predictive power with regard to each specific e_i, which is the conditional probability $P(e_i|h\&k)$ that e_i is true if h and k are true. As we do not only desire that our theories are (probably) true, but also that they are maximally informative, we will generally prefer theories that perform well on all four dimensions.

From Swinburne's brief definitions of the notion of scope, we may conclude that his notion embraces the first three dimensions defined in Chapter 9.1, above. For he says that the scope of a theory is greater 'in so far as it purports to apply to more and more objects and to tell you more and more about them'. Elsewhere, he stresses that the scope of a theory is greater to the extent that the values it predicts are more exact.[61] Let me call the rich notion of scope so defined the *three-dimensional* notion, in contrast to my own one-dimensional concept of scope. However, in his argument to the effect that empirical background knowledge is irrelevant for determining the prior probability of theism, Swinburne cannot use the term 'scope' in this three-dimensional sense. For in this sense, it would be *obviously false* to say that theism is a hypothesis 'of enormous scope' since it 'purports to explain everything logically contingent (apart from itself)'.

The reason is that even if theism as a theory could predict a few phenomena, such as the existence of the human species, it would not yield any detailed, or quantified, or exact predictions concerning these few phenomena, such as when the human species evolved, or how many human beings there will be in the year 2050. Indeed, if the scope of theism is 'enormous' in some vague collective sense, because theism purports to explain the existence of the universe as a whole, its scope is minimal in a distributive sense, since the vast majority of data in the universe cannot be predicted by theism at all. Theism can predict, for example, neither that God chose the path of evolution to produce human beings, nor that the present universe is as big and old as it is. Furthermore, it can neither predict that horrifying entities such as black holes and asteroids exist, which might eventually cause the destruction of our solar system or the Earth, nor that predators and lethal parasites evolved, nor that the dinosaurs would become extinct. If these data and many others were relevant to theism at all, they would rather be disconfirmations than confirmations.

In short, if theism has any predictive power at all, its scope in the three-dimensional sense is Lilliputian, because it performs below par on the three dimensions of

---

[61] Swinburne (EG), p. 55. Cf. (EJ), p. 82: 'The scope of a hypothesis is a matter of how much it tells us (whether correctly or not) about the world – whether it tells us just about our planet, or about all the planets of the solar system, or about all the planets in the universe; or whether it purports to predict exact values of many variables or only approximate values of few variables'. Cf. also (BT), p. 12.

predictive power that I distinguished above. Even worse, theism had better perform badly on these dimensions in order to avoid the risk of God-of-the-gaps, as I argued in Chapter 10. Sophisticated theists will avoid deriving any predictions from theism which they do not know already are true, since otherwise theism will run the risk of being disconfirmed.[62] Indeed, theists will only purport to explain a few phenomena that are either 'too big' or 'too odd' for science to explain in order to avoid rivalry with scientific explanations. For all these reasons, one commits a fallacy of division if starting from the premise that theism is a theory of the widest possible scope, since it purports to explain the existence of the universe, one concludes that its scope is so complete that there cannot be any background knowledge which should be taken into account in determining theism's prior probability.[63]

The fallacy is best analysed as a fallacy of division of the first type. Even if theism would predict the existence of the universe as a whole, its predictive power does not include predictions of each and every part of the universe in great detail and with a high degree of precision. Perhaps there is also another fallacy of ambiguity involved. Theists might conflate the proposition that *God* (by definition) caused or permitted directly or indirectly everything contingent apart from him, with the very different proposition that the predictive power of our human *theory of theism* has a universal scope. Since we can only conclude legitimately that God exists if the posterior probability of our theory of theism is higher than some threshold value, it is the predictive power of theism that we should take into account if we want to argue for the irrelevance of background knowledge for determining its prior probability, and not the proposition that God created everything apart from himself.

One has to admit, then, that Swinburne's global argument to the effect that background knowledge is irrelevant for assessing the prior probability of theism fails.[64] However, theists might reply that there is a more local argument to the same effect. As we have seen, the truth of theism is very improbable in the light of specific items of background knowledge. If theists are able to argue that *all* these specific items are situated within the explanatory scope of theism, they will belong to *e* and not to *k*, so that they will not influence negatively theism's prior probability. For example, Swinburne argues at length that if God wants to create animate beings with significant but limited free choice, these humanly free agents, as he calls them, need bodies and situatedness in a material world in order to exercise their limited freedom in a significant manner.[65] Hence, the fact that human mental faculties do not exist on Earth without human bodies is explained by theism. Consequently, it is not a piece of background knowledge that can be used in assessing theism's prior probability.

---

[62] Accordingly, Swinburne stresses that theism 'does not yield predictions such that we can know only tomorrow, and not today, whether they succeed' (EG), p. 70.
[63] In his article 'Bayes, God, and the Multiverse' (to be published in the volume of proceedings of the conference on Formal Perspectives on the Epistemology of Religion, which took place in June 2009 at the University of Louvain, Belgium), Swinburne argues that what rules out the need of fit with background knowledge is merely the 'size of the field' which the theory purports to explain (see note 5 of his article), and this is the one-dimensional scope of the theory in my sense. If this size were universal, the criterion of fit with background knowledge would be irrelevant. But this argument is also unconvincing, because, in principle, all knowledge that the theory cannot explain may be used to evaluate its prior probability.
[64] Cf. for another argument to the same effect: Smith (1998), pp. 93–6.
[65] Swinburne (EG), pp. 123–31: the section 'Humanly Free Agents Need Bodies'.

Unfortunately for theists, however, such a local argument is not available with regard to all relevant items of empirical background knowledge. Swinburne seems to hold that apart from creating humans, it is equally likely that God also created semi-divine beings of limited power but perfect freedom.[66] Since the freedom of these demi-gods is perfect, they do not need a body in order to exercise it significantly. But, as I said above, there is no empirical evidence whatsoever for the existence of demi-gods. Moreover, we know from brain research on animals that more sophisticated mental powers and functions depend on more complex brain structures. Hence, the empirical hypothesis that more advanced mental powers cannot exist without more complex brain structures is well supported by the evidence, quite apart from the existence of human beings. Since theism explains neither this evidence nor the hypothesis supported by it, these empirical items constitute background knowledge for theism. Given this background knowledge, we should expect that the immense mental powers of God, such as his omniscience and mental omnipotence, are supported by a monstrously big and extremely complex brain, and cannot exist without a body, if they can exist at all. In other words, even local arguments will not suffice to show that there are no devastating objections against theism on the basis of well-confirmed empirical background knowledge.

11.7 THE SINGULAR SIGNIFICANCE OF SIMPLICITY

Suppose, however, that Swinburne's argument against the relevance of empirical background knowledge to theism's prior probability were valid, since the three-dimensional scope of theism really contained 'everything logically contingent (apart from itself)', that is, 'the universe and all its characteristics', or at least all pertinent pieces of empirical background knowledge.[67] Then the prior probability of theism preceding all empirical tests would be identical with its intrinsic probability, if at least the objectivist conception of this prior is correct. According to Swinburne, the intrinsic probability of a theory can be determined objectively on the basis of its scope and its simplicity.[68] If this can be achieved without taking empirical background knowledge into account, what will be the result? Is the prior probability of theism then quite large, or vanishingly small, or perhaps indeterminate? Or should we admit that we simply don't know?

It is easy to see that the larger the three-dimensional scope of a theory, the smaller will be its prior probability. As Swinburne says, '[t]he greater the scope of a hypothesis, the less it is likely to be true'.[69] Let us assume, for example, that we can measure the scope of a theory by counting the number of singular empirical statements p_{1-n} derivable from it, requiring that these statements are both logically and

---

[66] Swinburne (EG), p. 119.
[67] Swinburne (EG), pp. 66 and 108.
[68] Swinburne (EG), pp. 53, 67–72; (BT), p. 12; (EJ), pp. 82, 105, 110–19.
[69] Swinburne (EJ), p. 82, cf. 86. Cf. also (EG), p. 55: 'Clearly the more you assert, the more likely you are to make a mistake'. In order to keep things relatively simple, I abstract here from differences in strength of the predictive power of a theory h with regard to a specific e, assuming that each e is entailed by h.

probabilistically independent from each other.[70] Let us also assume that we can determine in each case the prior probability that such an empirical statement p_i is true. If theory h_1 merely entails p_1, and theory h_2 entails both p_1 and p_2, whereas $P(p_1|k)$ and $P(p_2|k)$ are each equal to ½, the prior probability of h_2 as measured by its scope will be merely ½ of the prior probability of h_1. In other words, to the extent that the intrinsic probability of a theory is determined by its scope, it diminishes drastically when its scope increases.

Accordingly, if the scope of theism really is universal, it seems to follow that its intrinsic probability must be near to zero. This conclusion will be unacceptable to theists, however, since the few empirical arguments for theism they can adduce will probably not enable them to overcome such a low prior probability. How, then, can they avoid this conclusion? In *The Existence of God*, Richard Swinburne argues that the intrinsic probability of theism is greater than that of any alternative theory which is able to explain the same observations. His arguments to this effect aim at establishing the following conclusions. (1) The intrinsic probability of a theory is also determined objectively by its simplicity. (2) Simplicity outweighs by far wideness of scope for determining the intrinsic probability of a theory. (3) Theism is a very simple hypothesis, much simpler than its rivals. Hence, the intrinsic probability of theism is very high compared to other hypotheses about what there is.

Before concluding this chapter, I shall now briefly examine these three claims. They are of a singular significance to Swinburne's argumentative strategy for theism, because it cannot succeed without them.

11.8 SIMPLICITY AS A CRITERION FOR THEORY CHOICE

What should we think of the first conclusion (1) Swinburne aims at establishing? Nobody contests that scientists often use simplicity as one of the criteria for theory choice. Suppose we plot the results of a number of measurements concerning two interrelated magnitudes as points in a coordinate system. To the question how these magnitudes are related in general, we may answer by drawing a line, which represents a mathematical function. In principle, infinitely many lines can be drawn that fit the data quite well, but make different predictions concerning new data. Usually, this *curve-fitting problem* is resolved by applying criteria of simplicity and empirical adequacy, among others. A simple curve (a curve representing a simple function), or a curve explained by a simple model, is easier to handle, but often it fits the data points less adequately. Scientists arrive at a solution by weighing these criteria of simplicity and goodness-of-fit (among other criteria) against each other.

The criterion of empirical adequacy is truth-conducive, because it is concerned with the truth of the predictions that can be derived from the hypothesis. If this were the only criterion relevant to the curve-fitting problem, scientists would always prefer

---

[70] Measuring the scope or empirical import of a theory is notoriously difficult, and here I am making a number of idealizations. One is that the scope can be determined without taking other data or theories into account. Another is that that the singular empirical statements derivable from the theory are logically simple and probabilistically independent. Finally, I am assuming for the sake of argument that the number of logically simple empirical statements derivable from the theory is finite.

curves that pass exactly through the data points.[71] But in fact they do not do this. They use the criterion of simplicity as well, often preferring simple curves that do not fit the data points perfectly. Because this is the case, however, it is not obvious that the criterion of simplicity as used by scientists in solving curve-fitting problems is also and always truth-conducive, as Swinburne claims.[72] If scientists often prefer simpler curves that do not fit the data as adequately as more complex ones, they might do so for pragmatic reasons, for example, since such a use of simplicity removes us somewhat from the truth concerning the data obtained.[73] There is no reason to assume that the criterion of simplicity suddenly becomes indicative of the truth when it is used to choose between the (infinitely) many curves that fit the present data equally well.[74]

Yet natural theologians who insist that we can apply the criterion of simplicity in order to determine the prior probability of theism, must hold that simplicity is truth-conducive, and not merely of pragmatic interest. For the prior probability of a theory is the logical probability that the theory is true before one considers the evidence. How can theists argue for this contention? In *Epistemic Justification*, Swinburne avers that one cannot justify any further the principle that among theories performing equally well in all other respects, 'a simple one is more probably true'. Since any empirical justification of the principle relies on it, the principle must be a priori. And since it does not follow 'from some more obvious a priori principle [...] it must be a fundamental a priori principle'.[75] But this is unconvincing. In the case of many measures of the simplicity of a theory, there is no intelligible connection between simplicity and truth, and many prominent philosophers of science hold that simplicity is not a truth-conducive criterion. For example, Van Fraassen endorses an outspoken view on this issue: 'it is surely absurd to think that the world is more likely to be simple than complicated (unless one has certain metaphysical or theological views not usually accepted as legitimate factors in scientific inference)'.[76]

If one studies in detail the uses of simplicity by scientists as a criterion for the choice of theories or models, the topic of simplicity turns out to be of baffling complexity.[77] One might distinguish three dimensions of this problem of simplicity in the philosophy of science. One dimension (a) concerns the question as to how various uses of the simplicity criterion are justified. Do we prefer simple theories merely for aesthetic reasons, or on pragmatic grounds, because such theories are easier to grasp, to handle

---

[71] I am abstracting here from problems of random noise in the data and from criteria of theoretical plausibility.

[72] Swinburne (EG), pp. 58–9 concludes without further argument from the fact that simplicity is used as a criterion for resolving curve-fitting problems that '*Simplex sigillum veri* ("The simple is the sign of the true")'. Cf. also Swinburne (EJ), pp. 83–102.

[73] That criteria of simplicity are used in theory-choice for pragmatic reasons and are not truth-conducive has been argued by many philosophers of science, such as Newton-Smith (1981), p. 231, and Van Fraassen (1980), pp. 87ff.

[74] One might argue, however, that simplicity is relevant to the expected degree of predictive accuracy concerning future data (cf. the main text, below). But this cannot hold for theism, which according to Swinburne 'does not yield predictions such that we can know only tomorrow, and not today, whether they succeed' (EG, p. 70).

[75] Swinburne (EJ), p. 102.

[76] Van Fraassen (1980), p. 90. Cf. Foley (1993), p. 212: 'We cannot simply assume that the world is likely to be simple'. Foley rather assumes that 'there is no rationale, or at least no non-question begging one, for thinking that simplicity is a mark of truth' (p. 213).

[77] For an overview of these problems, cf. Sober (2002b).

in calculations, and to remember? Or should we prefer simpler theories because they are better testable or falsifiable, as Karl Popper argued?[78] Or, finally, do we select simpler theories because their simplicity is somehow related to truth?

In this latter case, there are once again many possibilities. Hans Reichenbach once argued that always choosing the simplest curve in a curve-fitting problem would enable us to approximate continually the true curve when we collect more and more data.[79] According to Kelly and Glymour, always choosing the simplest hypothesis in a series of theory choices is the shortest route to the true theory, because it minimizes the number of required mind-changes.[80] Others have attempted to justify specific criteria of simplicity by a meta-induction over the history of science.[81] And according to Hirotugu Akaike, one should prefer a simpler hypothesis over more complex ones, which all explain the same set of data, because it has a smaller estimated distance to a hypothesized but otherwise unknown true theory.[82] However, although these authors establish links between the simplicity of hypotheses on the one hand and our epistemic objective of having true beliefs and avoiding false beliefs on the other hand, none of them thinks that the simpler of two hypotheses, all else being equal, is more likely to be true, as Richard Swinburne holds. They merely argue that always choosing the simplest hypothesis is a means to the end of *ultimately* finding the true hypothesis.

A second dimension (b) of the problem is how the simplicity of hypotheses or theories should be measured. Again, there are many possibilities here, which have been spelled out in detail by philosophers of science and information theorists. Should we measure the simplicity of theories in some standard formulation syntactically, by measuring the length of the shortest sentence that expresses them, as has been proposed by Jorma Rissanen? Or should we prefer semantic measures of simplicity, such as the number of probabilistically independent empirical claims made by a theory, or the number of assumptions, or the number of adjustable parameters? Some of these measures are clearly probabilistic indicators of the truth of a theory, whereas others are not. Another problem here is that different measures for simplicity often lead to opposite results.

Two examples discussed in detail by Elliott Sober are instructive in this last respect.[83] Psychological egoism is a simpler theory concerning the motives of human actions than motivational pluralism, since the former allows for one type of ultimate motives only. But if psychological egoists have to explain an altruistic action, their explanation will be more complex than that of the motivational pluralist, since they have to postulate more motivating beliefs in the subject under investigation. Should we measure the simplicity of a theory *in abstracto*, without considering how it functions in explanations of specific phenomena? Or should we take these latter contexts into account? Clearly, these different options will often lead to opposite results.

---

[78] Popper (1959), Chapter VII.
[79] Reichenbach (1938), section 42. The problem with this convergence view is that almost any initial bias will eventually be swamped by incoming information, so that convergence fails to support the use of simplicity.
[80] Kelly and Glymour (2004), and Kelly (2004).
[81] Cf. McAllister (1996).
[82] Cf. Sober (2008), pp. 82–96.
[83] Cf. Sober (2002b).

Another example of diverging outcomes resulting from different measures for simplicity is concerned with nested models. Should we measure the simplicity of a model by counting its assumptions, or rather by counting its adjustable parameters? In the case of nested models, these two measures lead to conflicting conclusions. For example, if one counts adjustable parameters (a, b, c), the hypothesis that x and y are linearly related as $y = a + bx$ is simpler than the hypothesis that they are parabolically related as $y = a + bx + cx^2$. But the former hypothesis is more complex than the latter if one takes paucity of assumptions as the measure of simplicity, since $y = a + bx$ is equivalent to $y = a + bx + cx^2$ in conjunction with the independent assumption that $c = 0$.

The third dimension (c) of the problem of simplicity is how one should balance simplicity considerations and other criteria for theory choice, such as predictive power, empirical adequacy, fit with background knowledge, avoidance of ad-hocness, prospects for theoretical unification, and so on. How should sacrifices in simplicity be traded off against gains on some other criterion, and vice versa? Of course, if one has to choose between 'scientific theories of equal scope, fitting equally well with background evidence, and yielding the same data with the same degree of logical probability', that is, between theories equally good in all other respects, then the criterion of simplicity may be decisive for one's theory choice.[84] But in fact the theories one compares are almost never equally good in all other respects. Can one say in general, for example, that simplicity always outweighs by far wideness of scope for determining the intrinsic probability of a theory, as Swinburne argues (see under point 2, the next section)?

In order to cut the Gordian knot of these complexities, theists such as Richard Swinburne have to propose measures of simplicity that are uncontested indicators of truth, and not merely of pragmatic interest. For they want to argue that because of theism's great simplicity, its intrinsic logical probability is 'very high' if compared 'to other hypotheses about what there is'.[85] These measures must be such that theists can also maintain their other contentions about theism, like the claim that theism has a very wide scope. It will not be easy to satisfy these two requirements, as is shown by the following considerations. If one takes as a measure of simplicity the number of logically independent simple singular statements about what there is made or entailed by a theory, this measure is clearly an indicator of truth. If a theory makes or entails no such claims at all, because it consists of tautologies only, the prior probability of its truth will be 1. With each independent extra empirical claim p_i added, the prior probability of the theory decreases quickly, since it will be multiplied by the prior probability of p_i, which is between 0 and 1, as I argued above. Swinburne suggests such a truth-indicating measure of simplicity where he says that '[t]he simplicity of a theory, in my view, is a matter of it postulating few (logically independent) entities, few properties of entities, few kinds of entities, few kinds of properties [. . .]'.[86] Let us call this *measure*$_t$, where the subscript 't' stands for 'truth-indicator'.

---

[84] This is what Swinburne argues in (EJ), pp. 83ff.
[85] Swinburne (EG), p. 109.
[86] Swinburne (EG), p. 53. In the passage I left out, Swinburne adds measures that are not clearly truth-indicators, such as 'properties more readily observable, few separate laws with few terms relating few variables', etc.

In his arguments concerning the intrinsic probability of theism, Swinburne makes the idealizing assumption that the scope of theism can be determined a priori, without taking any background knowledge into account. Endorsing this idealization, we should measure the simplicity of theism inclusive of its scope. For if this assumption is correct, all statements about what there is that belongs to its scope are somehow implied by theism alone. This means, however, that if the scope of theism really is universal, since theism purports 'to explain the universe and all its characteristics', the theory of theism must be hugely complex, if assessed by measure$_t$, and not at all simple. In other words, the contention that theism is a very simple theory if measured by this truth-indicating gauge, squarely contradicts the claim that theism has an enormous scope.

Theists might respond by rejecting the idealizing assumption that the scope of theism can be determined without taking any background knowledge into account. Indeed, as we have seen in Chapter 9, even Swinburne holds that theism on its own has no predictive power whatsoever, so that its scope is non-existent. In order to derive predictions from theism, we allegedly must use the realm of objective ethical truths as relevant background knowledge. Now theism in itself might be a very simple theory. It merely says that God exists, where 'God' is defined by a small number of properties. As such, theism will seem to be much more simple than each variety of polytheism, for example.

Suppose, however, that both theism and a variety of polytheism are proposed as explanations of the existence of the universe and all its characteristics. Suppose also that they really have this explanatory power, if at least theism is conjoined with the set of ethical truths. Should we now prefer theism to polytheism, since it is more likely to be true because of its simplicity? What we should rather do, of course, is compare the simplicity of this variety of polytheism to the simplicity of theism conjoined with the set of ethical truths. But Swinburne does not do this. Neither is it clear how it should be done. So we come to the conclusion that we are ignorant concerning the simplicity of theism as an explanatory theory, that is, the simplicity of theism in conjunction with the realm of ethical truths.

11.9 WEIGHING SCOPE AND SIMPLICITY

Let us now investigate the second conclusion (2), which Swinburne aims at establishing, that the criterion of simplicity outweighs by far wideness of scope for determining the intrinsic probability of a theory. If the intrinsic probability of a theory is determined by its scope and by its simplicity, as Swinburne holds, how should these two criteria be traded-off against each other?

We prefer theories of greater three-dimensional scope because they are more informative. If truth-conducive measures of simplicity are used, we might prefer simpler theories because they are less likely to be false. Hence, we may say that each of these two criteria for theory choice serves a different objective of the epistemic community: to maximize informativeness and to minimize the risk of falsehood. Clearly, the two criteria pull in opposite directions. If they were both indicators of the truth or intrinsic probability of a theory, it would be contradictory to claim that theism both has an enormous scope and is very simple, as we saw above. For this would mean that its intrinsic probability is both very low and very high.

From earlier sections of this chapter, the reader will have gathered that there are at least two possible escapes from this contradiction. One is to admit that the three-dimensional scope of theism is Lilliputian instead of enormous. If theism has any predictive power at all, we may derive a few predictions from it, which we already know succeed. But the vast majority of data concerning the past, present, or future cannot be derived from theism. The second escape is to admit that theism has predictive power only if it is conjoined with a realm of objective ethical truths, so that we should apply our measures of simplicity to this conjunction. In that case, theism as an explanatory theory turns out to be much less simple than it may appear at first sight.

However, neither of these two escapes is available to a theist such as Richard Swinburne. The first escape would imply a confession that his argument for the irrelevance of background knowledge to the prior probability of theism contains a fallacy of ambiguity (cf. section 11.6, above). The second way out would mean that it is not convincing to prefer (mono)theism to polytheism because it is simpler and hence more likely to be true, assuming that these religious views perform equally well as explanations of the universe on all other accounts. I suppose that this is why Swinburne prefers a third type of escape. He argues that the criterion of scope is much less important than the criterion of simplicity for determining the intrinsic probability of a theory. But how can this be shown, if one uses truth-indicative measures for both scope and simplicity?

Swinburne puts forward a general argument and illustrates it by the example of classical mechanics. He starts by admitting that 'a theory's intrinsic probability is diminished in so far as its scope is great'. As a result, a theory about all metals has a lower intrinsic probability than a theory about copper only, for example. Vice versa, the intrinsic probability of a theory is greater to the extent that its scope is smaller. Then he argues very briefly as follows:

> But typically, if a theory loses scope, it loses simplicity too, because any restriction of scope is often arbitrary and complicating [...]. For this reason I do not think that the criterion of small scope is of great importance in determining prior probability...[87]

However, this argument is a *non sequitur*. If both scope and simplicity are truth-indicators, which pull in opposite directions, a restriction of scope increases the intrinsic probability of a theory, whereas an increase of complexity reduces it. If all restrictions of the scope of a theory were accompanied by compensating increases of complexity, a larger-scope version of the theory might a priori be as likely to be true as a smaller-scope version, and we would always prefer the former to the latter. For under this condition, the former version will have the very same intrinsic probability as the latter, whereas its informativeness is greater. But this conclusion does not show that the criterion of scope is of no great importance in determining prior probability. For according to the scenario imagined here, two versions of a theory with different scopes can have the same intrinsic probability only if the version with the smaller scope is

---

[87] Swinburne (EG), pp. 55–6.

more complex. This presupposes that a decrease in scope augments intrinsic probability unless it is compensated by an increase in complexity. Instead of showing that the criterion of scope is of less importance than the criterion of simplicity for determining the intrinsic probability of theories, the argument rather presupposes that both criteria are of equal importance. And this is how it should be with regard to the intrinsic probability of theories, if both of these criteria are really based upon truth-indicating measures.

Whether it is really the case that restrictions of scope are always accompanied by an increase in complexity, is another matter. Indeed, there is no reason to assume that a theory about all metals is necessarily simpler than a theory about copper, for example. More importantly, it seems contradictory to assume that a restriction of scope is accompanied by an increase of complexity if one really uses a purely truth-indicative measure of simplicity, such as the number of logically independent singular empirical statements derivable from the theory. For one cannot suppose that this number is both reduced by a restriction of scope and, simultaneously, increased by an increase in complexity. Hence, the argument also contains a fallacy of ambiguity concerning the notion of simplicity, which is used in a non-truth-conducive sense in the premise of the argument. But if simplicity is relevant to the intrinsic probability of a theory, a truth-conducive sense of simplicity should be used, as Swinburne does in his conclusion.

Having put forward this invalid argument, Swinburne purports to illustrate its conclusion by the example of classical mechanics. As he says, '[t]he fact that overall it [i.e. Newton's theory of motion] was judged enormously probable illustrates my point that the criterion of scope is of far less importance than the other criteria'.[88] But this claim also involves a fallacy of ambiguity. The point Swinburne wanted to make is that the criterion of scope is much less important than the criterion of simplicity for determining the intrinsic probability of a theory. However, what Swinburne's discussion of Newton's classical mechanics shows is a different point, to wit, that in spite of its great scope, its posterior probability and scientific value were judged to be high on three other criteria: its impressive empirical adequacy, its relative simplicity, and the fact that the phenomena it predicted could not be expected on the basis of rival theories, such as Descartes' laws of motion and his vortex theory.[89] The fact that these three factors favoured Newton's mechanics over its competitors does not support or illustrate Swinburne's contention that the criterion of scope is of less importance than the criterion of simplicity for the intrinsic probability of a theory. So we come to the conclusion that Swinburne does not succeed in resolving the contradiction implied by his claim that theism both is very simple and has an enormous scope.

---

[88] Swinburne (EG), p. 57. Cf. pp. 108–9: 'I argued earlier that it is clear from examples that the great simplicity of a wide hypothesis outweighs by far its wideness of scope in determining intrinsic probability'.
[89] As I argued above, the victory of classical mechanics should be analysed by using the law of likelihood instead of Bayes' theorem.

11.10 IS THEISM SIMPLER THAN ITS RIVALS?

Theists might react to this conclusion by weakening the claim involved. Believers in theism do not need to pretend, so they might aver, that their theory both has an enormous scope and is very simple. What they need to argue is merely that theism (in conjunction with the set of objective moral truths) is a simpler hypothesis than each rival theory of equal scope, fitting equally well with background knowledge (if any), and predicting the same data with the same degree of logical probability.[90] This is the third claim (3) that I intend to discuss.

If this contention can be substantiated, and if simplicity is a seal of the truth (*simplex veri sigillum*), then the truth of theism is more probable than that of each of its rivals. Although this result will not enable theists to assert that the truth of theism is probable to some determinate degree, it will at least justify their preference for theism over all rival religions and other competitors.

But can this contention be substantiated? Given the complexities of simplicity discussed above, some awkward problems emerge. First, there are in principle infinitely many rivals of theism, of which humanity has never thought. How can we compare the simplicity of these unknown theories with that of theism conjoined with the set of objective moral truths? Second, it is rarely the case that competing theories have exactly the same scope, fit equally well with background knowledge (if any), and predict the same data with the same degree of logical probability. What should we do, for instance, if a rival of theism predicts data with more precision and with a higher degree of logical probability, but is slightly less simple? How should we then trade-off the criteria of predictive precision, predictive strength, and simplicity? Finally, even if we compare theism with a specific rival theory that we know quite well, insurmountable difficulties may arise. It might be difficult, for example, to find truth-conducive measures of simplicity that can be applied to both theories equally well, if, say, a rival of theism can be formulated mathematically whereas theism cannot.

In order to study these problems in some detail, we have to leave behind the preliminary conceptual issues of this chapter and focus on a particular example of a probabilistic argument for the existence of God: the cosmological argument (Chapter 12). Surveying the conceptual problems I raised, however, we may conclude that the prospects for calculating with sufficient objectivity a specific posterior probability of theism, which is sufficiently high for justifying the belief that God exists, are not very promising. Not only is the prior probability of theism extremely low given our background knowledge concerning animal and human brain research, which shows that mental functions never exist without a neural basis. As we saw, Swinburne did not succeed in ruling out the relevance of this background knowledge. But in view of the problems of determining the prior probability of the catch-all hypothesis $\sim h$, and hence of determining the prior probability of e given k, it also seems impossible to do more than determine the relative posterior probability of theism compared to that of a specific rival hypothesis, given evidence e, by applying the law of likelihood. However, doing so cannot in principle yield the results theists such as Richard Swinburne are aiming at, that is, to show that the probability of God's existence is higher than ½.

---

[90] Cf. Swinburne (EJ), p. 83, and (EG), p. 109: 'The intrinsic probability of theism is, relative to other hypotheses about what there is, very high, because of the great simplicity of the hypothesis of theism'.

12

Cosmological Arguments

Arguments for the existence of God that use as an empirical premise the fact that there is a universe, or the fact that something exists rather than nothing, or some other very unspecific empirical fact, are called cosmological arguments. In this chapter we shall focus on the following question: do theists succeed in formulating a correct C-inductive cosmological argument for theism? In other words, is there a cosmological argument that makes theism more probable than it would be otherwise?

Many different cosmological arguments have been advanced in the literature on natural theology. Indeed, cosmological arguments belong to the oldest arguments adduced for the existence of a god, and we find one version already in Plato's *Laws* and another is suggested in Paul's letter to the Romans (I.18–20).[1] We may classify cosmological arguments by distinguishing different types, and I shall briefly argue that inductive cosmological arguments to the best explanation for the existence of God are *prima facie* more promising than deductive cosmological arguments.

In the main sections of this chapter, I discuss three inductive cosmological arguments put forward by Richard Swinburne in *The Existence of God*: the synchronic causal cosmological argument, and two diachronic causal cosmological arguments: the argument from the infinite series of cosmic time-slices as a whole, and the argument from the Big Bang to God. My conclusion will be the following. Even if one assumes that the criterion of simplicity for theory-choice is truth-conducive and crucial for determining the prior probability of theism and its rivals, Swinburne's cosmological arguments do not raise the probability that theism is true.

12.1 A CLASSIFICATION OF COSMOLOGICAL ARGUMENTS

Three parameters may be used in classifying cosmological arguments. First, we distinguish different types according to their logical form. Are the arguments deductive or inductive? If inductive, are they instances of enumerative induction, arguments by analogy, or rather arguments to the best explanation, for example? If the latter, do they apply the law of likelihood or the theorem of Bayes? And what, exactly, is their intended logical force? Does an argument merely aim at making theism more probable

---

[1] Plato, *Laws*, pp. 893–9.

than it would be otherwise, so that it is meant as a correct C-inductive argument? Or is it claimed that the argument makes theism more probably true than false, so that it purports to be a correct P-inductive argument? Or, finally, does it perhaps aim at raising the probability of theism higher above a threshold of $1/2$?

Second, not all cosmological arguments refer to the same empirical data as facts to be explained. Modal cosmological arguments start from the fact that there is a contingent entity or event and assume that, ultimately, contingent entities or events must be caused by something that exists necessarily. Non-modal arguments start from a different explanandum. In their factual premise it is stated, for example, that the present universe exists, that there is something rather than nothing, or that there is change in the universe. Then it is argued that, ultimately, such a fact cannot be explained by a scientific explanation, so that, if it can be explained ultimately at all, there must be a personal explanation, which posits God.

Finally, cosmological arguments can be classified on the basis of the universal premises they use. Since the logical form of the factual premise, such as 'there is a contingent entity' or 'our present universe exists', is not universal but singular, and the conclusion that God exists is singular as well, at least one universal premise is needed for deductive cosmological arguments. In modal arguments this universal premise often is a version of the Principle of Sufficient Reason. Non-modal cosmological arguments typically rely on the premise that an infinite regress of causes backwards in time is impossible. Inductive arguments also have to employ some general principle because of the problem of under-determination. Since the same set of empirical data can inductively support equally well indefinitely many different explanatory hypotheses, criteria are needed for preferring one to the others. As we have seen in Chapter 11.8, Richard Swinburne claims that the criterion of simplicity primarily fulfils this role with regard to theism.

Critics of deductive cosmological arguments often conclude that no version is sound. They hold, in other words, that each deductive cosmological argument either is deductively invalid or has at least one false premise. Richard Swinburne argues that since the empirical premise of cosmological arguments reports an evident facet of experience, '[t]here is no doubt about the truth of statements that report that they hold'. It seems equally evident to him 'that no argument from any of such starting points to the existence of God is deductively valid'. His reason is that 'if an argument from, for example, the existence of a complex physical universe to the existence of God were deductively valid, then it would be incoherent to assert that a complex physical universe exists and God does not exist. There would be a hidden contradiction buried in such co-assertions'. Since this is implausible and has not been shown, Swinburne concludes that theists have to abandon the hope of inventing a sound deductive cosmological argument. Instead, apologetic philosophers of religion should settle for inductive versions.[2]

But one cannot dismiss deductive cosmological arguments so easily. Admittedly, a deductive argument is invalid if one can deny the conclusion without contradicting the conjunction of *all* its premises. In the case of deductive cosmological arguments, however, these conjunctions contain more premises than the mere report of an evident facet of experience, such as the existence of our complex physical universe. As I argued, deductive cosmological arguments need at least one universal premise, and, given this

---

[2] Swinburne (EG), pp. 136–7; cf. p. 155.

premise, many of them are deductively valid. Hence, if one wants to write off all deductive cosmological arguments, one should establish that each of them either is invalid or essentially relies on a false universal premise.

In fact, it seems that this has been shown in the existing philosophical literature for all known deductive cosmological arguments. Let me give two examples only. The standard modal cosmological argument can be put succinctly as follows:

1. A contingent entity exists (that is, an entity of which we can suppose without contradiction that it does not exist), or a contingent event occurs.
2. Each contingent entity or event has a sufficient cause.
3. Contingent entities or events alone cannot constitute, ultimately, a sufficient cause for the existence of a contingent entity or the occurrence of a contingent event.
4. Therefore, at least one necessary entity or event exists (that is, an entity or event of which we cannot suppose without contradiction that it does not exist or occur). And because it exists necessarily, it does not stand in need of an explanation.

Clearly, the singular premise (1) states an evident facet of experience, so that it is obviously true. But the two universal premises (2, 3) are also needed to deduce validly the conclusion (4), if we assume as self-evident that every thing or event is either (logically) necessary or contingent. The conjunction of (2) and (3) amounts to one of the versions of the Principle of Sufficient Reason, which Leibniz once defended.[3] But why should one accept (2) and (3)? The causal principle (2) is often rejected because, it is said, quantum-mechanical considerations limit its applicability.[4] But this objection is debatable, since it assumes that quantum-mechanical indeterminacy is an ontological and not merely an epistemological matter.

What one should repudiate is premise (3), since causal explanations cannot but refer to causes that exist or occur contingently. If one explains causally an event E with reference to a cause C, what one means is that, *ceteris paribus*, if C had not occurred, E would not have occurred either, assuming that there is no causal redundancy. Hence, it is essential to the very meaning of the word 'cause' that we can always suppose without contradiction that a cause C did not occur. All causal explanations must connect contingent facts or events. And indeed, it is obvious that a contingent complex of causes can be sufficient for producing a contingent entity or event, so that (3) is false.

A second example of a cosmological argument that contains a false universal premise runs as follows:

1. This event in the universe is fully or partially caused by earlier events. The same holds for other events. They are caused by causal chains going backwards in time.
2. Infinite causal regresses are impossible.
3. Therefore, there must have been a first cause of each causal chain.

This first-cause argument had great appeal to mathematically minded philosophers of the past. The reason was that the concept of infinity seemed to imply a number of

---

[3] Cf. Leibniz (1697), p. 487: 'Therefore, since there must be an ultimate root in something which has metaphysical necessity, and since there is no reason for an existing thing except in another existing thing, there must necessarily exist some one being of metaphysical necessity, or a being to whose essence belongs existence'.
[4] Cf., for example, Grünbaum (2004), pp. 566–7 and 574–5.

contradictions, the most obvious of which is as follows. Take the series of natural numbers, which is infinite in the sense that there is no largest positive integer.[5] Intuitively, one will think that the set of odd numbers must be merely half as large as the set of all natural numbers. And yet the set of all odd numbers is also equally large as, or equivalent to, the set of all natural numbers, in the sense that one can pair all the members of the second set to members of the first in a one-to-one correspondence, and vice versa. So the set of odd numbers is both as large as, and only half as large as, the set of natural numbers. Such apparent contradictions led many philosophers to endorse the universal premise (2) of the first-cause argument, for if the notion of infinity implies contradictions, it cannot be true that there are infinite series of events.

However, as Georg Cantor showed during the last quarter of the nineteenth century, these paradoxes of infinity arise because one mistakenly conceives of infinite sets on the model of finite sets. In order to avoid such confusions, infinite sets should be defined explicitly as sets that can be paired in a one-to-one correspondence with a proper subset. Each set is a subset of itself, because by definition a set A is a subset of a set B if and only if all members of A are also members of B. But a proper subset of B is defined as a set A, all the members of which are members of B, whereas not every member of B is also a member of A. Clearly, finite sets are never equivalent to proper subsets of themselves, whereas infinite sets are. After Cantor's elimination of the traditional paradoxes of infinity, it seemed to mathematically informed philosophers that the first-cause cosmological argument had been refuted conclusively. There was no longer good reason to endorse its universal premise (2), even though Cantor's set theory generated a number of new paradoxes.

During the twentieth century, however, the first-cause argument was revived in slightly different versions on two grounds, one empirical and one a priori. As has been noted before, the empirical success of hot Big Bang theory in relativistic cosmology convinced Pope Pius XII and many others that in fact the universe must have had a first state in time, which cannot be explained scientifically. Does Big Bang theory not open up perspectives for a personal explanation of the first state of the universe by the hypothesis of theism? This question, raised in Chapters 10 and 11, will be discussed further in sections 12.6–12.9.

A second reason for reviving the first-cause argument has been developed with fervour by the philosopher-theologian William Lane Craig, and his works on the so-called *kalam* cosmological argument have generated a vast literature. Craig argues that although the concept of infinity is not contradictory as long as one interprets infinite sets as potentially infinite – for instance, we may always construct a positive integer larger than a given natural number – the idea of an *actually existing* infinity leads to unacceptable paradoxes.[6] Accordingly, Craig substituted for the second premise of the first-cause argument the following three claims:

---

[5] If we call the number of elements in a set A its cardinal number $n(A)$, infinite sets may also be defined as sets so large that their cardinal number is not found among the whole numbers (0 or the natural numbers).

[6] Cf. 'The Kalam Cosmological Argument', in Craig (2002), pp. 92–113.

2a. An actually infinite set cannot exist.
2b. If there were beginningless causal chains backwards in time, they would be actually infinite sets.
2c. Therefore, beginningless causal chains backwards in time cannot exist.[7]

Although this argument is a deductively valid *modus tollendo tollens*, premises (2a) and (2b) are both problematic. To begin, the word 'actual' stands in need of clarification. In classical, non-intuitionistic mathematics, infinite sets are treated as actual in the sense that they are regarded as completed totalities, independently of any human process of construction.[8] If one considers a set as actual only if its elements have been counted explicitly by humans, actually infinite sets are impossible. But it does not follow that there cannot be an infinity of objects or states existing in time and space. With regard to infinities of such naturally existing entities or events, Aristotle distinguished already between potential and actual infinities. According to an Aristotelian definition of 'actual', an infinite set of natural things or events exists actually if and only if all its elements exist simultaneously.[9] Given this meaning of 'actual', premise (2b) is false.

Craig's arguments for premise (2a) have been refuted by many critics. Craig asks us, for instance, to imagine Hilbert's Grand Hotel, a thought experiment that the German mathematician David Hilbert invented in order to illustrate the paradoxes of infinity. Since the Grand Hotel has infinitely many rooms, a proper subset of its rooms, say those with even numbers, can be paired in a one-to-one correspondence with all its rooms. Suppose that there are infinitely many reliable bookings for rooms, so that, one might think, all rooms of the hotel are occupied. What happens if a new guest asks for accommodation? If the hotel had only a finite number of rooms and was fully booked, the proprietor would say, 'Sorry, we cannot accommodate a new guest, since all our rooms are occupied'. But if the hotel has infinitely many rooms, there is no problem. Since the infinitely many bookings may be paired in a one-to-one correspondence with a proper subset of the set of all rooms, the infinitely many guests who booked a room may be lodged in even-numbered rooms, for example, so that infinitely many odd-numbered rooms are available. Hence, in the case of Hilbert's Grand Hotel, the statement that 'there are as many bookings as there are rooms' is not equivalent to the statement 'we cannot accommodate a new guest'. Indeed, infinitely many new guests can be accommodated.

Craig cleverly plays with these traditional paradoxes of infinity. Surely there can be no hotel that is both fully booked and has rooms available for new guests? Is it not clear, then, that this thought experiment leads to paradoxical results? But by doing so he does not succeed in showing 'in general that it is impossible for an actually infinite number of things to exist'.[10] His results seem paradoxical merely because in fact there are no hotels with infinitely many rooms, or libraries with infinitely many books. Such things are physically impossible on Earth, so that it is perhaps psychologically

---

[7] Cf. Craig (2002), p. 94 (with some modifications).
[8] Cf. Kleene (1952), p. 48.
[9] Cf. Aristotle, *Metaphysics* IX, Chapter vi. According to Aristotle, 'Infinite does not exist potentially in the sense that it will ever exist separately in actuality' (ibidem, 1048a; cf. *Physics* III, Chapters iv–viii).
[10] Craig (2002), p. 96.

impossible to imagine them. But this does not show that the existence of infinite series of causes in time is logically impossible.[11] By illustrating the traditional paradoxes of infinity by examples of things that in fact can only be finite, such as hotels, libraries, people writing a diary, or our counting things in time, he is seducing the reader to conceive of an infinite set on the model of a finite set. As Cantor showed, it is precisely this implicit model that generates the traditional illusion of paradoxes. By mobilizing our background knowledge about finite things when asking us to think through the implications of infinity, Craig has 'unfairly loaded the dice', as one critic observes.[12]

Apart from the fact that all known deductive versions of the cosmological argument rely on a false universal premise (or at least on a universal premise that is deeply problematic), there is a second reason to prefer inductive versions to deductive ones. As we saw with regard to the two examples I discussed, deductive versions of the cosmological argument can merely conclude that there is at least one necessary being, or that there is at least one first cause. A wide logical gap yawns between this conclusion and (mono)-theism. In order to show that there is at most one first cause or necessary being, and that this cause or being is identical to God, one needs many more premises, which may be problematic. Inductive arguments to the effect that theism is the best explanation for the existence of the universe are not confronted by this problem. If theism predicts the existence of the present universe with some probability, and if it can be shown that theism is a better explanation of this fact than its rivals, we have arrived at the conclusion of theism immediately. There is no further gap to be bridged.

This is why I shall now discuss the three inductive versions of the cosmological argument developed by Richard Swinburne in *The Existence of God*. Since they are intended as C-inductive arguments to the best explanation, they are all causal arguments. The general facet of experience that these arguments purport to explain is the existence of our (present) complex physical universe. Indeed, according to Swinburne, '[t]here is a complexity, particularity, and finitude about the universe that cries out for explanation, which God does not have'.[13]

Swinburne uses the theorem of Bayes in order to argue that these inductive cosmological arguments make the existence of God more probable than it is prior to considering the existence of the universe as evidence. Holding that empirical background knowledge is irrelevant to the prior probability of theism, he specifies the content of the background knowledge k for determining this prior probability $P(h|k)$ as consisting of purely conceptual knowledge only, such as logic and mathematics. Because we may distinguish between synchronic and diachronic causes, there are synchronic and diachronic causal cosmological arguments. And since the history of the universe backwards in time is either finite or infinite, the diachronic arguments may take either of these possibilities into account. After some preliminary considerations concerning cosmological scenarios, I shall first consider the diachronic argument

---

[11] Cf. Sobel (2004), p. 187.
[12] Cf. Everitt (2004), pp. 64–5. Cf. Smith, in Craig and Smith (1993), p. 85. For other criticisms, cf. Swinburne (EG), p. 138, note 10. The same criticism applies to arguments according to which 'the infinite cannot be traversed', because 'traversing' typically means to pass across some finite stretch. Cf. Oderberg (2002), pp. 307–20.
[13] Swinburne (EG), p. 150.

Cosmological Arguments 227

from a temporally infinite universe, then the synchronic argument, and finally the argument from the Big Bang to God.

12.2 THREE COSMOLOGICAL SCENARIOS

By definition, a correct C-inductive argument for theism is an argument that raises the probability that theism is true above the value it had before the evidence was adduced. In more formal terms, if h is theism, k the background knowledge, and e the evidence, a C-inductive or confirming argument is correct if and only if $P(h|e\&k) > P(h|k)$, and this is the case if and only if the relevance criterion $P(e|h\&k) > P(e|k)$ is met.[14] Clearly, correct confirming arguments may be stronger or weaker. But what is the empirical evidence e in the case of cosmological arguments?

According to Swinburne, the 'starting points of cosmological arguments are evident facets of experience', because the truth of the factual statement e has to be certain.[15] In the inductive cosmological arguments that he develops, this starting point e is specified as 'the existence of a complex physical universe'.[16] We might wonder, however, how one should spell out this notion of a complex physical universe in order to make sure on the one hand that its existence is an evident facet of experience, and on the other hand that one's cosmological arguments avoid the risk of God-of-the-gaps. In other words, how should we specify the e of inductive cosmological arguments in such a way that it is both obviously true and 'too big for science to explain' (cf. Chapter 10.6)?

We are acquainted merely with a minute spatio-temporal part of the complex physical universe in which we live. The testimony of travellers and scientists provides us with some reliable knowledge of other spatio-temporal regions. Swinburne defines a physical universe as 'a physical object consisting of physical objects spatially related to each other and to no other physical object'.[17] As he says, this definition does not rule out the logical possibility of other universes apart from our universe. If that is so, however, the explanandum e of his cosmological arguments cannot be specified merely as 'the existence of *a* complex physical universe'. The universe should be *our* universe, for otherwise its existence will not be an evident facet of experience. For the same reason, the explanandum e should be defined even more precisely as *the present time-slice of* (the known part of) our universe, postulating that each time-slice or state of the universe has a fixed finite duration, such as one year, and that there is a metric which enables us to determine these finite unities of time throughout the universe.[18]

Let us now assume that it makes sense to ask what caused our present time-slice or state S_0 of the universe diachronically.[19] What is the best explanation of this immense

---

[14] Cf. Chapter 9.1, above, and Swinburne (EG), p. 70.
[15] Swinburne (EG), p. 136.
[16] Swinburne (EG), p. 133.
[17] Swinburne (EG), p. 133. The second clause of this definition is somewhat unclear. It means, presumably, that no physical object which is part of the universe can be spatially related to another object that is not part of the universe.
[18] On the problem of metrics in cosmology, cf. Brian Pitts (2008), pp. 681–2.
[19] One might do so assuming that 'a cause of the occurrence of a collection of states is any collection of the causes of each [. . .], which are not members of the former collection' (Swinburne (EG), p. 141).

collection of items and events, summarized by *e*? The proper answer to this question will not be that God caused it diachronically. On the contrary, one will have to say, roughly, that it was caused by the preceding time-slice S_{-1} of the universe according to laws of nature *L*. For although the theistic hypothesis may be much simpler than a secular explanation that refers to this preceding time-slice S_{-1} of the universe, the predictive power of the latter explanation vastly exceeds that of the former. The same holds, in its turn and *mutatis mutandis*, for the best explanation of this preceding time-slice S_{-1}: it is caused by time-slice S_{-2} according to the laws of nature, and so on. Where, then, does God enter into the picture of diachronic explanations of our present universe? Which *e* really is 'too *big* to be fitted into any pattern of scientific explanation'?[20]

If the present state of the physical universe is our explanandum *e*, we have explained it sufficiently with reference to the preceding state as its collective diachronic cause. There is no need to refer to earlier states of the universe. One might think that a causal explanation is somehow defective as long as the posited causes are not causally explained in their turn. For example, a protagonist in David Hume's *Dialogues Concerning Natural Religion* objects to the hypothesis of a god causing our well-ordered material universe, that this postulated entity would itself need an explanation. But Swinburne correctly points out that those who raise this objection commit what he calls the *completist fallacy*. We cannot require that in order to explain adequately an event with reference to a cause, this cause should also be explained causally, and so on ad infinitum, since on that condition nothing whatsoever would ever be explained adequately.[21]

Hence, if only the present state of our universe is the explanandum *e* of the diachronic inductive cosmological argument, there is no room for a theistic hypothesis. But at this point natural theologians implicitly broaden the explanandum. Not only the present time-slice of the universe should be explained causally, but also the preceding time-slice, and its preceding time-slice, and so on. The price one pays for broadening the explanandum in this way, so that it will contain each past time-slice of the universe, is high, since the larger explanandum is no longer an 'evident facet of experience'. However, let us assume that we want to explain not only the present state of the existing universe, but also each of its earlier states.

The regression of explanations of time-slice S_{-n} of the universe by the preceding time-slice $S_{-(n+1)}$ leads to an 'interesting question', as Swinburne calls it: the question about whether the universe is of finite or infinite age. Have there been merely finitely many time-slices of equal duration up to the present, or rather infinitely many?[22] In the old days, this issue was a cherished topic of religious revelation or metaphysical speculation, although Kant argued that it is beyond the scope of our epistemic faculties.[23] But since Einstein and De Sitter started to apply the theory of general relativity in cosmology, and hot Big Bang theories were proposed, many of us decided

---

[20] Cf. Swinburne (EG), p. 74 (Swinburne's italics).
[21] Swinburne (EG), pp. 76–7.
[22] Swinburne (EG), p. 138.
[23] Kant, KdrV (*Critique of Pure Reason*), B432–560.

that the question concerning the finite or infinite age of our universe has to be settled by science, if it can be settled at all, and Swinburne agrees.[24]

This decision creates a formidable challenge for theists who want to advance a cosmological argument that does not run the risk of God-of-the-gaps, as I argued in Chapter 10.6. They cannot base their cosmological argument merely on the cosmological scenario that most scientists nowadays endorse. On the contrary, they have to argue that *whatever* scenario will turn out to be the most probable one from a scientific point of view, there is room for a theistic explanation of the universe. Accordingly, Swinburne argues that:

[i]f we confine ourselves to scientific explanation, it will now follow that the existence of the universe (for as long as it has existed, *whether a finite or an infinite time*) has no explanation.[25]

What must be meant is that the existence of the universe has no *possible* scientific explanation, whether it is finitely or infinitely old, so that *in principle* it is 'too big' for science to explain. And if there is no possible scientific explanation, the theist will conclude that there is room for a non-scientific, that is, a personal explanation, which posits God's creative act as the ultimate cause of the existing universe.

But at first sight, these conclusions do not follow. We have to consider at least three possible diachronic scenarios. *Scenario 1* is that the universe is infinitely old and that, in principle, each state of the universe can be explained scientifically by its preceding state. According to *scenario 2*, the universe is infinitely old, but the possibility of empirically testable scientific explanations is restricted. For example, it might be the case that empirically confirmed science cannot explain the development of the universe before Planck time, that is, before 10^{-43} seconds after the postulated Big Bang, but that nevertheless our universe is infinitely old.[26] Finally (*scenario 3*), the universe might be finitely old only, in such a way that there is a first temporal state of the universe, which is not caused by an earlier state of the universe.[27]

It will seem that only in this third scenario is there room for theism as a (very indirect) diachronic explanation of the presently existing universe. For if we go backwards in time from this initial explanandum to ever earlier states of the universe that cause the later ones, we will eventually arrive at a first time-slice of the universe, which had no physical cause, so that it may perhaps have had another, personal cause. In the other two scenarios, there is no first temporal state of the universe, so that there seems to be no need or room for theism as a causal hypothesis to the best explanation. What, then, are Swinburne's arguments for concluding that *in all three scenarios* there is room for a personal explanation, because allegedly the existence of the universe has no scientific explanation, *whichever* scenario obtains?

---

[24] Cf. Swinburne (EG), pp. 138–9: 'We do not know whether the universe has a finite or an infinite age, but science may be able to show us which is the more probable'.

[25] Swinburne (EG), p. 140 (my italics).

[26] Cf. Ellis (2007), § 6.1: 'Did a start to the universe happen? If so, what was its nature? [...] Tested physics cannot give a decisive answer; it is possible that *testable* physics also cannot do so' (p. 1235, Ellis' italics).

[27] In fact, there are more possible scenarios, which I shall not discuss here, such as the scenario of a finite age of the universe without there being a first temporal state or beginning of the universe. Cf. Brian Pitts (2008), pp. 679–80.

As a start, Swinburne attempts to reduce the second scenario to the third. Let us formulate all three scenarios in terms of current hot Big Bang cosmology. Edwin Hubble established in 1929 that our universe is expanding, as is shown by the fact that the greater the distance of a galaxy from us, the greater the galaxy's redshift (the Hubble law). Hence, a reconstruction of the past of our universe in conformity to well-confirmed laws of physics L, among which is general relativity theory, leads to the conclusion that if we reversed the arrow of time, our universe would be contracting ever more. Boldly extrapolating this contraction for some 13.7 billion years backwards in time on the assumption that we can use a metric that is isotropic, we arrive at an extremely dense and hot quantum state of the universe, which one might call state-P, that is, the state of the universe at Planck time. An even bolder extrapolation to a point 10^{-43} seconds earlier leads to a limiting state-S of the universe, in which matter and energy are packed into a point with infinite density, infinite temperature, and infinite curvature, which is called the Big Bang singularity. But we cannot apply our well-confirmed laws of physics L before Planck time, that is, before 10^{-43} seconds after the Big Bang singularity.[28]

We may now spell out the three scenarios mentioned above as follows. According to scenario 1, physicists will succeed in inventing and empirically confirming new theories, such as a string theory or a quantum gravity theory, by means of which it will be possible to explain the very dense quantum state-P of the universe at Planck time as caused by earlier states, and so on, ad infinitum. Cosmological speculations concerning a bouncing universe may suggest this scenario. In scenario 1, the universe is infinitely old, and each time-slice of the universe can be explained, at least in principle and to some extent, with reference to an earlier time-slice and scientific laws.[29]

According to scenario 2, physicists will never succeed in developing and empirically confirming such new theories, although the universe is infinitely old. If so, there was a beginning to the universe-as-governed-by-known-laws L, and we can have no scientific knowledge of anything earlier.[30] Let us suppose that this beginning, or the first state of the universe-as-we-know-it, is the universe at Planck time. Although in this second scenario the question as to what caused state-P of the universe is in fact 'too big' for science to answer, it does not seem to follow that there is room for a theistic explanation of any state of the universe. For according to this scenario, the universe is infinitely old, even though human science cannot discover earlier states than the state of the universe at Planck time. Hence, we may assume that each state or time-slice, including state-P, is caused by an earlier state.

Scenario 3, finally, is the standard hot Big Bang scenario, according to which (it seems that) there is a first state of the universe, at the beginning of which space-time

---

[28] Cf. Kaufmann (1994), p. 530.
[29] Cf., for example, Bojowald (2008), pp. 31–2. *In fact*, such explanations may be impossible, however, because '[t]he universe before the big bang could have been fluctuating very differently than it did afterward, and those details did not survive the bounce. The universe, in short, has a tragic case of forgetfulness. It may have existed before the big bang, but quantum effects during the bounce wiped out almost all traces of this prehistory' (p. 32).
[30] Swinburne (EG), p. 140.

and matter came into existence. One might think that this first state is state-S of the cosmic singularity. But according to many physicists, state-S is an idealization, which can have no physical reality. What should we do if extrapolating the inverted expansion of the universe backwards according to the well-confirmed laws of present-day physics L leads to a physically impossible state?

It is at this point that Swinburne reduces scenario 2 to scenario 3. The crucial passage reads as follows:

> In so far as science shows that the fundamental laws of nature operating today are L, and that extrapolating L backwards leads to a physically impossible state, we have to conclude that there was a beginning to the universe-governed-by-today's-laws and that we can have no knowledge of anything earlier than that. There might have been a physical universe governed by quite different laws, or there might have been no universe at all. But it is always simpler to postulate nothing rather than something; and so, in the absence of observable data made probable by the hypothesis that quite different non-fundamental laws were operating in the past, the hypothesis that the universe came into existence a finite time ago will remain the more probable hypothesis.[31]

What Swinburne argues here is that for reasons of simplicity one should never adopt scenario 2. Depending on the progress of science, we should either endorse scenario 1, if scientists succeed in developing well-confirmed fundamental laws and theories that enable us to extrapolate the history of the universe backwards forever, or we should endorse scenario 3, stipulating that the universe came into existence at state-P of Planck time.

There is no doubt that scenario 3 seems to offer the best prospects for cosmological arguments. The fourth Lateran Council proclaimed in 1215 as a doctrine of faith that the world has had a first temporal state or beginning, and the traditional doctrine that God created the world *ex nihilo* at some point in the past seems to be confirmed by scenario 3.[32] But what should we think of Swinburne's reduction of the second scenario to the third? My view would be that in this case, considerations of simplicity are overruled by considerations of background knowledge. We have ample experience of the fact that normally physical states are caused by earlier states, *even when we have not figured out the laws according to which this happens*. I suggest that given this background knowledge, we should accept as highly probable the hypothesis that state-P of the universe is caused by an earlier state, and so on, even if discovering the laws according to which this happened will turn out to be impossible. In other words, if state-S of the universe is physically impossible, scenario 2 is more plausible than scenario 3.

12.3 FROM AN INFINITELY OLD UNIVERSE TO GOD

Let us now focus on scenario 1 (and 2), and wonder whether Swinburne succeeds in establishing that even in the case of an infinitely old universe, there is room for a diachronic causal explanation of the existence of the universe by the hypothesis of

---

[31] Swinburne (EG), p. 140.
[32] Cf. McMullin (1981), pp. 29, 54, and 55; Brian Pitts (2008), p. 682.

theism. As I said, there seems to be no need or explanatory room for such a theistic explanation, because according to this scenario each state of the universe is caused by an earlier state. Surely one would violate Ockham's razor or the principle of simplicity if one postulated God as a diachronic cause of such a state as well, or perhaps of each state, so that there would be causal overdetermination. In order to resolve this impasse for the theist, Swinburne offers a number of arguments, the first of which I rebutted in Chapter 10.6.

As we saw, Swinburne establishes what I called the *principle of collective causality*, which says that 'in so far as a finite collection of states has a cause, it has its cause outside the collection'.[33] That is, concerning a diachronic series of causal states, one might say that if the present state a is caused by state b and b is caused by c, c is the cause of the collection $\{b+a\}$, and so on. Hence, the set of time-slices of the universe $\{b+a\}$ is caused diachronically by time-slice c, where c causes b directly and causes a indirectly. This principle holds only for all finite diachronic sets of causal states up to the present, for in each of such sets there is a first state in time, which is not explained by other members of the set. Swinburne correctly concludes that 'if the universe is of finite age, and [. . .] scientific causality alone operates, the whole collection of past states will have no cause and so no explanation'.[34] But to our surprise, Swinburne continues:

The same result follows if the universe is infinitely old (and so its history consists of the occurrence of an infinite collection of past states each lasting for the same finite time). The whole infinite series will have no full explanation at all, for there will be no causes of members of the series lying outside the series. In that case, the existence of the universe over infinite time will be an inexplicable brute fact. There will be an explanation (in terms of laws) of why, once existent, it continues to exist. But what will be inexplicable is its existence at all throughout infinite time. The existence of a complex physical universe over finite or infinite time is something "too big" for science to explain.[35]

As I pointed out in Chapter 10.6, Swinburne's argument commits a fallacy of ambiguity if this conclusion is meant to follow from the principle of collective causality. For this principle holds merely for finite sets of diachronic causal states. Only in the case of such finite sets is there a first temporal state of the universe that lacks a causal explanation with reference to an earlier state. But in the cited passage, I cannot detect any further argument, unless it is hidden in the word 'whole' (see below, section 12.4).

On the next page, Swinburne quotes with approval Leibniz, who argued in his tract *On the Ultimate Origin of Things* of 1697 that if our present copy of the book of the elements of geometry has been copied from an earlier copy, whereas each copy has been copied from an even earlier one and so ad infinitum, we would not have found a sufficient reason for its existence. Leibniz then argues that '[w]hat is true of the books is true also of the different states of the world', so that 'however far you go back to earlier states, you will never find in those states a full reason why there should be any world

---

[33] Swinburne (EG), p. 142.
[34] Swinburne (EG), p. 142.
[35] Swinburne (EG), p. 142.

Cosmological Arguments

rather than none, and why it should be such as it is'. Like Leibniz, Swinburne concludes without any critical comments 'that the existence of the universe over finite or infinite time would be, if only scientific explanation is allowed, a brute inexplicable fact', so that 'there is the possibility of an explanation of that existence in personal terms'.[36]

It is surprising that Swinburne takes the quote from Leibniz seriously and agrees with it, because it predates Cantor by two centuries. As turned out to be the case in Craig's arguments, Leibniz misleadingly uses as a paradigm for understanding infinite sets an example of a set that in fact can only be finite, to wit, the diachronic series of copies up to the present of a book that once was written. Moreover, the analogy between copies of books and time-slices of the universe suggests mistakenly that just as a book must have been created originally by an author, this must have been the case with the universe. But is our universe really like a book in this respect? That has to be shown, whereas Leibniz begs the question by assuming it without further argument.

We come to the following conclusion. Swinburne has not succeeded in arguing convincingly that if the universe is infinitely old, the present time-slice of the universe cannot but be explained, ultimately, by a personal explanation. On the contrary, if the universe is infinitely old, there is no explanandum *e* with regard to which one might propose theism as a best (ultimate) explanation, since 'no moment of time would be early enough' to bring about the existence of the universe.[37]

12.4 EXPLAINING 'THE WHOLE'

At this point, theists usually reply that, even though in the case of an infinitely old universe each time-slice can be explained causally by its preceding time-slice, this infinite series of time-slices *as a whole* is not yet explained. As Swinburne says, the 'whole infinite series will have no [...] explanation at all'. Although we saw that Swinburne's argument for this thesis with reference to the principle of collective causality is fallacious, we might wonder whether the thesis can be defended on other grounds. In order to do so, one would have to refute the standard objection to this thesis, which Swinburne quotes from Hume's *Dialogues*: 'But the *whole*, you say, wants a cause. I answer that the uniting of several parts into a whole [...] is performed merely by an arbitrary act of the mind, and has no influence on the nature of things.'[38] If Hume is right here, proposing *the whole* as an extra explanandum *e* apart from each of its parts is illegitimate.

However, Hume's maxim that if each of the members of a set is explained causally, the whole is explained causally, is true for some cases and false for others. If we want to explain why a whole bunch of letters is in a letterbox, we will say that each of them was thrown into the box separately. By explaining this for each letter, we have *eo ipso* explained it for the whole set of letters, and there is no meaningful explanandum left. But if we want to explain why there is a flock of sheep in the meadow, to say that the

---

[36] Swinburne (EG), p. 143.
[37] Cf. Chapter 10.6, above, and Swinburne (CT), p. 265.
[38] Swinburne (EG), p. 140; jo. Hume (1779), Part 9, p. 217. Cf. also Gale (1991), pp. 252ff., for an instructive discussion of this issue.

farmer let each sheep into the meadow will not be a sufficient explanation. We also have to explain why the sheep form a flock, that is, a specific type of structured whole. In this example, there are two types of explanation of this extra fact. We might investigate which proximate mechanisms in the brains of sheep cause them to flock together. And we might reconstruct the evolutionary history of sheep, in order to show why these proximate mechanisms were once adaptive, so that ancestors having them produced more offspring than ancestors that lacked them.

We might say, then, that if the members of a set do not form a structured whole, the collection of explanations of each will amount to an explanation of the whole set, as Hume claimed. If, on the contrary, the members do form a structured whole, there is room for an explanation of this structure, apart from the explanations of each member. How should we classify the infinite set of all time-slices up to the present of an infinitely old universe? Do these time-slices form a structured whole or not? And if the former, will it make sense to posit God as a possible explanation of this structure?

In a scientific explanation, we not only refer to causes or initial conditions in order to account for a phenomenon. We also use laws of nature. As Swinburne says, we not only specify *what* brought the phenomenon about, that is, its causes, but also *why* these causes brought the phenomenon about. 'To explain the occurrence of the high tide', for example, 'is to state what brought about the tide – the moon, water, and the rest of the earth being in such-and-such locations at such-and-such times, and why the moon, etc. had that effect – because of the inverse square law of attraction acting between all bodies'. For this reason, Swinburne distinguishes sharply between two components in the scientific explanation of an event E: the *what* that made E happen, that is, its causes, and the *why*, that is, the laws according to which the causes brought about E.[39]

It follows that the series of time-slices of the universe is a diachronically structured whole, at least if each (element of each) time-slice causes (each element of) its successor in conformity with laws of nature. The laws of nature describe this diachronic structure. But does it follow that the structure is an explanandum for which we might propose theism as a possible explanation? In other words, can we say meaningfully that the structured 'whole' of the infinite series of time-slices up to the present of an infinitely old universe is caused by God?

Swinburne answers this question in the affirmative. He claims that even if each of the infinitely many time-slices of the universe has a full explanation of a scientific kind,

that leaves open the possibility that there may be a deeper explanation, one in which the explanatory factors themselves are explained by a personal cause acting at the time when they act. In particular, the operation of the laws of nature may be due to an external cause whose action, together with the previous state of affairs, provides a complete explanation of any given

---

[39] Swinburne (EG), pp. 23–4. Swinburne writes: 'the "what" that made E happen and the "why" that made E happen' (p. 24). But this is misleading, because laws of nature are not causes. They do not 'make' things happen. Perhaps it is also misleading to refer to laws of nature as the 'why', for these laws are not reasons for which things behave as they behave. Cf. Rundle (2004), p. 39: 'To be told that the movement exhibits lawlike rather than random behaviour is, of course, to be told something about the character of the movement, but it is not to be told *why* the body moves as it does' (Rundle's italics).

state. This will be the case if a person G brings it about by a basic action at each period of time that the laws of nature L operate; and so brings it about that $S_{-(n+1)}$ brings about S_{-n}.[40]

However, there is no room for such a 'deeper explanation', because this passage contains two fatal confusions. First, there is a confusion between the *what* and the *why* in a causal explanation. The *why* in scientific causal explanations consists of laws of nature. These laws of nature neither are causes of phenomena, nor can be caused themselves, for it does not make sense to ask, for example: 'what caused the inverse square law of attraction?' Only causes, that is, the *what* in scientific explanations, can be caused by other causes. But if theists posit God in order to explain phenomena, they are arguing 'to a causal explanation of the phenomena in terms of the intentional action of a person', that is, to a *cause*, as Swinburne stresses elsewhere.[41] So one cannot meaningfully propose God or God's action as a cause of the laws of nature in the scenario of an infinitely old universe. That specific fundamental laws of nature hold in such a scenario is not an explanandum that can be explained by postulating a supernatural *cause*.[42]

In other words, the quoted passage contains a fallacy of ambiguity concerning the expressions 'explanatory factors' and 'the operation of the laws of nature'. That laws of nature are operating may be explained in terms of deeper laws of nature, so that a *why* may explain another *why*. And the existence of causes may be explained by the existence of deeper or earlier causes. However, the expression 'that laws of nature are operating' is ambiguous between the *why* and the *what*. For one might also say that the inverse square law of nature is operating at a specific time in such a way that there is a high tide because the Moon and the Earth are in a specific relative position. What one then means is that the latter phenomenon causes the former *according to* the law. But one cannot infer, as Swinburne does, that there may be yet another cause of the fact that the inverse square law is operating, apart from the positions of the Earth and the Moon, to wit, God. For by so arguing, either one supposes that laws of nature may be caused, which is nonsensical, or one violates Ockham's razor by positing causal overdetermination of phenomena without necessity.

Someone might object that laws may very well be caused. Do we not say that kings or governments cause laws to hold in a specific country by promulgating them? Indeed, to promulgate a law is by definition to cause a law to be brought into effect by means of an official public declaration. But if one raises this objection, one commits a second confusion, which may be implicit in the quote from Swinburne as well. One should sharply distinguish between laws in the legal sense and laws of nature. The first are normative whereas the second are descriptive. If legal laws are violated, the trespasser may be punished, whereas violations of proposed laws of nature will motivate natural scientists to modify their theories. Although we can meaningfully

---

[40] Swinburne (EG), p. 142. I have changed the subscripts in the quote in order to harmonize them with my notation. Cf. (EG), p. 107, note 10: 'The far more plausible view [compared to occasionalism] advocated in the text is that God causes the operation of scientific causality'.

[41] Swinburne (EG), p. 23, and *passim*.

[42] One might object that according to the SPL (substances-powers-liabilities) account of scientific laws, creating laws of nature *is* possible in principle, because it is *nothing but* creating kinds of substances with specific powers and liabilities. But in that case, God must have created these substances at some point in time, which is a possibility ruled out in the scenario of an infinitely old universe. Cf. for the SPL account, which Swinburne prefers: (EG), pp. 33–5, 162–4.

ask for the causes of the fact that a specific legal law holds in a country, we cannot meaningfully ask for the causes of the fact that specific fundamental laws of nature are true.

We have to conclude again that if the universe is infinitely old, there is no room for the hypothesis that God caused the existence of the universe. As Hume already wrote: 'How can any thing, that exists from eternity, have a cause; since that relation implies a priority in time and a beginning of existence?'[43]

12.5 THE SYNCHRONIC COSMOLOGICAL ARGUMENT

According to most monotheistic religions, God not only *created* the universe at some point in the past, but also *sustains* its existence at each moment. Many Christian philosophers have thought so as well, such as Thomas Aquinas, Descartes, Leibniz, and Berkeley. Might the theist not propose God as a *synchronic* or *sustaining* cause of each time-slice of the universe, even if the series of time-slices up to the present is infinite? In this manner, theists who want to avoid the risk of God-of-the-gaps might meet the requirement that their cosmological argument holds, irrespective of which cosmological scenario will ultimately be preferred by scientific experts.

As is clear from the last quote in the preceding section, Richard Swinburne suggests that there is room for postulating God as a synchronic sustaining cause of the universe, even if it is infinitely old. As he says, there may be a 'deeper explanation, one in which the explanatory factors themselves are explained by a personal cause acting at the time when they act'.[44] If there were such a deeper explanation, God would ensure 'by an intention continuing over infinite time that the whole infinite series of states exists', so that 'the power of the universe to continue its existence' would depend on a person who keeps the universe in being.[45] I have argued that if this deeper explanation is interpreted as providing a cause for the fact that laws of nature hold or are operative, there is no legitimate explanandum, since it cannot be said meaningfully of laws of nature that they are caused. But may we not provide a more charitable interpretation of this 'deeper explanation'? Can we not say that there must be a sustaining cause of the fact that things continue to exist, which is operative at the very time at which they exist? In other words, is the continued existence of things not a meaningful explanandum, for which we might postulate God as a synchronic sustaining cause?

There is no doubt that we can meaningfully ask for a sustaining cause of specific phenomena in some contexts. If a certain state or entity always continues to obtain or exist for a limited time, but now is going on much longer, we might ask what causes its continued existence. Or if there is a disintegrating cause, which acts upon a certain state or thing, we might wonder what counters or inhibits this cause, if the state or thing continues to exist. For example, a vegetable may usually rot within a week, but now it remains fresh for a month. What sustains its fresh condition? No doubt, preservatives were added. Some phenomena even require a sustaining cause for their continued existence, irrespective of how long they last. An example is the tone

---

[43] Hume (1779), Part 9, p. 217.
[44] Swinburne (EG), p. 142.
[45] Swinburne (EG), pp. 143, 144.

produced by a violoncellist, although in this case the sustaining cause is identical with the producing cause. But there is no doubt either that in many other cases we think it queer, or a symptom of misunderstanding, if one asks for a sustaining cause of a continued existence. For example, if someone enquires why a body continues to move through space in a straight line at the same speed, the correct answer within the framework of classical mechanics is that no cause is acting on it at all, as is implied by the law of inertia. Or if one wonders why a proton continues to exist when it is not at the end of its lifetime and nothing causes it to decay, the correct answer is, again, that no positive cause is needed for its continued existence.

It is false, then, that in order to explain the continued existence of the universe, we have to postulate some sustaining cause for everything at each moment. As Bede Rundle argues at length, 'no form of causation, divine or otherwise, is in general required to ensure persistence in being'.[46] And if a sustaining cause of some specific phenomenon is required, we will look for it within the universe. It is not very informative to say that our vegetable stayed fresh because of God's will, for example. What we have to remember is that speaking about the universe just means speaking about the set of all things spatio-temporally related to each other. If some of these things need a sustaining cause in order to stay in a certain state, or in order to continue their existence, this cause will be another thing or process in the universe, and not God.

At this point we may confront the proponent of the synchronic cosmological argument with a dilemma. With regard to some specific phenomenon, it is either generally acknowledged that it needs a sustaining cause or it is not. In the first case, the sustaining cause will be found within the universe, such as the preservatives for vegetables, or the violoncellist for a lasting tone. To posit God as an extra sustaining cause in such cases would violate the principle of simplicity or Ockham's razor, because one would postulate causal overdetermination without necessity. In the second case, critics of theism will object that God is postulated as an explanans for a non-existing explanandum.

A doctrine of double causality will not help the apologist either. According to such a doctrine, there are two types of causality. Scientific investigations can only discover secondary causes of phenomena, whereas God is the primary sustaining cause of everything. Even if secondary sustaining causes were superfluous, such as in the case of a stone or a pen which continues to exist, God would be needed as a primary sustaining cause. But how could one establish or make probable by an inductive argument to the best explanation that God is such a sustaining cause for everything at each moment? In other words, how can the fact that something continues to exist yield a good C-inductive argument for theism? One would have to meet the relevance condition spelled out above. That is, one would have to show that $P(e|h\&k) > P(e|k)$, where h is the hypothesis of theism and e is the empirical fact that something, such as my pen, did not cease to exist a moment ago.

But why would we esteem it more probable that my pen continues to exist if God exists than if God does not exist? Nobody thinks it likely that a pen turns into nothingness within the next second. Even if the pen is burned or run over by a

---

[46] Rundle (2004), p. 93.

truck, it does not disappear into nothingness. On the contrary, we are absolutely sure that this will never happen if God does not exist. But if God exists and is omnipotent, he might annihilate (in the strict sense of turning into *nihil*) my pen now. How should we discover that he does not want to do so, just in order to show me his omnipotence? It follows that the relevance condition is not satisfied for the hypothesis of God as a sustaining cause.

According to traditional Christian theologians, each entity would immediately fall into nothingness unless God sustained it in being.[47] We might call this the Principle of the Natural Collapse into Nothingness (PNCN).[48] If this were the case, the universe would indeed stand in need of a synchronic sustaining cause at each moment. But why should we assume that an abyss of nothingness threatens each entity at every moment of its existence, and endorse the PNCN? As long as no convincing arguments for this devilish assumption are put forward, we should reject it out of hand. So we come to the same conclusion as at the end of the last section. Both in the case of the diachronic cosmological argument for an infinitely old universe and in the case of the synchronic cosmological argument, there is no meaningful explanandum on which the argument can be based.

12.6 FROM THE BIG BANG TO GOD

It may have become superfluous by now to discuss the question as to whether there is a correct C-inductive argument from the Big Bang to God. Because the argument from the universe to God as a sustaining cause is unconvincing, theists who want to avoid the threat of God-of-the-gaps have to hold that *both* in scenario 1 of an infinitely old universe *and* in scenario 3 of a finitely old universe, 'there is the possibility of a person *G* being the ultimate cause' of its existence.[49] But such a possibility turned out to be absent in scenario 1. As a consequence, Swinburne's overall cosmological argument fails, as do all other cosmological arguments that aim at avoiding the God-of-the-gaps risk.

Yet there may be theists, inspired by Pope Pius XII, who want to bet on the ultimate tenability of standard Big Bang theory in order to exploit it in a cosmological argument for theism. For it may seem that according to standard Big Bang theory, there must have been a first temporal state or beginning of the universe which has not been caused by an earlier time-slice of the universe. Hence, there seems to be room for God's creative action as a diachronic cause of the universe *only* in scenario 3. If that is

---

[47] Cf. Thomas Aquinas, *Summa Theologiae*, I, Q. 104, Art. 1. Cf. Rundle (2004), p. 87, and Grünbaum (2004), p. 569.

[48] This principle is somewhat stronger than the principle of an ontological Spontaneity of Nothingness (SoN), which, according to Grünbaum (2000, 2004), is presupposed by the question why there is something rather than nothing. Swinburne says that he rejects a metaphysical version of SoN although he accepts an epistemological version (Cf. Swinburne (2005), p. 920). But Grünbaum (2005) replied correctly that (the metaphysical version of) SoN 'is being peremptorily taken for granted' by Swinburne (p. 930).

[49] Swinburne (EG), p. 145.

so, must theists not attempt to argue from the Big Bang to God? But this is a risky strategy for at least three reasons.[50]

First, the extrapolation from an expanding universe backwards in time to a Big Bang singularity on the basis of the general theory of relativity (GTR) is a bold and somewhat dubious one. Not only does the empirical success of GTR in weak- and medium-strength gravitational fields provide little support for the reliability of the theory in very strong gravitational fields, such as near the postulated Big Bang.[51] Standard textbooks on GTR also stress that 'at the extreme conditions very near the big bang singularity one expects that quantum effects will become important, and the predictions of classical GTR are expected to break down'.[52]

Secondly, given the underdetermination of theories by the data, (infinitely) many alternative theories are possible in principle, which explain the present data equally well as GTR, but do not lead to the extrapolation of a Big Bang singularity. In the case of arguments from the Big Bang to God, the risk of God-of-the-gaps arguments is very serious indeed, even if physicists have not yet succeeded in developing a well-confirmed alternative theory.[53]

Finally, there is a pragmatic reason for theists to abandon this argumentative strategy. Arguments from the Big Bang to God 'depend crucially on various highly technical premises which most people cannot even entertain, much less evaluate'. If we intend to rely on experts in this area, most of us will discover that we 'cannot even reliably identify relevant experts'. Should one's belief in God really 'depend on which factor ordering for the Hamiltonian constraint is correct in quantum gravity', as one author asks ironically?[54]

Notwithstanding these considerations, I shall discuss Richard Swinburne's C-inductive argument from the Big Bang to God. Since Swinburne thinks that there is room for a personal explanation of the universe whether it is finitely or infinitely old, he develops his cosmological argument with the existence of a complex universe as the explanandum e.[55] However, because there turned out to be a legitimate explanandum for neither a synchronic argument, nor for a diachronic argument if the universe is infinitely old, we have to modify somewhat Swinburne's formulations.[56] We may reconstruct his cosmological argument for the scenario of a finitely old universe as follows:

1. There has been a first temporal state or beginning of the universe, which is the explanandum e.
2. There is no scientific explanation of this first state, because it is 'too big' for science to explain.
3. Hence, there is room for a personal explanation, which postulates (the creative act of) a person G as the cause of this first state, and thereby as an ultimate diachronic cause of there being a universe at all.

---

[50] Cf. Brian Pitts (2008).
[51] Brian Pitts (2008), p. 696.
[52] Wald (1984), p. 100; quoted by Brian Pitts (2008), p. 695.
[53] Brian Pitts (2008), pp. 695–7.
[54] The quotes in this paragraph are from Brian Pitts (2008), pp. 697–8.
[55] Swinburne (EG), pp. 145–52.
[56] Cf. Swinburne (EG), pp. 145–52: *The Argument to God*.

4. Since no background knowledge is relevant to hypotheses about the ultimate cause of the universe, we have to use the criterion of simplicity for deciding which candidates for G have the highest prior probability (assuming that they have the same predictive power).
5. The theistic hypothesis of God defined as an omniscient, omnipotent, perfectly free, and perfectly good person is simpler than any other religious hypothesis. In particular, it is simpler than polytheism. It also has a higher predictive power than polytheism.
6. Hence, 'the choice is between the [first temporal state of the] universe as stopping point and God as stopping point' of diachronic causal explanation. In other words, we may specify P(e|k) in the present application of Bayes' theorem as the probability that the first state of the universe exists uncaused, that is, 'without anything else having brought it about'.[57]
7. Although the assumed first temporal state of the universe is much less complex than the now existing universe, which has 'a complexity, particularity, and finitude [...] that cries out for explanation', 'it had to have a certain kind of complexity built into it if there was to result a complex physical universe'.[58]
8. The hypothesis that God exists is 'an extremely simple supposition', since 'the postulation of a God of infinite power, knowledge, and freedom is the postulation of the simplest kind of person that there could be'. In particular, God is unextended and the divine properties are simple because 'they are properties of infinite degree'.[59]
9. Hence (from 7 and 8), 'the existence of [the first state of] the universe is less simple, and so less to be expected a priori than the existence of God', even if 'it is vastly improbable a priori that there would be anything at all'.[60] In other words, $P(e|k)$ as defined in (6) is low, whereas $P(h|k)$ is much higher (h is the hypothesis that God exists).
10. As we saw in Chapter 9, above, Swinburne argues that the probability that God creates humanly free agents is roughly ½, and that, since humanly free agents need bodies and a material world in order to make significant moral choices, the probability that God creates a material world for this reason alone is no less than ½. In any case, Swinburne states that 'the logical probability that, if there is a God, there will be a physical universe is quite high'.[61] He assumes without further argument that $P(e|h\&k)$ is also quite high if e is the first temporal state of the universe.
11. It follows (from 9 and 10) that there is a good C-inductive cosmological argument from the first temporal state of the universe (e) to the existence of God (h), since the relevance condition $P(e|h\&k) > P(e|k)$ is satisfied.[62]

Swinburne summarizes his argument in simple words as follows: 'There is quite a chance that, if there is a God, he will make something of the finitude and complexity

---

[57] Swinburne (EG), pp. 147 and 149. The addition between square brackets is mine.
[58] Swinburne (EG), p. 150.
[59] Swinburne (EG), pp. 150–1.　　[60] Swinburne (EG), p. 151.
[61] Swinburne (EG), p. 151.　　[62] Swinburne (EG), pp. 151–2.

of a universe. It is very unlikely that a universe would exist uncaused, but rather more likely that God would exist uncaused. Hence, the argument from the existence of the universe to the existence of God is a good C-inductive argument.'[63]

Let us start our discussion of this reconstructed argument from the Big Bang to God by solving two problems of interpretation. First, what role in Swinburne's summary is played by the phrase 'it is [...] rather more likely that God would exist uncaused'? We might formalize the sentence in which this phrase occurs by the formula $P(e|k) \ll P(h|k)$, where e is the first temporal state of the universe, h the hypothesis of theism, and k merely conceptual or tautological background knowledge. But there is no need for this claim in a C-inductive argument. In order to yield a correct C-inductive argument, it suffices that the relevance condition $P(e|h\&k) > P(e|k)$ be satisfied. My interpretation is that the phrase plays a role in Swinburne's subsidiary argument to the effect that $P(e|k)$ is quite low, so that it is in any case lower than the probability of roughly ½ attributed by Swinburne to $P(e|h\&k)$. In my reconstruction, this argument is summarized in steps 7–9.

Far more important is the second interpretative problem. What is the best candidate for the first temporal state of the universe postulated in premise (1), that is, for the explanandum e of the cosmological argument from the Big Bang to God? Should we decide that it is the cosmological singularity, the state-S when the entire universe was like the centre of a black hole, characterized by infinite curvature, in which space and time are all tangled up? Or should we rather decide that it is state-P of the universe at Planck time, when the universe began to behave roughly in accordance with the laws of physics we endorse today?

As we saw, Swinburne prefers the second option, arguing that we should assume that state-P is not caused by any earlier state of the universe, because 'it is always simpler to postulate nothing rather than something'.[64] But as I argued in section 12.2, state-P is an implausible candidate for a *first* state of the universe, a state that is not caused by earlier states. Furthermore, and as a consequence, if theism is proposed as an explanation of state-P, it will enter into a competition with rival explanations that are developed by physicists, such as theories of a bouncing universe. State-P of the universe clearly does not yield an e that *in principle* is 'too big' for science to explain.

For these reasons, theists may seem to be well advised to take state-S of the cosmological singularity as the first temporal state of the universe, if at least they want to argue from a finitely old universe to God. For if at time $t = 0$ of the Big Bang singularity, space-time itself came into existence, there can be no physical cause of the universe at $t = 0$. Should one not conclude that only the singularity is in principle 'too big' for science to explain, and that, consequently, there is room for a personal explanation, which posits the creative act of a person G (steps 1–3 in my summary of Swinburne's argument)?

Yet there are some embarrassing problems as well if one specifies e as state-S of the cosmological singularity. First, many physicists and philosophers, Swinburne included, believe that the singularity must be seen as a theoretical fiction, a mere idealization, and not as a physical reality that once existed. How can the entire mass of the universe

---

[63] Swinburne (EG), p. 152.
[64] Swinburne (EG), p. 140.

have been compressed into a pointlike space with infinite curvature and zero volume? Density is defined as the ratio of mass to unit volume. It follows that the density at the cosmic singularity is $n/0$, where n is the enormous but finite number of kilograms of mass in the universe. It has been argued that this density is infinite, not in Cantor's sense that it can be mapped onto a transfinite set, but because it is impermissible to divide by zero.[65] But does this notion of infinite density make sense, if dividing by zero is not defined for real numbers?[66] And if it makes sense, is state-S of an infinite density not physically impossible, as Swinburne apparently thinks?

When we assume, however, that state-S of the Big Bang singularity is not conceptually or physically impossible, a second problem emerges. The great advantage of specifying explanandum *e* of the cosmological argument as the Big Bang singularity seemed to be that this state is the very *beginning* of the present universe, so that *in principle* physics cannot causally explain it. But in standard handbooks of cosmology, the argument that *physics* cannot explain state-S often merges into an argument that the very question as to the cause of state-S is altogether *meaningless*. If that is the case, however, state-S of the Big Bang singularity is not only 'too big' for science to explain. It would also be impossible for theology to explain the singularity, because it simply *would not make sense* to say that it was caused diachronically by God's creative act.

In the fourth edition of Kaufmann's cosmology textbook, we read for example:

> At the moment of the Big Bang, a state of infinite density filled the universe [...]. Thus, we cannot use the laws of physics to tell us exactly what happened at the moment of the Big Bang. And we certainly cannot use science to tell us what existed before the Big Bang. These things are fundamentally unknowable. The phrases "*before* the Big Bang" or "at the *moment* of the Big Bang" are meaningless, because time did not really exist until *after* that moment.[67]

This quote shows that it is not easy to formulate coherently what Kaufmann wants to say. In the first half of the passage he uses the very expressions which he claims at the end are meaningless. But that it can be formulated coherently is shown by a philosopher such as Bede Rundle.

As Rundle argues, the 'notion of something's having had a beginning concerns a state of affairs which is intelligible only as occurring *within* the universe'.[68] The same point holds for the concept of an initial event. Although standard Big Bang theory says that the universe is only finitely old, it is misleading to conclude that there must have been a beginning or initial event of the universe, so that there is room for God as a diachronic cause of this beginning or initial event. We cannot conceive of a (spatio-)temporal void into which the Big Bang exploded, because the Big Bang was the unfolding of space-time itself. Hence, we 'are spared from having to fathom a mystery

---

[65] Quentin Smith, 'Atheism, Theism, and Big Bang Cosmology', reprinted in Martin and Monnier (2006), p. 52: 'Since it is impermissible to divide by zero, the ratio of mass to unit volume has no meaningful and measurable value and *in this sense is infinite*' (author's italics). Cf. ibidem, pp. 50–3 and 71–3 for Quentin Smith's criticisms of the idea that the cosmic singularity must be seen as a theoretical fiction. Cf. also Craig and Smith (1993), *passim*.
[66] Cf. Suppes (1957), §8.5, for an introduction to the vexing problem of defining division by zero.
[67] Kaufmann (1994), p. 530.
[68] Rundle (2004), pp. 120–1 (my italics).

on the far side of the Big Bang, since there is no far side'.[69] It follows that if the singularity cannot have a physical diachronic cause, this is because it can have no diachronic cause whatsoever. Hence, lemma (3) of my summary of Swinburne's cosmological argument does not follow from premises (1) and (2), whereas premise (1) is misleading.[70]

A third and final reason to reject cosmological arguments that specify e as state-S of the Big Bang singularity is that this singularity cannot qualify as an 'evident facet of experience' with which cosmological arguments have to begin. Even if the singularity is a physically possible state, many alternative cosmological views have been proposed, according to which this state has not in fact occurred. For example, according to quantum gravity proposals including the Hartle–Hawking no-boundary conditions, space-time is finitely old, whereas there is no singularity.[71]

From these initial considerations concerning the question as to how e has to be specified in cosmological arguments from the Big Bang to God, many readers will conclude that there is not much prospect for such arguments. If the explanandum e is state-P of the universe at Planck time, there is no good reason for assuming that it is not caused by earlier physical states, even if physicists will not be able to reconstruct such earlier states in a reliable way. For this reason, theists should prefer to specify e as the Big Bang singularity. But as we saw, it is dubious whether state-S is physically possible, whether one can meaningfully say that it has been caused, and whether it has in fact existed.

However, let us now suppose for the sake of argument that there really has been a Big Bang singularity (premise 1), and that it makes sense to ask for its cause. Let us also suppose that there is no physical cause of state-S of the universe (premise 2), and that, consequently, there is room for a G as the personal cause of this singularity (lemma 3).[72] In this case, it will be possible in principle to stage a correct C-inductive cosmological argument from the cosmic singularity (e) to the existence of God (h) if and only if the relevance condition $P(e|h\&k) > P(e|k)$ is met. But is this condition satisfied? Is the cosmic singularity really more to be expected if theism is true than it would be otherwise (cf. lemma 9–11)? We shall now discuss the two terms of the relevance condition, respectively.

12.7 THE PROBABILITY OF THE SINGULARITY GIVEN THEISM

How should we determine the value of the likelihood $P(e|h\&k)$? According to Richard Swinburne, there is a logical probability of roughly ½ that God wants to

---

[69] Rundle (2004), p. 118.
[70] Theists will react to this second problem by claiming that apart from the space-time framework we know by experience, there is another type of time, metaphysical time, within which God's act of creation occurred. But we can have no idea what is meant by 'metaphysical time', since our ordinary notion of time is informed by experience, whereas we have no experience of such a metaphysical time.
[71] For a popular account of this proposal, see Hawking (1988), pp. 143–9.
[72] I also assume here, for the sake of argument, that there is a meaningful conception of a bodiless G, or number of G's, and that it is meaningful to say that a bodiless person caused something physical to exist. The first assumption was contested in Chapter 7, above, whereas the second is denied on good grounds by Rundle (2004), pp. 8–10, 27–8, 74–80, 93, 148–58.

create humanly free agents. Because these creatures need to have complex bodies in order to make morally significant free choices, the probability that God will create a complex physical world is no less than ½ for this reason only, so that $P(e|h\&k) \geq$ ½ if e is a complex physical world.[73] As I stated in lemma 10 of the summary (section 12.6), Swinburne assumes without further argument that $P(e|h\&k)$ is also quite high if e is the Big Bang singularity. But unfortunately, this does not follow. As Quentin Smith has argued, in this case $P(e|h\&k) = 0$, or very near to zero, so that the last sentence of lemma (10) states an assumption which is false.[74]

The crucial premise of Smith's argument is that the cosmic singularity is inherently chaotic, so that it 'would [...] emit all configurations of particles with equal probability'.[75] Hence, if theists were right in believing that the creation of humans was the main reason for God to create a physical universe in the first place, God would never have attempted to create humans indirectly by causing the cosmic singularity as the first state of the universe. It follows that standard Big Bang theory is inconsistent with the hypothesis that God exists, is omnipotent, and wanted to create the human species (let i stand for this last conjunct). In other words, if e is the cosmic singularity, $P(e|h\&i\&k) = 0$, whereas $P(e|h\&k)$ is nugatory.

In his article 'Atheism, Theism, and Big Bang Cosmology' of 1993 and in subsequent papers, Quentin Smith has defended this 'cosmological argument for God's non-existence' against many objections. Theists will protest, for example, that after having created the cosmic singularity, God might have intervened in the developing universe in order to ensure the genesis of man. But as Smith correctly replies, this would be in conflict with God's rationality and omnipotence. If God really wanted to create man, it would be 'a sign of incompetent planning to create as the first natural state something that requires immediate supernatural intervention to ensure that it leads to the desired result'.[76] Also, this scenario of 'guided evolution' violates the requirement that one should avoid God-of-the-gaps arguments.

In order to refute another objection, Smith dives into the logic of counterfactuals. Could one not refute his argument by saying that if standard Big Bang theory is true, the cosmic singularity *in fact* gave rise to the human species, albeit accidentally and very indirectly? Would God then not have known before creation that if he produced the singularity, humans would ultimately be the accidental result, since God is omniscient? Smith replies to this objection that current theories of counterfactuals require for the possible truth of a counterfactual either a law of nature on which it is based, or an actual world with which the relevant possible world can be compared. Since the first condition is not met in the case of the Big Bang singularity, whereas the second is not satisfied before creation, God cannot have known before creation that the following counterfactual was true: 'if I create the cosmic singularity, man will eventually and accidentally evolve after some 13.7 billion years'.[77]

---

[73] Swinburne (EG), p. 151; jo. (EG), Chapter 6.
[74] At least, this is so according to Quentin Smith if one includes in $h\&k$ God's intention to create humans.
[75] Quentin Smith, 'Atheism, Theism, and Big Bang Cosmology', in Craig and Smith (1993). I quote from the reprint in Martin and Monnier (2006), p. 43. The quoted sentence comes from S. W. Hawking (1976), p. 2460.
[76] Smith in Martin and Monnier (2006), p. 47.
[77] Smith in Martin and Monnier (2006), pp. 54–6 and 73–5.

Theists might disagree with present philosophical theories concerning counterfactuals, and argue on the basis of a new theory, yet to be invented, that God can and will have known before creation that the singularity would eventually and accidentally lead to the evolution of the human species. In that case, there is no *contradiction* between standard Big Bang theory and the claim that God exists and intended to create humanity by causing the cosmic singularity as a reliable mechanism that would ultimately produce our species. But as far as we can see, it remains extremely improbable that if God wanted to create humans, he chose to do so by creating the cosmic singularity in the first place. No omnipotent craftsman would opt for such a risky and long-winded detour.[78] In other words, if the predictive power $P(e|h\&i\&k)$ in this case is not zero, it will be negligibly small, and $P(e|h\&k)$ will be nugatory as well.

Swinburne seems to endorse this assessment in another context. As he says,

> There is no doubt that the theory that the universe began at a point is simpler and so intrinsically more probable than any particular theory that it began with many substances, so much simpler that I suggest that it is more probable than the disjunction of all theories claiming that it began from or always contained many substances.
>
> But, if it did begin from an unextended point, the simplest theory of such a beginning would seem to be that that point would have had no power to produce extended substances.[79]

Whereas the first sentence of this passage may be interpreted as expressing yet another reason for preferring the Big Bang singularity to state-P of the universe as the explanandum *e* of a cosmological argument from the Big Bang to God, the second sentence seems to confirm the view that if *e* is this singularity, $P(e|h\&k)$ is very low indeed.

Swinburne is mistaken, then, in thinking that if there is a probability of roughly ½ that God created a complex universe containing the human species, it follows that there also is a probability of ½ that God did so by creating the Big Bang singularity. On the contrary, this probability should be considered negligible or at least very small, so that it will be difficult to demonstrate that the relevance condition $P(e|h\&k) > P(e|k)$ is satisfied. It is met only if $P(e|k)$ is even lower than $P(e|h\&k)$. Can this be shown?[80]

12.8 THE PRIOR PROBABILITY OF THE SINGULARITY

The prior probability of the Big Bang singularity can be spelled out as follows: $P(e|k) = P(e|h\&k).P(h|k) + P(e|\sim h\&k).P(\sim h|k)$, *h* being theism, *e* the singularity, and $\sim h$ the hypothesis that theism is false.[81] As we saw in Chapter 11.5, $\sim h$ is a

---

[78] I am assuming here, with Swinburne, that our moral intuitions give access to God's intentions. If not, the hypothesis of theism will barely have any predictive power, as I argued in Chapter 9. So it will not do to object that God's intentions are inscrutable.

[79] Swinburne (EG), pp. 163–4. Cf. also p. 106: 'The simplest scientific starting point [of an explanation of the existence of the universe] would be an unextended point'.

[80] Cf. for an argument to the effect that $P(e|h\&k)$ is nought if *h* = theism and if *e* = *our* universe: Clark (1989), §§ 3 and 4.

[81] Cf. Swinburne (EG), p. 72.

catch-all hypothesis, which comprises all possible alternatives to theism. If we form the set S including theism and all its possible rivals, so that S is both exclusive (each member is incompatible with all others) and exhaustive (one member must be true), $P(e|k)$ can be expanded as follows: $P(e | k) = P(e|h_t \& k) \cdot P(h_t|k) + P(e|h_1 \& k) \cdot P(h_1|k) + P(e|h_2 \& k) \cdot P(h_2|k) + P(e|h_3 \& k) \cdot P(h_3|k) + \ldots$, where h_t stands for theism and 'h_1', 'h_2', 'h_3', etc. stand for rivals to theism.

What are the members of S apart from the hypothesis of theism, and how many members does S have? We assumed for the sake of argument that there can be no physical cause of the cosmological singularity, so that S does not contain any physical theory. The set S will include all religious theories according to which 'some person G' (as Swinburne says), or more than one god, caused the cosmic singularity. Clearly, there are infinitely many possible religious theories, since polytheistic religions may posit any number n of gods for $n > 1$. It will also be possible to construct an infinite number of monotheistic theories incompatible with theism, assuming, for instance, that there are infinitely many degrees of knowledgeableness between complete ignorance and omniscience, and attributing a different degree to each alternative unique god. Finally, the set S will include the theory that the existence of the Big Bang singularity was an ultimate brute fact, which had no cause whatsoever. Let me call this latter theory the hypothesis of a *Generatio Spontanea Mundi*, abbreviated as GSM.

As we shall see, Swinburne attempts to argue that all members of set S apart from theism and GSM can be discarded, so that $P(e|\sim h \& k)$ should be interpreted as the probability that the Big Bang singularity occurred 'as a brute, inexplicable fact'. Accordingly, 'the choice is' between the cosmic singularity 'as a stopping point' of all diachronic explanations and 'God as a stopping point' (see lemma 6 in section 12.6).[82] He uses two types of reasons for eliminating all rivals of theism and GSM: one type is concerned with their prior probabilities and the other with their explanatory powers (or 'likelihoods'). Let me briefly discuss his arguments, beginning with the problem of priors. How should we determine the prior probabilities of each of an infinite number of alternative theories?

According to a traditional Bayesian answer, we should adopt the Principle of Indifference, which says that before considering any evidence about which member of an exclusive and exhaustive set of hypotheses is true, one must assign them equal priors, unless background knowledge indicates otherwise. Because Swinburne assumes that in the case of explaining the cosmic singularity, no empirical background knowledge can be relevant, it follows that all priors are equal.[83] And because S has infinitely many members, whereas according to the Principle of Countable Additivity, the priors of all exclusive and exhaustive alternatives must add up to 1, the prior of each must be less than any finite number, that is, either 0, or infinitesimal if one allows infinitesimals.[84] Since it is usually assumed that only

---

[82] Cf. Swinburne (EG), p. 147.

[83] As I implied in the introduction to this chapter, I assume here for the sake of argument that Swinburne's elimination of background knowledge is successful (cf. Chapter 11.6 for my argument to the contrary).

[84] Cf. Swinburne (EG), p. 332, note 1, and (EJ), Additional Note G.

contradictions have a prior of 0, Swinburne prefers to use infinitesimal numbers in other contexts.[85]

However, this outcome is unacceptable, both in itself and for Swinburne in the context of natural theology. It is unacceptable in itself because of well-known problems concerning the Principle of Indifference, which arise since there are many ways to divide the logical possibility space into parts.[86] And Swinburne cannot endorse it because if the prior probability of theism were infinitesimal, there would not be much hope that one can ever show on the basis of all evidence pro and contra, that the posterior probability of theism exceeds ½, as he intends to argue. How can one avoid the assumption that if S has infinitely many members, the priors of each of them are either infinitesimal or 0?

As we saw in Chapter 11.6–9, Swinburne holds that if there is no relevant empirical background knowledge, the prior probability of a theory before any evidence can be determined mainly by its simplicity, because (allegedly) simplicity is a mark of truth.[87] Furthermore, he claims that 'infinite' properties such as omniscience or infinite velocity are simpler than finite properties.[88] On these grounds, he argues that the prior probability of theism is higher than that of its rivals, and even higher than the sum of the priors of many rivals:

[...] it may still be argued that, although any particular hypothesis of a finite god or many finite gods is less probable than theism, still there are so many possible hypotheses of finite gods with different degrees of power, and knowledge and different numbers of gods, that surely it is more probable that one of these is correct than that theism is correct. But [...] consideration of the weight we give to simplicity in other areas of inductive inquiry suggests that we normally give it such weight that a really simple hypothesis is intrinsically more probable than a disjunction of many more complex hypotheses.[89]

---

[85] Swinburne (EJ), p. 244: 'So we should adopt the Principle of Countable Additivity as an axiom additional to the Axiom of Finite Additivity, and, if necessary, use infinitesimal numbers in our calculations'. In (ICT), he preferred the opposite view, according to which the prior of each of infinitely many equiprobable hypotheses is 0 (p. 71). But then, all posteriors calculated on the basis of Bayes' theorem would be 0 as well. Cf. Bradley (2002), p. 388, who does not notice that Swinburne changed his view between ICT and EJ.

[86] Cf. Sober (2008), pp. 27–8. For example, one might consider the monotheism according to which there is one non-omniscient god who knows the truth value of between one billion and two billion propositions as one theory, but one also might spell out one billion different monotheisms, each of which claims that their god knows the truth value of a specific number of propositions between one billion and two billion. Of course, these problems are particularly pertinent if one assumes that S has a finite number of members.

[87] Cf. Swinburne (EG), pp. 145–6. Cf. p. 109: 'The intrinsic probability of theism is, relative to other hypotheses about what there is, very high, because of the great simplicity of the hypothesis of theism'.

[88] Swinburne (EG), p. 55.

[89] Swinburne (EG), p. 146. Of course, this passage should not be interpreted as saying that the prior of theism is greater than the sum of the priors of *all* its mutually exclusive rivals contained in the catch-all set $\sim h$. For in that case, the prior probability of theism would be greater than ½, because $P(h|k) + P(\sim h|k) = 1$, and no empirical evidence would be needed in order to conclude that the posterior probability of theism is greater than ½. Cf. Gutenson (1997) for this uncharitable interpretation.

Furthermore, Swinburne claims that the predictive power of theism is higher than that of polytheism if one adds to the explanandum of a complex universe the (alleged) fact that the universe is governed throughout space and time by the same natural laws. For if this temporal order in the world is to be explained by many gods, 'some explanation is required for how and why they cooperate', which makes polytheism even less simple. However, if these gods do not cooperate, 'we would expect to see characteristic marks of the handiwork of different deities', such as different laws of nature obtaining in different parts of the universe.[90] From these considerations concerning simplicity and predictive power, Swinburne concludes that 'the choice is between the universe as stopping point and God as stopping point' of all explanations (lemma 6).[91]

Does Swinburne succeed in eliminating all members of set S except theism and GSM, so that we may 'regard $P(e|\sim h\&k)$ as the probability that there be a physical universe without anything else having brought it about'?[92] I do not think so, for the following reasons. First, Swinburne's argument to the effect that the predictive power of theism is greater than that of polytheisms with regard to a law-governed universe is inapplicable if e is the cosmological singularity, which is inherently chaotic. As we saw, the predictive power of theism with regard to the Big Bang singularity is negligible, and there is no reason to assume that the predictive power of each of its religious rivals is even smaller. Can we discard all rivals of theism except GSM by arguing that on the criterion of simplicity, the prior probability of theism is greater than the priors of each of its religious competitors, as Swinburne avers? This has been contested both for bad and for good reasons.

It is a bad reason, for example, to say that the (truth-conducive) simplicity of a theory can be measured by the number of *kinds* of entities posited by a theory, but not by the number of *individual* entities of each kind, so that one cannot assign a higher prior to theism than to all polytheisms for reasons of simplicity.[93] Why should Ockham's razor, according to which *entia non sunt multiplicanda praeter necessitatem*, not apply to existence claims concerning individual entities? If one makes many such existence claims, the risk of falsehood is greater than if one makes one existence claim only, as theism does. When one admits that simplicity may be a seal of truth, or at least a property of theories that diminishes the risk of falsehood, there is no reason to restrict that admission to the number of *kinds* postulated by a theory.

Far more problematic is Swinburne's application of the simplicity criterion to properties and magnitudes. In order to eliminate all monotheistic rivals to theism, he has to argue that theism is somehow simpler than each of its conceivable monotheistic alternatives. An essential part of his argument runs as follows:

(a) Hypotheses 'attributing infinite values of properties to objects are simpler than ones attributing large finite values'.[94]

---

[90] Swinburne (EG), p. 147. [91] Swinburne (EG), p. 147.
[92] Swinburne (EG), p. 149.
[93] Bradley (2002), p. 389: 'D.C. Williams long since pointed out that the actual number of entities postulated by a theory, what he calls its "gross tonnage", is irrelevant to the question of simplicity'. Cf. Williams (1966), p. 133. Apparently, Bradley assumes that all gods are of one kind.
[94] Swinburne (EG), p. 55.

(b) Some properties that theism attributes to God, such as omnipotence and omniscience, have infinite values.[95]
(c) No other monotheistic theory can attribute infinite values to these properties or posit a person that is simpler than God in other respects, or equally simple.
(d) So, theism is simpler, and therefore a priori more likely to be true, than its monotheistic rivals. As Swinburne says, theism 'postulates the simplest kind of person that there could be'.[96]

There are several objections to this argument, which show that it is unconvincing. For example, why should one endorse premise (a) and also assume that this type of simplicity is truth-conducive, as is needed for deriving (d)? According to Swinburne, attributing infinite values is simpler than attributing large finite values for two reasons. First, a formulation of a law is 'mathematically simpler than another in so far as the latter uses terms defined by terms used in the former but not vice versa'.[97] He then argues that, accordingly, 'hypotheses attributing infinite values of properties to objects are simpler than ones attributing large finite values', because 'we can understand, for example, the notion of an infinite velocity [...] without needing to know what the googleplex is ($10^{10^{10}}$)'.[98] But this argument is a *non sequitur*, since the inverse is also true: we can understand the googolplex without understanding a notion of mathematical infinity. Furthermore, understanding mathematical infinity has turned out to be very difficult in view of the contradictions that seemed to inhere in this concept. Finally, there is no good reason to think that ease or simplicity of understanding is related to the probability that certain properties (large ones or infinite ones) in fact exist. In other words, the measure of simplicity consisting in ease of understanding does not seem to be truth-conducive (cf. Chapter 11.8).

Swinburne's second reason for claiming that attributing infinite properties is simpler (in a truth-conducive sense) than finite ones is that 'scientific practice shows this preference for infinite values over large finite values of a property'.[99] His favourite example is that scientists preferred to assume for reasons of simplicity that the speed of light is infinite until data were found that refuted this hypothesis.[100] But historical analysis shows that this was not the case.[101] Empedocles argued, for example, that light was something in motion, and for that reason must take some time to travel. For Aristotle, light was not a movement but rather the actualization of the potential transparency of air and other media, so that the question of its speed could not arise. According to Alhazen, who deeply influenced medieval optics, the propagation of light takes time, even though this is hidden to our senses.[102] And whereas Descartes

---

[95] Swinburne (EG), pp. 97–8.
[96] Swinburne (EG), p. 97.
[97] Swinburne (EG), p. 54.
[98] Swinburne (EG), p. 55; 'googleplex' is Swinburne's spelling. Usually, the googolplex is defined as 10 to the power (10 to the power 100).
[99] Swinburne (EG), p. 55.
[100] Swinburne (EG), p. 55, cf. p. 97. Cf. for a critical analysis of Swinburne's other examples, Bradley (2002), pp. 391–4. Incontestable empirical evidence for a finite speed of light was first collected by Ole Rømer in 1676, by noting discrepancies in the apparent period of planet Jupiter's moon Io.
[101] Cf. MacKay and Oldford (2000).
[102] Bradley (2002), p. 393, with reference to Crombie (1979), I, p. 114.

did in fact hold that the speed of light is infinite, he did not do so for reasons of simplicity. Rather, his doctrine of the instantaneous propagation of light fitted in with his mechanistic theory that light is a form of pressure, which is propagated instantaneously, in the same way that the pressure exerted at one end of a stick is instantaneously present at the other end.[103]

Of course, attributing infinite values to magnitudes might be simpler than attributing finite values for computational reasons, but again there is no reason to assume that such a measure of simplicity is truth-conducive. Hence, if premise (a) is true, it is not true in the sense needed for deriving the conclusion (d). With regard to premise (b), one might wonder whether the divine properties of omnipotence and omniscience are 'infinite' in the required mathematical sense. This has been denied by Descartes, for example.[104] However that may be, we can safely conclude that one cannot eliminate all rivals of theism and GSM from set *S* by arguing that theism is the simplest possible theory because it attributes infinite properties to God.

It follows from this conclusion, however, that we cannot determine the value of $P(e|k)$, so that we cannot show that the relevance condition $P(e|h\&k) > P(e|k)$ is satisfied. Who can exclude that apart from theism there is another monotheistic hypothesis, which explains the Big Bang singularity better than theism does? In order to show that this is more than a shot in the dark, I shall now briefly sketch such a hypothesis. It is inspired by David Hume's suggestion that our present universe, with its mixture of good and evil, is better explained by a morally indifferent god than by the God of theism.[105] Let us abbreviate 'morally indifferent god' by MIG, and wonder whether theism is, or is not, superior to the MIG-hypothesis.[106]

The context of Hume's suggestion is very different from the present argumentative framework. Hume wondered whether one may correctly infer God's goodness from morally mixed worldly phenomena by an analogical design argument, whereas we are considering the question as to which of two rival hypotheses, theism or the MIG-hypothesis, provides a better explanation of the Big Bang singularity. Furthermore, whereas Hume supposed that his morally indifferent god is neither omniscient nor omnipotent, I shall stipulate that MIG is both, and also perfectly free, in order to refute premise (c) of the argument summarized above. But this stipulation seems to raise a problem concerning the coherence of the MIG-hypothesis. Can one suppose

---

[103] Descartes, *Le monde ou Traité de la lumière*, AT xi, p. 99.

[104] Descartes, *Réponses aux premières objections*, AT ix, pp. 89–90; *Principia Philosophiae* I, §§ 26–7. Quentin Smith (1998) points out that Swinburne uses the term 'infinite' in at least three different senses: (a) of the first transfinite cardinal, aleph-zero, (b) of the instantaneous propagation of light, and (c) of the maximum degree of a degreed property, so that his argument for God's simplicity given the 'infinity' of God's properties commits a fallacy of equivocation (pp. 92–3). And if God knows all numbers, his omniscience is infinite in (d) Cantor's 'absolute' sense (the number of all transfinite cardinals). Furthermore, Gwiazda (2009) wonders whether God's properties can be called 'infinite' given all the restrictions of these properties Swinburne introduced in *The Coherence of Theism* in order to make theism coherent. Accordingly, Swinburne's argument that God is the simplest possible person because of the infinity of his properties contains some fallacies of ambiguity.

[105] Hume (1779), part x–xi.

[106] My MIG hypothesis is akin to the epicurean hypothesis proposed by Bradley (2007), but developed somewhat differently in order to avoid Swinburne's criticisms of Bradley in Swinburne (2007).

without contradiction that there is a god who is omniscient, omnipotent, perfectly free, *and* morally indifferent?

The answer to this question depends on whether we use a formal or a substantial notion of morality. According to a formal notion, which Swinburne adopts, the morally best action is the one for which there is an *overriding reason*, and there is an overriding reason for doing an action if it is *overall better* than the available alternatives.[107] Because God is omniscient, he will know all reasons for doing or not doing actions, so that he will be able to determine in each case for which action there is an overriding reason, if any. Because God is perfectly free, he is not subject to *akrasia*, and hence will always choose to do the actions for which there is an overriding reason. Because God is omnipotent, he will always perform these actions. Consequently, the hypothesis of a morally indifferent god is self-contradictory if one uses this formal notion of morality.[108]

According to a more substantial notion of morality, however, acts are morally good if they instantiate paradigmatic moral values such as courage, charity, gratitude, humility, temperance, curiosity, perseverance, generosity, and so on. Clearly, the answer to the question which acts of a subject are overall better than the alternatives, so that there is an overriding reason for doing them, will depend upon the specific values that the subject does or should endorse, and upon the interrelations between these values. But as I argued in Chapter 9.3, there is no good reason to assume that God has the same moral values as we humans have, or is a norm-subject for the same moral norms. Indeed, none of the moral values I just mentioned *makes sense* for a god having properties such as omnipotence or omniscience, and who is unique in his kind instead of being a social animal.

It follows that although it is contradictory to suppose that an omniscient, omnipotent, and perfectly free god is morally indifferent in Swinburne's formal sense, it is coherent and even plausible to suppose that such a god is indifferent to the moral values we humans cherish. Consequently, such a god will be morally indifferent in our substantial sense of morality. Now suppose that there is a probability of ½ that MIG will create something and an equal probability that MIG will not create at all. Suppose also that MIG highly values chaotic things, because she does not want to be bored. Then it follows that whereas the prior probability of the MIG-hypothesis as determined by Swinburne's measures of simplicity is equal to the prior probability of theism, its predictive power with regard to the Big Bang singularity may be greater, since this singularity is inherently chaotic.

We must conclude again that Swinburne fails to eliminate all rivals to theism from set *S*, apart from the hypothesis of a *Generatio Spontanea Mundi* (GSM). Consequently, $P(e|k)$ cannot be calculated, so that there is no correct C-inductive argument from the Big Bang singularity to God. The best thing theists can do, it seems, is to argue that theism is better than some specific rival, such as polytheistic Hinduism, by using the law of likelihood (cf. Chapter 11.4–5). Unfortunately, however, the theistic hypothesis cannot even prevail over the MIG-hypothesis in this manner, so that Hume was right, after all.

---

[107] Cf. Swinburne (EG), p. 101; Swinburne (2007), pp. 274–5.
[108] Swinburne (2007), p. 275.

12.9 GOD OR THE BIG BANG AS A BRUTE FACT?

Suppose, however, that the attempt to eliminate from set S all rivals of both theism and the theory of *Generatio Spontanea Mundi* (GSM) was successful. Then the choice is between God's *creatio ex nihilo* and the Big Bang singularity as the ultimate stopping point of all explanations. Which option should we prefer? Is there a correct C-inductive argument for theism if $P(e|k)$ is the probability of the singularity as a brute fact?

According to Richard Swinburne, there is such a C-inductive cosmological argument, and its force is considerable. As we saw, he argues that the logical probability that God creates a complex universe apt for humans is 'quite high', that is, at least ½ (lemma 10 of the summary in section 12.6).[109] He also argues that the existence of the universe, or rather of the Big Bang singularity, as a brute fact 'is not a priori very probable at all'.[110] Hence, the relevance condition $P(e|h\&k) > P(e|k)$ is met in a substantial way if $P(e|k)$ is the prior probability of the universe existing as a brute fact, so that $P(h|e\&k)$ will substantially exceed $P(h|k)$. I argued above, however, that the value of $P(e|h\&k)$ would be much smaller than ½ if e were the Big Bang singularity. How should the value of $P(e|k)$ be determined if e is the existence of the singularity as a brute fact and k is nothing but tautological or conceptual background knowledge?

I hold that this value cannot be determined non-arbitrarily, so that there is no convincing C-inductive cosmological argument from the Big Bang singularity to God. As Richard Dawkins says with regard to applications of Bayes' theorem in the philosophy of religion, the 'GIGO principle (Garbage In, Garbage Out) is applicable here', because the input of the theorem is subjectively judged.[111] Swinburne thinks otherwise. He argues on the following two grounds that the existence of the Big Bang singularity as a brute fact 'is not a priori very probable': (1) 'because (it may well seem) it is vastly improbable a priori that there would be anything at all', and (2) 'because, if there is anything, it is more likely to be God than an uncaused [singularity]'.[112]

It is unclear how one can justify the first reason, which seems to be an arbitrary assumption.[113] What should we think about the second? As we would expect, Swinburne argues for (2) on the basis of the simplicity criterion. This is clear from lemma (9), according to which 'the existence of [the first state of] the universe is less simple, and so less to be expected a priori than the existence of God'. Is this true if the first state of the universe is the cosmic singularity?

There are several good reasons for answering this question in the negative. First, it is striking that Swinburne in lemmas (7–9) no longer speaks about the simplicity of *theories*, but rather about the simplicity of *entities*. With regard to theories, there may

---

[109] Swinburne (EG), p. 151.
[110] Swinburne (EG), p. 151.
[111] Dawkins (2006), p. 106.
[112] Swinburne (EG), p. 151. I substituted 'singularity' for 'complex physical universe'.
[113] Cf. the discussion between Adolf Grünbaum and Swinburne on what Grünbaum calls the ontological Spontaneity of Nothingness assumption (SoN) in Grünbaum (2004), Swinburne (2005), and Grünbaum (2005).

be measures of simplicity that are truth-conducive. But why should we think that simpler entities are more likely to exist than more complex entities? I do not see any good reason for endorsing this claim.

Second, we must wonder whether it makes sense to compare the cosmic singularity and God in terms of simplicity. The term 'simple' belongs to what Peter Geach once called (logically) *attributive* adjectives, which he contrasted with (logically) *predicative* adjectives. In phrases of the form 'an $A\ N$' (where 'A' is an adjective and 'N' a noun), the adjective is logically predicative if the predication can be split up without loss of meaning into a pair of predications 'is an N' and 'is A', as can be done with 'x is a red book'. Otherwise, the adjective is logically attributive, as in 'x is a big flea'.[114] Geach argued that 'good' and 'bad' are attributive adjectives in this sense, because if one calls something good or bad, one always explicitly or implicitly conceives of it as an N, such as a good shotgun, a good walking stick, or a good cricketer. The reason is that only by specifying the noun do we know which standards or criteria are used to determine whether the thing or person is good, and to what degree.

Similarly, it is only by specifying the noun N that we know which criteria are relevant for determining whether something is simple or complex, and these criteria are different for different N's. Chess is a complex game, even though there are maximally 32 pieces on a board, whereas we would consider a building consisting of only 32 bricks or stones as a very simple structure. This is the reason why we talk nonsense if we want to compare things of very different kinds in terms of simplicity or complexity. Questions such as 'what is more complex, your job or the city of Rome?' do not make sense. Since God is defined as a pure and bodiless spirit, whereas the cosmological singularity, if real, is a physical entity, one cannot meaningfully ask whether God is simpler or more complex than the singularity. As a consequence, lemma (9) is neither true nor false, because it is meaningless.

One might attempt to solve this problem of meaninglessness by constructing a common measure of simplicity, which holds for entities of different kinds. If Wittgenstein's *Tractatus Logico-Philosophicus* were correct in claiming that there are logically simple propositions, which correspond to simple facts if they are true, we might take as a measure of simplicity for comparing spirits with physical entities the number of simple propositions a spirit believes or knows to be true, and the number of simple facts which hold with regard to the physical entity. Admittedly, we know that this measure will not work, since there are no logically simple propositions, as Wittgenstein discovered after having written the *Tractatus*.[115] But let us suppose it worked. Would we then be justified in concluding that God is a simpler entity than the cosmic singularity?

As we saw, Swinburne argues that God is a very simple entity, because God is a spirit of infinite power, knowledge, and freedom. Since God is an unextended object, whereas the divine properties fit together closely, he is 'the simplest kind of person that there could be'.[116] But the cosmic singularity at the start of the Big Bang is also defined by a number of infinities, being a pointlike state of infinite density and infinite temperature, characterized by an infinite curvature of space-time. If both God and

---

[114] Geach (1967), p. 64.
[115] Cf. Hacker (1986), Chapter V.
[116] Swinburne (EG), pp. 150–1 and Chapter 5, pp. 96ff.

the singularity are infinite in various respects, it will seem difficult to argue that God is simpler than the cosmic singularity.[117]

Indeed, there is a compelling argument to the opposite effect, if we apply the Tractatarian measure of simplicity. Since God is omniscient, he must have known everything about the cosmic singularity. In other words, the number of simple propositions God believes to be true about the singularity must equal the number of simple facts holding of the singularity. But as a supreme spirit, God also has perfect self-knowledge. As a consequence, the content of God's knowledge is always more complex than things existing apart from God, such as the cosmic singularity. And this conclusion holds also concerning our universe as it is now, with its 'complexity, particularity, and finitude', which, according to Swinburne, 'cries out for explanation'.[118] Since God, being omniscient, knows everything about it, his mind's content must be at least as complex as the universe. And because God also knows everything about all past stages of the universe, his omniscience is vastly more complex than the universe as it exists now.

To this line of argument, Swinburne might object that we should not conceive of God's omniscience on the model of human *propositional* knowledge. Allegedly, the knowledge of God is more like the knowledge we humans have when we inspect a complex situation at a single glance. God can know everything in one simple mental act.[119] But this reply is unsatisfactory for two reasons. If one takes this line, one *eo ipso* rejects the common *Tractatus* measure of simplicity, which allowed us to compare God and the cosmic singularity under the aspect of complexity. As a result, this intended comparison slides back into meaninglessness, because 'simple' is an attributive adjective.

Moreover, if we cannot attribute ordinary propositional knowledge to God, many things believers say of him become meaningless as well. What might it mean to say, for example, that God knows us to be sinful creatures, or that he knows the objective truths of ethics? What is worse, it now has no sense to say that there is a probability of ½ that God intended to create humanly free agents and a physical universe, if it can no longer make sense to attribute any propositional attitude to God. As a result, the endeavour of applying Bayes' theorem and staging arguments to the best explanation in order to make probable the existence of God, becomes unintelligible.

At the end of this long chapter, we come to the following conclusions. Since there (probably) are no sound deductive cosmological arguments, inductive cosmological arguments to the best explanation seemed to be more promising. But neither for the argument from an infinitely old universe to God, nor for the synchronic causal argument to God as a sustaining cause, was there a valid explanandum. It follows that inductive cosmological arguments cannot be immune from the risk

---

[117] Swinburne seems to admit this in Chapter 14 of EG, where he writes: 'Let me allow for the sake of argument (despite my doubts on this) that the unextended point hypothesis is as simple as theism' (p. 336).

[118] Swinburne (EG), p. 150.

[119] Swinburne (2010), pp. 18–19): 'So I think that the best analogy for God's beliefs are the beliefs we acquire when we look at a scene before our eyes. Merely by looking we acquire innumerable beliefs about what objects there are, where they are, and what they look like. We are aware of these beliefs but not as linguistic entities [. . .]. So too God's wider fused pre-linguistic state of belief is one integrated state of himself. It does not consist of separate items within himself but it is a property of himself'.

of God-of-the-gaps, because one cannot hold that these arguments are correct irrespective of which cosmological scenario obtains.

Furthermore, for various reasons it cannot be shown that the relevance condition is satisfied in the case of the cosmological argument from the Big Bang singularity to God. And finally, theists cannot claim convincingly that preferring God as the ultimate stopping point of all explanations is more satisfactory than accepting the Big Bang singularity as the ultimate stopping point, because allegedly the former is simpler than the latter, so that its existence is a priori more probable. If God and the pointlike Big Bang singularity can be compared at all in terms of simplicity, the former must be vastly more complex than the latter.

13

Arguments from Order to Design

As Richard Swinburne suggests, '[a] *priori* the existence of anything at all logically contingent, even God, may seem vastly improbable, or at least not very probable'.[1] What the apologetic natural theologian has to show, then, is that this presumed low probability of the existence of God prior to the examination of empirical evidence is raised sufficiently for a legitimate belief that God exists by the addition of C-inductive or confirming arguments supporting theism, even if one takes all evidence against the existence of God into account. In his book *The Existence of God* Swinburne considers some eleven arguments in such a cumulative case strategy. He claims that for most pieces of empirical evidence e_n which he adduces, where $n = 1, \ldots 11$, $P(h|e_n \& k) > P(h|k)$, so that most of these arguments are correct and indeed good C-inductive arguments for the existence of God.[2]

In Chapter 12, we found Swinburne's cosmological arguments (e_1) wanting in this respect. It is the aim of Chapters 13 and 14 to investigate which of the remaining inductive arguments based upon evidence e_{2-9} are indeed good C-inductive arguments for (or against) theism (I discussed the evidence e_{10} of miracles in Chapter 10.2–5). In the present chapter, two global design arguments will be examined, the argument from the law-governed nature of the universe and the argument from fine-tuning. But, Swinburne stresses, '[t]he crucial issue [. . .] is whether $P(h|e_{1-11} \& k) > \frac{1}{2}$', that is, whether the addition of all C-inductive arguments raises the probability that theism is true to above $\frac{1}{2}$.[3] In other words, does the cumulative case strategy yield a good P-inductive argument for the existence of God? Is his existence more probable than not, given the total evidence? In Chapter 15, I shall investigate this decisive issue, and also discuss Swinburne's argument from religious experience (e_{11}).

---

[1] Swinburne (EG), p. 151.
[2] Cf. Swinburne (EG), p. 17. C-inductive arguments are correct iff $P(h|e \& k) > P(h|k)$, and they also are 'good' arguments iff their premises are 'known to be true by those who dispute about the conclusion', as Swinburne says (EG), pp. 6–7.
[3] Swinburne (EG), p. 17. In his text, Swinburne writes somewhat misleadingly: 'whether $P(h|e_{11} \& k) > 1/2$'. I substituted e_{1-11} for e_{11}.

13.1 THE ARGUMENT FROM TEMPORAL ORDER

Whereas cosmological arguments start from the empirical premise that something, or our universe, or a first temporal state of our universe, exists or has existed, design arguments typically begin with an empirical premise that is more specific. Design arguments are based on evidence of order or functionality in empirical phenomena. Their advocates claim that this order or functionality is best explained by the hypothesis that it is designed and produced by some intelligent being.[4] Traditionally, these arguments are called 'arguments from design'. As has often been observed, however, the label 'arguments *to* design' would be more suitable, since the concepts of design and of a designer occur in the conclusion of the arguments but not in their incontestable empirical premise.

One may distinguish between global and local design arguments. In global design arguments, the empirical evidence adduced is an overall feature of the universe, which is either evidently given in experience or established beyond reasonable doubt by scientific research. Local design arguments start from specific phenomena within the universe. As we have seen in Chapter 10, sophisticated theists will attempt to avoid the risk of God-of-the-gaps, so that the evidence for theism must be either 'too big' or 'too odd' for science to explain. Global design arguments are more promising in this respect than local ones. Since local arguments are based upon the evidence of specific natural phenomena, these phenomena may be explained in principle by discovering their causal genesis within the universe. This is why, I suppose, Richard Swinburne focuses primarily on global design arguments, or teleological arguments, as he calls them.[5] The first argument of this kind that he proposes, is the argument from temporal order.

In general, temporal order is defined as the global feature of the universe that there are regular successions of events which can be codified in relatively simple laws of nature. How should the explanandum *e* of temporal order be specified in such a way that a God-of-the-gaps explanation is ruled out, because *e* is too big for science to explain? One cannot describe this explanandum in terms of a *specific* set of laws of nature, for each specific set might be explained scientifically by a set of more fundamental laws. There is no good reason to suppose that there must be a set of most fundamental laws of nature, which future science cannot explain in principle.

Swinburne suggests that this problem may be solved by saying that 'although there is an explanation of the operation of any finite subseries' of such an infinite series of sets of ever more fundamental laws, 'there is no explanation of the operation of the whole series'.[6] In other words, he argues that an infinite series of sets of ever more

---

[4] With regard to cosmological arguments, Swinburne claimed that no deductive version could be valid: (EG), p. 136. On similar grounds, he holds that no deductive design argument can be valid: (EG), p. 155: 'Although the existence of order may be good evidence of a designer, it is surely compatible with the non-existence of one – it is hardly a logically necessary truth that all order is brought about by a person'. But as I argued in Chapter 12.1, whether a deductive design argument is valid will also depend upon the other premises used. I would argue that in all cases of deductive design arguments, at least one of the other premises is false, so that they are not sound, although they may be deductively valid.

[5] Swinburne (EG), p. 153. He discusses local design arguments under other labels, such as arguments from consciousness or from providence.

[6] Swinburne (EG), p. 158, footnote 3. Cf. for this move to 'the whole' also Chapter 12.4, above.

fundamental specific laws of nature might be an explanandum *e* that is too big for science to explain, so that the risk of a God-of-the-gaps can be avoided. But this is not correct, since if each set of laws is explained scientifically by a set of more fundamental laws, the infinite collection of all these explanations *eo ipso* amounts to an explanation of the whole infinite series of sets of laws, so that there is no explanandum left for theism. As I argued in Chapter 10.6, the most promising solution (if any) to this problem is perhaps to propose some vague description of the explanandum *e*, and this is what Swinburne often does.[7] He specifies *e* as the temporal 'orderliness of the universe', that is, 'its conformity to formulae, to simple, formulable, scientific laws'. And he claims that this temporal orderliness of the universe 'is a very striking fact about it', since '[t]he universe might so naturally have been chaotic'.[8]

By definition, the argument from temporal order is a correct C-inductive argument iff $P(h|e \& k) > P(h|k)$, where *e* is temporal order, *h* is the hypothesis of theism, and *k* includes the existence of the universe. As we saw, this is the case if, and to the extent that, the relevance condition $P(e|h \& k) > P(e|k)$, or, alternatively, $P(e|h \& k) > P(e| \sim h \& k)$, is met. What Swinburne has to show is, then, that the global feature of temporal order in the universe is more likely if God exists (*h*) and created the universe, than if God does not exist ($\sim h$). Although the negation $\sim h$ of theism is a catch-all hypothesis, which consists of the set of all possible explanations for the temporal order in the universe apart from theism, Swinburne does not consider any rival explanation. One reason is that the explanandum of temporal order, being defined very vaguely, allegedly is too big for science to explain, because each scientific explanation will presuppose some kind of temporal order in the universe.

This latter point might be illustrated by attempts to explain the temporal order in our universe by a multiverse hypothesis. According to such a hypothesis, there are many universes within the multiverse, and our universe is one of them. Exuberant advocates of a multiverse hypothesis claim that *all logically possible* universes in fact exist.[9] Some of these universes are completely chaotic, whereas others are law-governed. We live in a universe that is temporally ordered, since organisms cannot exist in a completely chaotic universe. But the fact that our universe is temporally ordered does not need a special explanation if the exuberant multiverse hypothesis is correct. The reason is that our universe is just one of the existing possible universes, some of which are chaotic. In other words, if all logically possible universes exist, the fact that our temporally ordered universe exists has probability 1, so that any further explanation of its existence is superfluous.

However, one might confront the proponents of a multiverse hypothesis with the following dilemma. Either there are convincing empirical reasons apart from the temporal order in our universe for thinking that many universes exist, or there are not. In the first case, specific empirical data in our universe will be best explained by the hypothesis that a long time ago there was a universe split in the multiverse, at which other universes budded off from our universe. But in that case, there must be fundamental laws of nature which explain both these empirical data and the budding-off events, so that the multiverse as a whole is temporally ordered. Hence, in this first

---

[7] In fact, Swinburne builds his argument 'on the not implausible assumption that there are most fundamental laws of nature' ((EG), p. 158, note 3), and it is this alleged fact 'that these are fundamental laws of nature', which is 'something "too big" for science itself to explain' (p. 160).
[8] Swinburne (EG), p. 154.   [9] Cf. David Lewis (1986a), *passim*.

case the multiverse hypothesis presupposes the vague feature of global temporal order that had to be explained.

In the second case, the multiverse hypothesis has no empirical confirmation other than the temporal order in our universe. But then 'it would be the height of irrationality to postulate innumerable universes' just to explain the temporal order in our universe, since it is much simpler to do so 'by postulating just one additional entity – God', as Swinburne observes.[10] In other words, the multiverse hypothesis 'represents about the most extreme negation of "Ockham's razor" that one could imagine'.[11] This use of the simplicity criterion or Ockham's razor is unobjectionable, if at least the notion of God makes sense and theism has the relevant predictive power. It conforms to Newton's first rule of reasoning in empirical science, according to which '[w]e are to admit no more causes of natural things than such as are both true and sufficient to explain their appearances'.[12]

It seems to follow that we do not need to consider any possible scientific hypothesis as a rival to the theistic explanation of the temporal order in our universe. This temporal order cannot be explained either by a so-called 'weak anthropic principle', according to which it is no wonder that we discover a temporal order in the universe, since without it we humans would not exist. This purported explanation confuses effects with causes, or conditions of discovery with the causes of these conditions. Of course it is true that a temporal order is a necessary causal condition for the existence of humans. But that does not explain causally why this temporal order exists.[13]

In assessing Swinburne's teleological argument from temporal order, we should not forget two caveats I argued for in the preceding chapter. First, it does not make sense to say that God caused laws of nature, or that God caused the fact that laws of nature are operating, except when this means that God *created a universe* governed by such laws (Chapter 12.4).[14] As a consequence, Swinburne's teleological arguments will turn out to depend essentially on his cosmological arguments, and the former cannot have any force if the latter do not have any force.[15] Furthermore, we have seen (Chapter 12.8)

---

[10] Swinburne (EG), p. 165.   [11] Davis (1987), p. 143.
[12] Newton (1729), Vol. II, p. 398.
[13] Cf. Swinburne (EG), pp. 156–7. I do not think either that in the case of temporal order in our universe (or of fine-tuning) the appearance of *e* is due to an observational selection effect (OSE), as is the case in the following example (invented by Eddington). From the fact *e* that all the fish we catch from a lake are more than 10 inches long, we might conclude that this evidence better supports hypothesis h_1 that all the fish in the lake are more than 10 inches long than it supports h_2 that only half of the fish in the lake are more than 10 inches long. However, if we caught the fish by using a net that cannot catch fish smaller than 10 inches (=k), the evidence does not favour either of these hypotheses over the other, because $P(e|h_1 \& k) = P(e|h_2 \& k) = 1$. In contradistinction to Sober (2003) and Sober (2004), I hold that the evidence of temporal order in the universe (or of the fine-tuning of our universe) is not due to an OSE, because in contradistinction to the fish example we cannot imagine a method of observation (for any possible observer) of our universe via which it would not exhibit instances of temporal order (or of fine-tuning), such as the sequence of days and nights on Earth, for example. Although it is true that we could not exist in a universe without temporal order (or fine-tuning), the evidence of temporal order (or of fine-tuning) is not due to, or produced by, our existence. Cf. Weisberg (2005) for other criticisms of Sober's view.
[14] The hypothesis that God first created a chaotic universe and at a later time induced temporal order into it, is difficult to reconcile with the theistic definition of God as omnipotent and omniscient. Surely, we should not imagine God as a mediocre craftsman.
[15] Incidentally, one might wonder whether teleological arguments can have any C-inductive force if the cosmological arguments on which they depend *are* good C-inductive arguments. Does the

that Swinburne's attempt to eliminate all rival religious explanations by using the criterion of simplicity is unconvincing, at least for the case of rival monotheisms. But here I shall assume for the sake of argument that there are no religious rivals of theism, which might explain the feature of temporal order better than, or as well as, theism.

Given this assumption, the relevance condition $P(e|h\&k) > P(e|k)$ will be met by Swinburne's first teleological argument if it can be shown that the global feature of temporal order in our universe is more likely on the hypothesis of theism than it is if the existence of our universe is an ultimate brute fact. In other words, what Swinburne has to show is both that

(1) 'it is a priori improbable that a Godless universe would be governed by simple laws', so that $P(e|k)$ is low,

and that

(2) 'there is quite a significant probability that a God-created universe would be governed by simple laws', so that $P(e|h\&k)$ is substantial.

If these two things can be established, the operation of (relatively simple) laws of nature would be 'evidence – one strand of a cumulative argument – for the existence of God'.[16]

13.2 THE ARGUMENT FROM TEMPORAL ORDER EVALUATED

As we shall see presently, Swinburne does not succeed in demonstrating these two points (1, 2). Yet his arguments are often misunderstood, and objections have been launched against them that are not completely convincing as they stand.

Swinburne's first probandum (1), according to which 'it is a priori improbable that a Godless universe would be governed by simple laws', may also be stated as follows: it is a priori probable that a Godless universe is completely chaotic. The reason for this near-equivalence is the following. If one accepts (like Swinburne) that simplicity is a truth-conducive criterion for theory choice, one will have to hold that the existence of a Godless universe governed by *simple* laws is a priori much more probable than the

---

evidence of the teleological arguments contain something new compared to the evidence of the cosmological arguments? It does not, if according to the cosmological argument God created a universe in order to enable life, and, in particular, in order to make possible significant free choices of humanly free agents, as Swinburne argues. Monton (2006, p. 421) correctly observes that from the cosmological argument, 'I already knew that, if God existed, God would have to choose to actualize our life-permitting universe from among a sea of non-life-permitting universes [. . .]. The fine-tuning evidence doesn't change any of that, and hence the fine-tuning evidence doesn't change my probability for the existence of God'. According to Monton, this 'is the strongest reply one can give to the fine-tuning argument' (ibidem). The same would hold for the argument from temporal order.

[16] Swinburne (EG), p. 166. Of course, Swinburne does not assume that one can give precise numerical values to these probabilities; cf. (EG), p. 17: 'In using the symbols of confirmation theory I do not assume that an expression of the form $P(p|q)$ always has an exact numerical value. It may merely have relations of greater or less value to other probabilities [. . .]'; cf. p. 68, and (EJ), p. 62: 'Inductive probability (unlike statistical or physical probability) does not normally have an exact numerical value'.

existence of a Godless universe governed by *complex* laws.[17] Hence, it will be difficult to show that the existence of a universe governed by simple laws is *even more* probable on the assumption that God created it, so that the relevance criterion would be satisfied. However, if one can establish that a Godless universe not governed by any laws at all, so that it is *completely chaotic*, is a priori more probable than the existence of a Godless universe governed by simple laws, the prospects that the relevance criterion will be satisfied are somewhat better.

It has been objected that one cannot even hope to establish this greater prior probability of a chaotic universe, since the notion of a chaotic universe is conceptually incoherent. Nicholas Everitt argues for this contention as follows. Kinds of stuff or objects are identified in part by their causal powers. For example, bread typically nourishes human beings, fish can swim, and diamonds can scratch sapphires. Furthermore, speaking about the powers of objects is a way of referring to regularities in their behaviour. So it is logically impossible that the universe contains kinds of stuff or objects and yet is entirely chaotic, since this supposition is conceptually incoherent. Everitt concludes that 'there cannot be a *separate* argument from order', since '[f]or there to be a material universe at all is for there to be an *orderly* material universe'. In other words, 'Swinburne is wrong to say that "the universe might so naturally have been chaotic"'.[18]

This objection would be convincing if it also were a conceptual truth that no universe can exist that does not contain *kinds* of stuff or *types* of objects. However, the supposition that there is a universe, in the sense of a spatio-temporal continuum, which is not empty, even though its non-emptiness cannot be specified by saying that it contains certain *kinds* of stuff or *types* of objects, is not *obviously* self-contradictory. Let us suppose, then, that it is conceptually possible that a universe is entirely chaotic. If so, Swinburne's first probandum (1), according to which 'it is a priori improbable that a Godless universe would be governed by simple laws', because a completely chaotic Godless universe is a priori more probable, is not conceptually incoherent.

But how can one argue for this alleged a priori improbability? As we have seen in Chapter 11.8–10, Swinburne holds that considerations of simplicity are decisive here,

---

[17] Cf. Swinburne (EG), p. 161: 'a very simple theory [of the universe] is [intrinsically] more probable than a disjunction of many more complex theories'. This is the case if one applies the simplicity criterion to *individual* theories of the universe or to individual possible universes. But there is another way of applying the simplicity criterion, which Swinburne also employs, that yields the opposite result. Because (allegedly) 'there are a very large number of complex ways' in which a universe can be temporally ordered, whereas there are only a few simple ways, Swinburne says that the prior probability that the universe is ordered by simple laws is much smaller than the prior probability that it is ordered by complex laws (ibidem, p. 162). But if the simplicity criterion is used in this manner to *sets* of theories of the universe, it should also be applied to sets of religious theories. Since the set of polytheisms is infinitely large, whereas the set of theisms contains one member only, this application of the simplicity criterion would lead to the conclusion that the prior probability of theism is much smaller than the prior probability that one or another polytheistic theory is true. Similarly, the prior probability of theism would be much smaller than the prior probability that one of the infinitely many monotheistic theories that posit one finite god is true (cf. Wynn (1993), pp. 329–33, for this objection of internal inconsistency, and Chapter 12.8, above). Because it leads to this internal inconsistency in *The Existence of God*, I shall not discuss this second way of applying the simplicity criterion here (cf. also section 13.4, below).

[18] Everitt (2004), p. 89 (Everitt's italics).

so that his argument should be built upon such considerations only.[19] Let us now see whether he succeeds in making plausible his first probandum (1), assuming for the sake of argument that *simplex veri sigillum*. What Swinburne has to show is that the assumption of a *chaotic* Godless universe is somehow simpler, and hence more likely to be true, than the assumption of a Godless universe *governed by simple laws*. What are his arguments for this contention? Are these arguments convincing?

Swinburne provides three arguments, each adapted to one particular way of interpreting laws of nature. On the Humean account of natural laws as developed by David Lewis, laws are extrapolations from past behaviours of kinds of objects to their future behaviour, which belong to the simplest and most adequate system of such extrapolations.[20] Swinburne argues correctly that 'in a Godless universe on the Humean theory of laws of nature there is no more fundamental explanation of the coincidence in the ways in which objects behave'.[21] But does this argument show that the assumption of a chaotic Godless universe is somehow simpler than the assumption of a Godless universe governed by simple laws? Not at all, although this was the first probandum. If this is what Swinburne wanted to show, his argument is an *ignoratio elenchi*.

Furthermore, on Swinburne's own measures of simplicity, the assumption of a universe governed by simple laws is simpler than that of a universe governed by complex laws. But if one wants to compare the former with the assumption of a completely chaotic universe in terms of simplicity, how should one proceed? Is positing a completely chaotic universe even more complex than positing a universe governed by complex laws? Or is it simpler than positing a universe governed by simple laws? We have no intuitions here, and this fact can be explained easily. Since a completely chaotic universe can neither contain kinds of objects nor be governed by laws, the comparison in terms of simplicity between a chaotic universe and a universe governed by simple laws does not make sense. The reason is that there is no possible common measure of simplicity (cf. Chapter 12.9).

For example, one cannot say that a chaotic universe is simpler because it contains fewer kinds of objects than a universe governed by simple laws, because the very notion of kinds of objects does not make sense with regard to a completely chaotic universe. One cannot argue either that a chaotic universe is simpler because it contains less individual entities, since counting entities is possible only if they are somehow individuated as entities of certain kinds, as Frege convincingly argued.[22] Finally, one cannot say that a completely chaotic universe is simpler than a universe governed by simple laws because the laws of the former universe are even simpler than the laws of the latter, since there are no laws at all that govern a completely chaotic universe. It follows that one cannot succeed in arguing convincingly for the first probandum on the Humean conception of laws of nature.

According to a second conception, defended by David Armstrong and others, laws of nature are logically contingent relations between universals, which are instantiated

---

[19] Cf. Swinburne (EG), p. 161: 'Among theories of the universe as a whole (which will thus have equal scope), simplicity is the sole indicator of intrinsic probability'. Cf. p. 163: 'Simplicity is the sole relevant a priori criterion'.
[20] Swinburne (EG), pp. 27–31, and Lewis, 'A Subjectivist's Guide to Objective Chance', in Lewis (1986).
[21] Swinburne (EG), p. 160.   [22] Frege (1884), §§ 46 ff.

by entities and events in the universe.²³ Swinburne suggests without further argument 'that a universe without connections between universals would be simpler than one with connections'. Accordingly, 'it would be very probable that there would be no connections between universals at all – that the universe would be chaotic'.²⁴ His reason is, I guess, that if one merely asserts that there are instantiated universals, one claims less than if one also asserts that there are logically contingent relations between these universals, which constitute the physical necessity of natural laws. The first assertion is both simpler and more likely to be true than the second, since if one claims less, one runs a smaller risk of advancing a falsehood. But unfortunately, this argument falls prey to Everitt's objection. If one thinks of a universe as containing objects instantiating universals, these objects belong to kinds. And no universe containing kinds of objects can be entirely chaotic, since kinds of objects are partly defined in terms of regularities.

Finally, laws of nature can be conceived of as regularities in the causal powers and liabilities of individual substances in nature. The generality of laws of nature, then, is explained by the contingent fact that substances fall into kinds, so that all entities of the same kind to some extent have the same powers and liabilities.²⁵ Like the Humean account, this substances-powers-and-liabilities or S-P-L account of laws of nature, which Swinburne prefers, 'raises the question of why so many substances have similar powers and liabilities to each other'.²⁶ On the S-P-L account this question can be answered in terms of 'the causal ancestry of substances'.²⁷

At this point, Swinburne discusses three possible cosmological scenarios.²⁸ First (a), if the universe is infinitely old, the fact that each time-slice of the universe contains substances falling into a number of kinds can be explained by the fact that this also was the case in the preceding time-slice. Second (b), if the universe is finitely old and started with a first state that is 'literally pointlike', it would be simplest to suppose that this pointlike singularity 'would have had no power to produce extended substances' at all, or if it had, 'that it would have the power to produce just one extended substance'.²⁹ Finally (c), if the universe began with a very dense but not pointlike state, this state 'must itself have consisted of a very large number of substances of very few kinds' if it 'was to give rise to our present universe of very few kinds of substance'.³⁰ Swinburne concludes that in each of these three scenarios, the claim that the universe once consisted of many substances falling into relatively few kinds with identical powers and liabilities is a priori 'a theory of a very improbable coincidence', and that '[s]uch a coincidence cries out for explanation in terms of some single common source with the power to produce it'.³¹

Again, we should wonder whether Swinburne's argument is not an *ignoratio elenchi*. Does he really show that the supposition of a chaotic Godless universe is a priori more probable than the supposition of a Godless universe governed by simple laws? What Swinburne says concerning the first and the third scenario (a, c) does not show this.

---

[23] Cf. Swinburne (EG), pp. 32–3; and Armstrong (1983, 1997).
[24] Swinburne (EG), p. 161.   [25] Swinburne (EG), pp. 33–4.
[26] Swinburne (EG), p. 34.   [27] Swinburne (EG), p. 162.
[28] Swinburne (EG), pp. 162–4. I discuss them here in another order than Swinburne does.
[29] Swinburne (EG), p. 164.   [30] Swinburne (EG), p. 163.
[31] Swinburne (EG), p. 164.

The argument that in each of these scenarios the limited number of kinds of substance in our present universe is explained by saying that this was also the case in an earlier time-slice of the universe, does not yield the conclusion that the assumption of a completely chaotic universe is simpler, and for that reason 'cries out for an explanation' less loudly. What should we think, then, of the second scenario (b), according to which our universe is finitely old and evolved from a pointlike singularity? Does Swinburne argue and indeed show that in this scenario a completely chaotic universe is a priori simpler and therefore more to be expected than a universe consisting of a limited number of kinds of objects?

Swinburne makes three compelling observations concerning the second scenario (b). First, he says that according to his 'assessment of the present state of cosmology', an 'evolution from a very dense state is more probable than evolution from an infinitely dense state'.[32] I share this assessment, and as I said in Chapter 12.2, it is implausible to assume that such a postulated very dense state was not caused by earlier states of the universe. Second, as Swinburne observes correctly (if at least we assume that simplicity is truth-conducive), 'the theory that the universe began at a point is simpler and so intrinsically more probable than any particular theory that it began with many substances'. He even suggests that it is 'so much simpler [. . .] that it is more probable than the disjunction of all theories claiming that it began from or always contained many substances'.[33] It follows that if we abstract from scientific a posteriori reasons for preferring one of the three scenarios to the other two, we should prefer the second scenario (b) for reasons of simplicity. Finally, Swinburne claims that given a Big Bang theory postulating an infinitely dense first state of the universe, the genesis of a universe like ours, which contains many substances falling into relatively few kinds, is highly unlikely. As he says, if the universe 'did begin from an unextended point, the simplest theory of such a beginning would seem to be that that point would have had no power to produce extended substances'.[34] What he means by this is perhaps that such a point would probably produce a completely chaotic universe, since, as Everitt argued, the presence of substances conceptually implies the presence of order.

So we may conclude that Swinburne has demonstrated his first probandum (1) for the second scenario (b) only, if he has demonstrated it at all. Assuming that the universe started with an infinitely dense pointlike Big Bang, a chaotic universe may be more likely than a universe containing many substances falling in relatively few kinds. Expressed in a formula, we might say that $P(e|h_p \& k)$ is low, where e is the evidence of temporal order, and h_p is the hypothesis of a pointlike Big Bang. Can we now argue that the evidence of temporal order favours the hypothesis of theism over the hypothesis of a pointlike Big Bang, so that the relevance condition $P(e|h \& k) > P(e|k)$ is satisfied at least for the case that $P(e|k)$ is specified as $P(e|h_p \& k)$?

At this point it is crucial to see that the teleological argument from temporal order cannot be considered separately from the cosmological argument. What the theist

---

[32] Swinburne (EG), p. 163.   [33] Swinburne (EG), pp. 163–4.
[34] Swinburne (EG), p. 164. Cf. the arguments concerning the Big Bang singularity put forward by Quentin Smith (Chapter 12.7, above).

wants to argue is not that at some point in time God transformed a pre-existing chaotic universe into a temporally ordered universe, but rather that God created a universe which was temporally ordered from the very start.[35] So let us try to connect the upshot of our discussion so far to the results of Chapter 12 on the cosmological argument. As we saw in that chapter, there is no room for a hypothesis of divine creation of the universe if the universe is infinitely old. Moreover, there turned out to be no legitimate explanandum that might be explained by the hypothesis of God as a sustaining synchronic cause of the universe. It follows that both an inductive cosmological argument and an inductive argument from temporal order should start from a postulated first temporal state of the universe, so that these arguments always depend upon the contingent results of scientific cosmology. In other words, neither of these arguments can avoid the risk of God-of-the-gaps.

How should one specify the postulated first state of the universe e, in order to propose theism as its best explanation? As we saw in Chapter 12.6, one might either opt for a pointlike singularity as this first state (scenario b), or for some very dense but not infinitely dense state (scenario c). Let us investigate briefly what the predictive power of theism $P(e|h\&k)$ is for e in these two alternative scenarios (cf. probandum 2). On the assumption that there is a probability of roughly $1/2$ that God wanted to create humanly free agents, Swinburne argued that there is a probability of at least $1/2$ that God wanted to create a physical universe. The reason is that humanly free agents will only be able to make significant free choices if they are incarnated and live in a material world. If this is correct, the probability that God wanted to create a temporally ordered universe must also be at least $1/2$, since only if the material world is temporally ordered in a relatively simple way, it will enable humanly free agents to live, to learn from experience, to build their characters, and to make significant free choices.[36] If God created a temporally ordered world by creating a singularity in the first place, from which this world would evolve, this 'leads us to expect that God will bring about an initial singularity of the right kind'.[37]

At this point theists face the following dilemma. Assuming that God wanted to create humans, they have to argue that God probably created the temporally ordered world either by creating a pointlike singularity from which the world evolved (b), or by creating a very dense singularity or state from which the world evolved (c). As I argued in Chapter 12.7 with reference to Quentin Smith, however, the first option is excessively unlikely, if not contradictory, because the pointlike cosmic singularity is inherently chaotic. Hence, if God wanted to create a temporally ordered universe, he would not have done so by creating a pointlike singularity in the first place. In other words, if the theist opts for this first horn of the dilemma, the predictive power of theism $P(e|h\&k)$, where e is the pointlike singularity, will be near to zero, so that probandum (2) is false and (probably) the relevance condition will not be met.

---

[35] Cf. Swinburne (EG), p. 166: 'It is not possible to treat a teleological argument in complete isolation from the cosmological argument. We cannot ask how probable the premiss of the teleological argument makes theism, independently of the premiss of the cosmological argument, for the premiss of the teleological argument entails the premiss of the cosmological argument'.
[36] Swinburne (EG), pp. 158–60; and 164–5, jo. pp. 151–2.
[37] Swinburne (EG), p. 165.

Hence, we have to prefer the second horn of the dilemma, and must suppose that the explanandum *e* of the argument from temporal order is a very dense state of the universe, which is temporally ordered. As we saw, Swinburne did not succeed in arguing that such a state is a priori less probable, because less simple, than an entirely chaotic state. What he succeeded in arguing was merely that an ordered state is very unlikely on the assumption that it is caused by a pointlike cosmic singularity. We should admit to Swinburne that such a state *e* is indeed more likely on the assumption that it is caused by God. But does this imply that the relevance condition $P(e|h\&k) > P(e|k)$ is met, so that the argument from temporal order is a correct C-inductive argument for the existence of God if at least the universe is finitely old?

Not at all. What has been shown is merely that $P(e|h_t\&k) > P(e|h_p\&k)$, where h_t is the hypothesis of theism and h_p is the hypothesis of a pointlike singularity. That is, the probability of a temporally ordered state of the universe on the assumption that God caused it is greater than the probability of this state on the assumption that it was caused by a pointlike singularity. But if the hypothesis of theism has to be compared with *one* scientific hypothesis, it should be compared with *all* proposed scientific hypotheses in order to show that $P(e|h\&k) > P(e|k)$. Let us not forget that $P(e|k)$ implicitly contains $\sim h$, and that $\sim h$ is a catch-all hypothesis (cf. Chapter 11.5). In order to show that the relevance condition is met, theists have to argue that the early dense state *e* of the universe is more likely on the hypothesis of theism than it is on many possible scientific hypotheses, which explain this state with reference to earlier states of the universe, such as in bouncing universe theories. Since theists cannot show this, I conclude that the argument from temporal order to God is not a correct C-inductive argument.

13.3 ARGUMENTS FROM SPATIAL ORDER

Apart from temporal order, we also find spatial order in the universe. On the basis of the geocentric cosmology current before Copernicus, Kepler, and Galileo, it was often thought that a harmonious spatial order characterized the universe or cosmos as a whole, and global design arguments from spatial order were popular. But the assumption of a harmonious global spatial order in the universe became less plausible within the framework of the universe as conceived of by Newton and his followers. How can we know that there is overall spatial order rather than overall chaos in a universe, if the universe is infinitely extended in time and space? This is one reason why after Newton, local design arguments from spatial order tended to prevail in the literature.

The empirical premises of these arguments usually refer to the functionality and complexity of organisms such as animals and their functional parts, like eyes or wings. Local design arguments were often construed as arguments from analogy. Is the functionality and complexity of an organism such as a tiger or an anteater not similar to the functionality and complexity of a watch, for example?[38] Should we not infer

---

[38] Cf. Paley (1802), p. 7. A similar argument can be found in Ray (1691) and Nieuwentijt (1715), for example.

that since watches are produced by intelligent designers, this has been the case for the first tiger and anteater that ever existed as well, because similar effects must be produced by similar causes? Surely the designer of these latter complex and well-adapted animals must be more intelligent and powerful than human artisans. It was concluded that God must have been the designer and creator of the first specimens of each biological species, so that natural theology confirms what *Genesis* says.

However, as Hume showed in his justly celebrated *Dialogues Concerning Natural Religion*, design arguments for theism are weak if they are construed as arguments from analogy.[39] First, the analogy between animals and artefacts is a very partial one only. The order that we find in nature does not much resemble the order inherent in artefacts. Furthermore, human artefacts such as watches are produced intentionally, whereas plants and animals come into being by natural procreation. Why should we assume that once upon a time this was different? Second, even if there were a convincing and strong analogy between organisms and artefacts in certain respects, the analogy argument would not support the conclusion that the first specimens of each biological species were created by God as traditionally defined. For typically, human artefacts are designed and put together by many craftsmen, so that polytheism would be a more natural conclusion of the local design argument by analogy than monotheism. Furthermore, human craftsmen never create *ex nihilo*, hence we should not suppose that gods do so, if at least we intend to argue by analogy. And since all designers we have experience of are bodily beings, we should not suppose that the designer of organisms is bodiless. Finally, there is no reason to think that the designer was omnipotent or omniscient, since one should never postulate more in a cause than is needed to explain the effect, if at least knowledge of the effect is our only source of knowledge concerning the cause.[40]

Because local design arguments from spatial order are weak if they are construed as arguments from analogy, they can better be seen as likelihood arguments. If the existence of complex and well-adapted functional organisms is more likely on the theistic hypothesis h that God exists than it is if we assume ($\sim h$) that God does not exist, h is better confirmed than $\sim h$ by the existence of biological organisms.[41] Because $\sim h$ is a catch-all hypothesis, we should specify which rivals of theism are considered in a local design argument.[42] Traditionally, the only rival hypothesis seriously examined

---

[39] Cf. Hume (1779), *passim*. Cf. Gaskin (1976) for a defence of Hume's criticisms against objections by Swinburne.

[40] Oppy (2006a), pp. 174–87, argues that it is historically more adequate to reconstruct Paley's Watchmaker Argument as a deductive argument, which relies on the premise that '[t]here are cases in which the presence of function and suitability of constitution to function makes it inevitable that we infer to intelligent design' (p. 181). The watch is proposed as an example of this premise, and biological organisms are other examples. In this reconstruction, the argument is also very weak, as Oppy points out. For it is not the presence of function and suitability of constitution to function that makes it inevitable that we infer to intelligent design in the case of watches (and other artefacts), but rather our background knowledge about the origins of the materials watches are made of and about the process of manufacturing them. Since we lack such background knowledge concerning organisms, their complex functionality does not make the inference to intelligent design 'inevitable'.

[41] Of course, such a reconstruction presupposes that the problem of predictive power of theism can be solved. Cf. Chapter 9, above, for discussion. Cf. Sober (2008), §§ 2.4–2.14 for the likelihood reconstruction of design arguments.

[42] Cf. Sober (2008), §§2.4–2.12.

was the hypothesis of classical atomists that the first specimens of each biological kind have emerged by chance, that is, from random movements of atoms. Many philosophers were convinced that the existence of well-adapted complex organisms favours theism over this hypothesis of chance, since the likelihood that God created them was thought to be much higher than the likelihood that they emerged from completely random processes.[43]

The intellectual situation changed drastically, of course, when Darwin and Wallace put forward their theory of descent with modification, as Darwin called it. Since there is some random variation in the inheritable traits of the individuals that belong to a species, growers and breeders can produce different varieties of a species within a number of generations by selecting the individuals that are allowed to procreate. Similarly, since organisms in nature produce more offspring than can survive, there is a struggle for life and a mechanism of natural selection, which favours those individuals in a population whose traits happen to be best adapted to the environment. And because the distinction between varieties merely differs in degree from the distinction between species, we may conclude that over time different species emerged from varieties by natural selection, as varieties are produced under domestication from the offspring of the same ancestors by artificial selection.

As an explanation of the origin of species, the theory of natural selection is vastly superior to the theistic hypothesis of special creation, so Darwin argued in more than twenty passages of *The Origin of Species*. For example, the hypothesis that new species emerged by descent with modification from an older species accounts better both for the data of the fossil record and for the geographical distribution of species than the theory that God created separately the first individuals of all species that now exist. The same holds for evidence such as imperfect adaptations and the degenerated remnants of anatomical forms that were once adaptive. Furthermore, the theory of evolution provided a mechanism that explains the emergence of ever more complex adapted organisms from simpler organisms in the course of history. And it is a progressive research programme in the sense that it generated an immense number of fruitful problems and hypotheses for further research. After the theory of evolution had been reconstructed in the 1920s on the basis of Mendelian models of inheritance by scientists such as J. B. S. Haldane, R. A. Fisher, and Sewall Wright, all other biological disciplines were gradually integrated with evolutionary theory, and empirical evidence for its correctness continues to accumulate.

Consequently, if the local design argument from spatial order in organisms is construed as a likelihood argument, it is refuted by the theory of evolution.[44] It

---

[43] Cf. Sober (2003), pp. 28ff. and Sober (2008), §§2.4ff. One may wonder whether the likelihood of the design hypothesis with regard to the adaptive complexity of organisms is really greater than that of the chance hypothesis if one assumes that the universe is infinitely old, and that there is a finite number of particles in the universe, and a finite number of possible states of the universe, which never comes to a standstill. If so, it is very probable that a state of the universe containing complex organisms that procreate will be produced *sometime*, and it is inevitable that we observe such a state of the universe, because we are such organisms.

[44] This is so if we assume for the sake of argument that theism has some predictive power. Cf. Chapter 9. According to Sober (2008), §2.12, the lack of independently attested information about God's intentions is the 'Achilles heel' of design arguments construed as likelihood arguments. In spite of this justified criticism, we may reconstruct the way in which the theory of evolution refuted the organismic design argument as follows. Assume that the traditional assumptions theologians made

turned out to be a God-of-the-gaps argument, which has been disposed of by scientific progress. Although so-called Intelligent Design theorists such as Michael Behe have attempted to revive the local design argument from biological spatial order by claiming that there are 'irreducibly complex' phenomena in nature, such as the bacterial flagellum or the blood-clotting system, the genesis of which cannot be explained in principle by the theory of evolution, critics have proposed evolutionary scenarios that might explain these phenomena.[45] As I said before, local design arguments are too risky for theists, since they are threatened by the trap of God-of-the-gaps.

This is why, I suppose, Richard Swinburne opts for the strategy of transforming the local design argument from spatial order into a global one. Although the prospects for global design arguments from spatial order were dim within the Newtonian framework of a spatio-temporally infinite universe, they seem at first sight to be more promising on the basis of modern relativistic cosmology. According to Swinburne, it is a 'mistake' that the theory of evolution 'led to the virtual disappearance [. . .] from popular apologetic' of the design argument from spatial order. The reason is that the argument 'can easily be reconstructed in a form that does not rely on the premises shown to be false by Darwin'. Swinburne summarizes this 'easy reconstruction' as follows:

The basic mistake of those who regarded Darwin's discoveries as destructive of the argument from spatial order is that they ignored the fact that only certain processes acting on a certain kind of inorganic matter would have produced human bodies (and animals and plants), and that it is *a priori* improbable that the processes and initial matter would be of the right kind, but that this is to be expected if theism is true.[46]

The accusation that those who regarded Darwin's discoveries as destructive of the argument from spatial order made a 'mistake' is considered to be disingenuous by some critics.[47] For the argument refuted by the theory of evolution is a *local* design argument from spatial order as exhibited in the adaptive complexity of organisms, whereas Swinburne's reconstructed argument is a very different and *global* design argument. This argument, which is usually called the argument from fine-tuning, is based upon a number of scientific results, many of which were still unknown in Darwin's time, and it is dubious whether it should be called a design argument from spatial order. But of course the accusation is correct with regard to those writers, if any, who claim that Darwin's theory of evolution refuted *each and every* possible design argument.

As we saw, Swinburne has 'no wish to challenge Darwin's account' of how complex adapted organisms evolve from simpler ones.[48] Equally, he will admit that empirical

---

concerning God's intentions are correct, and add these intentions to the theistic hypothesis. Then the biological evidence (as summarized by Darwin in *The Origin of Species*, for example) clearly favours the theory of evolution over theism.

[45] Cf. Behe, 'The modern intelligent design hypothesis: breaking rules', in Manson (2003), Chapter 15. For criticisms, cf. for example Sarkar (2007), Chapter 6, and Oppy (2006a), pp. 187–200.
[46] Swinburne (EG), p. 168.
[47] Cf., for example, Everitt (2004), p. 105.   [48] Swinburne (EG), p. 171.

problems such as how chemical elements were produced by thermonuclear reactions in the cores of massive stars, or how life originated on Earth, must be left to the scientist. In order to avoid the trap of God-of-the-gaps, he wants to focus on phenomena that are too big for science to explain. However, the data referred to in the empirical premises of the argument from fine-tuning do not lie open to view. Rather, they are produced by complex scientific research, so that a risk of God-of-the-gaps cannot be entirely excluded. Indeed, some arguments put forward by scientists for the conclusion that the universe is fine-tuned for life, or even for human life, are criticized by leading physicists, such as Steven Weinberg or Alan Guth.[49] In the next two sections, I shall discuss Swinburne's C-inductive versions of the argument from fine-tuning, which are propped up by an argument from beauty.

13.4 THE FINE-TUNING ARGUMENT FROM LOGICALLY POSSIBLE UNIVERSES

Arguments from fine-tuning are based on the claim that life as we know it, and in particular human beings, can only evolve in a universe if the initial conditions of the evolution of this universe are of very specific kinds, and if the constants in many laws of nature obtaining in that universe have values lying in intervals that are quite narrow compared to some postulated range of theoretically possible values. If the very first initial conditions of our universe are indeed of these kinds, and if the constants in the laws of nature holding for our universe fall within these narrow intervals, the universe is said to be 'fine-tuned for' (human) life. In arguments from fine-tuning it is assumed that many such facts are well established by scientific research. Then it is argued that a priori these facts of fine-tuning are very improbable if God does not exist, whereas they are much more probable if God exists.[50] Consequently, if e stands for the facts of fine-tuning and h for theism, $P(e|h\&k) >> P(e|\sim h\&k)$, so that '[w]e have here a powerful C-inductive argument for the existence of God', as Swinburne concludes.[51] But why are the facts of fine-tuning thought to be a priori very improbable if there is no god? And are they really (much) more probable on the assumption that God exists?

Since there are many of these (alleged) facts, which can be understood properly only by those who possess the relevant scientific knowledge, I shall merely mention some of them as examples without spelling them out in detail. Furthermore, I shall assume for the sake of argument that sufficiently many facts of fine-tuning have been adequately established by scientists, that is, by doing the necessary calculations and by selecting ranges of theoretically possible values, compared to which the intervals of values that permit the development of (human) life must be very small.

One well-known example is the value of the cosmological constant. If this value were large and positive, space would expand at such a great rate that most objects in the universe would fly apart, whereas if it were large and negative, the universe would collapse. Only if the current value of the cosmological constant is near to zero will

---

[49] Cf. Robin Collins, 'Evidence for Fine-Tuning', in Manson (2003), p. 178.
[50] I discuss here the Bayesian version of the Fine-Tuning Argument only. For a critical review of other versions, cf. Bradley (2001).
[51] Swinburne (EG), p. 189.

there be a universe in which galaxies can be formed and stable objects can exist. Should we not conclude, then, that the value of the cosmological constant is fine-tuned for (human) life, or, at least, for stable objects in the universe?

Another example is gravity. If the strength of gravity on Earth were a billion times greater, all land-based organisms of our size would be squashed by the gravitational pull of our planet. Furthermore, if the force of gravity were increased by a factor of more than 3,000, no stars could exist with a lifetime of more than one billion years, so that there would not be much time for the evolution of life on planets, and humans would not have evolved. On the other hand, if the strength of gravity were zero or negative, no stars, planets, or other solid bodies could exist at all. It may seem that the interval within which the value of this strength should lie in order to enable human life to develop is still quite large, say between 0 and $3 \times 10^3 G_0$. But if we suppose that the range of theoretically possible values of constants is the total range of strengths of the forces in nature, the interval within which the strength of gravity should lie in order to enable life to develop is relatively small indeed. For the total range of strengths of forces in nature is 0 to $10^{40} G_0$, where G_0 denotes the strength of gravity and $10^{40} G_0$ is the strength of the strong force. Accordingly, the interval of 0 to $3 \times 10^3 G_0$ is merely one part in 10^{36} of the total range of forces.

Analogous results may be calculated for many other constants of nature, such as the strong force and the electromagnetic force, small changes which would probably prevent the production of carbon or oxygen in stars, so that the evolution of carbon-based life as we know it would be very unlikely.[52] One may conclude that only a relatively small subset of all logically possible constants in the laws of physics is compatible with the existence of life, and of human life in particular.

Similar considerations apply to the initial conditions of the evolving universe. For example, the early universe must contain triggers of fluctuations, which will later evolve into galaxies, and these triggers or ripples must be of the right size so that structures form without collapsing into black holes. Furthermore, if background radiation never dropped below 3,000 K, matter would be always ionized, so that molecules of life could never emerge. And life can only originate if chemical elements are formed in thermonuclear processes within stars, and are then distributed through space by supernovae explosions. So the size and age of the universe must be large enough for second-generation stars to come into existence, whereas we can easily imagine universes that expanded and recollapsed with a total lifetime of, say, 100,000 years. Probably, our universe must be more than 13 billion years old for intelligent life to evolve.[53] Again, we may conclude that only a small subset of all logically possible initial conditions of the evolution of the universe is compatible with the existence of (human) life.

Can we legitimately conclude that these facts of fine-tuning are a priori very improbable, or even 'immensely improbable', if God does not exist?[54] Swinburne offers two arguments to this effect, an argument from possible universes and an

---

[52] Cf. for a more detailed discussion of these examples Robin Collins, 'Evidence for Fine-Tuning', in Manson (2003), pp. 178–99, on which Swinburne's account of fine-tuning in (EG), pp. 172–82, is based. Cf. also Collins (2009).
[53] Cf. Ellis (2007), §9.1.
[54] Cf. Swinburne (EG), p. 179: 'tuning is a priori immensely improbable'.

argument from complexity. According to the argument from possible universes, we should compare the set of life-permitting possible universes, to which our universe belongs, with the reference-set of all *logically* possible universes. The probability that if God does not exist, any universe will be fine-tuned for (human) life, will now be equal to 'that proportion of logically possible universes' that are conducive to the evolution of (human) life.[55] Because this proportion is very small, it is concluded that the prior probability of a life-permitting universe is very low. Reformulated in terms of comprehensive theories of a universe, the argument reads as follows: 'the prior probability (in a Godless universe), that a universe will be tuned is a function [...] of all the possible theories and boundary conditions there could be for any universe at all'. Swinburne assumes that this probability is small indeed, although he adds that it 'is not [...] within my ability to calculate this value, nor – I suspect – within the ability of any present-day mathematician'.[56]

Is this argument from a small proportion to a low probability a valid one? In my view, the argument from possible universes is unconvincing, because it lands theists in the following dilemma. On the one hand, we might assume, like Swinburne, that the reference-set in the argument from possible universes is the set of all *logically* possible theories and boundary conditions there could be for any universe at all. This set is generated, inter alia, by supposing that each constant in each possible law of nature can take any value in the real number interval $[0, \infty]$. On the other hand, one might assume that the reference-set is somehow limited, for example by supposing that constants may take values within some finite interval of natural numbers. Both horns of this dilemma create insuperable difficulties for theists.

If the reference-set is the set of all logically possible theories, as Swinburne assumes, serious problems arise with regard to the notion of probability. For example, in order to make this argument from fine-tuning work, we have to assume for each constant a flat probability distribution over the range of its possible values, that is, we have to assume that equal segments of values in the interval have the same fixed positive probability, however small.[57] Since the interval is infinite, these supposed probabilities do not conform to the principle of countable additivity: the probabilities of all alternative and mutually exclusive segments of values do not add up to one.[58] But within the framework of the argument from possible universes, speaking of probabilities makes no sense unless the sum of the probabilities of all possible disjoint alternatives is equal to one.[59]

Furthermore, if one considers all logically possible universes as a reference class in order to argue that the existence of a fine-tuned universe is much less probable than the existence of a universe that is not fine-tuned, because the value-intervals for life-permitting universes are much smaller than the value-intervals for not-life-permitting

---

[55] Swinburne (EG), p. 182.   [56] Swinburne (EG), p. 182.

[57] Cf. Sober (2008), pp. 27–8, for the well-known problems with this 'principle of indifference'. Cf. also Manson (2000), pp. 345ff.

[58] Cf. for this argument Oppy (2006a), p. 206. According to standard probability theory, *any* set with finite measure, however large, will have a probability of zero if it is considered in relation to a reference class with infinite measure. But if $P(e|k) = 0$, we cannot use Bayes' theorem in order to calculate a posterior probability of h, so that Swinburne's argument collapses.

[59] Cf. also McGrew et al., 'Probabilities and the Fine-Tuning Argument', in Manson (2003), p. 203.

universes, the objection will be raised that different intervals of real numbers are equinumerous, since they all contain infinitely many real numbers. Hence, if one considers all logically possible universes, one cannot argue that the existence of a universe that belongs to the class of life-permitting universes is less probable than the existence of a universe that belongs to the class of not-life-permitting universes.[60]

Finally, if the limited interval of life-permitting values of a constant has to be compared with the infinite real number range as a reference-set, *each* limited interval will have a probability of zero, no matter how large the interval is. If this is what fine-tuning means, it is guaranteed gratis and for free, so to say. And as soon as this has been established for one constant in a law of nature, further examples of fine-tuning are superfluous, because they cannot make it less likely that a universe is life-permitting.[61] For these and other reasons, the first horn of the dilemma is unattractive for theists. They should not assume, like Swinburne, that the reference-set in the fine-tuning argument from possible universes is the set of all *logically* possible universes, so that the interval of possible values of constants is the real number range.

According to the second horn, a constant in a law of nature can take values only in a finite interval $[L, H]$ of natural numbers, for example, where L is a finite lower limit and H a finite upper limit. But now the fine-tuning argument will be open to the objection that the choice of L and H is arbitrary. Whereas the proponent of the fine-tuning argument will choose values of L and H such that a universe fine-tuned for life is a priori very improbable, the critic will prefer values such that a universe fine-tuned for life is quite probable.[62] In general, one might say that it is impossible to assign in a non-arbitrary manner well-defined probabilities to the different possible universes, since as far as we know there is one universe only.[63]

Readers may not be impressed by these technical objections to Swinburne's first fine-tuning argument. For example, the problem concerning the principle of countable additivity may perhaps be resolved by allowing infinitesimal numbers, or by arguing that comparative probability judgements can make sense even when the probabilities for the exclusive and exhaustive alternative possibilities do not add up to one.[64] It is clear, however, that concerning these issues the burden of proof rests on the proponents of the fine-tuning argument. They have to show that a meaningful notion of probability can be used in the fine-tuning argument, construed as an argument from logically possible universes. Perhaps they will be able to demonstrate this by new research in probability theory. Yet there is a strong reason why Swinburne (or any other theist) should not rely on this first argument, since it leads to the following internal inconsistency.[65]

If the intrinsic probability of a life-permitting universe should be determined by comparing the set of life-permitting universes to the set of all logically possible

---

[60] Cf. Colyvan et al. (2005). Bradley (2001) criticizes the assumption of the fine-tuning argument that the set of life-permitting universes has to be contrasted with all sets of not-life-permitting universes taken together (pp. 458–61).
[61] Cf. Manson (2000), pp. 346–7.
[62] Cf. again Oppy (2006a), p. 206, and Manson (2000), pp. 347–8.
[63] Cf. Manson (2000), pp. 347–8.
[64] Cf. Swinburne (EG), p. 178, note 25, jo. Swinburne (EJ), p. 244; and Monton (2006), pp. 409–10.
[65] Cf. footnote 17 to this chapter.

universes, as Swinburne holds, one should also determine the intrinsic probability of theism by comparing the set of theisms to the set of all possible religious theories. But the former set has only one member, theism itself, whereas the latter set is infinite in at least two dimensions. There is not only an infinity of possible polytheistic theories, since polytheisms may posit any number n of gods for $n > 1$. There also is an infinite set of possible monotheisms, which each postulate one god who has some finite amount of knowledge and power (cf. Chapter 12.8, above). If the prior probability of theism had to be established by comparing these sets, and if prior probabilities could be established by such arguments, one would have to conclude that both some polytheistic theory, and one or another monotheistic theory that posits some finite god, are much more probable a priori than theism. But this conclusion would be fatal for any inductive natural theology that aims at establishing theism. Let us disregard Swinburne's fine-tuning argument from possible universes for these reasons and focus on his second argument.

13.5 THE FINE-TUNING ARGUMENT FROM COMPLEXITY

According to Swinburne, we can objectively estimate the prior (intrinsic) probability of a hypothesis (mainly) on the basis of criteria of simplicity. So he argues, secondly, that if God does not exist, the prior probability of the hypothesis that our universe is fine-tuned for human life is very low, because such a universe must be quite complex.[66] He stages this fine-tuning argument as an argument from human bodies, which have sense organs, a complex brain structure with an information processor and a memory bank, and so on.[67] He then argues that in order for these organs to function, the world must be very complex indeed. Given 'the considerable a priori weight of simplicity', it allegedly follows that 'in a Godless universe it is a priori improbable that any one universe will be tuned so as to yield human bodies'. As a consequence, $P(e|\sim h \& k)$ is very low, where e is the existence of a universe fine-tuned for human bodies, h is theism, and k is the fact that a universe exists and conforms to natural laws.[68] Is this fine-tuning argument from complexity more convincing than the fine-tuning argument from possible universes?

As we did with respect to Swinburne's argument from temporal order, we have to stress here that the fine-tuning argument cannot be considered separately from the cosmological argument. If God has fine-tuned the universe for human life, God must have done so at some point in the history of the universe. And if one assumes, as Swinburne does, that the evolution of the universe and of life on Earth is the domain of scientific investigations and not of theology, the fine-tuning argument from complexity must presuppose that there has been a first state of the universe, which was fine-tuned for the existence of human life or of human bodies.

This consideration has two implications. First, in the argument to the effect that $P(e|\sim h \& k)$ is very low, and indeed much lower than $P(e|h \& k)$, 'e' should not

---

[66] Swinburne (EG), pp. 182ff.
[67] Swinburne (EG), pp. 168–9.
[68] Swinburne (EG), p. 184.

stand for our present complex universe containing human bodies, but for a postulated initial state of this universe, which according to Big Bang cosmology was dramatically less complex than its present state. Accordingly, the argument from complexity is much weaker than Swinburne pretends. Second, in order to assess the argument to the effect that $P(e|\sim h\&k)$ is very low, and that $P(e|\sim h\&k)$ is much lower than $P(e|h\&k)$, we have to consider the three cosmological scenarios discussed above (section 13.2), to wit, (a) that the universe is infinitely old, (b) that the universe began to exist some 13.7 billion years ago with a pointlike singularity, and (c) that it started some 13.7 billion years ago with a very dense quantum state.

Swinburne admits that 'life is much more likely to evolve at some time in the course of the history of our universe if it has an infinite past than if it has only a finite past'.[69] But the more important point with regard to the first scenario (a) of an infinitely old universe is a different one. Since in this scenario no moment in time is early enough for God to create the universe, no moment of time is early enough for him either to fine-tune the universe for life. As has been argued in Chapter 12.3, a divine creation of the universe is excluded if the universe is backwardly eternal. Hence, for this scenario (a), if 'e' denotes fine-tuning at the moment of creation, $P(e|h\&k)$ is not defined, and, a fortiori, cannot be compared with $P(e|\sim h\&k)$.

With regard to the second scenario (b), according to which the universe started with a pointlike singularity, there are three problems for Swinburne's argument to the effect that $P(e|h\&k) >> P(e|\sim h\&k)$. First of all, if e is the pointlike singularity, the initial conditions of the universe cannot be fine-tuned for life at all, since as we saw in Chapter 12.7, the pointlike singularity is intrinsically chaotic. Second, we concluded for this reason that in this case $P(e|h\&k)$ will be very low. If the probability that God wanted to create humans is roughly $1/2$, the probability that he intended to do so by creating a pointlike singularity in the first place will be (near to) zero. Finally, since the pointlike singularity is the simplest physical object one can imagine, one cannot argue for a low value of $P(e|\sim h\&k)$ by an argument from complexity. Hence, the argument from fine-tuning fails for scenario (b) as well, even if it is merely intended as a C-inductive argument.

It is no wonder, then, that Swinburne prefers view (c), that fine-tuning of initial conditions occurred in a state of the universe 'immediately after the time of the Big Bang'.[70] We may assume that at that time, the universe was not yet very complex, so that an argument from complexity will not suffice to show that if God does not exist, fine-tuning is a priori very improbable. Are the prospects any better for arguing that $P(e|h\&k)$ is very high, so that $P(e|h\&k) > P(e|\sim h\&k)$, if e is this fine-tuned early state of the universe? There are at least three serious objections to such an argument.

First, since $\sim h$ is a catch-all hypothesis, the likelihood of theism has to be compared with the likelihoods of all scientific explanations of this fine-tuned early state of the universe, such as theories of a bouncing universe which is infinitely old, or the theory of cosmological natural selection defended by L. Smolin.[71] In other words, in this case it is not true that facts of fine-tuning are in principle '"too big" for science itself to

---

[69] Swinburne (EG), p. 180.
[70] Swinburne (EG), p. 172.
[71] Smolin (1992).

explain', as Swinburne claims.[72] Since none of these scientific explanations has been worked out in detail, $P(e|{\sim}h\&k)$ cannot be established. This first objection is already fatal.

Second, as Swinburne admits, the argument from fine-tuning can have strength only if the fine-tuned early state of our universe will lead to the evolution of humans 'not merely possibly but with significant probability'.[73] He avers that there is 'a very considerable, but not unanimous, scientific view that the laws and initial conditions of our universe make it very probable that human life will evolve in more than one place in the universe'.[74] However, as he admits, this view is contested, and some scientists claim that in spite of fine-tuning and of the estimated huge number of planets in the universe, the probability that there is life on one of them is nearly nil.[75] Clearly, the fine-tuning of the universe for human life is merely a necessary condition for its evolution, and it may be that in spite of fine-tuning, the emergence of human life is still very improbable. Again, we have to conclude that the fine-tuning argument risks being a God-of-the-gaps argument.

Finally, in order to determine the value of $P(e|h\&k)$, where e is a very dense early fine-tuned state of the universe, we need more independently attested information about God's intentions than Swinburne's assumption, according to which there is a probability of roughly $\frac{1}{2}$ that God wanted to create human beings. In particular, we need to know what reason God might have had for taking the excessively long evolutionary detour via the Big Bang. As Swinburne says, if God's 'only aim in creating a universe was to populate it with human beings, there would indeed be no point in producing them by a long evolutionary process'.[76] Since God is omnipotent, he could have created a universe with humans (and plants and animals) at one instant, or within one week.

13.6 THE ARGUMENT FROM BEAUTY

What reasons can God have had for preferring the long evolutionary route of the history of the cosmos and of life, if he wanted to create the human species? Theists should not answer this question by the traditional bromide that God's intentions are inscrutable for us.[77] Such a move would annihilate the predictive power of theism, and thereby destroy the prospects of natural theology. Instead, they should come up with convincing reasons for God to take the evolutionary detour, assuming that he probably wanted to create humans in the first place. Otherwise, they cannot but assign a very low value to $P(e|h\&k)$ if e is a dense and fine-tuned early state of the universe.

---

[72] Cf. Swinburne (EG), p. 172. Of course, an explanation of fine-tuning in some state of the universe by assuming fine-tuning in an earlier state of the universe will raise the issue as to how this latter fine-tuning should be explained. But in the scenario of an infinitely old bouncing universe, all such questions can be answered with reference to fine-tuning in ever earlier states, so that in principle the issue of fine-tuning is not 'too big' for science to explain.
[73] Swinburne (EG), p. 189.
[74] Swinburne (EG), p. 189.
[75] Cf., for example, Ward and Brownlee (2000); Gonzalez et al. (2001); Gonzalez and Richards (2004).
[76] Swinburne (EG), p. 188.
[77] Cf., for example, Behe (2006), p. 223.

Swinburne's answer is that 'there are other good features of the universe that God has good reason to bring about'. One is the beauty of the inanimate universe. Another is the beauty of plants and animals. Swinburne even claims that 'evolution is a very beautiful process'.[78] And it is also good that animals exist because of their ability to have pleasant sensations, and so on. Swinburne concludes that given all these good things, 'it is not too surprising that God should take the long (by our timescale) evolutionary route to produce human bodies'.[79]

I do not think that these (attributions of) reasons (to God) are convincing. Even if beauty were a species-independent property of things, there is no reason to assume that the universe as a whole is beautiful for any spectator, whether God or us.[80] Both beauty and ugliness are local properties of individual phenomena. Some landscapes or buildings are beautiful and others are hideous. Some human beings are beautiful and others are ugly. The same holds for phenomena in the inanimate universe. Asteroids or lunar craters depicted on photographs and pictorial reconstructions of mass extinctions may strike us as excessively sinister. Nobody will call dark matter or dark energy beautiful, whereas according to many present-day cosmologists dark matter and dark energy may account for the vast majority of the mass in the observable universe. Furthermore, apart from pleasant sensations, animals also have unpleasant ones. No one has ever demonstrated that, on balance, animals have more pleasant than unpleasant sensations, or that, on balance, there are more beautiful than ugly or horrifying phenomena in the universe. However, if the argument concerning beauty and pleasant sensations is unconvincing, the value of $P(e|h\&k)$ is very low if e is a dense and fine-tuned early state of the universe. I conclude that for none of the three possible cosmological scenarios (a–c), the argument from fine-tuning is a correct and a good C-inductive argument for the existence of God.

This conclusion is reinforced by many other features of our universe that are revealed by science. If the universe were an artefact created by God mainly in order to host humans and other living species, the universe would be very different from the way it in fact is. For example, one would not expect that in this designed universe, billions of years would have to pass before life could develop, and that life is an excessively rare phenomenon given the spatial magnitude of the universe.[81] Even in our solar system life is scarce indeed, since most planets cannot harbour it. Furthermore, one would expect clear symptoms of design and engineering in the universe, such as stars arranged in accord with aesthetic principles, or meaningful messages conveyed by the architecture of mountain ranges. But we find nothing

---

[78] Swinburne (1983), p. 389. One wonders, of course, whether this is meant to hold also for the mass extinction of dinosaurs, for instance, which was probably triggered by a bolide impact near the Yucatán Peninsula approximately 65.5 million years ago. Without this mass extinction, humans would never have evolved.

[79] Swinburne (EG), pp. 188–9.

[80] As Swinburne admits, the argument from beauty 'needs [. . .] an objectivist understanding of the aesthetic value of the universe, in order to have significant strength' ((EG), p. 190).

[81] Taking the speed of light (186,000 miles per second) as a measure of distance, the following data provide an impression of this spatial magnitude. Light takes about 4.2 hours to travel from the Sun to Neptune. Our solar system is a miniscule spot in the Milky Way, which comprises at least two hundred billion stars and measures roughly 90,000 light years in diameter. And the 'co-moving' distance from Earth to the most distant galaxies we can observe is now about 47,000,000,000 light years.

of the sort.[82] As I said at the end of Chapter 10, the idea that the universe is fine-tuned for (human) life is as absurd as if a fly residing in the palace of Versailles thought that the palace was built especially for it. Given the difference of relative dimensions between the universe and us on the one hand, and the palace and the fly on the other hand, even this is a massive understatement.

---

[82] This is argued in detail by Cordry (2006). As he says, 'When we try to find design in nature, especially on the large scale, we are befuddled. Nature does not appear to be arranged in accord with aesthetic principles; nor does it appear to contain any meaningful messages' (p. 274).

14

Other Inductive Arguments

Up to this point, we examined four of the eleven C-inductive arguments that Richard Swinburne considers in his cumulative case strategy, which results in a P-inductive argument for the existence of God. In these four arguments, the adduced evidence e was respectively the existence of the universe (e_1), the fact that the universe is temporally ordered according to relatively simple laws of nature (e_2), the alleged facts of fine-tuning of the early universe for human (bodily) life (e_3), and the beauty of the universe (e_4). I have argued that the existence of e_4 has not been established, whereas in the first three arguments the relevance condition $P(e|h\&k) > P(e|\sim h\&k)$ is not satisfied, so that they fall short of being either *good* or *correct* C-inductions.[1] In the present chapter, I shall briefly examine six of the remaining inductive arguments.

14.1 THE ARGUMENT FROM CONSCIOUSNESS: THE EVIDENCE

In the argument from consciousness, the evidence (e_5) is defined as 'the existence of souls with mental states connected to brain states' in regular ways.[2] As usual, h is the hypothesis of theism, whereas k contains the empirical premises e_{1-4} of Swinburne's first four arguments, apart from tautological or conceptual background knowledge.

In order to show that the relevance criterion is satisfied, Swinburne argues on the one hand that $P(e_5|\sim h\&k)$ is 'very low'.[3] This conclusion is supposed to follow from two premises. First, (m) 'we have every reason to believe that there can be no scientific theory and no scientific laws correlating brain states with souls and their states'.[4] Hence, 'it is immensely improbable that there could be a scientific explanation of the connections' between kinds of brain event and kinds of mental event, so that (very probably) e_5 is in principle 'too odd' for science to explain.[5] Second, (n) these (causal) connections in both directions are 'so detailed and specific that it is most improbable

---

[1] A C-inductive or confirming argument is called *correct* if the relevance condition is satisfied, that is, if the premises make the conclusion more probable than it would otherwise be, and a correct C-inductive argument is a *good* argument if the premises are known to be true by those who dispute about the conclusion. Cf. Swinburne (EG), pp. 6–7.
[2] Swinburne (EG), p. 210.
[3] Swinburne (EG), p. 210.
[4] Swinburne (EG), p. 210. I substituted 'no' for 'so' in the phrase 'and so scientific laws'.
[5] Swinburne (EG), p. 209.

that they would occur without an explanation'.[6] The conclusion (p) that $P(e_5|\sim h\&k)$ is 'very low' then follows on the assumption (o) that the catch-all $\sim h$ merely contains the hypothesis that the mind–brain connections occur as brute facts without any explanation. The catch-all $\sim h$ should not also comprise, for example, logically possible scientific explanations nobody thought of, or rival religious hypotheses, because in those cases we can have no idea about the value of $P(e_5|\sim h\&k)$.

On the other hand, Swinburne argues that $P(e_5|h\&k)$ has 'a moderate value', because God has good reasons for creating humans (and higher animals) having souls, the states of which are connected to their brain states in regular ways. If all this is convincing, the argument from consciousness is a correct C-inductive argument for the existence of God, because the relevance condition $P(e_5|h\&k) > P(e_5|\sim h\&k)$ is met.[7] According to Swinburne, it is 'a powerful argument to which philosophers have not given nearly enough attention'.[8]

In order to evaluate this fifth argument, we first have to investigate whether the evidence (e_5) has been established with sufficient certainty. The alleged evidence consists of two parts: (1) the existence of souls as substances and (2) the existence of a myriad of causal interactions between mental states (or events, properties, etc.) and brain states (or events, properties, etc.).

In his attempt to show that (1) souls exist as substances, Swinburne uses the science fiction argument from personal identity across brain transplants, which I already examined and rejected in Chapter 7.6.[9] From the story of the mad surgeon it allegedly follows that 'my survival consists in the continuing of [. . .] my soul', so that 'a human's continuing existence does not logically entail [. . .] the existence of that body at all'.[10] Furthermore, Swinburne thinks it likely that all vertebrates have mental life, and in consequence, souls, so that 'we must postulate a cat-soul that is the essential part of the cat'.[11] And he regards his argument in favour of substance dualism as 'inescapable'.[12] But as I put forward three decisive objections against this argument in Chapter 7.6, we should conclude that Swinburne has not established the first half (1) of the alleged evidence (e_5).[13] And since most philosophers and scientists reject substance dualism for many good reasons (cf. Chapter 7.2), e_5 cannot play a role in a 'good' C-inductive argument for the existence of God, that is, an argument that

---

[6] Swinburne (EG), p. 209.   [7] Swinburne (EG), pp. 210–11.
[8] Swinburne (EG), p. 193.
[9] Swinburne (EG), pp. 197–9. In Chapter 7.6, I analysed the versions in (CT) and (ES).
[10] Swinburne (EG), p. 199.
[11] Swinburne (EG), pp. 200–1.
[12] Swinburne (EG), p. 199. Cf. also (ES), Chapters 8–9. According to (ES), p. ix, Swinburne's arguments for the existence of the soul are 'ones of immense strength'.
[13] In oral discussions, Swinburne objected to this latter argument that it exemplifies 'dogmatic Wittgensteineanism'. But one cannot exclude a priori that Wittgenstein was *sometimes* right. Surely our everyday (and our legal) concept of personal identity breaks down for situations like the split-brain experiment, at least if we assume that the experiment does not simply kill the person. In order to deal with such thought experiments, we would have to establish new rules for the notion of personal identity, and these new rules should perhaps admit of *partial* personal identity. So we might say in this case that after the split-brain operation both P_1 and P_2 will be partially identical with me. Again, it would not follow that personal identity consists in the continued identity of a soul (cf. the preceding note).

satisfies the relevance condition *and* the premises of which are accepted by those who dispute about the conclusion.[14]

What should we think of the second half (2) of the alleged evidence (e_5)? Without any doubt, there are myriad connections between our mental life and states of and happenings in our brains. Swinburne assumes that these connections are causal connections.[15] But in the philosophy of psychology there are many other views concerning the connections between brain states or events and our mental life. According to some, mental events, states, or properties of persons supervene on complex events in, or states of, the brain, whereas there are many different explications of this notion of supervenience.[16] According to others, there is some kind of identity relation, such as a type or token identity. I shall not try to establish here which of the many possible positions on this complicated issue is correct. But it is not self-evident that we should interpret the correlations between what happens in our brains and our mental life as causal relations, so that it is not clear that the second half (2) of e_5 exists either.

Furthermore, if we allow for a more liberal interpretation of the second half of e_5, so that any type of connections between brain events and mental events might constitute the relevant evidence, it will be difficult to establish premise (m) of the conclusion (p) that $P(e_5|\sim h\&k)$ is 'very low'. According to this premise (m), it is immensely improbable that there could be a scientific explanation of the connections between kinds of brain event and kinds of mental event, so that there is room for a theistic explanation. But if a type-identity theory were correct, for example, scientific explanations of specific brain events would *eo ipso* be scientific explanations of the mental events which are identical with these brain events, so that the mental–brain correlations would not be 'too odd' for science to explain. Hence, we may conclude that either it has not been (sufficiently) established that the second half (2) of evidence e_5 exists, or it will be very difficult to establish premise (m) of conclusion (p).[17] In short, Swinburne's argument from consciousness cannot be a *good* C-inductive argument, because the evidence has not been established beyond reasonable doubt.

14.2 THE ARGUMENT FROM MEASUREMENT

We now have to investigate whether the argument from consciousness is a *correct* C-inductive argument, that is, whether it satisfies the relevance condition. For reasons

---

[14] Swinburne calls a C-inductive argument 'correct' if it satisfies the relevance condition, and he calls it a 'good' C-inductive argument if the premises are 'known to be true by those who dispute about the conclusion' (Swinburne (EG), pp. 6–7).

[15] Cf. Swinburne (EG), p. 201: 'The connection between one soul and one brain that gets established is a causal one. It is events in this particular brain that cause events in this particular soul, and events in this particular soul that cause events in this particular brain; this is what the connection between this brain and this soul amounts to'.

[16] Cf. Kim (1998) for an introductory discussion.

[17] Swinburne argues against mind/brain identity theories in (ES), Chapter 3, and *passim*. I cannot discuss his arguments here. However, since the evidence e in Swinburne's Bayesian arguments must consist of evident facets of experience, or at least of undisputed facts, whereas there is no consensus among philosophers concerning the correct view on the relations between mental states or events and brain states or events, we cannot but conclude that the second half (2) of e_5 has not been established if the mind–brain connections are interpreted as causal connections.

of space, I shall restrict my evaluation mainly to premise (m) for the conclusion (p) that $P(e_5 | \sim h \& k)$ is 'very low'.

In my summary of this premise, I used two different quotations. The first (m-1) reads: 'we have every reason to believe that there can be no scientific theory and no scientific laws correlating brain states with souls and their states'.[18] This may sound plausible, because it is perhaps difficult to imagine a scientific methodology by means of which one can investigate souls. But since the existence of souls has not been established, as I argued, we should prefer the second quote (m-2) for discussion: 'it is immensely improbable that there could be a scientific explanation of the connections' between kinds of brain event and kinds of mental event, so that (very probably) e_5 (its second half) is in principle 'too odd' for science to explain.[19]

Clearly, the words 'could be' in this latter citation should be interpreted as 'could be, in principle'. The question is not whether present-day humanity is able to develop a scientific explanation of these connections, but whether this is possible in principle, for an omniscient scientist, so to say. In other words, the real question is whether mental events (or states, properties, etc.) and their relations to brain events (or states, properties, etc.) can arise naturally in the universe, that is, without a divine or other supernatural intervention. Only if this is not possible *in principle* is one justified to conclude that these correlations are 'too odd' for science to explain, so that there is room for a theistic explanation. But how can one establish such a very strong premise?

There are three strands in Swinburne's overall argument to this effect, which I shall discuss briefly: an argument from measurement, an argument against reductionism, and an argument against evolutionary explanations. Swinburne distinguishes between two different types of explananda within the second half (2) of e_5. First (2a), there are myriads of causal connections between brain states and mental phenomena, such as that I experience something blue or have a strong desire to drink tea whenever specific things are happening in my brain. Second (2b), we have to explain which types of brains give rise to consciousness or mental life at all and why. Swinburne assumes that in principle scientists might be able to establish (2a) 'a long list of such causal connections in humans', and also to establish (2b) a list of 'which primitive brains give rise to consciousness – that is, to souls', even though there are many epistemological problems, such as the difficulty that we can only attribute mental life to animals on the basis of their behaviour and facial expressions.[20] But of course, making lists of such connections does not amount to explaining them.

The argument from measurement purports to establish the conclusion that quite probably, the explananda (2a, 2b) 'lie utterly beyond the range of successful scientific explanation'.[21] The reason is that a successful scientific explanation would have to consist in the derivability of (2a, 2b) 'from a theory consisting of a few relatively simple laws that fit together'. As Swinburne says, this theory would need to explain both (2b) 'why the formation of a brain of a complexity as great as or greater than that of a certain animal [...] gives rise to consciousness' and (2a) 'why brain events give rise to

---

[18] Swinburne (EG), p. 210.
[19] Swinburne (EG), p. 209.
[20] Swinburne (EG), p. 202.
[21] Cf. Swinburne (EG), p. 209.

the particular mental events they do'.[22] But such a theory is inconceivable in this case, since relatively simple laws about a large domain of phenomena are only possible if these phenomena are measurable on scales, whereas mental phenomena such as thoughts 'do not differ from each other along scales'.[23]

This argument from measurement is undoubtedly correct if its conclusion is merely that the explananda (2a, 2b) cannot be explained scientifically by a theory consisting of a few relatively simple laws that fit together. But the argument is unconvincing if it is meant to establish the much stronger conclusion that in principle there cannot be a successful scientific explanation of the explananda (2a, 2b) at all (= premise m). The reason is that we acknowledge more paradigmatic models of successful scientific explanation than the explanation of phenomena by deriving their descriptions from theories consisting of a few relatively simple laws (and from initial conditions, etc). One other acknowledged model is explanation by microreduction, another is the explanation of specific traits of organisms by reconstructing their evolutionary history. This is why, I suppose, Swinburne discusses both of these alternative models as well, and argues that they cannot be relevant to the explanation of the explananda (2a, 2b). But there are also models that he does not discuss, such as the model of explanation by providing possible or actual mechanisms.[24]

14.3 MICROREDUCTION

Swinburne's argument for the irrelevance of microreduction is similar to the one John Searle uses for the irreducibility of consciousness.[25] It may seem that in paradigmatic examples of microreduction, such as the (partial) reduction of thermodynamics to statistical mechanics, or the reduction of optics to electromagnetism, qualities such as heat or colour were explained by integrating them into theories about very different, measurable qualities, such as mean velocities of particles or mathematical properties of electromagnetic waves. If so, one may wonder why a similar microreduction might not be possible in principle for mental properties, states, or events.

However, according to the argument proposed by Swinburne and Searle, these paradigmatic microreductions could be successful only because it was assumed that in reality the reduced phenomena were not as they appeared to be. A distinction was made between the reduced *primary qualities* of heat and colour, which were conceived of as measurable physical properties or dispositions based upon them, and irreducible *secondary qualities*, which were conceived of as sensory properties or 'sensations' of felt hotness or of seen colours 'in the mind'. If these past paradigmatic microreductions could be successful only by projecting the properties-as-experienced-by-us into the realm of the mental, this apparently rules out in principle 'any final success in integrating the world of the mind and the world of physics', as Swinburne concludes. For 'when you come to face the problem of the mental events themselves [...], you

---

[22] Swinburne (EG), p. 203. [23] Swinburne (EG), p. 204.
[24] Cf. Machamer et al. (2000). [25] Cf. Searle (1992), pp. 118–24.

cannot distinguish between them and their underlying causes and only explain the latter'.[26]

I do not think, however, that microreductions such as the (partial) reduction of thermodynamics to the kinetic theory of gases should be interpreted in terms of the philosophical distinction between primary and secondary qualities, or that such reductions can be successful only by 'siphoning off' experienced qualities such as thermal ones 'to the world of the mental'.[27] In fact, this is a strongly misleading metaphysical interpretation of such reductions, which originated in the seventeenth century.[28] Take, for example, the temperature in my room at a given moment. This is both an objective property of the air in the room, which can be measured indirectly by a mercury thermometer, and it is a property we can experience by a faculty of feeling. The range of temperatures we humans can feel is very limited, and the laws of thermodynamics also apply to temperatures far beyond that range. But this does not imply that we should 'siphon off' the felt temperature to the world of the mental.[29]

On the contrary, it is this objective property called temperature of which the physicist will say that it is in fact identical with the mean kinetic energy of the molecules that constitute the air in my room. Such a 'bridge law' states the identity of two objective properties, one of which we can perceive to some extent. Of course our feeling or perception of the temperature in the room is a mental experience, but *what* we experience is an objective property, which should not be confused with the perception of it. In other words, one should not identify the 'felt hotness' of the air in the room with the 'sensation' or feeling of hotness.[30] Consequently, the argument of Searle and Swinburne against the possibility of micro-reducing the mental is unconvincing, so that it does not contribute to showing that explananda (2a, 2b) are 'too odd' for science to explain.

14.4 EVOLUTIONARY EXPLANATIONS

Yet another scientific model for explaining explananda (2a, 2b) is that of an evolutionary explanation of traits. Suppose that mental phenomena are emergent properties, states, or events, which cannot exist without a neuronal basis. Let us assume that at a time in the past a specific neuronal configuration of great complexity resulted from genetic mutations in one individual animal, which, if stimulated in specific ways, gave rise to particular mental phenomena, such as being in pain. Let us admit for the sake of argument that this first instance of a brain–mental correlation was a brute fact, which

---

[26] Swinburne (EG), p. 206.
[27] This terminology is used by Swinburne (EG), p. 206.
[28] Descartes stipulated, for example, that matter can only have mathematical properties, so that all other properties we attribute to material objects, such as colours or temperature, must be subjective impressions in our minds, which we project onto these objects. There is no good reason whatsoever to endorse this metaphysical view.
[29] Indeed, it has been argued that such a 'siphoning off' cannot be formulated meaningfully. Cf. Bennett and Hacker (2003), pp. 128–35.
[30] As Swinburne does on pp. 205–6 of (EG), like many philosophers since the seventeenth century.

cannot be explained by science. What science can explain by the usual mechanisms of natural selection, however, is that this specific brain–mental correlation occurred more and more frequently in later generations of the population, if at least it was inherited by offspring and provided some adaptive advantage for the animal.

In his discussion of evolutionary explanations, Swinburne admits that they might take care in part of explanandum (2a), the myriads of connections between brain states and mental phenomena. As he says, a Darwinian theory of selection might explain 'why, having once appeared in evolutionary history, conscious animals survived' and proliferated, since consciousness might provide many advantages in the struggle for survival and procreation.[31] Similarly, the theory of natural selection might explain why the brain events connected to a specific belief are causally connected to events in or facets of the outside world, since 'animals with beliefs are more likely to survive if their beliefs are largely true'. Inversely, the theory is also able to explain why brain events causally connected to specific facets of the outside world are also connected to specific beliefs, and this is explanandum (2a). Furthermore, it 'is highly probable that natural selection will ensure survival of those organisms and only those organisms that use correct criteria of inductive inference'.[32]

As Swinburne observes, however, the theory of natural selection is merely a theory of elimination. Whereas it explains why some variants procreate whereas others were not fitted for survival, it does not account for the fact that these variants 'were thrown up in the first place'.[33] In the case of physical variants such as longer necks of giraffes, this problem may be solved by the theory of random genetic mutations. But Swinburne holds that this is impossible with regard to the problem of 'why some physical state causes and sustains the existence of souls with such mental properties as beliefs, desires, purposes, thoughts, and sensations, causally connected in a regular way with brain states'.[34] In other words, he holds that in principle there can be no secular and evolutionary explanation of explananda (1) and (2b).

As we have seen, however, the existence of substantial souls, that is, of explanandum (1) has not been established. And if mental phenomena are emergent properties, states, or events, it is *only the first instance* of a specific brain–mental connection (2a, 2b) that cannot be explained by the theory of evolution. The proliferation over time of such a brain–mental connection within a population is accounted for by the usual evolutionary mechanisms.

Let us suppose for the sake of argument, then, that such a first instance of a specific brain–mental connection cannot be explained by science at all. Does it follow that this first instance is very unlikely if God does not exist, and (much) more likely if God exists, as Swinburne will conclude? I do not think so, for two reasons. First, we might assume that initially many different specific brain–mental connections are thrown up by random mutations. As is the case with traits produced by random mutations in organisms, most of these brain–mental connections will be counterproductive, so that they will perish with their bearers. But once in a while, a random mutation will produce a brain–mental connection that is adaptive, so that it will proliferate in the

---

[31] Swinburne (EG), p. 207. [32] Swinburne (EG), p. 208.
[33] Swinburne (EG), p. 207. [34] Swinburne (EG), p. 207.

population over time. Given the large number of different brain–mental connections that are thrown up by random mutations, it is not at all unlikely that once in a while there is an adaptive one.

Second, we should wonder whether it is more likely that God produced the first instance of this specific brain–mental connection, which then spread through the population because of the usual evolutionary mechanisms. At this point we will have to raise *mutatis mutandis* the traditional objections of Darwin and later evolutionists against theories of special creation. If an omniscient, omnipotent, and perfectly good god created these first instances of specific brain–mental connections, we will expect that these connections are perfect ones, or at least optimal in some sense. But this is not the case. Take, for example, the brain–mental connection between specific brain states that are caused when injury-detectors are triggered on the one hand, and sensations of pain on the other hand.

As Michael Tooley has argued, these brain–mental connections are badly designed in at least four ways. First, the injury-detectors are not sensitive to numerous life-threatening bodily changes. Second, the detectors often produce intense pain when there is no bodily condition that seriously threatens our health, such as in cases of migraine. Third, we cannot switch off the injury-detectors whenever the pain sensations merely increase our misery instead of providing us with a useful warning of bodily damage. Finally, the system often produces levels of pain that are 'unbearably intense', whereas this is not needed for alerting us to bodily damage. Tooley even argues, following Hume, that the very use of pain in order to alert animals and humans to bodily damage is already a design fault, because less unpleasant other means are available that would work equally well.[35]

We should conclude that Swinburne's argument from consciousness is not a *correct* C-inductive argument because the relevance condition $P(e_5|h\&k) > P(e_5|\sim h\&k)$ is not satisfied with regard to specific examples of brain–mental connections. Indeed, some of these specific connections are not at all probable on the hypothesis that God exists, because they involve serious design faults, as I illustrated for the case of pain. Furthermore, the argument from consciousness turned out not to be a *good* C-inductive argument either, because its factual premises are not accepted by many of those who dispute about the conclusion.

14.5 ARGUMENTS FROM MORAL TRUTH AND MORAL AWARENESS

Traditional theists interpreted the so-called 'voice' of our conscience as the voice of God. Consequently, they considered the facts of human moral awareness as very strong evidence for God's existence. The argument from moral awareness is still popular today, even among a number of prominent scientists who are also religious believers. For example, in his popular book *The Language of God*, the geneticist Francis Collins defends the view that our awareness of what he calls 'The Moral Law' cannot

---

[35] Plantinga and Tooley (2008), pp. 111–12.

be explained by secular science or scholarship, and that it yields in fact the best reason for believing in God.[36]

If one endorses a theory of moral objectivism, discussed in Chapter 9.3, above, one has to distinguish between two possible inductive arguments from morality to God. The first argument is an argument from moral truth, in which the (alleged) fact that there are (objective, species-independent) moral truths functions as the explanandum (e_6). The second argument is an argument from moral awareness, in which the fact that we are aware of moral truths is the explanandum (e_7).[37] For two reasons, I shall discuss here the argument from moral awareness only, and interpret the explanandum of this argument liberally, so that it does not imply or presuppose a doctrine of moral objectivism.

First, I criticized Swinburne's arguments for his moral objectivism and for the existence of species-independent moral truths in Chapter 9.3. As I said, his view that moral norms are not species-relative is implausible given our background knowledge from socio-biology. More in particular, I claimed that the fundamental norms of human morality should be understood as typical for intellectually advanced gregarious higher animals, so that the explanandum (e_6) as conceived of by Swinburne does not exist. Second, Swinburne confesses, after having analysed one unconvincing Kant-inspired argument from moral truth, that he is 'too pessimistic' about the prospects of developing a convincing C-inductive argument of this kind for devoting more time to the task.[38] For these reasons, I confine myself to an analysis of the argument from moral awareness.

According to Swinburne, the argument from moral awareness is 'one more good C-inductive argument for the existence of God'. This is because 'there is no great probability that moral awareness will occur in a Godless universe', whereas 'God has significant reason to bring about conscious beings with moral awareness', so that the relevance condition $P(e_7|h\&k) > P(e_7|\sim h\&k)$ is satisfied.[39] In other words, Swinburne holds that in this case the likelihood of the theistic explanation is greater than that of alternative, secular explanations, covered by the catch-all hypothesis $\sim h$.

The reader can by now guess why, according to Swinburne, God had good reasons to bring about moral awareness in us. He holds that there is a probability of roughly $\frac{1}{2}$ that God wanted to create humanly free agents, who are able to make significant choices. He also holds that choices can only be 'significant' for beings with moral awareness. Admittedly, any conscious agent that can choose to do any action must see this action in some way as a good action. But if 'good' merely means here that the action produces pleasant sensations, or that the agent wants to perform the action because it serves his or her interests, the choice is not 'significant' in Swinburne's sense. We might say, then, that by definition humanly free agents can only make 'significant' choices if they have moral awareness. Moreover, choices would not be free and significant if humans were not tempted to do bad things or did not know that these

---

[36] Collins (2006), Chapter 1.
[37] Cf. Swinburne (EG), p. 212.
[38] Swinburne (EG), p. 215.
[39] Swinburne (EG), p. 218.

things are bad. Consequently, the likelihood that God causes moral awareness in human beings must be at least $1/2$.[40]

But why does Swinburne think that the probability of moral awareness in humans is lower than $1/2$ if God does not exist? Let me first mention the reasons for which Francis Collins claims that the existence of our moral awareness cannot be explained by secular science.[41] Collins focuses on cases of really unselfish altruism, which, he claims, can neither be explained by the usual models of reciprocal altruism or tit-for-tat, as developed by R. Trivers and later evolutionary game theorists, nor by models of kin-selection first proposed by W. D. Hamilton.[42] Take, for example, the courageous actions of Oskar Schindler during the Second World War, by which more than one thousand Jews were saved from the extermination camps. Or take the case of Mother Teresa, who won the Nobel Peace Prize in 1979 because, following her religious vocation, she devoted her life to helping the poor in the slums of Calcutta, and founded her order of The Missionaries of Charity. Collins thinks that although these examples of courageous and unselfish altruism may be rare, a scrupulous analysis of the inner voice of our conscience will reveal that all of us are motivated to practise this type of altruistic love or *agapè*, although we frequently attempt to ignore our inner voice. What explains such a fact of moral awareness?

In *The Descent of Man*, Charles Darwin already acknowledged that this type of unselfish altruism creates a problem for the theory of natural selection. As he said with regard to the example of courageous warriors:

> He who was ready to sacrifice his life, as many a savage has been, rather than betray his comrades, would often leave no offspring to inherit his noble nature. The bravest men, who were always willing to come to the front in war, and who freely risked their lives for others, would on an average perish in larger numbers than other men. Therefore it hardly seems probable, that the number of man gifted with such virtues, or that the standard of their excellence, could be increased through natural selection, that is, by the survival of the fittest.[43]

We might formulate the problem somewhat more sharply than Darwin did by introducing the modern notion of *evolutionary altruism*, according to which an action is evolutionarily altruistic if and only if it involves a cost in terms of reproductive fitness to the donor and confers a fitness benefit on the recipient.[44] Given this definition, it will seem a mystery how the trait of evolutionary altruism, once originated by mutations, can spread and be stabilized in a population. Paradoxically, one might say that each possible solution to this problem with regard to a specific altruistic trait shows that the trait was not altruistic in the evolutionary sense after all, if at least one defines fitness costs and benefits as costs and benefits in the long run and at the genetic level.[45] But this does not imply that the trait is not genuinely altruistic in the vernacular sense.

Darwin already proposed a number of such solutions and many contemporary evolutionary biologists have proposed others. In order to claim, as Collins does, that

---

[40] Swinburne (EG), pp. 215–18. [41] Collins (2006), Chapter 1.
[42] Cf. Trivers (1971) and Hamilton (1964). [43] Darwin (1871), pp. 155–6.
[44] Cf. on the main differences between the ordinary or vernacular notion of altruism and the concept of evolutionary altruism: Sober (1988). Many altruistic actions in the vernacular sense are not altruistic in the evolutionary sense, and vice versa.
[45] Cf. for an instructive discussion of the problem of evolutionary altruism: Carter (2005).

no possible solution will work, one has to refute all logically possible evolutionary explanations, and this is a difficult task. In other words, Collins' argument to the effect that real (evolutionary) altruism cannot be explained by secular science, so that we must resort to a theistic explanation of our moral awareness, is an *argumentum ad ignorantiam*. For example, Collins argues that the solution of group selection, already proposed by Darwin, does not work because the majority of modern evolutionary biologists agree that group selection never occurs in nature. But a model of group selection that solves the altruism problem has been proposed by Elliott Sober and David Sloan Wilson, whereas even E. O. Wilson now accepts the correctness of so-called multi-level selection theory, according to which the mechanisms of natural selection may work at many different levels, including the level of groups of organisms.[46]

This is why, I guess, Richard Swinburne does not argue that altruism is an insoluble problem for the theory of natural selection. As he writes, '[i]f genetic mutations produce creatures naturally inclined to behave altruistically towards others of their community [...], then there may well be a good Darwinian explanation for their survival'.[47] Instead, Swinburne argues that our moral awareness includes moral beliefs, and that 'while God will give some creatures moral beliefs as features essential to their being humanly free agents, there is otherwise no particular reason why whatever processes give creatures beliefs should give them moral beliefs'. He claims that '[t]his is shown by the fact that, as far as we can tell, there are many species of animals that are naturally inclined to help others of their species, and yet do not have moral beliefs'.[48] The point of the argument seems to be that whereas once 'some mutation gave to creatures of some community such moral beliefs or the ability to acquire them, such beliefs would reinforce any selective advantage possessed by a natural inclination to altruistic behaviour', secular science and scholarship will not be able to explain what mutations or other factors cause such beliefs to arise in the first place.[49]

But this is a weak argument indeed. In the paradigmatic sense of 'belief', creatures can only have beliefs (beyond the most rudimentary ones) if they have a language, and the evolution of linguistic dispositions and of languages can be studied and explained in detail by biological anthropology and historical linguistics. Once humans acquire linguistic skills, it is to be expected that they gradually develop concepts both of phenomena in the world and of features of their own behaviour, such as acts of altruism. Initially, the proximate mechanisms that triggered these acts of altruism were merely the emotional systems that, in more rudimentary forms, are also present in other gregarious animals. According to behavioural biologists such as Frans de Waal, these emotional systems and capacities are the elementary 'building blocks' of human morality.[50] However, as both Darwin and De Waal stress again and again, human morality is inconceivable without language and our superior intellect, which enable us to analyse the past and foresee the future, and to develop general moral rules, which reflect in part our genetically based emotional systems.

---

[46] Sober and Wilson (1998); Wilson and Wilson (2007).
[47] Swinburne (EG), p. 216. In note 7, Swinburne refers to Sober and Wilson (1998).
[48] Swinburne (EG), p. 217.
[49] Swinburne (EG), p. 217.
[50] Cf., for example, De Waal et al. (2006) and De Waal (2005).

It is entirely to be expected, then, that human beings develop moral beliefs once they have acquired linguistic capacities. Both the evolutionary history of the emotional building blocks of human morality, and the later cultural developments of human morality can be studied and explained in detail by secular disciplines such as evolutionary and behavioural biology, biological anthropology, cultural anthropology, psychology, and historical research. The scientific and historical explanations of the origin of moral beliefs, both those that are shared by humanity at large and those that are typical for specific cultures in specific historical situations, are very detailed and instructive. Since all these explanations are part of the catch-all set $\sim h$, there is no reason whatsoever to think that in the case of Swinburne's argument from moral awareness, the relevance condition $P(e_7|h\&k) > P(e_7|\sim h\&k)$ is satisfied. On the contrary, in terms of the richness and precision of their three-dimensional scope, the likelihood of the secular explanations of human moral beliefs is much greater than that of the theistic explanation.[51]

14.6 THE ARGUMENT FROM PROVIDENCE

If there is a probability of $1/2$ that God intended to create humanly free agents who can make significant choices, it will be equally probable that God wanted to create a world in which such significant choices make a real difference, so that humans have responsibilities for themselves, for each other, and for specific aspects of the world. Swinburne calls a world in which there are opportunities for significant and effective choices *providential*. Accordingly, the C-inductive argument in which the explanandum (e_8) is the collection of those aspects of the world that give us opportunities for significant effective choices, is called the *Argument from Providence*. Swinburne avows that he does not know of anyone who has collected these aspects and put them 'in the form of precise argument for the existence of God'.[52] Since many of these aspects are often considered evils, the argument from providence prepares Swinburne's solution to the problem of evil, which I shall discuss in sections 14.7ff.

In Chapter 10 of *The Existence of God*, Swinburne describes a great many features of the world that create or enhance the range of our choices and responsibilities. For example, we have to provide for ourselves as bodily beings. But the range of our choices in this respect 'is greatly extended by our living in an environment where geography provides dangers, food is limited, there are predators, and other humans compete with us for the things that will satisfy our desires'.[53] Furthermore, human procreation and division of labour require cooperation, and 'it is good that we should have the opportunity to cooperate when cooperation matters'.[54] But our responsibility for other human beings will be greatly enhanced if we also have the power and opportunity to harm them, and 'to deprive them of what they badly need'.[55] In a similar vein, Swinburne lists many other features of the human condition which are

---

[51] Swinburne might reply that the universe is fine-tuned for humans-with-moral-beliefs. But if the argument from moral awareness turns out to be yet another fine-tuning argument, it will be open to the objections I raised in Chapter 13.4–6.
[52] Swinburne (EG), p. 219. [53] Swinburne (EG), p. 220.
[54] Swinburne (EG), p. 223. [55] Swinburne (EG), p. 225.

usually considered bad, and argues that a good God must have intended to create them if he wanted to create humanly free agents in the first place, who are capable of significant and effective moral choices.

Critics of theism have often argued, for instance, that a really good God would not have created us as mortal beings, because death is generally considered a bad thing. But Swinburne provides five reasons for reversing this argument, since the fact that we are mortal enhances our freedom and responsibility in five ways. First, if we were immortal, we would be stuck with each other, whereas our mortality gives us the power to kill each other and ourselves. By creating us as mortal, God gave us a mark of profound trust, whereas by 'refusing them this power, a God would refuse to trust his creatures in a crucial respect'.[56] Second, in a world without death, 'supreme self-sacrifice and courage' would be impossible, as would be 'cheerfulness and patience in the face of absolute disaster'.[57] In the third place, if we lived on indefinitely, we could always undo our actions, whereas they matter more if the length of our life is limited. Fourth, in a world with birth but without natural death, the young 'would always be inhibited by the experience and influence of the aged', and this would limit their freedom.[58] Finally, although it is good according to Swinburne that we can harm other people, because this enhances our responsibility, it is also good that there is a limit to the suffering that agents can inflict on one another, and a 'natural death after a certain small finite number of years provides the limit to the period of suffering'.[59] He even thinks that this is '[t]he greatest value of death'.[60]

I leave it to the reader to decide whether these five reasons really justify Swinburne's eulogy of death. Here, I assume for the sake of argument that the probability of God wanting to create us as mortal beings is indeed $1/2$, and that this holds also for the other features of the world referred to by Swinburne in his argument from providence. Then the question arises whether the argument from providence is a correct C-inductive argument. In other words, is the probability that God exists greater than it would be without this argument, because the relevance condition $P(e_8|h\&k) > P(e_8|\sim h\&k)$ is satisfied? Let us again take death as an example. According to Swinburne, $P(e_8|h\&k)$ is 'significant' if e_8 is the phenomenon of human mortality and h is theism.[61] But which alternative explanations of human mortality should be examined? What are the members of the catch-all set $\sim h$?

Since in this phase of Swinburne's strategy of addition it is considered *in confesso* that humans are a product of biological evolution, I suppose that $\sim h$ should include biological explanations of the fact that humans are mortal. Human mortality has a very high probability (near to one) given the fact that our ancestors during the past four million years were all mortal. These ancestors evolved from even earlier ancestors, tree-dwelling primates, who date back from about 5 million to about 60 million years ago. Ultimately, modern mammals evolved from mammal-like reptiles, and these reptiles evolved from terrestrial amphibians, which in their turn evolved from lobe-finned fish ancestors, all of which were also mortal, and so on. Since it would be very implausible to argue, for example, that God intended to make these early lobe-finned fish mortal

---

[56] Swinburne (EG), p. 229. Cf. for these reasons also (PPE), pp. 212–15.
[57] Swinburne (EG), p. 229. [58] Swinburne (EG), p. 229.
[59] Swinburne (EG), p. 230. [60] Swinburne (EG), p. 229.
[61] Swinburne (EG), p. 234.

for the five reasons adduced above, that is, because God wanted to enhance their freedom and responsibility, it follows that if a biological explanation of our mortality is included in the catch-all hypothesis $\sim h$, the relevance condition is not satisfied. Hence, the argument from providence is not a correct C-inductive argument for the existence of God, if e is human mortality. Similar considerations will apply to all other features of the human condition that belong to explanandum e_8.

It is perhaps because such a rejoinder is to be expected that at the end of his tenth chapter, Swinburne suddenly decides to interpret the argument from providence as yet another fine-tuning argument. Instead of defining explanandum e_8 as the set of all features of the world that give humans opportunities for effective significant choices, he now defines e_8 as the fact that 'the laws and boundary conditions of the world are such as to endow humans with such natures and put them in such circumstances as to give them the sort of responsibility described in this chapter'.[62] This fact allegedly is 'too big' for science to explain. And since we might imagine many possible worlds in which we humans do not have all the opportunities described, such as a possible world in which all humans 'live in shells, unable to talk to each other or to hurt each other, merely making contact at the moment of reproduction', the probability that our world has these opportunities is greater if God exists than if God does not exist, given 'the diversity of possible worlds'.[63] One may wonder whether these possible worlds can be conceived of consistently. But this does not matter very much. If the argument from providence is an argument from fine-tuning, it fails because of the objections I adduced in Chapter 13.4–6 against this latter type of argument.

14.7 ARGUMENTS FROM EVIL

Only two of the many C-inductive arguments that Swinburne considers in his cumulative case are arguments to the effect that God *not* exist. Both of them are concerned with types of evil in the world, where the word 'evil' stands for all things we consider to be bad, and not only for cases of moral depravity. With regard to these two arguments, Swinburne attempts to show either that they do not have any force at all, or that their force is weak, so that in the final reckoning they are outweighed by arguments to the effect that God exists. Explanandum (e_{9a}) is the presence of those states of affairs in the world that are bad or evil whether or not God exists, whereas explanandum (e_{9b}) consists of states of affairs that are bad only if God exists.

Concerning the first category (e_{9a}), it is customary to distinguish between moral evils, brought about by intentional human actions or knowingly allowed to occur by human beings, including those morally bad actions and instances of negligence themselves, and natural evils, such as diseases or the misery caused by predators, parasites, earthquakes, tornadoes, volcanic eruptions, and other natural disasters. With regard to category (e_{9b}), Swinburne focuses on the fact that although God is generally thought to be a good father of humanity at large, if he exists God fails to make himself known to quite a number of humans, as is shown by the presence of many inculpable non-theists in the world today, and during the history of humanity.

---

[62] Swinburne (EG), p. 235. [63] Swinburne (EG), pp. 234 and 235.

I shall first discuss Swinburne's attempts to refute or weaken the force of arguments from e_{9a} to the effect that God does not exist, that is, his treatment of the traditional problem of evil (sections 14.8–10), and review his solution to the problem of *Deus absconditus* or of God's hiddenness in the last two sections (14.11–12) of this chapter.

With regard to traditional arguments from evil against God's existence, we should distinguish between two varieties, the logical and the inductive or evidential. According to the logical argument, there is an implicit contradiction between the thesis that natural and moral evils occur in the world on the one hand, and the claim that God exists on the other hand, if at least God is defined by the properties of omniscience, omnipotence, perfect freedom, infinite goodness, and of having created everything apart from himself. For surely an infinitely good God would eliminate all evils he knows of and is able to eliminate, whereas being omniscient and omnipotent, God knows of all evils and is able to remove all of them. Since the existence of evil is considered an established fact, it is concluded by *modus tollendo tollens* that God does not exist.

In spite of its initial appeal, however, the logical argument from evil is bound to fail. For in order to show that there is a contradiction, one has to argue for the logical impossibility of God having good justifying reasons for admitting evils in the world, and this is a heavy burden of proof. Let us assume, then, that the existence of all actual evils in the world is logically compatible with the existence of God, so that the likelihood $P(e_{9a}|h\&k)$ is not zero. Yet it may be argued that this evidence (e_{9a}) is much less likely on the hypothesis that God exists than on some other hypothesis, such as an atheistic hypothesis combined with our knowledge about the geology of Earth and the evolutionary history of life. If this is true, the existence of the actual evils in the world (e_{9a}) yields a good C-inductive or evidential argument from evil against the existence of God.

Confronted with the logical argument from evil, theists have a rather easy task. They merely have to tell us some global story that is possibly true in the sense of not being self-contradictory or logically incompatible with our existing knowledge, and which shows that the existing evil in the world is logically compatible with God's existence. The myth of paradise and the fall of man is such a story. It tells us that God's existence does not logically exclude all 'post-lapsarian' evils, whether moral or natural, because it is fancied that they are all due somehow to man's free decision to sin.[64] Refuting the evidential argument from evil is much more demanding for theists, however.[65] With regard to each particular occurrence of evil and to all existing evils taken together, it has to be shown that they are not less likely on the hypothesis that God exists than on some rival, atheistic hypothesis.

---

[64] Cf., for example, Van Inwagen (2006), pp. 84ff. Whether this is a plausible story is quite another matter, but this is not needed for defeating the logical argument from evil. Of course, the story has to be supplemented by another story explaining the pre-lapsarian natural evils. Cf. Van Inwagen (2006), Lecture 7.

[65] For this reason it is surprising that Van Inwagen says about the distinction between the logical and the evidential arguments from evil that 'this is not a distinction I find useful' (2006, p. 8). Unfortunately, he does not tell us why he does not find the distinction useful, and he uses it in earlier publications such as (1991).

14.8 THE HIGHER-ORDER GOODS DEFENCE

In order to defuse the evidential problem of evil, Swinburne purports to show with regard to each instance of existing moral or natural evil E_i, and concerning all the n existing evils E_{1-n} taken together, that four conditions are satisfied. Put in Swinburne's own words, these conditions are the following:

(1) 'it is not logically possible or morally permissible to bring about some good G except by allowing E (or an evil equally bad) to occur or by bringing it about';
(2) 'God also in fact brings about the good G';
(3) 'God must not wrong the sufferer by causing or permitting the evil';
(4) 'the expected value of allowing E to occur – given that God does bring about G – must be positive' so that 'it is probable that the good will outweigh any evil necessary to attain it'.[66]

In Chapter 11 of *The Existence of God*, Swinburne argues that these four conditions are satisfied by 'all the world's evils', at least if one adds to theism 'one or two further hypotheses'.[67] He also concludes that although the evidential argument from evil is indeed a good C-inductive argument against the existence of God (bare theism), it is not very strong, whereas it is not a good C-inductive argument at all against the ramified theism of Christianity, because the latter includes the hypothesis of heaven.[68] Incidentally, there is a striking difference between the spaces devoted to each of the four conditions in Swinburne's chapter. Although he admits that condition (4) concerning the quantity of evil in the world is 'the crux of the problem of evil', he merely uses three pages to discuss it, whereas he devotes nineteen pages to condition (1), only one paragraph to condition (2), and six pages to condition (3).[69] I shall now argue that contrary to what Swinburne contends, the evidential problem yields a very strong C-inductive argument against the claim that God exists.

It is not difficult to show that (condition 1) for each evil E, there *might* be some good G, which would not be logically possible without E. If, for example, having *knowledge* is always good, each individual instance of evil E yields the possibility of a good G, to wit, that some individual knows of E.[70] And since evils consist mostly of suffering by humans or animals, each evil E of which we know will enable us to feel *compassion* for the suffering individual involved, whereas feeling compassion is also considered to be a good, except by philosophers such as Friedrich Nietzsche. Since knowledge of E and compassion with someone really suffering from E are logically impossible unless E exists, condition (1) is satisfied for all evils, whether moral or

---

[66] Cf. for the quotes in these conditions: Swinburne (EG), pp. 237–8. The formulations are somewhat different in (PPE), pp. 11–14 and *passim*. In particular, condition (2) is formulated as 'God does everything else logically possible to bring about G' (p. 14, condition c), in order to make it consistent with condition (4). These details are not important for my arguments in the main text.
[67] Swinburne (EG), p. 238.
[68] Swinburne (EG), p. 266. Unfortunately, Swinburne does not distinguish clearly between the logical and the evidential argument from evil.
[69] Swinburne (EG), p. 263. Cf. also (PPE), p. 239.
[70] Cf. for Swinburne's elaborate 'argument from the need for knowledge': (EG), pp. 245–56.

natural.[71] This is the point of the so-called *higher-order goods defence*, where the label 'higher-order goods' is defined as 'goods the occurrence of which is logically impossible without the occurrence of some evil'. Swinburne discusses many other higher-order goods, such as courage in the face of danger, helping someone who suffers, and stoic endurance of pain.

However, it is unlikely that this higher-order goods defence will be able to meet conditions (2) and (4). In order to outweigh or absorb the evils that exist in the world (condition 4), the second-order goods must exist as well (condition 2). But is it really the case that for each individual instance of suffering E_i that occurred in the past or is occurring now, there in fact existed, or exists, or will exist in the universe, a higher-order good, such as knowledge of E_i or compassion because of E_i? In the only paragraph that Swinburne spends on condition (2), he does not even raise this crucial question.[72] And it seems that we must answer it in the negative, if we think for example of the innumerable cases of animal suffering that have occurred during evolutionary history, with its mechanism of the struggle for life, and given the presence of predators and parasites.[73] Or think about the animal suffering that must have occurred during the Cretaceous-Tertiary mass extinction some 65 million years ago, which was probably caused by an asteroid impact, and during which nearly all the existing species of dinosaurs perished, mainly from starvation.

We humans cannot know in detail the great majority of such individual instances of suffering, since we cannot be acquainted with each of the individual animals that have existed in the distant past, before humans evolved. We can merely have some vague general ideas about them, and these instances of knowledge will certainly not suffice to outweigh the evils of evolutionary history. Also, I do not know anyone who really feels compassion for the dinosaurs that perished during this mass extinction, although children may feel compassion for their toy dinosaurs. I suppose that most humans are simply not able to feel compassion for dinosaurs.

Theists might retort that if an animal suffers, it must know of its own suffering, for the simple reason that one cannot have pain without being aware of one's pain, whereas being aware is a (simple) type of knowledge. But of course, if pain is considered bad, the 'knowledge' consisting in one's awareness of pain cannot be considered a good, since being aware of one's pain and having it are one and the same thing. In short, condition (2) will very probably not be met by the higher-order goods defence if considered on its own.

---

[71] Strictly speaking, one might have compassion for a person because one thinks mistakenly that this person suffers, so that God might create the good of human compassion by inducing in us the illusion that others are suffering. According to Swinburne, however, inducing such illusions would not be morally permissible for God. This is why Swinburne includes in requirement (1) the condition that it is not morally permissible to bring about some good G except by allowing E to occur.

[72] Swinburne (EG), p. 257, first paragraph. The reason is that he considers the second-order goods of human knowledge and compassion to be results of free choice. Then, by giving us free choice, God has done 'everything else logically possible to bring about G', as condition (2) should be read in this context (cf. PPE, p. 14). But these second-order goods are not always the product of free choice. Quite often, we cannot but witness evils, and cannot but feel compassion, and quite often knowledge is beyond our epistemic scope, such as in the case of William Rowe's fawn trapped in a forest fire. Cf. Rowe (1979).

[73] Cf. Rowe (1979) and (2006).

There are four good reasons for thinking that this defence (on its own) cannot show either that condition (4) is satisfied. First of all, apart from second-order goods such as compassion for suffering, there are also second-order evils, such as malicious delight in someone else's misfortunes. The existing second-order goods have to outweigh not only the existing first-order evils but also these second-order evils, to the extent that they exist. Or, to put this point somewhat more precisely, since the second-order evils can only satisfy conditions (1) and (2) if they in fact give rise to third-order goods, such as the moral indignation one feels concerning the presence of malicious delight, it is these third-order goods that must absorb or outweigh the second-order evils. But is there really more compassion and good moral indignation in the world than suffering and unholy glee? It is not obvious that this is the case.

Second, it is only the second-order and third-order goods that are available for outweighing the evils in the world, if one wants to solve the evidential problem of evil. Most goods, such as being kind to someone, or being generous, or being useful to someone, do not logically presuppose evils, so that God could have created these goods without creating or permitting any evils. In other words, only those goods can be fed into the equation of condition (4) that also satisfy condition (1), and it is very unlikely that (2) in fact enough of them exist.[74]

There are two further problems for theists who want to show that condition (4) is satisfied by all existing evils in the world. In order to argue convincingly that for all E's and G's, the expected value of allowing E to occur is positive, they have to quantify both probabilities and valuations, since the expected value of an action or scenario, such as the scenario of causing or allowing E to occur, is the long-term average outcome of that scenario.[75] No theist has attempted to solve this problem of quantification, which is my third reason for thinking that the higher-order goods defence cannot succeed in showing that condition (4) is satisfied.

Finally, there is the problem of valuation itself. It is here that theists will meet the strongest opposition to the higher-order goods defence. For they will have to argue that the negative value of an existing and horrendous evil such as the Holocaust is outweighed by the positive value of all existing second-order goods made possible by the Holocaust, such as the compassion people feel for the victims, if taken together. Critics will object that the valuations of E and G required for such a result are unacceptable from a moral point of view. As John Stuart Mill once said, they will amount to a 'jesuitical defence of moral enormities'.[76] Surely, they will protest with indignation, *no* amount of present and future goods like compassion, knowledge, or remorse concerning the Holocaust will be able to outweigh the total amount of evil of the Holocaust itself.

And indeed, concerning their valuations, theists who endorse the higher-order goods defence will seem to commit a dubious inversion. Given that the Holocaust

---

[74] Of course, if one makes the higher-order-goods defence dependent on the free-will defence, one might argue again that God has done everything logically possible to bring about G. Cf. Swinburne (PPE), p. 14 and *passim*. But I am here considering the higher-order-goods defence on its own merits. Also, many higher-order goods are not within the ambit of our free decisions.

[75] In other words, the expected value of an action A is the sum of the values of its possible consequences, both positive and negative, each multiplied by the probability that the consequence occurs if A is done. Cf. Swinburne (PPE), p. 12, note 7.

[76] Mill (1878), p. 187, quoted by Van Inwagen (2006), p. 57.

occurred, it is urgently desirable that a maximum amount of second-order goods will be induced by it. But no one will claim that the occurrence of these second-order goods is so probable, and that these goods are so valuable, that God would be justified to cause the Holocaust, or allow it to happen, in order to produce them, as condition (4) requires. A similar argument applies to most second-order goods. For example, was it justified to cause or permit the AIDS epidemic, which induced the agonizing deaths of millions, in order to give to humanity the 'outweighing good' of the opportunity to investigate its causes and to prevent future instances of AIDS?[77] I conclude that if considered on its own, the higher-order goods defence is utterly unconvincing. There are too many and too serious evils in the world, both natural and moral, which seem to be pointless and gratuitous in spite of the higher-order goods defence.

14.9 THE FREE-WILL DEFENCE

Given the weakness of the higher-order goods defence if considered on its own, it has to be buttressed by other defences. With regard to moral evils, theists might supplement the higher-order goods defence by the free-will defence in the following manner. Concerning the Holocaust objection, they will admit that God would not be justified to cause or permit the Holocaust in order to produce the existing set of second-order goods that logically presuppose it. But, so they will object, God did not cause the Holocaust. All the horrendous events that, taken together, constitute the Holocaust were the product of free decisions taken by human beings, for which God cannot be held responsible. As Swinburne says:

> If humans are to have the free choice of bringing about good or evil, and the free choice thereby of gradually forming their characters, then it is logically necessary that there be the possibility of the occurrence of moral evil unprevented by God. If God normally intervened to stop our bad choices having their intended effects, we would not have significant responsibility for the world.[78]

The idea is, clearly, that human free will and human 'significant responsibility for the world' are very great goods, which, taken together with the higher-order goods discussed above, will outweigh the sum of existing evils in the world. But do they? Again, we have to check whether the four conditions listed above are met.

First of all, it is debatable whether free will is a higher-order good in the sense of condition (1), that its existence is logically impossible without the existence of moral evils caused by free decisions. Can we not imagine a logically possible world containing free agents who always decide for the good? Perhaps significant freedom requires logically that these agents have some temptation to do evil, as Swinburne argues, and surely this temptation is in itself a bad thing. But is it not logically possible that humans always freely decide to overcome these temptations? If God in his omniscience would know all 'true counterfactuals of creaturely freedom', and could partly pre-programme the outcomes of the free deliberations of created free agents, God could have created a better world than in fact exists, so that even the logical problem of evil is

---

[77] Cf. for this example Smith (1992), p. 350.
[78] Swinburne (EG), p. 238.

not resolved by the free-will defence.[79] Let us assume, however, that condition (1) is met because, among other reasons, the freedom to form our character is logically impossible unless we sometimes do bad things, as Swinburne argues. What, then, should we think of the remaining three conditions?

As has been argued by many authors, the free-will defence works and condition (2) is satisfied only if our free decisions are in fact not completely causally determined, and if compatibilism is false. If compatibilism were true and the world created by God were deterministic, God would be causally responsible for and involved in all the moral evils of the world because of his omniscience and omnipotence, so that the fact of moral evil in the world would entail that God does not exist. Incompatibilism is indeed Swinburne's view, but it is not easy to establish it convincingly.[80] Furthermore, one might argue that the value of human freedom would not be diminished much if compatibilism were true and if the world were deterministic. If so, could and should God not have created a world with compatibilist freedom, which is such that humans always choose to do the good? If compatibilism is in fact false, does this then not demonstrate that God does not exist, or at least diminish the probability that God exists?

Even more serious problems for the free-will defence arise with regard to conditions (3) and (4). One may wonder, for example, whether with regard to the Holocaust or concerning the lethal regime of the communist Khmer Rouge in Cambodia, condition (3) is satisfied, according to which 'God must not wrong the sufferer by causing or permitting the evil'. Even if God did not cause the Holocaust, he could have prevented it by interfering before the Nazis implemented their horrendous project. By not preventing it, God permitted the Holocaust. Can one really argue that by so doing, God did not wrong the more than six million massacred Jews and gypsies? The critic of theism will hold that in such cases, one cannot decently defend that 'it is a blessing for a human (or animal) if the possibility of his suffering makes possible the good for others of having the free choice of hurting or harming him', as Swinburne argues with regard to lesser evils.[81] Such a defence would be seen by many as yet another case of a 'jesuitical defence of moral depravity'.

Furthermore, theists have a predilection for the good-father metaphor. This metaphor seems to imply that in such cases God should have interfered. If so, the fact that God did not interfere disconfirms theism. Admittedly, '[i]f God normally intervened to stop our bad choices having their intended effects, we would not have significant responsibility for the world', as Swinburne argues.[82] But good fathers should sometimes interfere, and a perfectly good God should have intervened in Nazi Germany, the Stalinist Soviet Union, Maoist China, Cambodia, and Rwanda. Many Christians and Jews thought so after the Second World War and forsook their faith. As Swinburne says in another context, 'the parental analogy suggests that a good God will come to the help of the human race if communities go too far astray'.[83]

---

[79] For a technical discussion of this issue, cf. Plantinga (1974), Chapter 9, and Flint (1998), Chapters 2 and 4–7.
[80] Cf. Swinburne (RA), Chapter 3 and (ES) Chapter 13.
[81] Swinburne (EG), p. 260.
[82] Swinburne (EG), p. 238; cf. the quote in the main text above.
[83] Swinburne (EG), p. 276.

According to condition (4), 'the expected value of allowing E to occur – given that God does bring about G – must be positive' so that 'it is probable that the good will outweigh any evil necessary to attain it'. With regard to this condition, theists who endorse the free-will defence seem to be confronted by a dilemma. When God created our world, in which human beings with a libertarian free will would evolve, either he could neither influence nor foresee *at all* what amount of moral evil would result from their free decisions (assuming that this can be reconciled with God's omnipotence and omniscience), or he could influence and foresee this to some extent. In the first case, condition (4) is not satisfied, and God took an irresponsible risk by (indirectly) creating humanly free agents. As Everitt says, it 'would be a kind of cosmic Russian roulette'.[84] In the second case, God is co-responsible for the existence of moral evil after all, so that the free-will defence does not disculpate him.

Even if this dilemma is not fatal, theists will have great difficulties in showing that condition (4) is satisfied if G is free will and E are the resulting moral evils. They will have to cope again with the problems of valuation, quantification, and inversion that I discussed above. How highly should we value the existence of libertarian free will in itself, for example, to the extent that it really exists? A bad action such as killing someone is generally considered to be worse if it is done out of free deliberation than if it results from blind impulses, as is acknowledged by criminal law. Similarly, good actions are perhaps morally better if they are done out of free will than out of inclination, as Kant seems to have argued. Should we not conclude that the value of freedom in itself is neutral, so that even *one* evil action would outweigh the total value of human freedom? Surely, God could have pre-programmed humans in such a way that, being unfree, they would not commit any evil deed.[85] And how can it be shown that condition (4) is satisfied without a reliable quantification of valuations and probabilities? Finally, given the fact that humans can harm and kill each other, it certainly is a good that they can freely decide to avoid this. But is it not an unacceptable inversion to argue that God should give to humans the opportunity to kill each other *in order to* broaden their range of significant free choices?

14.10 HEAVEN AND THE VALUE OF FREEDOM

If it cannot be shown convincingly that all existing evils in the world, both if taken separately and if taken together, satisfy conditions (1–4), the evidential argument from evil is a strong C-inductive argument against the existence of God. How can theists avoid this conclusion? One possibility is to attribute an enormous value to significant libertarian freedom, but that does not solve the problem of natural evils. Another is to introduce an auxiliary hypothesis. Let me finally discuss these two options.

Although Swinburne expresses '*considerable initial* sympathy' with critics of theism who argue that there simply is too much prima facie pointless evil in the world for

---

[84] Everitt (2004), p. 249.
[85] The theist typically values freedom in itself very highly. Cf. Swinburne (EG), p. 257: 'The bad nature and bad effects of human free choices being so much worse than the bad nature and bad effects of instinctive animal reactions, the free nature of their choices is, I suggest, needed to justify a good God allowing them to cause such evil as they can'.

theism to be probable (condition 4), he values the existence of significant libertarian freedom so highly that usually it will outweigh both the existence of the natural evils which provide us with opportunities for significant choices, and the risk of moral evils which result from these choices.[86] As he says, if God had restricted considerably these natural evils and 'the harm that humans can do to each other', we would live in 'a toy-world; a world where things matter, but not very much'. And he adds that '[t]he objector is asking that God should not be willing to be generous and trust us with his world'.[87]

Nevertheless, Swinburne admits that conditions (3) and (4) may not always be met. Although God has great rights over us because he is the author of our being, so that he is morally justified in permitting many evils, he must on balance remain a benefactor for each of us (condition 3). It follows that if 'there are humans whose lives on earth are such that on balance it would have been better for them never to have lived, then God has an obligation to provide them with enough (in quantity and quality) of a good life after death'.[88] This 'crucial assumption of a compensatory life after death' may also be needed in order to show that condition (4) is justified. In other words, Swinburne proposes the hypothesis of heaven in order to answer those who think that '[n]o God ought to have allowed Hiroshima, the Holocaust, the Lisbon Earthquake, or the Black Death'.[89] Furthermore, he is 'inclined to suggest' that 'if God makes humans (and animals) suffer to the extent to which he does, albeit for good purposes, he would in virtue of his perfect goodness share our suffering himself', that is, become incarnated in Jesus, crucified, and resurrected.[90] Considering all these things, Swinburne concludes that 'while evil may provide a good C-inductive argument against the existence of God (bare theism)', 'it does not provide a good C-inductive argument against Christian theism', which includes these two auxiliary hypotheses.[91]

What does account for the differences between this conclusion and my own view that the evidential problem of evil yields a *strong* C-inductive argument against the claim that God exists? As we have seen, discrepancies in moral evaluations play a crucial role. One might think that such discrepancies of valuation can never be bridged, since they might be subjective to a large extent. But this is neither Swinburne's view, since he endorses a version of moral objectivism, nor mine, because I hold that moral values are typically species-relative (and perhaps to some extent culturally relative), but not relative to each individual.

So let us try to assess the moral evaluation by which Swinburne overrules his '*considerable initial* sympathy' with the inductive argument from evil.[92] As we saw, he claims that if God very greatly diminished the number and intensity of natural evils, and the harm that humans can do to each other, our 'world would be a toy-world; a world where things matter, but not very much', so that God would not 'trust us with' the world in which we in fact are living.[93] But is this really true? Let us take homicide

---

[86] Swinburne (EG), p. 263 (Swinburne's italics); cf. (PPE), p. 239: 'some initial sympathy'.
[87] Swinburne (EG), pp. 263–4.
[88] Swinburne (EG), pp. 261–2.
[89] Swinburne (EG), pp. 262 and 263ff.
[90] Swinburne (EG), pp. 264ff.; cf. RGI.
[91] Swinburne (EG), p. 266.
[92] Swinburne (EG), p. 263 (Swinburne's italics); (PPE), p. 239.
[93] Swinburne (EG), p. 264.

as an example. Suppose that we lived in a world in which human beings could not kill one another. Would that world be more of a 'toy-world' than the world in which we are living? And is it for that reason good and justified that God created a world in which animals and humans are able to kill each other?

Since we cannot properly evaluate facts without understanding them thoroughly, I take one specific type of homicide as an example. Two men get involved in an argument, which results in the one murdering the other, typically in view of many bystanders. Such killings are usually classified by criminologists as 'trivial altercations' or as 'meaningless violence'. But seen from an evolutionary point of view, they are not trivial or meaningless, since what really happens is a competition for status among males. That the competition takes this crude form is due to the contingent circumstances in which these men were born and are living: poor neighbourhoods, in which people have a relatively low life expectancy. In such environments, men often have to take extreme risks in order to obtain status and prospects for reproduction.[94]

In other circumstances, such as life in academia, the competition for status takes a different form, and males very rarely attempt to murder each other. Instead, they work hard and aim at publishing excellent articles or books. Should we really conclude that academics live more in a 'toy-world' than those who live in the criminal slums of Chicago, London, Shanghai, or Buenos Aires? That would not be my evaluation, and I hope it is not the considered view of the reader either. Hence, if men have to compete for status, it probably would have been better if God had created them without the capacity to kill each other, and similar arguments may be proposed for other types of homicide. As a consequence, the fact that humans are able to kill each other provides a strong C-inductive argument against the existence of God.

That many theists implicitly share this evaluation is shown by the fact that they believe in heaven. The same holds for Swinburne, since he proposes the auxiliary hypothesis of heaven or of 'a compensatory period of good life after death' in order to guarantee that God 'remains on balance a benefactor' for each of us (condition 3).[95] I suppose that in heaven humans do not lose their significant libertarian freedom. Indeed, as Swinburne says elsewhere, '[t]he range of free choice open to the Blessed could be enormous compared with the range of choice open to us on Earth'.[96] But since there is no moral evil in heaven, they always choose what is good. According to the usual representations of heaven, there is no natural evil in heaven either, and souls are not able to kill one another. However, if God created heaven, and if heaven really is a better place for us than Earth, as all orthodox monotheists believe, this implies that God could have created a better world than he in fact created, to wit, heaven only, without Earthly life.[97] This means that by proposing the auxiliary hypothesis of

---

[94] Cf. D. S. Wilson (2007), Chapter 14; Wilson and Daly (1997); cf. also Daly and Wilson (1988).
[95] Swinburne (EG), pp. 264 and 258.
[96] Cf. Swinburne (PPE), p. 251.
[97] As Swinburne says in (PPE), p. 249, '[t]his world-order is a very dangerous and costly experiment which its author might be expected to bring to halt one day'. But as I am suggesting, if God created heaven anyway, he should not have created the 'costly experiment' of our Earthly life. The theistic rejoinder that heaven is good only if its citizens have deserved it by their labours on Earth is convincing only if life on Earth exists anyway.

heaven, Swinburne *eo ipso* destroys the credibility of the evaluations that seemed to enable him to solve the evidential problem of evil.

From a methodological point of view, the auxiliary hypothesis of heaven is problematic as well. It is clearly ad hoc in the sense that it does not imply new predictions. No independent evidence is or can be adduced to support it, unless one accepts a specific revelation. And as Swinburne admits, the hypothesis complicates theism, so that according to his own standards the prior probability of theism is reduced if the hypothesis of heaven is added.[98] In the beginning of his chapter on the problem of evil, Swinburne rejected the traditional hypothesis of fallen angels who cause natural evils, for precisely these reasons. As he said:

> it does have the major problem that it saves theism from refutation by adding to it an extra hypothesis, a hypothesis for which there does not seem to me to be much independent evidence – the hypothesis that angels of this kind exist created by God and have limited power over the rest of God's creation. This hypothesis is not entailed by theism, nor does theism make it especially probable [. . .]. A hypothesis added to a theory complicates the theory and for that reason decreases its prior probability and so its posterior probability.[99]

On these same grounds, Swinburne should have rejected the hypothesis of heaven. Consequently, we have to stick to our conclusion that the amount of prima facie pointless evil existing in the world yields a strong C-inductive argument against the existence of God.

14.11 *DEUS ABSCONDITUS*

Many devoted believers belonging to theistic traditions have complained that God is insufficiently manifest to humanity. 'Truly, thou art a God who hidest thyself', we read in *Isaiah* (45.15), a text rendered in the *Vulgate* as 'Vere, tu es Deus absconditus, Deus Israel, Salvator'. The theme of *Deus Absconditus* or of God's hiddenness occurs again and again in the Christian tradition. For example, in his *Proslogion*, Anselm of Canterbury lamented 'You are my God and my Lord, and never have I seen You [. . .]. I was made in order to see You, and I have not yet accomplished what I was made for'.[100] Blaise Pascal, another Christian philosopher who ardently longed for God, wrote in his *Pensées*: 'I look around in every direction and all I see is darkness', so that 'I am in a pitiful state, where I have wished a hundred times over that, if there is a God supporting nature, she should unequivocally proclaim Him'.[101]

In some senses of 'hiddenness', it is unproblematic for a believer to say that God is hidden or concealed. Since God is defined as infinite in many respects, we will never be able to fathom his nature exhaustively, and as a consequence, we shall never detect the

---

[98] Swinburne (EG), pp. 262 and 264.
[99] Swinburne (EG), pp. 239–40.
[100] Anselm (1965), Ch. 1, pp. 110 and 111: 'Domine, deus meus es, et dominus meus es, et numquam te vidi [. . .]. Denique ad te videndum factus sum, et nondum feci propter quod factus sum'.
[101] Pascal (1966), fragment 429. The French text reads: 'Je regarde de toutes parts, et je ne vois partout qu'obscurité [. . .]; je suis dans un état à plaindre, et où j'ai souhaité cent fois que, si un Dieu la soutient, elle le marquât sans équivoque' (Pascal 1963, p. 555, fragment 429/*229*).

full pattern of his activities in the world. Moreover, since God is defined as bodiless, we cannot expect that we perceive him with our senses, in spite of his omnipresence. But what is problematic for a believer is the sense of hiddenness stressed by Anselm and Pascal, according to which God's very existence is concealed even to those who are longing to meet him. In other words, if there is any evidence for God's existence at all, the evidence is not very compelling. And this fact, which is acknowledged by many theists, yields a powerful argument to the effect that God does not exist. The argument has been developed convincingly by John L. Schellenberg in his book *Divine Hiddenness and Human Reason*. In a nutshell it runs as follows.[102]

If God exists and if he really created humanity, he will desire a reciprocal relationship of love to obtain between himself and each individual human being who is capable of it, because perfect love is an essential property of God. Indeed, according to Richard Swinburne, God created humanly free agents because 'God needs to share, to interact, to love'.[103] However, such a reciprocal relationship with a human being is logically impossible if this person does not believe that God exists. Accordingly, since he is a loving father of all humans, God will want to produce for each individual human being at the earliest possible stage of life sufficient evidence for his existence, such as a non-sensuous experience of God, which is approximately the same for everyone, and is described by all human beings in a similar way. Since God is omnipotent, he will produce such evidence for each human being. But in fact there is no compelling and universal evidence for God's existence of this or any other kind. Indeed, there were and are a great many reasonable and inculpable unbelievers, who honestly think that there is insufficient evidence for God's existence. From these premises, it follows by *modus tollendo tollens* that God does not exist. Or, if the argument is construed as an inductive one, it should be concluded that '[t]he weakness of evidence for theism [...] is itself evidence against it'.[104]

It may be misleading to speak of 'the argument from hiddenness', although I shall go on doing so, since this might suggest that one begs the question by assuming that God exists and intentionally withholds convincing evidence of his existence.[105] Better labels for the argument are 'the argument from God's love' or 'the meta-argument from insufficient evidence'. A particular merit of this argument, if convincing, is that it will settle the debate between intellectually responsible agnostics and atheists. Having examined conscientiously the available evidence and the arguments for and against God's existence, intellectually responsible agnostics conclude that there is epistemic parity in this case. Since they hold that given the total evidence it is equally probable that God exists and that he does not exist, if one can decide about these probabilities at all, they suspend judgement. But if the argument from hiddenness is sound, this very conclusion of epistemic parity entails that God does *not* exist. Hence, all intellectually responsible agnostics must convert to atheism. As Schellenberg says, the agnostic

---

[102] Schellenberg (2006). Cf. also Schellenberg (2005a).
[103] Swinburne (EG), p. 119; cf. PPE, pp. 111–15.
[104] Schellenberg (2006), p. 2. To those who object that religious belief should be a personal choice and does not need any evidence, one should point out that belief is involuntary. It is simply false that we can freely choose what to believe (the thesis of doxastic involuntariness).
[105] Schellenberg (2006), p. 5.

conclusion of epistemic parity 'must in their case be viewed as *itself* a consideration that tips the balance in favour of atheism'.[106]

For a believing theist such as Swinburne, who thinks that God created humans in order 'to interact' with them 'in love',[107] the existence of reasonable and inculpable unbelievers is a state of affairs that is bad or evil. But clearly, this state of affairs is bad only if God exists, so that it belongs to explanandum (e_{9b}) as defined in section 14.7. If we assume that God abhors favouritism and loves all human beings equally, the existence of *one* inculpable unbeliever would be sufficient to argue that God does not exist. But in fact, the extension of explanandum (e_{9b}) is far greater, and indeed, enormous (cf. Chapter 3.3, above).

As far as we know, there were no monotheists before Egyptian pharaohs such as Apophis and Akhnaton, whereas Judaism became monotheistic only in the seventh or sixth century BCE.[108] However, Homo sapiens evolved gradually between 400,000 and 200,000 years ago, and the earliest fossils of anatomically modern humans are over 100,000 years old. Hence, we may safely assume that during the first 200,000 years or so that humanity existed, there were no monotheists. Historians of religion and anthropologists have discovered many cultures in which the theistic concept of God was completely unknown, and today there are probably more than three billion living non-theists. Even Catholic theologians during Vatican II admitted 'that it is possible for a *normal adult to hold an explicit atheism for a long period of time* – even to his life's end – *without this implying moral blame* on the part of such an unbeliever'.[109] If God is conceived of as a good father, inculpable absence of belief that he exists on this massive scale must be seen as compelling evidence against his existence, since '[f]athers who absent themselves too much from their children are rightly judged less than adequately loving'.[110]

14.12 JUSTIFYING GOD'S HIDDENNESS

How can theists rebut this argument from hiddenness? If the argument is construed as a purely logical objection only, according to which the claim (a) that God exists and loves us is (implicitly) *inconsistent* with the fact (b) that reasonable unbelief in theism occurs, it may suffice to invent a story (c), so that (a) and (c) entail (b). In other words, one might argue that the assumption according to which there would be no reasonable unbelief if a perfectly loving God existed, may very well be false.[111]

But if the argument from hiddenness is construed as a C-inductive argument against the existence of God, the task of rebutting it is more demanding. Now it has

---

[106] Schellenberg (2006), p. 212 (Schellenberg's italics). Cf. Schellenberg (2005b) for replies to counterarguments to this claim.
[107] Swinburne (EG), p. 119.
[108] In the older sections of the Hebrew Bible, such as some of the Psalms, Yahweh, the god of Israel, is a member of a larger divine council, of which the god El is the head. Cf., for example, Psalm 82, and Mark Smith (2001), *passim*.
[109] Schellenberg (2006), p. 69, quoted from Karl Rahner (emphasis in the original).
[110] Swinburne (EG), p. 267 and (PPE), p. 204.
[111] That it is not easy to invent a relevant story which accomplishes this aim is shown by Schellenberg (2005a) and (2005b).

Other Inductive Arguments 305

to be shown that the existence of massive reasonable unbelief is not more likely if God does not exist than it is if God exists. In other words, theists have to argue that the relevance condition $P(e_9{}_b|\sim h\&k)>P(e_9{}_b|h\&k)$ is not satisfied for the argument from hiddenness, where h is the hypothesis of theism. But surely, the occurrence of massive inculpable unbelief, including the existence of non-theistic religions and of atheists, is at first sight *much* more probable if God does not exist than if God exists and loves each human being individually as a good father.

In *The Existence of God*, Swinburne attempts to meet this challenge by arguing that both for unbelievers (A) and for believers (B) there is in fact a higher-order good if reasonable unbelief exists. He also argues that this higher-order good is sufficient to outweigh the evil of reasonable unbelief, or 'honest agnosticism', as he calls it, and that it is morally permitted for God to allow reasonable unbelief 'for the period of our earthly life', if at least there is an afterlife.[112] In other words, he argues that with regard to the occurrence of reasonable unbelief as E, there are higher-order goods G that satisfy all four conditions (1–4) quoted in section 14.8, which should be met by an adequate refutation of the C-inductive argument from evil in general.

In case (A) of unbelievers, this higher-order good allegedly is the free and significant choice between good and evil. Indeed, Swinburne claims that 'it is not logically possible that God give us both a strong awareness of his presence and a free choice between good and evil', at least if certain other conditions are fulfilled, such as the condition that we have both a strong desire to be liked by good persons and the belief that God will not like us if we do bad things, or the condition that we have both a strong desire for our future well-being and the belief that probably God will not provide a good afterlife for bad people.[113] If at least some of these conditions obtain, and given a strong awareness of God's presence, we 'will inevitably do the good', so that 'there will be no free choice between good and bad'.[114] Furthermore, the existence of reasonable unbelief 'also makes possible a great good for the religious believer', Swinburne claims (case B). It 'allows the believer to have the awesome choice of helping or not helping the agnostic' to acquire belief, so that the latter will understand 'who is the source of his existence and of his ultimate well-being'.[115]

Regrettably, Swinburne does not explain how these two defences (A) and (B) can be reconciled, and it is precisely their combination that yields some serious problems. Suppose (a) that God gave many believers 'a strong awareness of his presence' by means of religious experiences, for example. Then, according to the first defence, they 'will inevitably do the good', so that they will have 'no free choice between good and bad', except if their temptation to do the bad is exceptionally strong. But how, assuming that their temptation is average, can Swinburne subsequently claim that these believers have 'the awesome choice of helping or not helping the agnostic'? Surely an 'awesome' choice must be a free choice, so that supposition (a) leads to a

---

[112] Cf. Swinburne (EG), pp. 267–72. Swinburne defines 'agnosticism' in this context as including atheism (p. 267). Cf. also Swinburne (PPE), Chapter 11, pp. 203–12.

[113] Swinburne (EG), p. 269 jo. p. 268; (PPE), pp. 205–6, where Swinburne speaks of 'a deep awareness of the presence of God' (p. 206).

[114] Swinburne (EG), p. 269. The conditions are spelled out somewhat more amply in (PPE), pp. 203–9.

[115] Swinburne (EG), p. 271.

contradiction. And indeed, if the defence with regard to (A) is credible, we cannot assume that God gave *any* believer on Earth 'a strong awareness of his presence', since he would thereby destroy their significant freedom, which, according to Swinburne, is one of the highest goods one might imagine.

We have to suppose, then, (b) that God gave theistic believers at best a moderate amount of evidence for his existence, an amount which on the one hand is sufficient to justify and cause their theistic belief, but on the other hand is insufficient to destroy their significant freedom.[116] Perhaps this is precisely the amount of evidence that Swinburne aims at providing by his book *The Existence of God*, the conclusion of which is that '[o]n our total evidence theism is more probable than not'.[117] However, if this is possible, a loving God surely would have given this very amount of evidence for his existence to *all* human beings capable of a loving relationship with him, and he would have done so early in their youth, so that Swinburne's defence ad (A) collapses and the argument from hiddenness stands unrebutted.

Should we then conclude, finally, (c) that even this moderate amount of evidence would destroy human freedom, so that God, in order to safeguard our freedom, did not provide *any* human being with evidence on which his existence is (just) more probable than not? This seems to be the only option left if one wants to avoid the contradiction and solve the problem of hiddenness. But unfortunately, this option would exclude a priori that the very project of natural theology in which Swinburne is engaged can be successful, and it entails that all theistic beliefs are unjustified. But if they are, how are believers supposed to help unbelievers to acquire the belief that theism is true?

Perhaps one might avoid these unwelcome implications of option (c) by arguing that human beings can only acquire this moderate amount of evidence for God's existence by engaging in the difficult, time-consuming, and cooperative intellectual research of natural theology. Both the opportunity of deciding freely to engage in such research and that of persevering in it would be great higher-order goods, made possible by God's initial hiddenness.[118] Furthermore, it might be argued that the goodness of these things 'depends on their being temporary', and that this also holds for the goodness of significant moral freedom. As Swinburne says, God's hiddenness 'allows the agnostic to make a more serious commitment to the good than he would be able to make if the presence of God were more obvious'. But when the agnostic 'has become committed to the good, the advantage of agnosticism in helping him to do it with great seriousness disappears'.[119] However, if this higher-order goods defence were convincing, it would be a mystery why God would allow many young infants to acquire a strong religious belief by education early in life. Indeed, such an education would be an evil, since according to Swinburne it would preclude the significant freedom of young people, which they need in order to become seriously committed to the good.

Apart from these problems of consistency, there are other strong objections to Swinburne's attempt to rebut the C-inductive argument from hiddenness. Because

---

[116] Cf. Swinburne (PPE), p. 207.
[117] Swinburne (EG), p. 342.
[118] Cf. Swinburne (PPE), pp. 210–12.
[119] Swinburne (EG), p. 271.

similar objections have been discussed extensively by Schellenberg, I can be brief about them.[120] Before I state the objections, Swinburne's rebuttal has to be elucidated in two respects. First, it presupposes a specific conception of the type of human freedom of which he claims that it has great value. What is so valuable, he thinks, is our having 'significant free choice in the sense of a free choice that can make real differences to things for good or ill', and 'the free choice thereby of gradually forming [our] characters'.[121] It is not only necessary for our having a significant free choice in this sense that we have libertarian free will, which is defined as the freedom to choose whether or not to bring about some effect, where the totality of causes that influence us do not totally determine how we will choose.[122] It is also logically necessary that we have temptations to do evil, since otherwise we would always do the good.[123] Because we have such temptations, whereas God lacks them, our significant freedom is very different from God's perfect freedom. Indeed, Swinburne claims that '[o]ur good desires have to be outweighed in their causal influence on us by our evil desires if we are to make a free choice in favour of the good'.[124]

Second, we have to wonder what Swinburne means when he says that 'a strong awareness' of God's presence would 'inhibit someone's ability to choose freely between good and evil'.[125] Since our temptations to do evil will not simply disappear because of such an awareness, he must mean that they are overwhelmed by it. This may perhaps inhibit our ability to make a free choice in favour of the good, if it is required for such a choice that our good desires are outweighed in their causal influence on us by our evil desires. But does it really follow that God would be justified in hiding himself from us in the sense that he does not even provide us with evidence sufficient for believing that he exists? I think that the following objections effectively undermine this conclusion.

First, there is the problem of inversion and evaluation I discussed with regard to the argument from evil. Given the fact that we are cursed with temptations to do wrong, it is a great good that we have the possibility of overcoming them by significant free choices. But does it follow that God would be justified in giving us such evil temptations in order to provide us with significant freedom and then hide himself in order not to destroy this freedom? I do not think so, and according to my own valuation such a god would be perverse. In other words, the great value of significant free choice is conditional and not absolute. Accordingly, it would be a great good if our temptations to do evil were outweighed by an awareness of God's presence, even if the laborious type of human freedom, which Swinburne values so highly, were lost. We would thereby acquire a superior type of freedom, which more greatly resembles God's divine freedom. Would God as a good father be justified in preventing this outweighing, either by hiding himself from us or by considerably reinforcing our evil temptations? Again, I think that only a wicked god would do so.

Second, what a rebuttal of the C-inductive argument from hiddenness should show is not that a specific higher-order good would be logically impossible given a 'strong

---

[120] Schellenberg (2006), pp. 115–30; Schellenberg (2005a) and (2005b).
[121] Swinburne (EG), p. 119 and p. 238.
[122] Partly quoted from Swinburne (EG), p. 113; cf. pp. 169–70.
[123] Swinburne (EG), pp. 238; cf. pp. 103–5, 218.
[124] Swinburne (EG), p. 268.   [125] Swinburne (EG), pp. 268, 269.

awareness' of God's presence or a 'deep conviction of the existence of God'.[126] Rather, it should be argued that the higher-order good is logically impossible given a situation in which the evidence available is *merely sufficient* for belief, in the sense that on the evidence it is probable that God exists. According to the argument from hiddenness, it is this 'probabilifying' situation only that God would and should create if he loves us.[127] Given that situation, each of us can decide to develop a loving relationship with God, which might lead to a 'deep conviction' of his existence later on. In other words, by focusing on a 'strong awareness' or 'deep conviction' of God's existence, arguing that in such situations significant freedom would be logically impossible, Swinburne makes things too easy for himself.

But third, even this has not been shown. As we saw above, significant human freedom is logically impossible given a strong awareness of God's existence, only if some further conditions are met. One condition is that we have both a strong desire to be liked by good persons and the belief that God will not like us if we do bad things. Another condition is that we have both a strong desire for our future well-being and the belief that probably God would not provide a good afterlife for bad people. But it may very well be the case that for many believers, none of these conditions is satisfied. A believer may think, for example, that God is all-forgiving, and that he will love us even if we sometimes do bad things. For was it not God who created us with temptations to do evil? Or the believer might endorse a Calvinist doctrine of predestination, and hold that whether after death we shall go to heaven or hell has been decided long before our birth. Or the believer may notice that many bad people are very successful in life. Why, then, should believers give up assured short-term gains in view of a hypothetical punishment in a distant future, that is, after death, since clearly God often does not punish sinners during this life?

Fourth, one might admit to Swinburne that given a strong awareness of God, a *specific type* of significant human freedom would become impossible in case these further conditions were met. If, for example, we have both a strong desire for our future well-being and the belief that probably God will not provide a good afterlife for bad people, all of us would do the good on most or all occasions. Yet this strong awareness of God under these conditions would create an even more significant type of human freedom. We would have the significant choice between on the one hand doing morally good things merely for prudential reasons, because we desire a future life in heaven instead of in hell, and on the other hand cultivating the ability to do morally good things for purely moral reasons. Because this significant freedom is of a higher type than the freedom that is allegedly destroyed by a strong awareness of God's presence, God will give each of us this strong awareness early in life if he really loves us and is like a good father. In other words, if God can accommodate even higher goods within a loving relationship with him, he will do so, assuming that he is like a good father and is longing for a loving relationship with us. And of course, given his omnipotence, God will always be able to accommodate higher goods within our relationship with him.[128]

---

[126] Swinburne (EG), pp. 268, 269; cf. (PPE), p. 206: 'a deep awareness'.
[127] Cf. Schellenberg (2006), p. 122.   [128] Cf. Schellenberg (2005b), pp. 288–96.

Fifth, Swinburne's defence concerning (A) clearly has testable empirical implications, to wit, that people who have a deep conviction of God's existence will not sin at all, if at least their temptation to do evil is average. In order to rule out the escape that deep believers who sin nevertheless have exceptionally strong temptations to do evil, we should engage in statistical research on large samples of unbelievers and believers. Are believers on average morally better people than unbelievers? I do not know of any research on the issue that yields this result. Moreover, the many cases of child abuse by Catholic priests, both in the United States and in Ireland, Germany, Belgium and The Netherlands, show that people who on average have a deeper conviction of God's existence than laymen are very well able to sin.[129]

Finally, even if our temptations to do evil were overwhelmed by God's presence, the free choice of gradually forming our characters would not be taken away. Instead of merely doing what is morally required, we might engage in acts that are supererogatory. As Schellenberg says, we still would have a significant choice between moral mediocrity and moving in the direction of divinity.[130]

We may conclude on the one hand that the existence of explanandum (e_{9b}) is not probable at all if God exists. Having examined many attempts to rebut the argument from hiddenness, Schellenberg comes to a similar conclusion. As he says, 'the prospects for a future counterargument that would remove this threat to theism and revive the possibility of belief must appear dim'.[131] On the other hand, if God does not exist, explanandum (e_{9b}) is exactly what we expect. A natural history of religion will explain how, given the tendency of children to resort to animistic and anthropomorphic explanations of natural phenomena, and given longings for deceased human beings, polytheistic religions evolved. Perhaps David Hume was right that a contest of praising of gods between peoples of different religions ultimately yielded as its unsurpassable limit the monotheistic notion of one infinite god.[132] But given the fact that there is insufficient evidence for the existence of any god, there always will be a diversity of religions and there always will be reasonable unbelievers and atheists. In short, if h is theism, $P(e_{9b}|{\sim}h \& k) \gg P(e_{9b}|h \& k)$, so that the argument from hiddenness is a very strong C-inductive argument against the existence of God.

---

[129] Cf., for example, the data on BishopAccountability.org, and the report of the Irish 'Commission to Inquire into Child Abuse', published on 20 May 2009, available on the Web.
[130] Schellenberg (2006), p. 130.
[131] Schellenberg (2006), p. 213.   [132] Cf. Chapter 4.4 above.

15

Religious Experience and the Burden of Proof

The *argument from religious experience* plays a pivotal role in the global strategy of Richard Swinburne's justification of theism. It is not merely yet another C-inductive or confirming argument for the existence of God. Its function is, rather, to transfer the burden of proof from the religious believer to the critic of religions. In this final chapter, I shall first explain why such a transfer is needed, and then investigate whether the argument from religious experience accomplishes it.

15.1 THE ARGUMENT FROM RELIGIOUS EXPERIENCE

Let me begin by situating the argument from religious experience in its context. In his book *The Existence of God*, Swinburne examines the bearing that eleven sets of evidence (e_{1-11}) have on the probability that theism is true. We found in Chapters 12–14 that his arguments based upon evidence e_{1-8}, which aim at increasing the probability of theism, fail to do so convincingly. Moreover, the arguments from evil (e_{9a}) and from hiddenness (e_{9b}) turned out to be good C-inductive arguments against the existence of God, notwithstanding Swinburne's attempts to rebut them.

Swinburne adduces two further sets of evidence in favour of theism. Evidence e_{10} consists of public historical events, which are either much disputed (e_{10a}) or not disputed at all (e_{10b}).[1] Public historical events of the first kind are miracles, which I discussed in Chapter 10.2–5. Taking the alleged bodily resurrection of Jesus as an example, I argued that typically the existing testimony of miracles can be accounted for better by a purely secular explanation than by postulating that the miracle really occurred. Accordingly, we should not assume that evidence e_{10a} exists.

Public historical events of the second kind (e_{10b}) form a heterogeneous set, the members of which have in common that religious believers interpret them as manifesting 'the hand of God'. Examples one might think of are the appearance of prophets in history, the winning of battles by one's own people, the growth of one's religion, unexpected recoveries of the sick, or petitionary prayers that seem to be answered. Since to the extent that they have really occurred, such events can in principle be accounted for equally well or better by purely secular explanations, they will not yield

---

[1] Swinburne (EG), Chapter 12.

correct C-inductive arguments for God's existence. These events simply are not 'too big' or 'too odd' for secular science or scholarship to explain. And because Swinburne holds that they provide at best 'a bit more evidence' for the existence of God, I shall not discuss them.[2]

The argument from religious experience is concerned with evidence set e_{11}, which consists of non-public historical events. It claims that in fact there have been many private experiences purportedly of God, and that for this reason it is very probable that God exists.[3] Experiences allegedly of God may be called 'private' for the following reason. Ordinary cases of perception are public because the perceived object is such that all persons rightly positioned in space and time, having normal sense organs and who observe attentively, are able to perceive it. But religious experiences purportedly of God are not like that. Their object is said to be such that it can *choose* whether it will be perceived by one person and not by another. Religious experiences are called 'private', then, because their object is not a public entity that can be perceived by all rightly positioned competent observers.[4]

Why and how does the argument from religious experience transfer the burden of proof to the unbeliever? And why is such a transfer of the *onus probandi* needed? In order to answer the latter question, we have to consider Swinburne's cumulative case strategy.

15.2 THE CUMULATIVE CASE STRATEGY

It is the ultimate aim of an inductivist strategy in the apologetic philosophy of religion to add up all C-inductive arguments for (and against) the existence of God in order to show that this cumulative case amounts to a good P-inductive argument. Theists want to establish ultimately that the probability of God's existence is higher than some threshold value, which Swinburne fixes at $1/2$, so that $P(h|e \& k) > P(\sim h|e \& k)$, where h is the hypothesis of theism, e the collection of all evidence $e_{1\text{-}11}$ for and against the existence of God, and k stands for tautological or conceptual background knowledge.[5] In other words, what theists want to argue is that given the total evidence, it is more probable than not that God exists.

The very idea of such a cumulative case strategy is one of the great merits of Swinburne's approach. As he says, the 'tendency to treat arguments for the existence of God in isolation from each other' is an 'unfortunate feature of recent philosophy of religion'.[6] Since arguments purporting to support a hypothesis may reinforce or weaken each other, they have to be studied in their mutual relationships. I pointed out in Chapter 13.5, for example, that fine-tuning arguments cannot survive the

---

[2] Swinburne (EG), p. 277. Swinburne's argument seems to be yet another fine-tuning argument. He claims that if these undisputed historical events occur, 'the laws governing human behaviour' must have the feature that they make it probable that they occur, whereas the fact that they have this feature is more probable if God exists than if he does not. Furthermore, Swinburne claims that this feature is 'too big' for science to explain, since the fact that the laws governing human behaviour have it 'derives from the fundamental laws of nature' (p. 277). I leave it to the reader to assess these arguments in the light of Chapter 13.4–5, above.
[3] Swinburne (EG), p. 293.   [4] Swinburne (EG), p. 297.
[5] Swinburne (EG), pp. 17 and 329.   [6] Swinburne (EG), p. 12.

demise of the cosmological argument. Furthermore, it may be the case that whereas each of the eleven sets of evidence Swinburne considers, if taken separately, favours some rival hypothesis over theism, theism is favoured over each of these alternative hypotheses by the total evidence.[7] The crucial issue with regard to theism in the apologetic philosophy of religion undoubtedly is the cumulative effect of all the arguments for and against God's existence.

Readers who endorse my criticisms of the C-inductive arguments discussed so far may be tempted to think that there is no point in investigating the cumulative case strategy as such. Given the fact that the examined C-inductive arguments in favour of God's existence are unconvincing, whereas those against God's existence are quite good, they will conclude that the strategy is shipwrecked anyway. But other readers may judge that some of my criticisms were mistaken or insufficient. Furthermore, even if the examined C-inductive arguments turn out to be very weak if taken individually, their combined force might be considerable. Hence, we have to investigate the cumulative case strategy. As we shall see now, this strategy is confronted by three serious problems, which might motivate theists to shift the burden of proof to unbelievers.

First, there is the problem of completeness. Have apologetic philosophers of religion really examined all available evidence pro and contra? In their apologetic zeal, they may have overlooked some pertinent pieces of evidence that plead against God's existence. Many scientific discoveries of recent centuries have convinced great numbers of educated people that in view of what we now know about the universe, the existence of a creator-god as conceived by theism is extremely unlikely. For example, if God's main intention in creating the world really was to produce 'in his image' humanly free agents, as traditional theists hold, one might expect that he would create a universe on a human scale like the one described in the book *Genesis* or by Aristotle, with the Earth occupying the centre of a spatially rather limited world.

The Copernican revolution already destroyed this cosy picture of the cosmos. But what modern cosmology reveals about the universe is even more startling and unexpected from a theistic point of view. The fact that the number of stars in the observable universe is so huge, probably somewhere between 10^{22} and 10^{24}, may perhaps be interpreted as testifying to God's grandeur. But it seems to be completely at odds with God's intentions, as far as they can be scrutinized by theologians, that about 80% of all matter in the visible universe is dark matter, which is undetectable by electromagnetic radiation, as cosmologists tend to conclude from various lines of evidence. What is all this dead rubbish good for?

Furthermore, why would God create an 'open' universe that will continue to expand indefinitely, and will finally endure a 'heat death' or even a 'Big Freeze' when it asymptotically approaches absolute zero temperature? This scenario is currently the most commonly accepted theory within the cosmological community. But if God's primary intention was to create man, as Swinburne seems to think, it is a mystery why God would create a universe that will exist for an infinite future time during which no life will be possible at all.

---

[7] Cf. Swinburne (EG), pp. 18–19.

The biological sciences yield many other data that are unexpected from a theistic point of view, and were never accounted for by theologians. Theists assume that God wanted to create animals and man, but why would he have created life on Earth in such a way that bacteria are most successful, both in terms of numbers, longevity, variation of habitats, total biomass, and seniority?[8] The history of life on Earth as it is revealed in ever more detail by evolutionary biology also is unlikely and baffling from a theistic perspective. Why would God have created life in such a way that a great many species had to perish before man could evolve? What can explain mass extinctions from a theistic point of view? These scientific discoveries and many others reveal facts that seem to disconfirm theism, if at least theism has any predictive power, so that the cumulative case P-inductive argument as proposed by Swinburne is incomplete.[9]

The second problem that confronts the cumulative case strategy in natural theology is the problem of the determination of likelihoods. By definition, an inductive argument supporting a hypothesis h is a correct C-inductive argument if and only if $P(h|e\&k) > P(h|k)$, and this is the case if and only if the relevance condition $P(e|h\&k) > P(e|k)$ is satisfied. However, as Swinburne admits, it is impossible in the philosophy of religion to assign exact numerical values to the terms of these inequalities, that is, to the likelihoods $P(e|h\&k)$ and $P(e|k)$. Indeed, in this field our estimates of conditional probabilities are rarely backed up by statistics or other methods of quantification. Accordingly, Swinburne does not assume in using the symbols of Bayesian confirmation theory 'that an expression of the form $P(p|q)$ always has an exact numerical value', or that one can determine the intervals of these factors. Indeed, it 'may merely have relations of greater or less value to other probabilities [...], without itself having a numerical value' at all.[10]

But this lack of numerical values of likelihoods may seem to prevent the cumulative case strategy from getting off the ground. For even if it were possible to assign non-arbitrarily and roughly some numerical value to the prior probability of theism $P(h|k)$ before considering any evidence, one would never arrive at the conclusion that $P(h|e_{1-11}\&k)$ has a specific numerical value, or a value within a given interval, or is greater than $1/2$, unless one knows with what factors the good C-inductive arguments based on evidence $e_1 - e_{11}$ increase the initial probability of theism. One simply cannot calculate with the Bayesian formula without plugging in numbers or numerical intervals.

Suppose for the sake of argument, however, that one is able to assign non-arbitrary numerical values to the C-inductions, claiming for example that each of them is so strong that it approximately doubles the relative prior probability of theism. Even this would not seem to help theists, however, unless they could assign non-arbitrarily a numerical value to the absolute prior probability of theism, that is, the probability prior to considering all empirical evidence. This numerical value should be such that

---

[8] Cf. Everitt (2004), p. 218.
[9] So-called sceptical theists will reply that our limited minds are simply unable to think of the good reasons God might have had for creating these things. But this sceptical thesis annuls the predictive power of theism (if any, cf. Chapter 9), so that an inductive natural theology becomes impossible.
[10] Swinburne (EG), pp. 17–18.

the initial improbability of theism is swamped by the empirical evidence adduced.[11] As Swinburne observes, '[*a*] *priori* the existence of anything at all logically contingent, even God, may seem vastly improbable, or at least not very probable'.[12] But what numerical value should we assign to the absolute prior (im)probability $P(h|k)$ that God exists?

If we say that this prior is 1/100, six good C-inductive arguments that each double the relative prior probability of theism would yield a correct P-inductive argument, according to which the probability that God exists is 64/100. However, if we fix the (im)probability of theism prior to all evidence at 1/1,000,000,000, one would need thirty C-inductive arguments that double the relative prior probabilities in order to yield a P-inductive argument that raises the probability of theism to above $^1/_2$. But is it not subjective and arbitrary which value one gives to the prior (im)probability of God's existence? This problem of assigning non-arbitrarily an absolute prior probability to theism is the third difficulty that a cumulative case strategy for theism has to face.[13]

The latter two difficulties may also arise for scientists who want to argue that the logical probability of some scientific theory is greater than $^1/_2$. One might object, however, that no scientist will ever attempt to establish the conclusion that the truth of a general theory such as quantum mechanics is more probable than not. In order to account for scientific progress, it is sufficient to reconstruct the probability assessments of scientists concerning general theories as based on the law of likelihood (cf. Chapter 11.4–5, above). What they seek to establish is merely that the sum of the available evidence strongly favours one theory over all other theories that have been proposed.

Swinburne admits that there are difficulties in establishing that a general scientific theory has a probability of above $^1/_2$. 'Is Quantum Theory more probable than not? Or is the General Theory of Relativity? The answer is in no way clear.'[14] Yet he thinks that 'the situation is by no means hopeless', and that 'given the difficulty in reaching a conclusion about whether any scientific theory is more probable than not, any difficulty about reaching a conclusion about whether theism is more probable than

---

[11] Cf. Earman (1992), Chapter 6.3–6 on the question as to whether this is feasible. Cf. on the issue of prior probabilities also Priest (1981).

[12] Swinburne (EG), p. 151.

[13] As Jeremy Gwiazda (2009) has pointed out, it would have been helpful if Swinburne had given examples of numerical values assigned to his probabilities for which his cumulative case argument might *work*. This is not easy at all. For instance, if we assume with Swinburne that $P(e|h\&k)$ is 1/2 or no less than 1/3 for evidence $e_{1\text{-}10}$ (EG, pp. 338–9; h is theism), and if 'the choice is between the universe as stopping point and God as stopping point' of explanation, as Swinburne argues in his chapter on The Cosmological Argument (EG, p. 147), there are no plausible numerical values that one can assign to $P(h|k)$ and to $P(e|k)$ so that the cumulative case might work, if at least we endorse Swinburne's criteria for intrinsic probability. The reason is that according to Swinburne, the universe has 'a complexity, particularity, and finitude [...] that cries out for explanation' (EG, p. 150), whereas he claims that theism is 'a very simple hypothesis indeed' (EG, pp. 96–7) and that God is a very simple entity (EG, p. 336: 'it seems impossible to conceive of anything simpler (and therefore a priori more probable) than the existence of God'). If it is true that 'the simple is more likely to exist than the complex' (EG, p. 109), as Swinburne claims, and if God is *much* simpler than the universe as described by $e_{1\text{-}10}$, $P(h|k)$ must be *much* greater than $P(e|k)$, say 10 or a hundred or a million times greater. But then it follows on Bayes' theorem that $P(h|e\&k)$ is 10 or more times 1/3, which is logically impossible, since probabilities must lie between 0 and 1 (inclusive). Cf. Swinburne (2011) for an answer to Gwiazda's objection.

[14] Swinburne (EG), p. 331.

Religious Experience and the Burden of Proof

not is not in any way to the special discredit of theism'.[15] This may be true, and we shall see in the next section how Swinburne tries to bypass the difficulties of quantification. But why should we not conclude that the very objective of Swinburne's cumulative case strategy is chimerical?

15.3 THE TACTIC OF AVOIDING NUMBERS

In his chapter on *The Balance of Probability*, Swinburne ably attempts to solve the second and third problems discussed in the preceding section (15.2) by avoiding quantification altogether, and by using judgements of relative probability only.[16] Is this attempt successful?

The structure of his argument is as follows. Bayes' theorem in the form Swinburne uses it reads:

$$P(h|e\&k) = \frac{P(e|h\&k).P(h|k)}{P(e|k)}$$

Since $P(e|k) = P(e|h\&k).P(h|k) + P(e|\sim h\&k).P(\sim h|k)$, the theorem can be rewritten as:

$$P(h|e\&k) = \frac{P(e|h\&k).P(h|k)}{P(e|h\&k).P(h|k) + P(e|\sim h\&k).P(\sim h|k)}$$

The numerator on the right-hand side of the theorem, $P(e|h\&k).P(h|k)$, is identical to the first term of the denominator. Let us call this term a, and let us call the second term of the denominator after the '+' sign b. If we can argue convincingly that b will not be greater than a, $a + b$ will maximally be equal to $2a$, so that $P(h|e\&k)$ will be at least $\frac{1}{2}$. This crucial conclusion can be reached purely by assigning relative probabilities, that is, by arguing that $P(e|\sim h\&k).P(\sim h|k)$ is not greater than $P(e|h\&k).P(h|k)$.

The format of Swinburne's argument for this latter claim is as follows. He treats the cumulative case of the C-inductive arguments based on evidence e_{1-10}, that is, all the evidence apart from that of religious experience, as one big cosmological argument, in which the universe to be explained has all the empirical characteristics e_{1-10} adduced as evidence for (or against) God's existence. Swinburne then argues that the catch-all set $\sim h$ contains only three alternatives to theism, to wit, polytheism (h_1), the hypothesis that the present universe is caused, ultimately, by an initial physical first state, whether everlasting or not (h_2), and the hypothesis that there is no explanation for the universe at all (h_3). If this set of mutually exclusive alternatives is exhaustive, $P(e|\sim h\&k).P(\sim h|k) = P(e|h_1\&k).P(h_1|k) + P(e|h_2\&k).P(h_2|k) + P(e|h_3\&k).P(h_3|k)$. Swinburne claims that this sum is not greater than $P(e|h\&k).P(h|k)$, because either

---
[15] Swinburne (EG), p. 333.
[16] Swinburne (EG), Chapter 14, pp. 336–42.

the prior probabilities of these alternatives to theism or their likelihoods are much lower than those of theism.[17]

Even if this latter claim were correct, however, the argument does not work. One reason is that Swinburne has not succeeded in eliminating other rival hypotheses to theism, such as the monotheistic hypothesis of an omniscient, omnipotent, and morally indifferent god (MIG), as I argued in Chapter 12.8. In other words, the catch-all $\sim h$ contains many more rivals to theism than the three hypotheses h_1, h_2, and h_3. Furthermore, by treating the cumulative case for evidence $e_{1\text{-}10}$ as one big cosmological argument, Swinburne's cumulative case for evidence $e_{1\text{-}10}$ is open to all the objections I raised to his cosmological arguments in Chapter 12.[18] For example, there is no room for the theistic hypothesis of God creating the universe if the universe is infinitely old, whereas $P(e|h\&k)$ is very low if e is a pointlike Big Bang.

Suppose, however, that Swinburne's cumulative case argument for evidence $e_{1\text{-}10}$ were correct and convincing. Then it would have been established that $P(h|e\&k)$ is at least $\frac{1}{2}$. But this conclusion falls short of what the theist wants to argue. The cumulative case or P-inductive argument aimed at showing that the existence of God is more probable than not, whereas the conclusion of Swinburne's argument says 'that it is something like as probable as not that theism is true, on the evidence so far considered'.[19] Instead of yielding theistic belief, it merely amounts to agnosticism. And as we have seen, agnosticism is a vulnerable position. The balance can easily be tipped against God's existence by showing that the cumulative case argument is incomplete, or by the argument from hiddenness (Chapter 14.11). Clearly, the problems of numerical determination and the problem of completeness would be avoided, or would be much less threatening, if the burden of proof were shifted to the critic of theism.

---

[17] Again, it is difficult to invent plausible assignments of numerical values to these probabilities, given Swinburne's assumptions, such that the argument works (cf. Gwiazda, 2009, pp. 5–6). If h (theism), h_1 (polytheism), h_2 (the hypothesis that the present universe is caused, ultimately, by an initial physical first state), and h_3 (the hypothesis that there is no explanation for the universe), are mutually exclusive and exhaustive (one and only one of them is true), it must be the case that $P(h|k) + P(h_1|k) + P(h_2|k) + P(h_3|k) = 1$. Swinburne claims that '[a] priori the existence of anything at all logically contingent, even God, may seem vastly improbable, or at least not very probable' (EG, p. 151), so that '$P(h|k)$ might be very small' (EG, p. 112). Let us assign, for example, the value of 0.001 to the 'very small' intrinsic probability that God exists. It follows that $P(h_1|k) + P(h_2|k) + P(h_3|k) = 1 - 0.001 = 0.999$. But if this is the case, how can it be that the sum $P(e|h_1\&k).P(h_1|k) + P(e|h_2\&k).P(h_2|k) + P(e|h_3\&k).P(h_3|k)$ is not greater than $P(e|h\&k).P(h|k)$, as Swinburne claims? Moreover, Swinburne also says that the prior probability of theism (h) is much higher than that of polytheism (h_1), and that $P(h_3|k)$ is 'infinitesimally low', because the assumption that our universe should have the characteristics $e_{1\text{-}10}$ 'without there being some explanation of this is beyond belief' (EG, p. 341). It follows that $P(h_2|k)$ must be near to 1. If so, however, how can Swinburne claim that '[e]ither $P(e|h_2\&k)$ is going to be much lower than $P(e|h\&k)$ or $P(h_2|k)$ is going to be much lower than $P(h|k)$' (EG, p. 341)? On the one hand, Swinburne's contentions imply that $P(h_1|k) + P(h_2|k) + P(h_3|k) = 0.999$, or at least is very high, whereas on the other hand he suggests that each of the values $P(h_1|k)$, $P(h_2|k)$, and $P(h_3|k)$ is (or might be) very low. As a consequence, it turns out to be impossible to provide a consistent numerical example for which Swinburne's argument works, if at least we endorse his premises.

[18] This is true for all the evidence explained by assuming that, ultimately, God created the world. It is not true for the evidence of miracles, if they exist, because miracles are God's interventions in an already existing universe.

[19] Swinburne (EG), p. 341. In his argument up to this point, Swinburne leaves out the evidence of religious experience.

This is precisely what Swinburne's argument from religious experience purports to achieve. It allegedly shows that 'unless the probability of theism on other evidence is *very low*, the testimony of many witnesses to experiences apparently of God suffices to make many of those experiences probably veridical'.[20] It follows that if the argument from religious experience is compelling, and if all the other C-inductive arguments have the force Swinburne attributes to them, so that the probability of theism given these other arguments is not 'very low', theism is more probable than not on the total evidence, as Swinburne concludes at the end of his book.[21] However, if shifting the *onus probandi* to the critic of theism is so crucial to the success of Swinburne's cumulative case strategy, we must investigate whether the argument from religious experience can justify such a shift.

15.4 TWO FUNDAMENTAL PRINCIPLES OF RATIONALITY

Swinburne's argument from religious experience is based on two maxims, which he regards as fundamental principles of rationality.[22] According to the first maxim, baptized the *Principle of Credulity*, one is justified in believing that what one takes oneself to perceive or to have perceived in the past is probably the case, unless there are defeating considerations.[23] To the extent that the principle applies to religious experiences, subjects of such an experience are justified (in the absence of defeating considerations) to believe what they take themselves to behold.

The second maxim is needed in order to give evidential weight to religious experiences even for those who have never had them, and to increase this weight for all of us. According to the *Principle of Testimony*, the experiences of others are (probably) as they report them, unless there are defeating considerations.[24] If we combine these two principles, we come to the following conclusion. It is rational to

---

[20] Swinburne (EG), p. 341 (my italics). [21] Swinburne (EG), p. 342.
[22] This is said explicitly of the first principle. Cf. Swinburne (EG), p. 306: 'The Principle of Credulity is a fundamental principle of rationality'.
[23] Swinburne (EG), pp. 303ff; cf. (EJ), pp. 141–50, and (ES), pp. 11–12. According to (EG), the principle applies to (apparent) perception and memory: 'I suggest that it is a principle of rationality that (in the absence of special considerations), if it seems (epistemically) to a subject that *x* is present (and has some characteristic), then probably *x* is present (and has that characteristic); what one seems to perceive is probably so. And similarly I suggest that (in the absence of special considerations) apparent memory is to be trusted' (p. 303). In (EJ), Swinburne applies the principle to basic beliefs that result from perception, memory, and to the 'deliverances of reason', such as the belief that 2 + 2 = 4. According to this latter book, the Principle of Credulity even 'claims that every proposition that a subject believes or is inclined to believe has (in so far as it is basic) in his noetic structure a probability corresponding to the strength of the belief or semi-belief or inclination to believe' (EJ, p. 141). As Swinburne notes, the label 'Principle of Credulity' is somewhat unfortunate, because Thomas Reid used it for what Swinburne calls the Principle of Testimony (EJ, p. 141, note 14). In the main text, I endorse the narrower interpretation of the Principle of Credulity suggested by (EG). If the broader interpretation of (EJ) is adopted, the applicability of the principle to religious experiences is unproblematic, but the principle itself should be rejected as a principle of rationality, because it implies counter-intuitively that all basic beliefs are properly basic. In particular, I reject Swinburne's view (EJ, p. 144) that 'what the Principle of Credulity affirms is that the very having of a basic belief, semi-belief, or inclination to believe *is itself evidence of the truth of that belief*, in proportion to its strength' (my italics). But I cannot argue for this rejection here.
[24] Swinburne (EG), p. 322.

believe that what others tell us that they perceived probably is the case, other things being equal. Both principles are fundamental, since all attempts to justify them by inductive procedures will be question begging. Without the first principle, we would 'quickly find ourselves in a sceptical bog, in which we know hardly anything', as Swinburne says, and the same holds to a large extent for the second principle, since 'most of our beliefs about the world are based on what others claim to have perceived'.[25]

The argument from religious experience runs in outline as follows. Swinburne first rejects two attempts to restrict the Principle of Credulity in such a way that its application to religious experiences is ruled out. From this he concludes that 'the principle stands' and that it applies to religious experiences.[26] If this is the case, religious experiences might have as great an evidential force as ordinary perceptual experiences, and this force may be great indeed. For example, I think it extremely unlikely that my friend Peter is in Paris, since he lives in England and rarely goes to France. Yet, when I see him walking on the Boulevard Saint-Germain and meet him there, my perceptual experience decisively outweighs the initial improbability on my background knowledge of his presence in Paris.

Swinburne argues next that of the four logically possible defeaters, which cancel conclusions to the effect that things are as a subject purportedly perceived them to be, none applies convincingly to (most) experiences purportedly of God. It follows that such experiences should be considered veridical, unless one can show that 'very, very probably God does not exist'.[27] Since 'the onus is on the atheist to do so', the argument from religious experience shifts the burden of proof to the unbeliever, which is what Swinburne needed.[28] He concludes that '[o]nly if the prior probability of the existence of God is very low indeed will this combined weight of testimony be insufficient to overcome it'.[29] In the remainder of this chapter, I shall review critically these two phases of Swinburne's argument from religious experience, and I shall also question whether perceptual religious experiences of God are possible in principle.

15.5 RELIGIOUS EXPERIENCE AND THE PRINCIPLE OF CREDULITY

Let us first investigate whether and to what extent the Principle of Credulity really applies to religious experiences, apart from conceivable defeating considerations. The expression 'religious experience' is vague, and it may cover 'experiences' of many different kinds, such as emotional feelings, realizations of religious significance, religious associations, and perceptual experiences. Clearly, if the Principle of Credulity applies to religious experiences at all, it does so only to religious experiences of a perceptual kind, because only perceptual experiences of God may yield epistemic

---

[25] Swinburne (EG), pp. 306 and 322.   [26] Swinburne (EG), p. 310.
[27] Swinburne (EG), p. 321. Strictly speaking, even if this has been shown, the resulting low probability of God's existence might be outweighed by the sheer number of witnesses who had religious experiences.
[28] Swinburne (EG), p. 319.
[29] Swinburne (EG), p. 326. Swinburne's argument from religious experience structurally resembles the argument put forward by C. D. Broad in his (1939).

support for the belief that God exists. Emotional experiences, such as an intense fear or love of God, are based upon the belief that God exists, so that they do not have any evidential value with respect to this belief. The same holds for realizations of religious significance, religious associations, living in some religiously motivated hope, and so on.[30]

Does Swinburne succeed in refuting attempts to restrict the Principle of Credulity in such a way that its application to *perceptual* religious experiences is ruled out as well? The first attempt discussed by Swinburne consists in saying that the Principle of Credulity itself stands in need of inductive justification, whereas the inductive justification is available in cases of sense-perception but not in cases of religious perceptual experience.[31] Put in this crude form, the attempt clearly cannot be successful. As I said in section 15.4, each inductive justification of the Principle of Credulity will be circular. In order to justify the principle inductively, we must rely on earlier perceptions, trusting that they were veridical, and we must assume that we recall these past perceptions correctly. But in doing so we implicitly rely on the Principle of Credulity, since this principle says that, in the absence of defeating considerations, we should trust that things are the way they perceptually seem to be, and also that apparent memory of our own experiences is to be trusted. The Principle of Credulity is 'a basic principle not further justifiable', as Swinburne correctly stresses.[32]

However, a subtler version of the first attempt would be more difficult to refute. One might argue that we cannot but start with some weak credulity concerning perceptions of a specific type of thing, which becomes reinforced gradually whenever multiple experiences of an object confirm each other. With regard to perceptions of public objects, our initial trust in a perception may be perceptually confirmed and reinforced in four ways. First, when we see a house or a tree, for example, we may explore it further by using the same sense-modality of vision. Visual perception often involves an activity of exploration, during which we walk around an object in order to see it from various angles. Second, we may use other modalities of sense-perception in exploring the same public object, such as smell or touch. Third, we may manipulate an object in order to investigate its causal interactions with other objects and with ourselves. Finally, we may test our perceptions by comparing them with those of other people who are in a position to perceive the object. And of course, the fact that so many past perceptions of public objects were confirmed in these ways, so that they clearly were veridical, makes it more probable that one's present perception of a public object is veridical as well.

Our initial credulity with regard to some perception purportedly of a public object may be cancelled if it turns out that one or more of these reinforcements are not available.[33] In such cases, we often conclude that we had not really perceived the object

---

[30] In Chapter 4, I rejected the claim that non-perceptual religious experiences that are directly caused by an internal instigation of the Holy Spirit can have any evidential force for well-educated and intelligent people living in the twenty-first century.

[31] Swinburne (EG), pp. 305–7.

[32] Swinburne (EG), p. 305. One might compare Swinburne's Principle of Credulity with William Alston's idea of a 'basic doxastic practice'.Cf. Alston (1988), section II, and Alston (1991), passim. Whereas Swinburne's principle is primarily individualistic, Alston's doxastic practices are defined as social ones. Cf. Alston (1991), p. 195.

[33] Cf. my discussion of the four defeaters, sections 15.9–11 below.

we seemed to perceive, and that we were prey to some perceptual illusion or even to a hallucination. Or we conclude that we are not really certain that we perceived the thing. How should we apply these considerations to religious perceptual experiences? It seems to me that there are only two options.

One option is that we do not apply the Principle of Credulity to religious perceptual experiences at all, since in their case typically none of the confirming reinforcements is available. For example, having had a perceptual experience purportedly of God, I cannot engage in actively exploring God perceptually, as I can do with human beings. As a consequence, there are no imaginable perceptual circumstances either in which an experience purportedly of God turns out to be an illusion or a delusion. One might argue that in order to apply the Principle of Credulity to experiences of a certain type, such disconfirming circumstances must be possible in principle.[34]

But this first option will seem dogmatic, or at least excessively prudent. As against it, theists will argue that which confirming (or disconfirming) experiences are available in principle will depend on the type of object that we are (purportedly) perceiving. In the case of sounds, for example, we cannot perceive them by sense modalities other than hearing, although we may explore their source by using other modalities, but we expect sounds to grow louder as we approach the source and weaker when we put our hands on our ears. In general, we might say that each particular type of object essentially implies a schematism of possible perceptions of one of its instances, which mutually confirm each other. So we should wonder which schematism of possible confirming reinforcements we may expect in the case of perceptions of God.

As Swinburne stresses, God is not a public object of perception. As we saw in section 15.1, perceptions of God, if they exist, will be 'private' in the sense that God is 'a normally invisible person' who can choose whom to cause to have a perceptual experience of him.[35] But this implies that in the case of God, there does not seem to be *any* standard schematism of possible further perceptions. And if we cannot expect in principle any confirming reinforcement once we have had a perceptual experience purportedly of God, the absence of such reinforcements cannot be used as an argument against our trust in this experience. For this reason, the first option is mistaken, so the theist will argue.

This leaves us with the second option, which is that in the absence of confirming experiences, we should merely have a weak adherence to the Principle of Credulity, or rather, a very weak trust in the veridicality of our perceptual experience. This implies that in the case of religious perceptual experiences, the additional probability conferred by the experience on the proposition that what is purportedly perceived is really the case, is not very high. Indeed, it will be much lower than in the case of well-confirmed sense-perceptions of public objects, so that our own religious perceptual experiences by themselves will never make it more probable than not that God exists, unless they are repeated a number of times and somehow confirm each other. From this it follows, however, that for most of us the argument from religious experience cannot transfer the burden of proof to the unbeliever, unless we also rely on the Principle of

---

[34] Rowe (1982), pp. 90–1.   [35] Swinburne (EG), p. 297.

Testimony and on the veridicality of numerous converging perceptual religious experiences of others. Why we should not do so, will become clear in later sections of this chapter.

15.6 A PROBLEM OF EPISTEMIC PARITY

The second attempt to restrict the application of the Principle of Credulity discussed by Swinburne may be introduced as follows. Clearly, the principle justifies trust in standard cases of sense-perception, and not in beliefs resulting from dreams, hallucinations, hearing voices while no speaker can be detected, non-sensory feelings, and so on.[36] Indeed, Swinburne introduces the Principle of Credulity with reference to examples such as seeing a table in front of you, or hearing someone giving a lecture in the room you are in.[37] Hence, the crucial issue to be discussed is whether there are, and indeed can be, perceptual experiences purportedly of God that are sufficiently similar to cases of ordinary sense-perception for the Principle of Credulity to be applicable. In other words, what has to be shown in order to apply the principle to specific religious experiences, is that there is a strong epistemic parity between paradigmatic cases of sense-perception on the one hand, and these religious experiences purportedly of God on the other hand.[38]

Showing this, in turn, may require a detailed phenomenological exploration of the variety of existing religious experiences. For there is no doubt that the Principle of Credulity does not justify trust in many of them, such as experiences purportedly of God during dreams, or interpreting the feeling of being overwhelmed by the beauty of a landscape as feeling God's presence. With regard to these feelings, for example, it is implausible to claim that their occurrence is *prima facie* a good reason to believe that God exists, since only those who already believe that God exists will endorse a religious interpretation. Unbelievers may have the very same feelings while not interpreting them in a religious sense.

But this is not the manner in which Swinburne discusses the second attempt. Instead, he argues that it is impossible to draw once and for all a line between real perceptual experiences on the one hand, and their interpretations on the other hand, so that the Principle of Credulity applies to the first and not to the second.[39] One cannot say, for example, that the Principle of Credulity holds only for sensible characteristics and relations, such as colours, sounds, warmth or cold, and common sensibles such as movement, rest, or magnitude, but not for recognizing a table as Victorian or a ship as Russian. The reason is that humans may learn to recognize perceptually many different things under many different descriptions, such as blue-dwarf stars or elliptical galaxies, whereas it often is not possible to explicate these

---

[36] Cf. my discussion of the two inductive defeaters, sections 15.9–10, below.
[37] Swinburne (EG), p. 303. Swinburne uses the expressions 'seeming to see' and 'seeming to hear' in order to cancel the success-connotation of verbs of sense-perception.
[38] Cf. Geivett (2003) on 'the parity thesis'.
[39] Swinburne defines 'real' perceptual experiences of x as follows: 'if its seeming that an object (or characteristic) x is present is grounds for supposing that it is without need for further justification, then you have a *real experience* of x' (EG, p. 308, Swinburne's italics).

descriptions purely in terms of sensible characteristics.[40] Swinburne concludes that this second attempt to limit the application of the Principle of Credulity fails. If, in general, it is not possible to draw a clear line between what is really perceived and what is merely a matter of interpretation, one cannot use such a line in order to exclude the application of the Principle of Credulity to religious experiences on the ground that the typical objects of religious experience are always merely a matter of interpretation.

However, Swinburne's rebuttal of the second attempt is not very satisfactory. To begin, we should introduce a distinction between two very different questions. The first is whether one actually engages in two separate activities when one concludes on the basis of the perception of a ship that it is a Russian ship: the activity of perceiving and the activity of interpreting. Swinburne is right, of course, that often no such distinction can be made. When we have learned what are the typical characteristics of Victorian tables or of Russian ships, we can see immediately that a table is Victorian or a ship Russian. The other question is whether one can justify one's judgement that there is a Russian ship in one's visual field to someone else just by saying: 'Look, don't you see that this ship is Russian?' In such a situation of justification, extensive background knowledge may have to be made explicit if one is talking to a non-expert. In this example, background knowledge concerning typically Russian ways of building ships will figure in a justification of one's perceptual judgement.

Clearly, the second attempt to exclude the application of the Principle of Credulity to religious perceptual experiences focuses on this second question of justification and not on the first, factual question. If someone claims to have perceived God, the judgement that what one perceived is correctly identified as God goes far beyond a phenomenological description of the perceptual contents, as I shall argue in section 15.7. Consequently, one has to adduce a vast amount of background knowledge in order to justify the judgement to someone else, and one cannot merely refer to the Principle of Credulity. But if that is true, religious experiences cannot have the role Swinburne attributes to them within the framework of his cumulative case strategy, to wit, that they transfer the burden of proof to the unbeliever.

Furthermore, even if it were true that one cannot exclude the applicability of the Principle of Credulity to religious perceptual experiences *in general* by drawing a line between perception and interpretation, this would not suffice to show that the principle is applicable to *specific* religious experiences. As I said, in order to show this, it has to be demonstrated that these religious experiences resemble experiences of sense-perception so closely that there is epistemic parity between them. But with regard to religious experiences purportedly of God, it seems to be excluded that there is such an epistemic parity for two reasons. First, God is not the kind of entity that can be perceived by senses at all, because he is defined as bodiless. How, then, can a religious experience purportedly of God resemble experiences of sense-perception sufficiently for the Principle of Credulity to be applicable? Second, experiences purportedly of God typically turn out to be such that what is really presented perceptually to the subject (if anything at all) is not God but something else, which is not identical with God.

---

[40] Cf. Swinburne (EG), pp. 307–10.

Before elaborating these two points in the next section, I introduce an important conceptual distinction and state the argument from religious experience in more detail.

Verbs of perceptual experience are achievement or success verbs.[41] The statement that a subject hears or sees or smells an object O typically entails that O exists. However, when formulating the factual premise of an argument from religious experience, this entailment has to be neutralized in order to avoid begging the question. In Swinburne's terminology, we have to describe the religious experience *internally*, by saying for example that a subject S had a visual experience purportedly of O, or took herself to be perceiving O, and not *externally*, so that the description of the experience entails that O exists or existed when and where it was seen.[42] A second reason for adopting an internal description of a religious perceptual experience is that a subject may perceive an object O without identifying it *as* (an) O. In religious experiences described internally, it is also specified *as what* the subject of the experience is identifying its object.[43]

Quite often, authors attempt to give an internal description by using phrases such as 'S seemed to see O' and 'O appeared to be F to S'. But this is misleading, since in everyday usage these phrases typically imply that O did *not* exist or was *not* where S thought to see it, and that O in fact was *not F*. In order to formulate the argument from religious experience correctly, all existential implications concerning O have to be neutralized, both positive and negative ones. As Edmund Husserl would have said, we have to describe the religious experiences phenomenologically, and *bracket* all existential implications concerning their intentional objects. In what follows, I shall adopt the convention of characterizing experiences internally as 'of-O experiences' or 'experiences purportedly of O', and externally as 'experiences of O'.[44]

The argument from religious experience to the existence of God may now be formulated as follows:

(1) Many people have, or have had, of-God experiences.
(2) These of-God experiences sufficiently resemble ordinary perceptual of-O experiences for the Principle of Credulity to apply to them. In other words, there is epistemic parity between of-God experiences and ordinary perceptual experiences.
(3) One is justified in believing that one's own of-God experiences are veridical experiences of God, unless there are defeating considerations (from 1, 2).
(4) The Principle of Testimony holds for of-God experiences.
(5) On the basis of the many of-God experiences of others one is justified in believing that God (very probably) exists, in absence of defeating considerations, even if one did not have an of-God experience oneself (from 3 and 4).

As I said above, premise (2) will not be true of numerous of-God experiences. For example, if I wake up and remember hearing a thundering voice in my dreams, saying to me, 'I am God almighty; walk before me and be blameless', nobody will think nowadays that the Principle of Credulity applies to this experience, since the principle

---

[41] Cf. Ryle (1949), pp. 125ff.   [42] Swinburne (EG), p. 294.
[43] Sometimes, subjects of religious experiences claim that in their religious experience the distinction between subject and object is annulled.
[44] I borrow this notation from Geivett (2003).

does not apply to dreams.[45] So we have to investigate whether there are any of-God experiences of which premise (2) is true. Indeed, we have to investigate whether there *can* be real of-God experiences at all.

15.7 FIVE KINDS OF RELIGIOUS EXPERIENCE

In *The Existence of God*, Swinburne proposes a five-fold classification of religious experiences, of which he claims correctly that it is both exclusive and exhaustive.[46] However, when he argues that the Principle of Credulity is applicable to religious experiences, he no longer differentiates between the five different kinds he distinguished earlier.[47] This is unfortunate, since it may be more legitimate to apply the principle to some of these kinds than to apply it to others. But in fact there are no real perceptual of-God experiences to which the Principle of Credulity is applicable, as I shall now argue.[48] That is, there are no perceptual experiences the purported object of which can really be identified as being God.[49]

Swinburne presents the first two kinds of religious experience as cases of perceiving a public thing *as* something, or perceiving something *in* something public. For example, I may perceive a bright star in the sky near to the Earth and identify it *as* Venus, or *in* seeing a man dressed in a certain way I may see John Smith.[50] The first kind of religious experience (a) is defined as purportedly perceiving a supernatural object *in* perceiving a perfectly ordinary and public non-religious object.

But unfortunately, the example of (a) provided by Swinburne, that someone looks at the night sky and suddenly sees it as God's handiwork, is not an example of a real perceptual of-God experience, since perceiving a product is not the same thing as perceiving the maker of that product.[51] Indeed, the example is not an instance of (a) purportedly perceiving a supernatural object in perceiving a perfectly ordinary and public non-religious object, since even if one thinks that the night sky is God's handiwork, it is not for that reason a supernatural object. In cases of perceiving x as y, or y in x, x and y have to be identical. But God and the night sky are not identical, if at least one endorses the theistic definition of God, and does not slip into Spinozism. Clearly, it is logically excluded that there are perceptual of-God experiences of type

---

[45] If one takes the scope of the Principle of Credulity to be much larger than Swinburne explicitly says in (EG), one might construe its inapplicability to dreams as resulting from the first two defeaters discussed below (sections 15.9–10).
[46] Swinburne (EG), pp. 298–303.
[47] Cf. Swinburne (EG), p. 298: 'it will be useful to classify the different kinds of religious experience. In due course I shall make similar points about all of them'.
[48] Let me stress again that I am interpreting the Principle of Credulity in the narrow sense defined in Swinburne's (EG), and not in the wide sense of (EJ).
[49] This thesis has been defended on other grounds by Zangwill (2004).
[50] Swinburne (EG), p. 298.
[51] There may be boundary cases. For example, we may say that I am hearing you when I am hearing spoken words of which I reliably believe that you are uttering them, and we may say that I am smelling you if I smell an odour of which I reliably believe that your body is giving it off. But this is only so when there are no problems of identification concerning the owner of the voice or the producer of the smells. This is different in God's case.

(a), since God as defined by theism is not identical with any publicly perceivable object.[52]

For the same reason, an of-God experience cannot instantiate kind (b) of religious experience, which is defined as an experience purportedly of a supernatural object which people have in perceiving an extra-ordinary public object. For example, if there really was a resurrection of Jesus with his Earthly body, the disciples may have perceived this very unusual public object suddenly turning up among them and eating some fish. Inspecting his wounds, and recognizing his features and his voice, the disciples then 'had the religious experience of taking the man to be the risen Jesus Christ'.[53] Again, this experience of a public object as a supernatural object is not a perceptual of-God experience, notwithstanding the mystery of the Trinity. For even if Jesus is in some sense identical with God, this is not the ordinary sense of identity that is required for perceiving x as y or perceiving y in x. Indeed, since allegedly Jesus has risen with his Earthly body, whereas God is defined as bodiless, assuming that Jesus and God are identical in this ordinary sense would lead to a contradiction.

The conclusion that of-God experiences can instantiate neither kind (a) nor kind (b) is in fact sufficient to refute premise (2) of the argument from religious experience as stated in section 15.6. Since the Principle of Credulity applies paradigmatically to cases of sense-perception, whereas what is perceived in these cases always is a public object, of-God experiences should instantiate kinds (a) and (b) of religious experiences in order to qualify for a legitimate and uncontested application of the principle. The three other kinds of religious experience discussed by Swinburne are too different from ordinary sense-perceptions to have epistemic parity with them, because they do not involve taking public phenomena religiously.

Using the jargon of sense-datum theories of perception, Swinburne distinguishes between these three other kinds as follows. Either (c) the religious experience involves having certain sensations, private to the subject, which may be described in terms of our normal sensory vocabulary for qualities perceptible by the five senses, such as colours, sounds, smells, etc. Or (d) the religious experience involves having certain sensations or feelings, which, however, are not so describable. Finally (e), there may be religious experiences that do not involve having any sensations or feelings at all.

As an example of (c), Swinburne mentions the story of *Matthew* 1.20–21, according to which Joseph dreamt that he saw an angel who said things to him. In the context of the argument from religious experience, this example is unfortunate, however, because one should not apply the Principle of Credulity to dreams after waking up. Better examples would be the experiences of those who tell us they have heard God's voice, or have had a vision of God. But now other difficulties arise.

First of all, if someone hears a voice not heard by others in the room, so that this voice is a purely private phenomenon as (c) requires, few of us will be inclined to apply the Principle of Credulity and infer that there must be a speaker present, who cannot be detected by others. Second, identifying a voice as the voice of a person P requires that one sees or has seen P speaking, or that one is told, directly or indirectly, that this

---

[52] One might extend the first set (a) of of-God experiences by allowing that being y is a monadic property of x, since we may perceive a person *as* kind, or courageous. But even in this extended sense there cannot be of-God experiences of kind (a).
[53] Swinburne (EG), p. 299.

is *P*'s voice by those who have seen *P* speaking. But in God's case, this is impossible, since God is immaterial, so that he cannot be seen in the ordinary sense of the term. It is a fortiori impossible if the voice is merely a subjective phenomenon, which cannot be perceived by others.

One may hear voices that others cannot hear. Indeed, this is a phenomenon that is typical of many psychological pathologies. But it seems out of the question that one be able to identify such a voice as God's voice in any trustworthy manner. Finally, the sensible qualities which one takes oneself to perceive if one has a religious experience of type (c) cannot be qualities of God because of God's immaterial nature. As a consequence, these (c)-experiences will never be real perceptual of-God experiences, but rather experiences purportedly of things allegedly produced by God. Even if the Principle of Credulity were applicable to them, the principle would only justify the conclusion that these products exist, such as a thundering voice or the pitch-black night sky. It would not at all justify the conclusion that God produces them.

For these reasons, it seems that if there are real of-God experiences, they must belong either to kind (d) or to kind (e). If any sensations or feelings are experienced by the subject at all, these cannot be sensations or feelings that one might describe in terms of the vocabulary we use for sensible qualities or phenomena. The examples to which Swinburne refers are either mystical experiences by means of an alleged sixth sense (case d), or mystical experiences during which the subject does not have any sensations (case e).

One subject, quoted by William James in *The Varieties of Religious Experience*, describes such an experience as follows:

> I have on a number of occasions felt that I had enjoyed a period of intimate communion with the divine. These meetings came unasked and unexpected, and seemed to consist merely in the temporary obliteration of the conventionalities which usually surround and cover my life.... Once it was when from the summit of a high mountain I looked over a gashed and corrugated landscape [...] when I could see nothing beneath me but a boundless expanse of white cloud [...]. What I felt on these occasions was a temporary loss of my own identity [...].[54]

Another subject wrote in a report sent to the Religious Experience Research Unit in Harris Manchester College, Oxford:

> Then, in a very gentle and gradual way, not with a shock at all, it began to dawn on me that I was not alone in the room. Someone else was there, located fairly precisely about two yards to my right front. Yet there was no sort of sensory hallucination. I neither saw him nor heard him in any sense of the words 'see' and 'hear', but there he was; I had no doubt about it. He seemed to be very good and very wise, full of sympathetic understanding, and most kindly disposed towards me.[55]

There are numerous reports of this kind, which fit into Swinburne's classes (d) or (e). But should they be considered real perceptual of-God experiences to which the Principle of Credulity applies? There are two problems with regard to all these religious experiences, and each of them justifies a negative answer to the question.

---

[54] James (1902), p. 84.
[55] Quoted from Alston (1991), p. 17, who quoted the report from Beardsworth (1977), p. 122.

First of all, since God is defined as bodiless, experiences will better qualify as of-God experiences to the extent that they differ more, phenomenologically speaking, from ordinary sense-perceptions. This is why they should belong to kinds (d) and (e). But since the Principle of Credulity applies to sense-perceptions primarily, the applicability of this principle is less justified to the extent that experiences differ more from ordinary sense-perceptions, phenomenologically speaking. In short, the more religious experiences qualify for being of-God experiences, the less the Principle of Credulity is applicable to them. Indeed, the feelings described in the first quotation cannot be considered perceptions at all, so that this type of religious experience can yield no epistemic support for the belief that God exists, and something similar holds for the experience described in the second quotation.

The other problem is even more serious. How can one know that a perceptual experience really qualifies as an of-God experience? It does so only if the subject is able in principle to identify the object of the experience as being God. Swinburne is right in arguing that we may learn to recognize objects of perception under many different descriptions. We may learn to identify a ship as Russian or a bright spot in the sky as Venus. But can we also learn to identify some peculiar non-sensory phenomenal content (case d), or no phenomenal content whatsoever (case e), as God or as a characteristic of God? In both cases (d) and (e), there seem to be insuperable difficulties.

For one thing, it is not clear what phenomenal features, if any, can belong to an immaterial being. Consequently, if specific perceptual experiences are taken by a subject as of-God experiences, they are given a much richer interpretation than the phenomenology of the experience by itself can support. The fact that the interpretation so radically transcends the perceived characteristics of what is purportedly experienced, can only be explained by assuming a heavy reliance on background beliefs. Furthermore, God is unique in his kind if monotheism is true, and learning to identify an object that is unique in its kind will barely be possible on the basis of the phenomenal properties of the object of a private experience.[56] Finally, God is defined as infinite in some respects, since he supposedly is omniscient, omnipotent, and absolutely free. But can one ever perceive that something infinite (in the relevant sense of 'infinite') is present, as contrasted with something that has the qualities of power, knowledge, and freedom in large but finite degrees?

Swinburne is optimistic in this respect. He writes:

Someone might well, through visual, auditory, tactual, etc. experience of recognizing persons of various degrees of power, knowledge, and freedom be able to recognize when he was in the presence of a person of unlimited power, knowledge, and freedom. Indeed, it is plausible to suppose that someone might be able to recognize extreme degrees of these qualities, even if he could not so easily recognize lesser degrees straight off without inductive justification.[57]

---

[56] Cf. McLeod (1993), pp. 157–61.
[57] Swinburne (EG), p. 307.

And elsewhere he adds:

> The description of God as the one and only omnipotent, omniscient, and perfectly free person may indeed suffice for someone to recognize him – by hearing his voice, or feeling his presence, or seeing his handiwork, or by some sixth sense.[58]

But it is difficult to understand how this optimism can be justified. First of all, in order to 'recognize a person' in the usual sense of this expression, that person has to be bodily present to us, whereas God is defined as an immaterial person. I argued in Chapter 7.2 that if one first attributes properties to God such as power, knowledge, or freedom, and then adds that God is bodiless, one negates a necessary condition for meaningfully attributing such properties to a person, so that the traditional theistic definition of God does not make sense. But let us suppose for the sake of argument that the definition makes sense, and that we can meaningfully attribute such personal properties to a being called 'God' even though he is defined as bodiless. Let us also suppose for the sake of argument that we can attribute a voice, which we hear, to an unknown person although we do not perceive this person. How, then, will we be able to know that this person has the defining properties of God, such as omniscience or omnipotence, so that he is God?

As Swinburne admits, we would need substantial inductive evidence in order to establish that a person possesses 'lesser' than 'extreme degrees of these qualities'. We can only discover that a person is quite knowledgeable about a topic by interviewing him or her for some time, at least if we are experts on the topic. Swinburne suggests that such inductive evidence is *not* needed in order to discover that, say, the voice we are hearing belongs to an omniscient person. But this suggestion is utterly counterintuitive. For in order to justify the claim that we are hearing the voice of an omniscient person, we would at least have to obtain satisfactory answers from that person to some difficult questions, such as questions concerning the solution to mathematical problems that mathematicians have not been able to resolve, or we would have to hear true answers to questions concerning personal matters about which we have not informed anybody, such as the PIN numbers of our credit cards. However, there are no reports of alleged of-God experiences in the vast literature on religious experience according to which the voice ascribed to God told the subject of the experience anything that could not be known or invented by the community to which the subject belonged.

On the basis of these considerations, we cannot avoid drawing the following three conclusions. First, there are no good reasons to suppose that any real perceptual of-God experiences have occurred among the multitude of religious experiences. If subjects claim to have had an of-God experience when they had an experience of one of the five kinds distinguished above, they claim much more than is warranted by the character of these experiences. Consequently, we should not endorse the first premise of the argument from religious experience as stated at the end of the previous section (15.6).[59]

---

[58] Swinburne (EG), pp. 318–19.
[59] I do not think that there are convincing answers to these objections in Alston (1991), who discusses the problem of the character of of-God experiences at length on pp. 43–63. Alston is 'inclined to think that what we know about sense perception provides a strong empirical argument against the possibility of God's directly appearing to *sensory* experience'. But he 'can't see that any

Second, if real of-God experiences have occurred, the Principle of Credulity would not be applicable to them, since such experiences would be too different from ordinary perceptual experiences for even some weak epistemic parity to obtain. Hence, the second premise should not be endorsed either. Finally, we should not apply the Principle of Testimony to religious experiences. According to this principle, we should believe what others tell us they have experienced, unless there are convincing defeaters. But it would be a mistake to apply this principle if the other persons are not in a position to know by experience what they are reporting, and this is the case with regard to religious experiences, as I argued. It follows that the burden of proof concerning theism cannot be shifted to the unbeliever by the argument from religious experience.

15.8 THE SWAMP OF GULLIBILISM

Let us now suppose for the sake of argument, however, that genuine of-God experiences are conceivable and do in fact occur. Let us also assume that the Principle of Credulity is applicable to them, and, indeed, that it applies to all religious experiences, as Swinburne argues, unless there are specific defeating considerations. What would these assumptions imply for our epistemic situation?

Critics of Swinburne have pointed out two problems that should be raised. First of all, if the Principle of Credulity can be used positively in order to justify on the basis of an experience the belief that God is present and therefore exists, one will expect that it can also be used negatively in order to justify on the basis of experiences the belief that God is absent. It may be that quite a few people have had an of-God experience. But many other people have sought God passionately and have not found him. They will have had grievous experiences purportedly of God's absence. '[A]re not experiences of the absence of God good grounds for the non-existence of God', as one critic argued? Indeed, if the positive Principle of Credulity applies to religious experiences, one might think that a negative Principle of Credulity should be applicable as well, saying that if a subject experiences purportedly that x is absent, then probably x is absent. Of course, experiences of God's absence might be defeated by showing that very, very probably, God exists. But the burden of proof would be shifted back to the theist.[60]

However, this objection to the argument from religious experience is problematic for a number of reasons. First, there is no symmetry between positive and negative applications of the Principle of Credulity even in cases of ordinary perception. When I (purportedly) see a paperclip lying on my desk, this is a sufficient ground for believing that there is a paperclip on my desk and, indeed, the perception yields knowledge. But if I do not see a paperclip, this does not show that it is absent from my desk. It may be hidden by books or papers. In order to have sufficient grounds for concluding that the clip is absent from the desk, I must have searched thoroughly. Furthermore, with regard to spatio-temporal particulars we usually cannot infer

---

empirical considerations count against the supposition that God presents Himself to our experience in a non-sensory fashion' (p. 59, note 49, Alston's italics). However, Alston does not solve the problem of identification raised above, for example.

[60] Martin (1986), pp. 82–3ff and (1990), Chapter 6, pp. 169–74.

non-existence from absence at a specific place. The particular paperclip I was looking for may have fallen off the desk, or it may have been swallowed by a vacuum cleaner.

One might reply that God is said to be omnipresent, so that in his case establishing absence at a specific place would indeed show his non-existence. But of course, this is not how God's omnipresence should be understood. Finally, the theist will hold, as we saw, that God is a person who can choose whom to cause to have an of-God experience, since he is 'a normally invisible person with the power of letting you, but not me, see him'.[61] He is not a public object perceptible by all people who have normal perceptual capacities and are in the right position to perceive it. Of-God experiences, if they exist, must be private experiences.[62] If this is so, the proper place to consider the evidential value of absence-of-God experiences is the argument from hiddenness, which I discussed in Chapter 14.11–12, and not the argument from religious experience.[63]

The second problem is more serious, since for monotheists it will risk being a *reductio ad absurdum*. If the Principle of Credulity really applies to religious experiences, we may find ourselves 'landed in the swamp of gullibilism'.[64] In the absence of defeating considerations, we must not only accept the existence of God, but also the existence of numerous Hindu deities, of flying saucers, of Martians who rape American ladies, of witches, of all kinds of demons and devils, of wood elves and goblins, etcetera, since with regard to all these things there are or were many people who claim(ed) to have experienced them. The burden of proof for the (mono)theist now is to show that in the case of each experience purportedly of such a queer thing, there are defeating considerations showing that the experience is illusory, whereas with regard to of-God experiences such defeating considerations do not apply. Although Swinburne attempts to establish the latter point, he does not extensively argue for the former.[65] Let me briefly review the four possible defeating considerations (1–4) discussed by him, which limit the application of the Principle of Credulity, in order to see whether this double burden of proof can be substantiated.

---

[61] Swinburne (EG), p. 297.
[62] Cf. Swinburne (EG), pp. 297–8: 'If S has the experience of its seeming to him that o is there, but, either because of o's choice or for some other reason, not every other attentive person rightly positioned and equipped would necessarily have the experience, then S has, I shall say, a private perception of o. If religious experiences are of anything – that is, are perceptions – they are normally private perceptions'.
[63] Cf. Swinburne (EG), p. 304: 'given that my rejection in Chapter 11 of "the argument from hiddenness" is correct, there are no good grounds for supposing that, if there is a God, necessarily the atheist would have experienced him'.
[64] Cf. Martin (1986), p. 92.
[65] Swinburne (EG), pp. 310–22. Swinburne answers to the objection of gullibilism that whereas the Principle of Credulity may well give an initial degree of probability to the existence of Hindu deities, flying saucers, and Martians, the positive arguments for theism, which are not paralleled in the case of Hindu deities, etc., rule out these deities, etc. (private correspondence). But this answer will not do if the probabilifying weight of religious experiences is as great as Swinburne claims and needs for shifting the burden of proof to the unbeliever. Indeed, in ordinary cases of perception, a real of-O perception outweighs strong arguments to the effect that O does not exist when and where it is perceived.

15.9 UNRELIABLE SUBJECTS OR CONDITIONS

Each of the defeating considerations (1–4) must be such that, if in a specific case it is added to the report of an of -O experience, it prevents the experience from making it probable that O was present. Two types of consideration are based on inductive inferences from the failure of similar claims, whereas the remaining two types are not so based. To begin, (1) an of -O experience will not make it probable that O is present if the experience is had by subjects or under conditions found unreliable in the past with regard to similar experiences. For example, the subject may be suffering from hallucinations, or the experience was had when the subject was drugged, or seriously ill, or ecstatic.

Swinburne dismisses this first defeater by declaring lapidarily that it is 'hardly generally available', since '[m]ost religious experiences are had by people who normally make reliable perceptual claims, and have not recently taken drugs'.[66] But this generalization is insufficiently substantiated by the available empirical evidence. Psychoactive drugs have been used for many thousands of years for religious purposes. The Aztecs called their hallucinogenic mushrooms 'God's flesh', the Pythia at Delphi got her states of divine possession by Apollo probably from methane, ethane, and ethylene, whereas psilocybin has been used to prompt mystical experiences during Protestant church services.[67] Furthermore, in all pre-modern cultures including ancient Greece, attacks of epilepsy were interpreted as manifestations of supernatural forces. Nietzsche diagnosed St Paul's conversion on the way to Damascus, during which Paul fell to the ground and heard a voice saying to him: 'Saul, Saul, why do you persecute me?', as an epileptic attack triggered by deep distress.[68] Nietzsche's diagnosis has been confirmed by modern research, which shows that a substantial number of founders of major religions, Paul and Muhammad among them, have been documented as showing symptoms of what is now diagnosed as epilepsy.[69] Furthermore, religious delusions are an important subtype of delusional experience in schizophrenia, whereas mood-congruent religious delusions are a feature of mania and depression. Many other mental afflictions can produce religious experiences, such as in patients with frontotemporal dementias who exhibit hyperreligiosity.[70] The fact that such subjects 'normally make reliable perceptual claims' when talking about their physical environment, does not refute the assertion that their perceptual claims concerning devils, demons, or deities are unreliable when made in trance, or shortly after an epileptic attack or a dream.

Having studied a large sample of religious experiences, William James concluded as follows:

---

[66] Swinburne (EG), p. 315.
[67] Cf. for the Aztecs: Wasson (1961) and Wasson et al. (1986). For the Oracle of Delphi, cf. Hale et al. (2003). For psilocybin in Protestant churches, cf. Doblin (1991).
[68] Jilek-Aall (1999), pp. 382–6. For Nietzsche's interpretation of Paul's conversion, see Nietzsche (1881) § 68 jo. the *New Testament*, Acts 9.3–6.
[69] Landsborough (1987). For Muhammed, cf. Saver and Rabin (1997).
[70] Saver and Rabin (1997).

Even more perhaps than other kinds of genius, religious leaders have been subject to abnormal psychical visitations. Invariably they have been creatures of exalted emotional sensibility. Often they have led a discordant inner life, and had melancholy during a part of their career. They have known no measure, been liable to obsessions and fixed ideas; and frequently they have fallen into trances, heard voices, seen visions, and presented all sorts of peculiarities which are ordinarily classed as pathological. Often, moreover, these pathological features in their career have helped to give them their religious authority and influence.[71]

Although James argues that '[i]f there were such a thing as inspiration from a higher realm, it might well be that the neurotic temperament would furnish the chief condition of the requisite receptivity', this is of no avail to Swinburne.[72] For the basis of his argument from religious experience is the Principle of Credulity, and not a speculative guess concerning mental disorders that might enhance receptivity for divine manifestations. However, even if Swinburne were right in dismissing the first type of defeating consideration, this would not help him very much. By dismissing it, he cannot avoid the swamp of gullibilism evoked above. Does he fare better with regard to the remaining three conceivable defeating considerations?

15.10 RELIGIOUS DIVERSITY AND RADICAL TRANSCENDENCE

The second defeater discussed by Swinburne consists in showing (2) 'that the perceptual claim was to have perceived an object of a certain kind in circumstances where similar perceptual claims have proved false', such as claims to have read something from a great distance or to have drank a specific wine which one had never tasted before.[73] There are two arguments against the reliability of of-God experiences that fall under this heading, the argument from religious diversity and the argument from radical transcendence. Again, Swinburne seriously underestimates the force of these objections.

To the argument from religious diversity he replies that contrary to what its proponents claim, experiences (apparently) of other deities are not (necessarily) in conflict with experiences (apparently) of God, because they are 'hardly experiences apparently of any person or state whose existence is incompatible with that of God'.[74] This is correct in itself, because in principle the omnipotent god of theism *could* have created nearly all the other deities that humanity has worshipped. But as an answer to the argument from religious diversity against theism the reply is an *ignoratio elenchi*. The point of this argument is that, if monotheism is true, all existence claims about other deities are false. Consequently, all religious experiences (apparently) of other deities must be illusory. It follows that all experiences (apparently) of God are indeed experiences (apparently) of 'an object of a certain kind in circumstances where similar perceptual claims have proved false', for God and other deities belong to the common kind of supernatural beings.[75] So it seems that theists are landed in a dilemma at this

---

[71] James (1902), p. 29.   [72] James (1902), p. 45.
[73] Swinburne (EG), p. 311.   [74] Swinburne (EG), p. 318.
[75] Swinburne (EG), p. 311. Swinburne might answer that all religious experiences (apparently) of other deities are mistaken interpretations of religious experiences that in reality are caused by God.

point. Either they have to convert to polytheism and live with the swamp of gullibility, or they have to abandon the argument from religious experience.

Swinburne suggests yet another remedy for the problem raised by the diversity of religions. Let us admit that by giving a particular description of the purported object of one religious experience, we may contradict descriptions of the purported objects of many other religious experiences. By claiming that one has perceived Poseidon, for example, one contradicts claims of Christian mystics that they have experienced God, or the implications of such contentions, since if (mono)theism is true, Poseidon does not exist. But, as Swinburne says, in such a case the subject of the religious experience 'need not [...] withdraw his original claim totally; he need only describe it in a less committed way – for example, claim to have been aware of some supernatural being'. In other words, such contradictions between the descriptions of purported objects of religious experiences mean 'only that we have a source of challenge to a particular detailed claim, not a source of scepticism about all claims of religious experience'.[76] However, even if this solution of an evasion into the undefined were acceptable, it would not help theists. For their argument from religious experiences purports to make it 'very, very likely', in the absence of defeating considerations, that *God* exists, and not merely that there is 'some supernatural being'. It is the former 'particular detailed claim' that they want to establish, and not the latter, less committed contention.

The argument from radical transcendence is equally damaging for theism. In contradistinction to the finite deities of polytheistic religions, God is radically transcendent to the universe. Because God is immaterial, no perceptual experience can present him directly to us, so that a proper perceptual experience of God *himself* is a priori impossible, as I argued in section 15.7. If God exists, we might at best 'perceive' him indirectly by perceiving certain perceptible effects that he allegedly produces, such as by hearing a voice that we attribute to him or by having a feeling of great joy, ultimate significance, or unity with the universe, which we interpret as caused by him. In such cases, we would perceive God in the same sense in which we may say that we 'perceive' a fire when we merely see some smoke. But in God's case, such indirect experiences have no clear evidential force, because the perceptible effects may be produced in many other ways, whereas we cannot establish any regular conjunction between the effects and God as a cause.

Swinburne merely responds to a weaker version of this argument, according to which 'those who make claims of religious experience have not had the kind of experience that is needed to make claims of this kind that are probably true'. The reason would be that one cannot recognize a person one never saw before unless one has previously been given a detailed description of his appearance. Swinburne answers, as we saw, that '[t]he description of God as the one and only omnipotent, omniscient, and perfectly free person may indeed suffice for someone to recognize him – by hearing his voice, or feeling his presence, or seeing his handiwork, or by some sixth sense'. Although Swinburne admits that '*some* mild suspicion is cast on a subject's

---

But this is not an attractive escape, not only because Swinburne rejects the experience/interpretation distinction, but also because it would turn God into a deceiver.

[76] Swinburne (EG), pp. 316–17.

claim to have recognized an agent with these qualities by the qualitative remoteness of his previous experiences from what he claims to have detected', he seems to overlook that if someone claims to have perceived something unextended and bodiless, that is, something *that cannot be perceived in principle* in the normal sense of the term, such a claim should be received with the greatest possible suspicion.[77] Indeed, it seems plausible to hold that the Principle of Credulity does not apply to such claims at all, as I argued in section 15.7.

We may conclude with regard to the second defeater discussed by Swinburne that it refutes his argument from religious experience. If God exists and monotheism is true, arguments from religious experience for the existence of all other deities are defeated by monotheism. But if these arguments are defeated, religious experience in general is highly unreliable, so that the argument from religious experience for monotheism is defeated as well. In short, if monotheism is true, the argument from religious experience for monotheism is a very bad argument. The evasion into the undefined does not help theists, as we saw, and God arguably is not the kind of thing that can be experienced perceptually at all.

15.11 PROBLEMS OF PRESENCE AND CAUSALITY

The third and fourth conceivable defeaters of arguments from religious experience are not based upon inductive inferences from the failure of similar perceptual claims. Since Swinburne endorses a causal theory of perception, according to which a subject truly perceives an object if and only if the subject has a perceptual experience purportedly of the object, and this experience is caused by its being present, he holds that we might defeat a perceptual claim concerning a given object either by showing that very, very probably it was *not present* where it was (purportedly) perceived, or by showing that even though it was present, its presence probably did *not cause* the experience.[78] There is a defeater of the third type (3) if it can be shown that the object very, very probably was not present, whereas defeaters of the fourth type (4) show that the object probably did not cause the perceptual experience. What should we think about the applicability of these defeaters in the case of perceptual experiences purportedly of God?

It is primarily the logic of the third defeater that shifts the burden of proof concerning theism to the unbeliever, if at least the Principle of Credulity applies to religious experiences. Swinburne correctly stresses that in order to launch this third defeater, it is not enough that on background evidence it is more probable than not that the object in question was not present where it was (purportedly) perceived. This is so because 'after all, most of the things that we think that we see are on background

---

[77] Swinburne (EG), pp. 318–19.
[78] Swinburne (EG), pp. 296 and 311–12. Nearly all authors on perception hold that there are further necessary conditions for veridical perception with regard to each particular sense, which are needed in order to exclude deviant causal chains, such as Strawson's condition for visual perception that the object has to be located in the subject's arc of vision (Strawson, 1974, pp. 79–80). If there are more necessary conditions for successful perception, there also will be more possible defeaters. Swinburne rejects such further conditions, however (EG, pp. 296–7, note 4). This implausible move is needed in order to rebut the fourth defeater, as we shall see below in the main text.

evidence less probable than not', as the example of Peter in Paris illustrated. Consequently, the defeater applies only if it can be made *very, very* probable that the object was not present, since only in that case does it outweigh the force of the perceptual experience itself. If perceptual experience did not normally have this force, we would be 'imprisoned within the circle of our existing beliefs', as Swinburne observes.[79] In the case of experiences purportedly of God, however, the defeater cannot consist in showing merely that God was not present at a particular place, since 'if there is a God, he is everywhere'. It follows that God 'is only not present if he does not exist'. Accordingly, Swinburne concludes that in order 'to use this challenge [...], you have to prove that very, very probably there is no God', and clearly 'the onus is on the atheist to do so'.[80]

Swinburne argues for a similar (albeit somewhat weaker) conclusion with regard to the fourth possible defeater (4), which would consist 'in showing that the religious experience probably had a cause other than its purported object'. Although this may be shown for experiences purportedly of lesser gods, he holds that such an argument from alternative explanations is a 'particularly awkward challenge' if applied to of-God experiences, because '[t]he mere fact that a religious experience apparently of God was brought about by natural processes has no tendency to show that it was not veridical'. The reason is, Swinburne avers, that since God is defined as the creator and sustainer of the universe, 'any causal processes at all that bring about my experience will have God among their causes' if God exists, whether it is an of-God experience or not. Furthermore, because God is omnipresent, he will always be at the place where one purportedly perceives him to be. Since Swinburne holds that apart from being of-God, the only other necessary conditions for an of-God perceptual experience to be a veridical perception of God are, first, that God belongs to the causal chain that brings about the experience, and, second, that God is where he is purportedly perceived to be, one can show that God did not cause an of-God experience only by showing that God does not exist. In other words, 'if there is a God, any experience that seems to be of God, will be genuine – will be of God'.[81]

Swinburne would have concluded from his discussion of the four defeaters that of-God experiences ought to be taken as veridical unless it can be shown that *very, very* probably God does not exist, if there had not been some considerations that somewhat weaken the force of these experiences. One consideration is that what the subjects claim to have recognized and identified, namely God, is qualitatively very remote from everything else they have experienced. For this reason, he somewhat modifies the 'very, very probably'. He concludes that of-God experiences ought to be taken as veridical unless it can be shown that it is 'significantly more probable than not that God does not exist'.[82]

If Swinburne's C-inductive arguments for the existence of God had the force he attributes to them, this conclusion might suffice not only to shift the burden of proof to the unbeliever, but also to make this burden unbearable. Moreover, Swinburne attempts to avoid the problem of gullibility by suggesting that arguments from

---

[79] Swinburne (EG), pp. 311–12.
[80] Swinburne (EG), p. 319.
[81] Swinburne (EG), p. 320. All quotes in the paragraph are from this page.
[82] Swinburne (EG), p. 321.

religious experience supporting the claim that some lesser gods exist, are open to the fourth challenge. Furthermore, 'they are far more open to the third challenge' to the effect that this lesser god does not exist, 'since there is no natural theology available to give some probability to the existence of supernatural beings other than one who sustains the whole universe'.[83]

Let us briefly evaluate these arguments in order to substantiate the final conclusion of this chapter. According to J. L. Mackie, the fourth defeater is a sufficient reason for rejecting all arguments from religious experience. As he says, 'we need not postulate any supernatural source or sources for these experiences, since they can be fully explained on purely natural grounds, by reference to otherwise familiar psychological processes and forces', if at least we add to the explanation elements of the cultural traditions in which the religious experiences occurred.[84]

Swinburne's rebuttal that such an argument from alternative explanations is a 'particularly awkward challenge' if applied to of-God experiences, since if he exists, God is both onmipresent and the ultimate cause of all causal chains in the universe, is unconvincing for two reasons. First, it does not take into account the problem of deviant causal chains. If a neurosurgeon N by influencing our brains via electrodes caused a visual of-N perceptual experience in us while our eyes are shut, nobody would say that we are seeing N, even though we have a visual of-N experience caused by N, and even though N is exactly where we purportedly see him to be. Similarly, even if God ultimately caused all causal chains in the universe, this would not imply that if someone has an of-God experience, he is experiencing God, although God is omnipresent and for that reason present where he is purportedly experienced. On the contrary, such a mode of causation should be considered deviant, since the subject purportedly perceives God directly, so to say, and not as an ultimate cause of all causal chains in the universe.[85]

Second, Swinburne's claim that 'if there is a God, any experience that seems to be of God, will be genuine – will be of God', which is entailed by his answer to the fourth defeater, is so strong that it amounts to a *reductio ad absurdum*. If this contention of an 'essential veridicality' of experiences purportedly of God were correct, even the of-God experiences of the Yorkshire Ripper, who murdered at least thirteen women and claimed to hear a divine voice telling him to kill, would be veridical.[86] So it cannot be the case that all of-God experiences are veridical simply because 'any causal processes at all that bring about my experience will have God among their causes', as Swinburne holds. Consequently, we should agree with Mackie that the argument from alternative explanations effectively undermines all arguments from religious experience.

With regard to the third possible defeater, we have to wonder whether arguments from religious experience for lesser gods are really 'far more open' to this challenge than the argument from religious experience for God. On the one hand, Swinburne

---

[83] Swinburne (EG), p. 321.
[84] Mackie (1982), p. 197.
[85] This conclusion holds, I presume, even though we cannot know the causal mechanism by which God would produce of-God experiences in us. Cf. on the problem of deviant causal chains in of-God experiences, Alston (1991), pp. 64–6.
[86] Levine (1990).

argues that because of reasons of simplicity, natural theology favours theism over the hypothesis that the universe is created by a consortium of gods. As a consequence, it will be easier to show that lesser gods do not exist than it is to show that God does not exist. But on the other hand, these lesser gods are not as qualitatively remote from ordinary objects of perception as is God. Accordingly, the argument from religious experience will have more force with regard to them, so that the task of defeating it will be more demanding. Should we not conclude that these two considerations are equally weighty, so that the force of the one cancels the force of the other? If so, Swinburne is landed in the swamp of gullibilism, unless there are other defeaters that apply to arguments from religious experience for lesser gods and not to the argument for God.

As we have seen, however, the opposite is the case. Whereas the first defeater will be equally strong with regard to both types of argument, the second defeater refutes the argument from religious experience for (mono)theism. But it does not apply to broadminded versions of polytheism, which accept the existence of any god or other supernatural entity that is supported by otherwise undefeated religious experiences. We have to conclude, then, that the argument from religious experience to God is defeated at least by two of the defeaters discussed, that is, by the second and the fourth, whereas the first defeater is much more serious than Swinburne thinks. And if it is not so defeated, theists cannot avoid the swamp of gullibilism, which is a *reductio ad absurdum* of the argument from religious experience for theism.

In earlier sections of this chapter, I argued that there are no real perceptual of-God experiences (section 15.7) and that, if there were, it is questionable to what extent the Principle of Credulity applies to them (sections 15.5–6). We now see that even if it is applicable in principle, this does not help theists. Shifting the burden of proof to the non-believer is accomplished by the logic of the third defeater only if no other defeaters apply to the argument from religious experience. But other defeaters do apply, of which the second is most decisive. It follows not only that the argument from religious experience is unconvincing, but also that Swinburne's cumulative case strategy is unsuccessful. In his book *The Existence of God*, Richard Swinburne fails to show that on our total evidence, God's existence is more probable than not. And since according to many experts, his cumulative case strategy yields the best pleading in support of theism available in the literature, it is unlikely that anyone else will succeed.

Conclusion

In the preface to his masterly classic *Pagans and Christians*, Robin Lane Fox declares that '[n]o generation can afford to ignore whether Christianity is true and, if it is not, why it has spread and persisted'. He continues by claiming that '[h]istorians have a particular contribution to make to these inquiries'.[1] Of course, one might affirm the same thing with regard to Islam and other major religions.

Strictly speaking, however, historical scholars will be able to make a greater contribution to the second enquiry ('why it has spread and persisted') than to the first ('whether Christianity is true'). Assuming that neither bare theism nor ramified Christian theism is true, historians will have to explain the surprising expansion of Christianity in the Roman Empire, for example, by reference to purely secular causal factors. If they succeed in doing so convincingly, their achievement may provide some slight evidence against the truth of the Christian creed. And if they fail, Christians will be tempted to conclude from this lack of success that specific historical developments could not occur without divine intervention. But in both cases, this evidence will not have a great impact on the balance of total evidence concerning the truth of Christian theism. The same holds, *mutatis mutandis*, with regard to other religious creeds.

If each generation of human beings has to enquire whether theism or some other religious creed is true, the proper discipline to engage in is the philosophy of religion, as I argued in Part I of this book. In particular, believers cannot endorse a religious creed conscientiously without relying on the apologetic arguments of natural theology (Chapters 1, 2). Similarly, when secular scholars of religion, such as historians, sociologists, or evolutionary psychologists, purport to explain the origin, functions, or rise and fall of religions, they cannot just assume that the creeds of these religions are *not* true without presupposing specific results of critical philosophy of religion. Clearly, then, the philosophy of religion in general, and natural theology in particular, has a priority as regards the issue of whether theism or any other religious creed is true. In this Conclusion, I summarize the results of my *Critique of Religious Reason* and define the epistemic stance we should adopt with regard to theism and other religious creeds.

The logical structure underlying Part I was that of a decision tree for the faithful, which I explained in the *Preface*. Concerning the contents of their creeds, such as the assertion that God or some other god exists, religious believers have to face the

---

[1] Lane Fox (1988), p. 8.

following three dilemmas. First of all, (1) they have to interpret such statements either (a) as claims to truth or (b) in a non-cognitive manner, so that no truth-claim is involved. As I argued briefly in the *Preface*, in Chapter 2.4, and in Chapter 6.4, the non-cognitive option (b) as elaborated by followers of Wittgenstein such as D. Z. Phillips, for example, is inadequate as an interpretation of most existing religious discourse, so that, at best, it can be seen as a proposal for a radical revision of the religious language game. Let us suppose for this reason that most religious believers in the world opt for alternative (a). Since according to option (b), credal statements do not involve claims to truth, believers who endorse this option do not disagree with atheists and other unbelievers in any philosophically interesting way.

If the statement that God exists is meant as a claim to truth (a), believers have to face a second dilemma (2). Either (c) a religious belief has to be backed up by reasons or evidence in order to be legitimate, or (d) such a backup is not or may not be needed. I argued in Chapters 3 and 4 that option (d), as elaborated by Alvin Plantinga under the banner of the *Reformed objection to natural theology*, is not available to 'sophisticated, aware, educated, turn-of-the-millennium people' who want to be conscientious and reasonable in endorsing their creed, although these people constitute the intended public of Plantinga's works.[2] It follows that our well-educated contemporaries can endorse a religious creed conscientiously only if (c) they rely on the arguments of natural theology, and in particular on what is called *positive apologetics*.

However, if conscientious religious believers have to take option (c), they are faced with a third and last dilemma (3) in our decision tree. Which methods of investigation and types of argument should they use in their positive apologetics in order to be epistemically rational in the relevant sense (cf. Chapter 5)? Should (e) these methods and types be completely unlike those employed by scientists or historians when they argue for the existence of theoretical entities, or postulate events in the past, or should (f) they rather be similar to the ones scientists and scholars employ? As I said in the *Preface* and argued in Chapter 6.3, both options imply great perils for religious believers. If they opt for (e) on the ground that God, Allah, or Ishvara is completely different from anything in the universe, but fail to spell out reliable and validated methods of investigation (as is always the case), they run the risk of forfeiting credibility in our age of science. Why would existence claims concerning God or gods be exempted from the obligation to justify them by using reliable and validated methods of investigation?

Of course, we should admit that methods of research at the lowest level of generality have to be adapted to the particular topic of investigation, as is the case in scientific and scholarly disciplines (cf. Chapter 6.1–3). With regard to gods, who are defined as persons of a kind, we should prefer methods of communication with them, for example. But at the highest level of generality and from a logical point of view, arguments for the existence of God or gods should be like arguments for the existence of theoretical entities in the sciences. If they opt for (f), however, religious believers will run the risk that their creed is strongly disconfirmed by empirical evidence, or eclipsed by scientific or other secular theories, which explain the adduced evidence better than their creed is able to do.

One of the reasons why I selected the works of Richard Swinburne as the 'toughest case' for a critical book on the philosophy of religion is that in my view, he offers the

---

[2] Cf. Plantinga (2000), p. 200; cf. p. 242.

most sophisticated solution to this third dilemma (e/f), which I called *The Tension* in Chapter 6.5. On the one hand, Swinburne interprets what he labels *bare theism* as a 'large-scale theory of the universe', which closely resembles 'large-scale scientific theories'.[3] And he declares that 'the structure of a cumulative case for theism' is 'the same as the structure of a cumulative case for any unobservable entity, such as a quark or a neutrino'.[4] By stressing these features of his natural theology, he aims at gaining credibility for theism in our age of science, and indeed for his Bayesian inductive method of substantiating it.

On the other hand, however, we saw that Swinburne tries to eliminate in several ways the risks of option (f), that theism will be disconfirmed empirically, or eliminated by secular rival theories, which explain the adduced phenomena better than theism does. He avers, for example, that theism 'does not yield predictions such that we can know only tomorrow, and not today, whether they succeed'.[5] Furthermore, although allegedly the theory of theism 'purports to explain everything logically contingent (apart from itself)', that is, 'all our empirical data', Swinburne also argues that the only empirical evidence that we should adduce in order to confirm theism consists of facts that are either 'too *odd*' or 'too *big*' for science to explain.[6] In this manner, he aims at avoiding so-called God-of-the-gaps arguments (cf. Chapter 10). Finally, he abstains from developing domain-specific methods of theological investigation at the lowest level of generality, such as the research project of showing by means of statistical methods that petitionary prayers said in the past have had positive results more often than not (cf. Chapter 6.3).

In short, Swinburne attempts both to circumvent the Scylla of option (e), that theism lacks credibility in our age of science, because the methods which theologians use in investigating its truth cannot be validated, and the Charybdis of option (f), that whenever natural theologians adduce specific evidence in order to confirm theism within the framework of a cumulative case, this very evidence might be explained better by a rival secular explanation. We may conclude with regard to the third dilemma (e/f) that instead of opting radically for one of its horns, Swinburne goes between them, and this is what all sophisticated theists will attempt to do. Bare theism as interpreted by Swinburne may closely resemble large-scale scientific theories in some respects, but it differs radically from such theories in other respects. For example, no scientist will take seriously a proposed fully developed theory, if it 'does not yield predictions such that we can know only tomorrow, and not today, whether they succeed'. And no scholar or scientist will engage in a field of research if there are no reliable domain-specific methods of investigation at the lowest level of generality (cf. Chapter 6.3).

This result of Part I does not imply, however, that there cannot be a convincing inductive cumulative case showing that the truth of theism is more probable than not on the total available and permissible evidence. Indeed, using the format of Bayesian confirmation theory, Swinburne claims that at the highest level of generality the method of his natural theology is exactly the same as the method scientists use implicitly when they assess the probability that a hypothesis is true. In other words,

---

[3] Swinburne (EG), p. 3.
[4] Swinburne (1983), p. 386.
[5] Swinburne (EG), p. 70.
[6] Swinburne (EG), pp. 66, 93, and 74–5.

at this abstract level the rationality of Swinburne's natural theology allegedly is identical with scientific rationality.

***

Lawyers often use the following argumentative strategy in courts. They first advance an argument for their most sweeping claim, which, if accepted, will leave their opponent empty-handed. Because they know that the other party will not endorse this argument, and some of the judges might also reject it, they then proceed by pleading for a less far-reaching position, and the procedure may be repeated many times. In Parts II and III this method is used again and again, and I shall call it the *Strategy of subsidiary arguments*.

Part II is devoted to the question whether theism really is an explanatory theory or hypothesis, which can be confirmed by empirical evidence. In Chapter 7, I argued that theists implicitly annul the very conditions for meaningfully applying psychological predicates to God by claiming that God is an incorporeal being. As a result, all these predicates, used in the very definition of 'God', have to be interpreted as irreducibly metaphorical or analogical. Furthermore, I argued that the notion of God as a bodiless person is incoherent unless one interprets the term 'person' analogically as well, and I criticized Swinburne's arguments for the coherence of this notion if taken literally.

Swinburne points out correctly that '[i]f theology uses too many words in analogical senses it will convey virtually nothing by what is says'.[7] Accordingly, we should conclude that theists do not succeed in giving any meaning to the word 'God', if indeed most of the terms by which the word is defined are used in an irreducibly analogical manner. Consequently, the question whether God exists cannot even arise.

Let us call this result *semantic atheism*. According to semantic atheism, theists do not succeed in giving any intelligible meaning to the word 'God' when they stipulate, for instance, that the proposition that God exists is logically equivalent to 'there exists necessarily a person without a body (i.e. a spirit) who necessarily is eternal, perfectly free, omnipotent, omniscient, perfectly good, and the creator of all things'.[8] If the word 'God' does not have an intelligible meaning and referential use, the coherence of theism cannot be demonstrated directly or indirectly, as I argued in Chapter 8.5–6, and theism cannot be a theory, which is open to empirical (dis)confirmation.

The term 'atheism' in this context has a strict or narrow sense. Semantic atheism merely claims that the traditional doctrine of theism, according to which God is a bodiless person with mental properties, lacks any intelligible meaning. It does not claim that all definitions of gods are meaningless. For example, the description of the Hindu-god Ganesha as a person with a human body, protruding belly, four (or more) arms, and an elephant's head is not meaningless, although it is very improbable that Ganesha really exists given our empirical knowledge about life on Earth. And if one stipulates that 'God' is a proper name of the Sun, without attributing any other properties to this god than pertain to the Sun according to scientists, the claim that God exists is not only meaningful but also true, although it does not tell us anything interesting or new.

---

[7] Swinburne (CT), p. 72.
[8] Swinburne (EG), p. 7.

Readers who are convinced by the arguments of Part I and of Chapters 7 and 8 of Part II will conclude that the decision tree described above does not contain any viable option for believers in theism. Since the tree provides an exhaustive inventory of all options for believers who want to be reasonable and conscientious, theists who are convinced by these arguments will have to give up either the truth claims concerning their creed, or their reasonableness. It is at this point that I applied the strategy of subsidiary arguments for the first time. In Chapter 9, I assumed counterfactually and for the sake of argument that theism is a meaningful theory, and investigated whether, if so, it can have any predictive power. Even if the notion of predictive power is defined in a minimalist sense, so that a theory has predictive power with regard to a piece of evidence e if and only if the relevance condition $P(e|h\&k)>P(e|k)$ is satisfied, I concluded that theism has no predictive power with regard to any existing e, because all attributions of values to the likelihood $P(e|h\&k)$ for all existing e's are completely arbitrary.[9] I criticized Richard Swinburne's attempt to solve this problem of the predictive power of theism by what I called the *moral access claim*, that is, by a very strong interpretation of meta-ethical objectivism or realism, and I confirmed my criticisms by analysing his attributions of creative intentions to God, which were exposed as anthropomorphic projections.

This result of Chapter 9 entails that all empirical arguments for and against theism based upon existing evidence come to nothing. It means, in more technical jargon, that no correct C-inductive or incrementally confirming argument for (or against) theism can be constructed on the basis of existing evidence. It does not follow, however, that we should be agnostic concerning the truth of theism. For if theism makes sense at all, the prior probability of its truth is low. Indeed, the hypothesis that there is a bodiless person (i.e. a spirit) with many mental powers and properties, if it were meaningful, would be very unlikely to be true, given the background of our growing scientific knowledge of the brain, which shows in ever greater detail that mental life cannot occur without brain processes.

As we saw in Chapter 11.6, Richard Swinburne attempts to eliminate this important objection against theism by arguing that no background knowledge can be relevant to the prior probability of theism, because the predictive scope of theism is universal. If theism really is a 'theory of everything' in the strict sense, all empirical data belong to the evidence to be explained by the theory, and there is no background knowledge left in the light of which one might assess the prior probability of theism. Swinburne concludes that '[i]t will not, therefore, be a disadvantage' to theism 'if it postulates a person in many ways rather unlike the embodied human persons so familiar to us'.[10] However, I argued in Chapter 11.6 that even on the assumption that theism has a global predictive power, Swinburne's argument for the irrelevance of background knowledge commits a fallacy of division. And if we accept the conclusion

---

[9] As I say in a footnote to Chapter 9, this does not imply that one might not *imagine* a piece of evidence e for which theism has some (or a strong) predictive power, such as the stars in the sky suddenly reconfigurating in order to form a Hebrew sentence saying 'I the Lord your God am a jealous God', or something like that. But such pieces of evidence do not occur, and, according to Swinburne, theism does not predict that they will occur. As we saw, he holds that theism 'does not yield predictions such that we can know only tomorrow, and not today, whether they succeed' (EG, p. 70).
[10] Swinburne (EG), p. 66.

of Chapter 9 that theism has no predictive power whatsoever concerning existing evidence, the premise of Swinburne's argument against the relevance of background knowledge is false, because theism has no predictive scope at all. In short, those who accept the conclusion of Chapter 9 should become atheists in a traditional sense.[11]

In contradistinction to semantic atheists, traditional atheists assume that the religious creed(s) which they reject are meaningful. One should make two distinctions between traditional atheistic positions. One is between narrow, local, or particular atheism on the one hand and broad, or global, or universal atheism on the other hand. Particular atheists merely reject theism as defined above, or some other specific religious doctrine, whereas universal atheists reject all existence claims concerning gods. Strictly monotheistic believers are particular atheists with regard to all gods except their preferred deity, such as God or Allah. The other distinction is between weak and strong atheism, and it is concerned with the scope of the logical negation 'a', the *alpha privans* in the term 'a-theism'. Weak atheists are those who do *not believe* that God, or some other god, exists, whereas strong atheists *believe* that God does *not* exist, or that some other god does not exist. Agnosticism is a form of weak atheism, whereas it is incompatible with strong atheism. Clearly, believing Jews, Christians, and Muslims are strong particular atheists with regard to many gods. Indeed, in the Roman Empire the term 'atheism' was used for Jews and Christians, because they wanted to worship their own god exclusively while denying the existence of everyone else's.

Using this terminology, we may say that if theism is meaningful, but does not have any predictive power regarding existing evidence, the argument from scientific background knowledge justifies strong particular traditional atheism with regard to theism as defined above. Although I have not argued this in the present book, arguments from background knowledge also justify strong universal atheism with regard to all traditional gods of which a meaningful description can be given. As I said, the existence of the god Ganesha is very improbable given our biological background knowledge, and something similar holds for all gods that humanity has worshipped, such as Wodan or Zeus. They may exist as cultural symbols, which we might love, but surely they do not exist as living persons at some place in the universe like Mount Olympus. All of us should adore Ganesha as a cultural symbol, because he is the patron of arts and sciences, the *deva* of intellect and wisdom. Unfortunately, monotheistic Jews, Christians, and Muslims lack such an enchanting god.

Combining the results of Chapters 7, 8, and 9, we should become what I call *disjunctive strong universal atheists*. Either religious believers have not succeeded in providing a meaningful characterization of their god(s), or the existence of this god or these gods is improbable given our scientific and scholarly background knowledge.

Anthony Kenny has objected to such a universal atheism that it is too strong, and that it cannot be substantiated. According to Kenny, the (universal) atheist 'says that

---

[11] According to Elliott Sober (2008), §2.12, the Achilles heel of design arguments construed as likelihood arguments is that we lack independently attested information about the intentions of the designer. It follows that empirical arguments both *for* and *against* God's existence come to nothing. But it does not follow that 'scientific theories are silent on the question whether God exists' (p. 188). The reason is that the prior probability of theism is very low given scientific background knowledge concerning the dependence of mental life on brain processes.

no matter what definition you choose, "God exists" is always false', so that 'atheism makes a much stronger claim than theism does'.[12] But these two contentions are mistaken. First of all, (mono)theists typically deny the existence of all gods except their own, so that their claim is at least as strong as the claim of the universal atheist. Second, no atheist will say that 'no matter what definition you choose, "God exists" is always false', for if I call my iPhone 'God', for example, there is no doubt that God, so defined, exists. What the disjunctive strong universal atheist claims is merely that with regard to all descriptions of gods that humanity has provided within the framework of known religions, we should conclude either that they are meaningless or that nothing corresponds to them in reality.

At the beginning of Chapter 10, I applied the strategy of subsidiary arguments for the second time. I assumed counterfactually and for the sake of argument not only that theism makes sense, but also that it has the substantial predictive power which Richard Swinburne attributes to it. He says, for example, that '[t]he theist argues from all the phenomena of experience, not from a small range of them'.[13] If that were really true, however, theism would be threatened by the Charybdis of option (f), described above, that whenever natural theologians adduce specific evidence in order to confirm theism inductively, this very same evidence might be explained better by a rival secular explanation, so that theism is disconfirmed. Swinburne attempts to eliminate this risk of so-called God-of-the-gaps arguments by the following prescription. Theists should adduce facts as evidence for theism only if these facts are either 'too *big*' or 'too *odd*' for secular science and scholarship to explain.

In Chapter 10, I investigated the extent to which this tactic of avoiding an epistemic contest between theism and secular science or scholarship can be successful. Taking the alleged miracle of Christ's resurrection with his Earthly mutilated body as an example of 'too odd', I argued that the tactic fails with regard to miracles. The testimony concerning Christ's bodily resurrection is defective for many reasons, and the existence of this testimony can be explained better by secular theories of cognitive dissonance, collaborative storytelling, and source amnesia, in conjunction with historical background knowledge, than by the hypothesis that the miracle really occurred (Chapter 10.2–5).

The risk of God-of-the-gaps arguments is not excluded either concerning the three main phenomena that allegedly are 'too big' for science to explain: the existence of the universe, the fact that the universe is governed by relatively simple laws of nature, and the fact that the universe is 'fine-tuned' for specific things, such as life (Chapter 10.6, and Chapters 12–13). If this is correct, and if theists really want to exclude a priori empirical disconfirmations of theism, which would occur if the evidence adduced for theism were better explained by rival secular theories, they should not attempt to confirm theism by empirical evidence at all. Again, we have to conclude that the decision tree sketched above does not contain any viable option for theists, if at least they want to avoid God-of-the-gaps arguments.

***

---

[12] Kenny (2006), Chapter 3 ('Why I am not an Atheist'), p. 21.
[13] Swinburne (EG), p. 71.

In Part III of the book, I examined critically the inductive cumulative case for theism. Since inductive arguments may mutually support or weaken each other, they should be studied with their mutual interconnections if one wants to assess their cumulative effect. In order to do justice to such a cumulative case, it is best to investigate the version developed by an individual philosopher who is very prominent in the field of natural theology. This is why I focused in Part III on the cumulative case for theism as elaborated by Richard Swinburne in the second, substantially revised edition of *The Existence of God* (2004). In discussing the C-inductive arguments for and against the existence of God I applied the strategy of subsidiary arguments many times, presupposing the two applications mentioned above.

Given the many detailed results of Chapters 11–14, believers should not nurse high hopes that a cumulative case for theism can succeed, even if one assumes counterfactually and for the sake of argument that theism is a meaningful theory and has predictive power. On the one hand, the arguments based upon evidence e_{1-8}, each of which aims at increasing the probability of theism, failed to do so convincingly. Either they are not *correct* C-inductive arguments for the existence of God because the relevance condition $P(e|h\&k) > P(e|k)$ is not satisfied, so that $P(h|e\&k)$ does not exceed $P(h|k)$, or they are not *good* C-inductive arguments for the existence of God, because the alleged evidence e has not been established. On the other hand, the arguments based upon evidence e_{9a-b}, i.e. the arguments from evil and God's hiddenness, turned out to be rather strong C-inductive arguments against the existence of God.

Evidence e_{10} does not modify this assessment. Swinburne admits that the evidence e_{10a} of miracles consists of public historical events which are 'much disputed'.[14] Consequently, one will expect that C-inductive arguments based upon such evidence will not be *good* C-inductive arguments by definition, since good arguments are defined as 'arguments from premisses known to be true by those who dispute about the conclusion'.[15] Indeed, we saw in Chapter 10.2–5 that the argument from Christ's bodily resurrection as developed in Swinburne's book *The Resurrection of God Incarnate* (2003), fails to be a good C-inductive argument for the existence of God. Swinburne also admits that undisputed historical events (e_{10b}), such as the rise of Christianity in the Roman Empire, or the military successes of early Islam, at best provide 'a bit more evidence for the existence of God' or Allah, so that they will not greatly influence the final balance of probability.[16] Everything considered, then, we should conclude that if theism were a meaningful theory and had predictive power, the total evidence apart from e_{11} would disconfirm theism. Since the prior probability of theism is low on background knowledge, we should become atheists.

Even if these things were otherwise, Swinburne's cumulative case for theism faces insuperable difficulties. In Chapter 15, I applied the strategy of subsidiary arguments again. It was assumed counterfactually and for the sake of argument that theism not only is a meaningful theory and has predictive power, but also that at least some of Swinburne's C-inductive arguments for the existence of God are correct and good C-inductive arguments. Even so, it seems to be unwarranted to conclude from such a

---

[14] Swinburne (EG), pp. 273–4.   [15] Swinburne (EG), pp. 6–7.
[16] Swinburne (EG), p. 277; cf. p. 292: 'a rather small addition to the probability of the existence of God'.

cumulative case that the existence of God is more probable than not, so that it amounts to a good P-inductive argument for theism. Since Swinburne cannot reliably quantify his probabilities, any quantitative input of the Bayesian formula will seem arbitrary, and the output will be arbitrary as well. As I argued in Chapter 15.3, Swinburne's attempt to solve these problems of quantification by avoiding numbers altogether and using relative probabilities only, fails for several reasons.

It is at this point in his campaign for theism that Swinburne uses his argument from religious experience. Clearly, perceptual experiences may drastically change the balance of probability concerning some hypothesis. Since the order of cards does not matter in Poker, 2,598,960 possible distinct hands of 5 cards can be dealt from a deck of 52 cards. Because there are four royal flushes (A, K, Q, J, 10), the probability that one of them is dealt at a time is roughly 0.00015%. I ardently desire that a royal flush is dealt to me, but I would be very surprised if it happened. When I turn my cards, however, lo and behold, I see that I got the royal flush of spades, so that an extreme improbability is changed into a certainty.

On this model of the possible probabilistic impact of perceptions, Swinburne argues that religious experiences drastically change the balance of probability. Indeed, they may shift the burden of proof concerning (a)theism to the unbeliever. As he says, 'unless the probability of theism on other evidence is *very* low, the testimony of many witnesses to experiences apparently of God suffices to make many of those experiences probably veridical'.[17] But does the occurrence of religious experiences (e_{11}) really shift the burden of proof to the unbeliever? In Chapter 15, I argued on several grounds that this is not the case.

In the final analysis, then, we should draw the following conclusions:

1. Theism is not a meaningful theory. So we should become particular semantic atheists.
2. If we assume for the sake of argument that theism is a meaningful theory, it has no predictive power with regard to any existing evidence. Because the truth of theism is improbable given the scientific background knowledge concerning the dependence of mental life on brain processes, we should become strong particular atheists with regard to theism.
3. If we assume for the sake of argument that theism not only is meaningful but also has predictive power, we should become strong particular atheists, because the empirical arguments against theism outweigh the arguments that support it, and theism is improbable on our background knowledge.

In short, we should become strong disjunctive particular atheists with regard to theism. And since I assume that either (1) or (2, 3) apply *mutatis mutandis* to all other gods that humanity has worshipped or still reveres, the ultimate conclusion of this book is that if we aim at being reasonable and intellectually conscientious, we should become strong disjunctive universal atheists.

[17] Swinburne (EG), p. 341 (my italics).

References

Abraham, William J. (1997). 'Revelation and Scripture'. Chapter 74 of *A Companion to Philosophy of Religion*, edited by Philip L. Quinn and Charles Taliaferro. Oxford: Blackwell.
Adler, Jonathan E. (1999). 'The Ethics of Belief: Off the Wrong Track'. *Midwest Studies in Philosophy*, XXIII: 267–85.
Allison, Dale C. (2005). *Resurrecting Jesus. The Earliest Christian Tradition and its Interpreters*. New York: T & T Clark International.
Alston, William P. (1987). 'Functionalism and Theological Language', in Thomas V. Morris, *The Concept of God*. Oxford: Oxford University Press.
Alston, William P. (1988). 'Religious Diversity and Perceptual Knowledge of God'. *Faith and Philosophy* 5: 433–48.
Alston, William P. (1989). 'Foley's Theory of Epistemic Rationality'. *Philosophy and Phenomenological Research* 50: 135–47.
Alston, William P. (1991). *Perceiving God. The Epistemology of Religious Experience*. Ithaca and London: Cornell University Press.
Alston, William P. (1997). 'Swinburne and Christian Theology'. *International Journal for Philosophy of Religion* 41: 35–57.
Alston, William P. (2005). 'Religious Language', Chapter 9 of *The Oxford Handbook of Philosophy of Religion*, edited by William J. Wainwright. Oxford: Oxford University Press, pp. 220–44.
Amundson, Ron, and George V. Lauder (1993–4). 'Function without Purpose: The Uses of Causal Role Function in Evolutionary Biology'. *Biology and Philosophy* 9: 443–69.
Anselm (1965). *St. Anselm's Proslogion*. Translated with an Introduction and Philosophical Commentary by M. J. Charlesworth. Oxford: Oxford University Press.
Aquinas, Thomas (1975). *Summa Contra Gentiles*. Book One: God. Translated with an Introduction and Notes by Anton C. Pegis, FRSC. Notre Dame: University of Notre Dame Press.
Aristotle (1933). *Metaphysics* I–IX, with an English translation by H. Tredennick. The Loeb Classical Library, Vol. 271. Cambridge, Mass.: Harvard University Press.
Aristotle (1934). *Nicomachean Ethics*, with an English translation by H. Rackham. Revised Edition. The Loeb Classical Library, Vol. 73. Cambridge, Mass.: Harvard University Press.
Aristotle (1935). *Metaphysics* X–XIV, with an English translation by H. Tredennick. The Loeb Classical Library, Vol. 287. Cambridge, Mass.: Harvard University Press.
Armstrong, D. M. (1983). *What is a Law of Nature?* Cambridge: Cambridge University Press.
Armstrong, D. M. (1997). *A World of States of Affairs*. Cambridge: Cambridge University Press.
Asch, S. E. (1956). 'Studies of Independence and Conformity'. *Psychological Monographs* 70, whole no. 416.
Audi, Robert (1998). *Epistemology. A Contemporary Introduction to the Theory of Knowledge*. London: Routledge.
Austin, J. L. (1962). *Sense and Sensibilia*, Reconstructed by G. J. Warnock from the Manuscript Notes. Oxford: Oxford University Press.
Ayer, A. J. (1936). *Language, Truth and Logic*. Harmondsworth: Penguin Books Ltd, 1971.
Baker, Deane-Peter (2007). *Tayloring Reformed Epistemology. Charles Taylor, Alvin Plantinga and the de jure challenge to Christian belief*. London: SCM Press.
Barth, Karl (1922). *Der Römerbrief*. Second Edition, Zürich: Theologischer Verlag, 1989.

Barth, Karl (1957). *Church Dogmatics* II. Edited by G. W. Bromley and T. F. Torrance. Edinburgh: T & T Clark.
Batteux, Charles (1769). *Histoire des causes premières, ou Exposition sommaire des pensées philosophiques sur les principes des êtres*. Paris: Saillant.
Bavelas, J. B., L. Coates, and T. Johnson (2000). 'Listeners as co-narrators'. *Journal of Personality and Social Psychology* 79: 941–52.
Beardsworth, Timothy (1977). *A Sense of Presence*. Oxford: Religious Experience Research Unit.
Behe, M. J. (2006). *Darwin's Black Box: The Biochemical Challenge to Evolution* (10th Anniversary Edition). New York: Simon and Schuster.
Bennett, M. R., and P. M. S. Hacker (2003). *Philosophical Foundations of Neuroscience*. Oxford: Blackwell Publishing.
Benson, Herbert, Jeffery A. Dusek, Jane B. Sherwood et al. (2006). 'Study of the Therapeutic Effects of Intercessory Prayer (STEP) in Cardiac Bypass Patients: A Multicenter Randomized Trial of Uncertainty and Certainty of Receiving Intercessory Prayer'. *American Heart Journal* 151: 934–42.
Block, N. J. (1981). 'Psychologism and Behaviorism'. *Philosophical Review* 90: 5–43.
Bloemendaal, P. F. (2006). *Grammars of Faith. A Critical Evaluation of D. Z. Phillips's Philosophy of Religion*. Leuven: Peeters.
Bojowald, Martin (2008). 'Follow the Bouncing Universe'. *Scientific American* 299: 28–33.
Bradley, M. C. (2001). 'The Fine-tuning Argument'. *Religious Studies* 37: 451–66.
Bradley, M. C. (2002). 'The Fine-tuning Argument: the Bayesian version'. *Religious Studies* 38: 375–404.
Bradley, M. C. (2007). 'Hume's Chief Objection to Natural Theology'. *Religious Studies* 43: 249–70.
Braithwaite, R. B. (1955). *An Empiricist's View of the Nature of Religious Belief*. Cambridge: Cambridge University Press.
Brian Pitts, J. (2008). 'Why the Big Bang Singularity Does Not Help the Kalām Cosmological Argument for Theism'. *British Journal for the Philosophy of Science* 59: 675–708.
Broad, C. D. (1939). 'Arguments for the Existence of God. II'. *The Journal of Theological Studies* 40: 157–67.
Carnap, Rudolf (1950). *Logical Foundations of Probability*. Chicago: University of Chicago Press. Second Edition 1962.
Carrazana, E. et al. (1999). 'Epilepsy and Religious Experiences: Voodoo Possession'. *Epilepsia* 40: 239–41.
Carrier, Richard C. (2006). 'Was Christianity Too Improbable to be False?' The Secular Web.
Carroll, Lewis (1895). 'What the Tortoise Said to Achilles'. *Mind*, n.s., 4: 278–80.
Carter, Alan (2005). 'Evolution and the Problem of Altruism'. *Philosophical Studies* 123: 213–30.
Casatelli, Christine (2006). 'Study Casts Doubt on Medicinal Use of Prayer'. *Science and Theology News* 6: 10–11.
Clark, Kelly James (1989). 'The Explanatory Power of Theism'. *Philosophy of Religion* 25: 129–46.
Collins, Francis S. (2006). *The Language of God. A Scientist Presents Evidence for Belief*. New York: Free Press.
Collins, Harry, and Trevor Pinch (1998). *The Golem. What You Should Know about Science*. Second Edition. Cambridge: Cambridge University Press.
Collins, Robin (2009). 'The Teleological Argument: an Exploration of the Fine-tuning of the Universe'. In W. L. Craig and J. P. Moreland (eds.), *The Blackwell Companion to Natural Theology*. Chichester: Wiley-Blackwell.
Colyvan, Mark, Jay L. Garfield, and Graham Priest (2005). 'Problems with the Argument from Fine Tuning'. *Synthese* 145: 325–38.

Conee, E., and R. Feldman (1998). 'The Generality Problem for Reliabilism'. *Philosophical Studies* 89: 1–29.
Congregatio Pro Doctrina Fidei (2000). *Declaratio Dominus Iesus, de Iesu Christi atque Ecclesiae unicitate et universalitate salvifica*. The Vatican.
Cook, Michael (2000). *The Koran. A Very Short Introduction*. Oxford: Oxford University Press.
Cordry, Benjamin S. (2006). 'Theism and the Philosophy of Nature'. *Religious Studies* 42: 273–90.
Cottingham, John, et al. (1988). *Descartes. Selected Philosophical Writings*. Cambridge: Cambridge University Press.
Coyne, Jerry A. (2006). 'Intelligent Design: The Faith That Dare Not Speak Its Name'. In John Brockman (ed.), *Intelligent Thought. Science versus the Intelligent Design Movement.* New York: Vintage Books.
Craig, William Lane (2002). *Philosophy of Religion. A Reader and Guide*. Edinburgh: Edinburgh University Press.
Craig, William Lane, and Quentin Smith (1993). *Theism, Atheism and Big Bang Cosmology*. Oxford: Oxford University Press.
Crombie, A. C. (1979). *From Augustine to Galileo*. London: Heinemann.
Curd, Martin (1996). 'Miracles as Violations of Laws of Nature'. Chapter 9 of *Faith, Freedom, and Rationality*, edited by D. Howard-Snyder and J. Jordan; pp. 171–83. Lanham, MD: Rowman and Littlefield.
Daly, Martin, and Margo Wilson (1988). *Homicide*. New York: Aldine de Gruyter.
Dancy, Jonathan (1986). 'Two Conceptions of Moral Realism'. *Proceedings of the Aristotelian Society* 60: 167–87 (Suppl.).
Darwin, Charles (1859). *On the Origin of Species By Means of Natural Selection*. Edited with an Introduction and Notes by Gillian Beer. Oxford: Oxford University Press, 1996.
Darwin, Charles (1871). *The Descent of Man, and Selection in Relation to Sex*. With an introduction by James Moore and Adrian Desmond. London: Penguin Books, 2004.
Davis, John Jefferson (1987). 'The Design Argument, Cosmic "Fine Tuning," and the Anthropic Principle'. *International Journal for Philosophy of Religion* 22: 139–50.
Dawkins, Richard (2006). *The God Delusion*. London: Bantam Press.
Decety, J. and P. L. Jackson (2004). 'The Functional Architecture of Human Empathy'. *Behavioral and Cognitive Neuroscience Reviews* 3: 71–100.
De Jonge, H. J. (1989). 'Ontstaan en ontwikkeling van het geloof in Jezus' opstanding'. *Té-èf.* Blad van de Faculteit der Godgeleerdheid van de Rijksuniversiteit te Leiden, 18, no. 3: 33–45.
De Jonge, H. J. (1991). 'Jesus' Death for Others and the Death of the Maccabean Martyrs', in idem, *Jewish Eschatology, Early Christian Christology and The Testaments of the Twelve Patriarchs. Collected Essays* (Supplements to Novum Testamentum 63). Leiden: Brill.
De Jonge, H. J. (2002). 'Visionary Experience and the Historical Origins of Christianity'. In R. Bieringer, V. Koperski, and B. Lataire (eds.), *Resurrection in the New Testament*. Festschrift J. Lambrecht. Leuven: University Press and Peeters, pp. 35–53.
Dennett, Daniel C. (2006). *Breaking the Spell. Religion as a Natural Phenomenon*. London: Penguin Books.
Descartes, René (AT). *Oeuvres de Descartes*, publiés par Charles Adam et Paul Tannery. Paris: Léopold Cerf, 1897–1913.
Deschner, Karlheinz (1996). *Abermals krähte der Hahn. Eine kritische Kirchengeschichte*. München: Btb Verlag.
De Waal, Frans (2005). *Our Inner Ape. The Best and Worst of Human Nature*. London: Granta Books.
De Waal, Frans, et al. (2006). *Primates and Philosophers. How Morality Evolved*. Princeton: Princeton University Press.

Dewhurst, Kenneth, and A. W. Beard (1970). 'Sudden Religious Conversions in Temporal Lobe Epilepsy'. *The British Journal of Psychiatry* 117: 497–507.
Doblin, R. (1991). 'Pahnke's "Good Friday Experiment": a Long-term Follow-up and Methodological Critique'. *Journal of Transpersonal Psychology* 23: 1–28.
Donellan, Keith S. (1966). 'Reference and Definite Descriptions'. *The Philosophical Review* 75: 281–304.
Douven, Igor (2006). 'Assertion, Knowledge, and Rational Credibility'. *The Philosophical Review* 115: 449–85.
Drake, Frank, and Dava Sobel (1992). *Is Anyone Out There? The Scientific Search for Extraterrestrial Intelligence.* New York: Delacorte Press.
Drummond, Henry (1896). *The Lowell Lectures on the Ascent of Man.* London: Hodder and Stoughton.
Dummett, Michael (1959). 'Truth'. In *Truth and Other Enigmas.* Cambridge, Mass.: Harvard University Press, 1978.
Durkheim, Émile (1912). *Les formes élémentaires de la vie religieuse.* 5e édition. Paris: Quadrige/Presses Universitaires de France, 2003.
Eagleton, Terry (2006). 'Lunging, Flailing, Mispunching'. Review of *The God Delusion* by Richard Dawkins. *London Review of Books*, 19 October 2006, pp. 32–4.
Earman, John (1992). *Bayes or Bust? A Critical Examination of Bayesian Confirmation Theory.* Cambridge, Mass.: MIT Press.
Earman, John (2000). *Hume's Abject Failure. The Argument against Miracles.* Oxford: Oxford University Press.
Edwards, D. and D. Middleton (1986). 'Joint Remembering: Constructing an Account of Shared Experience through Conversational Discourse'. *Discourse Processes* 9: 423–59.
Ehrman, Bart D. (2009). *Jesus, Interrupted. Revealing the Hidden Contradictions in the Bible (and Why We don't Know About Them).* New York: HarperCollins Publishers.
Ellis, B. (2001). *Scientific Essentialism.* Cambridge: Cambridge University Press.
Ellis, George F. R. (2007). 'Issues in the Philosophy of Cosmology', in J. Butterfield and J. Earman (eds.), *Philosophy of Physics*, pp. 1183–286. Amsterdam: Elsevier/North-Holland Publishing Company.
Everitt, Nicholas (2004). *The Non-existence of God.* London and New York: Routledge.
Festinger, Leon, et al. (1956). *When Prophesy Fails. A Social and Psychological Study of a Modern Group that Predicted the Destruction of the World.* Minneapolis: University of Minnesota Press.
Festinger, Leon (1962). *A Theory of Cognitive Dissonance.* Stanford: Stanford University Press.
Flew, Anthony (1976). *The Presumption of Atheism.* London: Elek/Pemberton.
Flint, Thomas P. (1998). *Divine Providence: The Molinist Account.* Ithaca, NY: Cornell University Press.
Foley, Richard (1987). *The Theory of Epistemic Rationality.* Cambridge, Mass.: Harvard University Press.
Foley, Richard (1993). 'What's to be Said for Simplicity?' *Philosophical Issues* 3: 209–24.
Forrest, P. (2002). Review of *Warranted Christian Belief* by Alvin Plantinga, *Australasian Journal of Philosophy* 80: 109–11.
Franklin, R. L. (1964). 'Some Sorts of Necessity'. *Sophia* 3: 15–24.
Frege, Gottlob (1884). *Die Grundlagen der Arithmetik. Eine logisch mathematische Untersuchung über den Begriff der Zahl.* Breslau: Verlag von Wilhelm Koebner.
Frege, Gottlob (1892). 'Über Sinn und Bedeutung'. *Zeitschrift für Philosophie und philosophische Kritik*, NF 100: 25–50.
Funk, Robert W., and the Jesus Seminar (1998). *The Acts of Jesus: The Search for the Authentic Deeds of Jesus.* New York: HarperCollins.

Gale, Richard M. (1991). *On the Nature and Existence of God.* Cambridge and New York: Cambridge University Press.
Gallese, V. (2001). 'The "Shared Manifold" Hypothesis: from Mirror Neurons to Empathy'. *Journal of Consciousness Studies* 8: 33–50.
Gallese, V. (2006). 'Intentional Attunement. A Neurophysiological Perspective on Social Cognition and its Disruption in Autism'. *Brain Research* 1079: 15–24.
Galton, Sir Francis (1872). 'Statistical Inquiries into the Efficacy of Prayer'. *Fortnightly Review* 12: 125–35.
Gaskin, J. C. A. (1976). 'The Design Argument: Hume's Critique of Poor Reason'. *Religious Studies* 12: 331–45.
Geach, P. T. (1967). 'Good and Evil', in *Theories of Ethics.* Oxford Readings in Philosophy; edited by Philippa Foot. Oxford: Oxford University Press, pp. 64–73.
Geivett, R. Douglas (2003). 'The Evidential Value of Religious Experience'. Chapter 9 of *The Rationality of Theism*, edited by Paul Copan and Paul K. Moser. London and New York: Routledge.
Gonzalez, Guillermo, D. Brownlee and P. D. Ward (2001). 'Refuges for Life in a Hostile Universe'. *Scientific American* 285, no. 4: 60–6.
Gonzalez, Guillermo, and Jay W. Richards (2004). *The Privileged Planet.* Washington: Regnery Publishing.
Gould, O. N., and R. A. Dixon (1993). 'How we Spent our Vacation: Collaborative Storytelling by Young and Old Adults'. *Psychology and Aging* 8: 10–17.
Gould, Stephen Jay (1989). *Wonderful Life: The Burgess Shale and the Nature of History.* New York: Norton.
Gould, Stephen Jay (1997). 'Nonoverlapping Magisteria'. *Natural History* 106: 16–22.
Gould, Stephen Jay (1999). *Rocks of Ages.* New York: Norton.
Gowler, David B. (2007). *What are They Saying about the Historical Jesus?* New York: Paulist Press.
Grünbaum, Adolf (2000). 'A New Critique of Theological Interpretations of Physical Cosmology'. *The British Journal for the Philosophy of Science* 51: 1–43.
Grünbaum, Adolf (2004). 'The Poverty of Theistic Cosmology'. *The British Journal for the Philosophy of Science* 55: 561–614.
Grünbaum, Adolf (2005). 'Rejoinder to Richard Swinburne's "Second Reply to Grünbaum"'. *The British Journal for the Philosophy of Science* 56: 927–38.
Gutenson, Charles E. (1997). 'What Swinburne Should have Concluded'. *Religious Studies* 33: 243–7.
Gwiazda, Jeremy (2009). 'Richard Swinburne, The Existence of God, and Exact Numerical Values'. Published online on 27 June 2009 and in *Philosophia* 38 (2010): 357–63.
Habermas, Gary (2001). 'The Late Twentieth-century Resurgence of Naturalistic Responses to Jesus' Resurrection'. *Trinity Journal* 22NS: 179–96.
Hacker, P. M. S. (1986). *Insight and Illusion. Themes in the Philosophy of Wittgenstein.* Revised Edition. Oxford: Clarendon Press.
Hacker, P. M. S. (1987). *Appearance and Reality. A Philosophical Investigation into Perception and Perceptual Qualities.* Oxford: Blackwell.
Hacker, P. M. S. (1990). *Wittgenstein, Meaning and Mind.* Volume 3 of *An Analytical Commentary on the Philosophical Investigations.* Oxford: Blackwell.
Hacker, P. M. S. (1996). *Wittgenstein's Place in Twentieth-century Analytic Philosophy.* Oxford: Blackwell.
Hacker, P. M. S. (2007). *Human Nature: The Categorial Framework.* Oxford: Blackwell.
Hacking, Ian (1965). *The Logic of Statistical Inference.* Cambridge: Cambridge University Press.
Hacking, Ian (1983). *Representing and Intervening. Introductory Topics in the Philosophy of Natural Science.* Cambridge: Cambridge University Press.

Hahn, Roger (2005). *Pierre Simon Laplace, 1749–1827. A Determined Scientist.* Cambridge, Mass.: Harvard University Press.

Hale, J. R., et al. (2003). 'Questioning the Delphic Oracle'. *Scientific American* 289, no. 2: 66–73.

Hallyn, Fernand (2000). *Metaphor and Analogy in the Sciences.* Dordrecht: Kluwer Academic Publishers.

Hamilton, W. D. (1964). 'The Genetic Evolution of Social Behaviour I and II'. *Journal of Theoretical Biology* 7: 1–16 and 17–52.

Hardin, C. D., and E. T. Higgins (1996). 'Shared Reality: How Social Verification Makes the Subjective Objective'. In R. M. Sorrentio and E. T. Higgins (eds.), *Handbook of Motivation and Cognition*, Vol. 3. New York: The Guilford Press, pp. 28–84.

Hare, R. M. (1989). *Essays in Ethical Theory.* Oxford: Oxford University Press.

Harris, William (1989). *Ancient Literacy.* Cambridge, Mass.: Harvard University Press.

Harrison, Jonathan (1973–4). 'The Embodiment of Mind, or What Use is Having a Body?' *Proceedings of the Aristotelian Society* 74: 33–55.

Hawking, S. W. (1976). 'Breakdown of Predictability in Gravitational Collapse'. *Physical Review* D14: 2460.

Hawking, Stephen W. (1988). *A Brief History of Time. From the Big Bang to Black Holes.* London: Bantam Books.

Heller, Mark (1995). 'The Simple Solution to the Problem of Generality'. *Noûs* 29: 501–15.

Hempel, Carl G. (1983). 'Valuation and Objectivity in Science', in R. S. Cohen and L. Laudan (eds.), *Physics, Philosophy, and Psychoanalysis. Essays in Honor of Adolf Grünbaum.* Dordrecht: Reidel, pp. 73–100.

Hezser, Catherine (2001). *Jewish Literacy in Roman Palestine.* Texts and Studies in Ancient Judaism 81. Tübingen: Mohr-Siebeck.

Hick, John H. (1961). 'Necessary Being'. *Scottish Journal of Theology* 14: 353–69.

Hick, John H. (1978). *The Center of Christianity.* San Francisco: Harper & Row.

Holder, Rodney D. (1998). 'Hume on Miracles: Bayesian Interpretation, Multiple Testimony, and the Existence of God'. *The British Journal for the Philosophy of Science* 49: 49–65.

Hooykaas, Reijer (1984). *G. J. Rheticus' Treatise on Holy Scripture and the Motion of the Earth.* Amsterdam: North-Holland Publishing Company.

Hornsby, Jennifer (1997). *Simple Mindedness. In Defense of Naive Naturalism in the Philosophy of Mind.* Cambridge, Mass.: Harvard University Press.

Howson, Colin, and Peter Urbach (1993). *Scientific Reasoning. The Bayesian Approach.* Second Edition. Chicago: Open Court.

Hume, David (1748). *An Enquiry concerning Human Understanding.* In *Enquiries* etc., reprinted from the posthumous edition of 1777, Third Edition by P. H. Nidditch. Oxford: Clarendon Press, 1975. References are to the marginal sections of this edition.

Hume, David (1757). *The Natural History of Religion.* Edited by A. Wayne Colver (in one volume with Hume 1779). Oxford: Oxford University Press, 1976.

Hume, David (1779). *Dialogues Concerning Natural Religion.* Edited by John Valdimir Price (in one volume with Hume 1757). Oxford: Oxford University Press, 1976.

Hyman, John (2010). 'Wittgenstein'. In Charles Taliaferro, Paul Draper, and Philip L. Quinn (eds.), *A Companion to the Philosophy of Religion.* Second Edition. Oxford: Blackwell.

Jackson, Patrick Wyse (2006). *The Chronologer's Quest. Episodes in the Search for the Age of the Earth.* Cambridge: Cambridge University Press.

James, William (1896). *The Will to Believe and Other Essays.* London: Longmans.

James, William (1902). *The Varieties of Religious Experience* (The Gifford Lectures delivered at Edinburgh 1901–2). Glasgow: Collins, Fount Paperbacks, 1977.

Jeffrey, Richard (1965). *The Logic of Decision.* Second Edition. Chicago: University of Chicago Press, 1983.

Jilek-Aall, Louise (1999). '*Morbus Sacer* in Africa: Some Religious Aspects of Epilepsy in Traditional Cultures'. *Epilepsia* 40: 382–6.
Josephus, Flavius (1966). *Jewish Antiquities*. Greek text and English translation, edited by H. St. S. Thackeray and R. Marcus. The Loeb Classical Library. Cambridge Mass.: Harvard University Press.
Joyce, Richard (2006). *The Evolution of Morality*. Cambridge, Mass.: The MIT Press.
Kant, Immanuel (KdrV). *Kritik der reinen Vernunft. Werke in zehn Bänden*, Band 3 and 4. Darmstadt: Wissenschaftliche Buchgesellschaft, 1968.
Katz, Leonard D. (ed) (2000). *Evolutionary Origins of Morality. Cross-Disciplinary Perspectives*. Bowling Green: Imprint Academic.
Kaufmann, William J., III (1994). *Universe*. Fourth Edition. New York: W. H. Freeman and Company.
Kelemen, Deborah (2004). 'Are Children "Intuitive Theists"?: Reasoning about Purpose and Design in Nature'. *Psychological Science* 15: 295–301.
Kelemen, Deborah, and C. Diyanni (2005). 'Intuitions about Origins: Purpose and Intelligent Design in Children's Reasoning about Nature.' *Journal of Cognition and Development* 6: 3–31.
Kelly, K. (2004). 'Justification as Truth-finding Efficiency: How Ockham's Razor Works'. *Minds and Machines* 14: 485–505.
Kelly, K. and C. Glymour (2004). 'Why Probability Does Not Capture the Logic of Scientific Justification', in C. Hitchcock (ed.), *Contemporary Debates in the Philosophy of Science*. Oxford: Blackwell, pp. 94–114.
Kenny, Anthony (1969). *The Five Ways. St. Thomas Aquinas' Proofs of God's Existence*. London: Routledge and Kegan Paul.
Kenny, Anthony (1992). *What is Faith? Essays in the Philosophy of Religion*. Oxford: Oxford University Press.
Kenny, Anthony (2004). *The Unknown God. Agnostic Essays*. London and New York: Continuum.
Kenny, Anthony (2006). *What I Believe*. London and New York: Continuum.
Kent, Jack (1999). *The Psychological Origins of the Resurrection Myth*. London: Open Gate.
Keynes, J. M. (1921). *A Treatise on Probability*. Cambridge: Cambridge University Press.
Keysers, C. and V. Gazzola (2007). 'Integrating Simulation and Theory of Mind: from Self to Social Cognition'. *Trends in Cognitive Sciences* 11: 194–6.
Kim, Jaegwon (1993). 'Mental Causation in a Physical World'. *Philosophical Issues* 3: 157–76.
Kim, Jaegwon (1998). *Mind in a Physical World*. Cambridge, Mass.: The MIT Press.
Kitcher, Philip (1998). 'Psychological Altruism, Evolutionary Origins, and Moral Rules'. *Philosophical Studies* 89: 283–316.
Kleene, Stephen C. (1952). *Introduction to Metamathematics*. 10th printing, 1971. Amsterdam/New York: North-Holland Publishing Company.
Kleinknecht, K. T. (1988). *Der leidende Gerechtfertigte: die alttestamentisch-jüdische Tradition vom 'leidenden Gerechten' und ihre Rezeption bei Paulus*. Tübingen: J. C. B. Mohr.
Koester, Helmut (1990). *Ancient Christian Gospels*. Harrisburg, PA: Trinity Press International.
Korsgaard, C. M. (1996). *The Sources of Normativity*. Cambridge: Cambridge University Press.
Kotre, J. (1995). *White Gloves: How We Create Ourselves Through Memory*. New York: The Free Press.
Krach, Helge (1996). *Cosmology and Controversy. The Historical Development of Two Theories of the Universe*. Princeton: Princeton University Press.
Kraus, R. M. (1987). 'The Role of the Listener: Addressee Influences on Message Formulation'. *Journal of Language and Social Psychology* 6: 81–98.
Kretzmann, Norman (1997). *The Metaphysics of Theism. Aquinas's Natural Theology in Summa contra gentiles I*. Oxford: Clarendon Press.

Kretzmann, Norman, and Eleonore Stump (1996). 'An Objection to Swinburne's Argument for Dualism'. *Faith and Philosophy* 13: 405–12.
Kripke, Saul (1971). 'Identity and Necessity', in Milton K. Munitz (ed.), *Identity and Individuation*. New York: New York University Press.
Kripke, Saul (1972). 'Naming and Necessity', in Donald Davidson and Gilbert Harman (eds.), *Semantics of Natural Language*. Dordrecht: D. Reidel.
Krucoff, Mitchell, et al. (2005). 'Music, Imagery, Touch, and Prayer as Adjuncts to Interventional Cardiac Care: the Monitoring and Actualisation of Noetic Trainings (MANTRA) II randomised study'. *The Lancet* 366: 211–17.
Kyburg, Henry (1961). *Probability and the Logic of Rational Belief*. Middletown, CT: Wesleyan University Press.
Lakatos, Imre (1976). *Proofs and Refutations. The Logic of Mathematical Discovery*. Edited by John Worrall and Elie Zahar, Cambridge: Cambridge University Press.
Lakatos, Imre (1978a). *The Methodology of Scientific Research Programmes. Philosophical Papers*, Volume 1, edited by John Worrall and Gregory Currie. Cambridge: Cambridge University Press.
Lakatos, Imre (1978b). *Mathematics, Science, and Epistemology. Philosophical Papers*, Volume 2, edited by John Worrall and Gregory Currie. Cambridge: Cambridge University Press.
Landsborough, D. (1987). 'St Paul and Temporal Lobe Epilepsy'. *Journal of Neurology, Neurosurgery, and Psychiatry* 50: 659–64.
Lane Fox, Robin (1988). *Pagans and Christians in the Mediterranean World from the Second Century AD to the Conversion of Constantine*. London: Penguin Books.
Laplace, Pierre Simon (1796). *Exposition du système du monde*, 2 vols. Paris: Cercle Social.
Laudan, Larry (1990). 'Aimless Epistemology?' *Studies in History and Philosophy of Science* 21: 315–22.
Lear, Jonathan (2006). *Radical Hope. Ethics in the Face of Cultural Devastation*. Cambridge, Mass.: Harvard University Press.
Leibniz, Gottfried Wilhelm (1697). 'On the Radical Origination of Things', translated by Leroy E. Loemker (ed.), *Gottfried Wilhelm Leibniz: Philosophical Papers and Letters*, Second Edition, pp. 486–91. Dordrecht (The Netherlands) and Boston: D. Reidel Publishing Company, 1969.
Levine, Michael P. (1990). 'If there is a God, any Experience which seems to be of God, will be Genuine'. *Religious Studies* 26: 207–17.
Lewis, David (1973). *Counterfactuals*. Oxford: Basil Blackwell.
Lewis, David (1986). *Philosophical Papers*. Oxford: Oxford University Press.
Lewis, David (1986a). *On the Plurality of Worlds*. New York: Basil Blackwell.
Lindberg, David C., and Ronald L. Numbers (eds.) (1986). *God and Nature. Historical Essays on the Encounter between Christianity and Science*. Berkeley: University of California Press.
Locke, John (1700). *An Essay Concerning Human Understanding*. Edited with an Introduction by Peter H. Nidditch. Oxford: Oxford University Press, 1975.
Loeb, Don (2007). 'The Argument from Moral Experience'. *Ethical Theory and Moral Practice* 10: 469–84.
Loftus, E. F. (1979). *Eyewitness Testimony*. Cambridge, M.A.: Harvard University Press.
Loftus, E. F. (2003). 'Our Changeable Memories: Legal and Practical Implications'. *Nature Reviews, Neuroscience* 4: 231–4.
Lowder, Jeffry J., and Robert M. Price (eds.) (2005). *The Empty Tomb: Jesus Beyond the Grave*. Buffalo: Prometheus Books.
Luck, Morgan (2005). 'Against the Possibility of Historical Evidence for Miracles'. *Sophia* 44: 7–23.
Lüdemann, G. and A. Ozen (1995). *What Really Happened to Jesus. A Historical Approach to the Resurrection*. London: SCM Press.

Machamer, P., L. Darden, and C. F. Craver (2000). 'Thinking about Mechanisms'. *Philosophy of Science* 67: 1–15.
MacKay, R. H., and R. W. Oldford (2000). 'Scientific Method, Statistical Method and the Speed of Light', *Statistical Science* 15: 254–78.
Mackie, J. L. (1969). 'The Relevance Criterion of Confirmation'. *British Journal for the Philosophy of Science* 20: 27–40.
Mackie, J. L. (1977). *Ethics. Inventing Right and Wrong.* Harmondsworth: Penguin Books.
Mackie, J. L. (1982). *The Miracle of Theism. Arguments for and against the Existence of God.* Oxford: Oxford University Press.
Mahler, Patrick (2006). 'The Concept of Inductive Probability'. *Erkenntnis* 65: 185–206.
Makinson, David (1965). 'The Paradox of the Preface'. *Analysis* 25: 205–7.
Manson, Neil A. (2000). 'There Is No Adequate Definition of "Fine-tuned for Life"'. *Inquiry* 43: 341–52.
Manson, Neil A. (ed.) (2003). *God and Design. The Teleological Argument and Modern Science.* London: Routledge.
Martin, Michael (1986). 'The Principle of Credulity and Religious Experience'. *Religious Studies* 22: 79–93.
Martin, Michael (1990). *Atheism. A Philosophical Justification.* Philadelphia: Temple University Press.
Martin, Michael, and Ricki Monnier (2006). *The Improbability of God.* Amherst, NY: Prometheus Books.
Mawson, T. J. (2002). 'God's Creation and Morality'. *Religious Studies* 38: 1–25.
McAllister, James W. (1996). *Beauty and Revolution in Science.* Ithaca: Cornell University Press.
McFague, Sallie (1982). *Metaphorical Theology.* Philadelphia: Fortress Press.
McGregor, I., and J. G. Holmes (1999). 'How Storytelling Shapes Memory and Impressions of Relationship Events over Time'. *Journal of Personality and Social Psychology* 76: 403–19.
McLeod, Mark (1993). *Rationality and Theistic Belief.* Ithaca, NY: Cornell University Press.
McMullin, E. (1981). 'How Should Cosmology Relate to Theology?' In A. R. Peacocke (ed.), *The Sciences and Theology in the Twentieth Century*, pp. 17–57. Notre Dame, IN: University of Notre Dame.
Mill, John Stuart (1878). *Three Essays on Religion.* New York: Henry Holt and Company.
Miller, Arthur I. (2000). 'Metaphor and Scientific Creativity', in F. Hallyn (2000) *Metaphor and Analogy in the Sciences.* Dordrecht: Kluwer Academic Publishers, pp. 148–64.
Monton, Bradley (2006). 'God, Fine-Tuning, and the Problem of Old Evidence'. *British Journal for the Philosophy of Science* 57: 405–24.
Morris, S. Conway (2003). *Life's Solution: Inevitable Humans in a Lonely Universe.* New York: Cambridge University Press.
Mumford, Stephen (2001). 'Miracles: Metaphysics and Modality'. *Religious Studies* 37: 191–202.
Mumford, Stephen (2005). 'Kinds, Essences, Powers'. *Ratio* (new series) 18: 420–36.
Newton, Sir Isaac (1729). *Philosophiae Naturalis Principia Mathematica.* Andrew Motte's translation of 1729, revised by Florian Cajori. Berkeley: University of California Press, 1962.
Newton-Smith, W. H. (1981). *The Rationality of Science.* London: Routledge and Kegan Paul.
Nietzsche, Friedrich (1881). *Morgenröte. Gedanken über moralischen Vorurteile.* Stuttgart: Alfred Kröner Verlag, 1964.
Nieuwentijt, Bernard (1715). *Het regt gebruik der wereldbeschouwingen, ter overtuiginge van ongodisten en ongelovigen (The Right Use of World Views, Designed to Convince Atheists and Unbelievers).* Amsterdam: Joannes Pauli, 1725.
Novick, N. R. (1997). 'Twice-told tales: Collaborative Narration of Familiar Stories'. *Language in Society* 26: 199–220.

Oderberg, David S. (2002). 'Transversal of the Infinite, the "Big Bang," and the *Kalam* Cosmological Argument'. *Philosophia Christi* 4: 303–34.

Oderberg, David S. (2007). 'The Cosmological Argument', in C. Meister and P. Copan (eds.), *The Routledge Companion to the Philosophy of Religion*. London: Routledge, Chapter 32.

Oppy, Graham (2006a). *Arguing about Gods*. New York: Cambridge University Press.

Oppy, Graham (2006b). *Philosophical Perspectives on Infinity*. New York: Cambridge University Press.

Otte, Richard (1996). 'Mackie's Treatment of Miracles'. *International Journal for Philosophy of Religion* 39: 151–8.

Paine, Thomas (1987). *The Thomas Paine Reader*, edited by Michael Foot and Isaac Kramnik. London: Penguin Books.

Paley, William (1802). *Natural Theology or Evidence of the Existence and Attributes of the Deity, Collected from the Appearances of Nature*. Edited with Introduction and Notes by Matthew D. Eddy and David Knight. Oxford: Oxford University Press, 2006.

Pannenberg, Wolfhart (1973). *Wissenschaftstheorie und Theologie*. Frankfurt a. M.: Surhkamp, 1987.

Parfit, Derek (1984). *Reasons and Persons*. Revised Edition. Oxford: Clarendon Press.

Parsons, Keith M. (1989). *God and the Burden of Proof. Plantinga, Swinburne, and the Analytic Defense of Theism*. Buffalo: Prometheus Books.

Pascal, Blaise (1963). *Oeuvres complètes*. Présentation et notes de Louis Lafuma. Paris: Éditions du Seuil.

Pascal, Blaise (1966). *Pensées*. Translated by A. J. Krailsheimer. Harmondsworth: Penguin.

Pasupathi, M. (2001). 'The Social Construction of the Personal Past and its Implications for Adult Development'. *Psychological Bulletin* 127: 651–72.

Paton, R. C. (1992). 'Towards a Metaphorical Biology'. *Biology and Philosophy* 7: 279–94.

Philipse, Herman (1990). 'The Absolute Network Theory of Language and Traditional Epistemology', *Inquiry* 33: 127–78.

Philipse, Herman (1998). *Heidegger's Philosophy of Being. A Critical Interpretation*. Princeton: Princeton University Press.

Phillips, D. Z. (2005). 'Wittgensteineanism. Logic, Reality and God'. Chapter 18 of *The Oxford Handbook of Philosophy of Religion*, edited by William J. Wainwright. Oxford: Oxford University Press.

Pius XII (1951). 'Una Ora'. *Acta Apostolicae Sedis – Commentarium Officiale* 44 (1952): 31–43.

Plantinga, Alvin (1965). 'The Free Will Defence', in Max Black (ed.), *Philosophy in America*. London: George Allen & Unwin, pp. 204–20.

Plantinga, Alvin (1974). *The Nature of Necessity*. Oxford: Oxford University Press.

Plantinga, Alvin (1982). 'The Reformed Objection to Natural Theology'. Reprinted from *Christian Scholar's Review* 11, no. 3: 187–98, in *Philosophy of Religion. Selected Readings*. Edited by Michael Peterson et al. Second Edition, Oxford: Oxford University Press, 2001.

Plantinga, Alvin (1983). 'Reason and Belief in God'. In *Faith and Rationality. Reason and Belief in God*, edited by Alvin Plantinga and Nicholas Wolterstorff. Notre Dame and London: University of Notre Dame Press, pp. 16–93.

Plantinga, Alvin (1986). 'The Foundations of Theism: A Reply'. *Faith and Philosophy* 3: 298–313. In part reprinted in Craig (2002), pp. 49–56.

Plantinga, Alvin (1993a). *Warrant: The Current Debate*. New York and Oxford: Oxford University Press.

Plantinga, Alvin (1993b). *Warrant and Proper Function*. New York and Oxford: Oxford University Press.

Plantinga, Alvin (1995). 'Pluralism: A Defense of Religious Exclusivism', in *The Rationality of Belief and the Plurality of Faith* (Thomas D. Senor (ed.), Cornell University Press, 1995).

Quoted from the reprint in *The Philosophical Challenge of Religious Diversity*, edited by Philip L. Quinn and Kevin Meeker. Oxford: Oxford University Press, 2000, pp. 172–207.
Plantinga, Alvin (2000). *Warranted Christian Belief.* New York and Oxford: Oxford University Press.
Plantinga, Alvin (2001). 'Rationality and Public Evidence: a Reply to Richard Swinburne'. *Religious Studies* 37: 215–22.
Plantinga, Alvin, and Michael Tooley (2008). *Knowledge of God.* Oxford: Blackwell Publishing.
Pollock, John (1974). *Knowledge and Justification.* Princeton: Princeton University Press.
Pope John Paul II (1998). Encyclical Letter *Fides et Ratio.* Rome: Libreria Editrice Vaticana.
Popkin, Richard H. (1979). *The History of Scepticism from Erasmus to Spinoza.* Berkeley: University of California Press.
Popper, Karl R. (1959). *The Logic of Scientific Discovery.* London: Hutchinson & Co.
Powell, Russell (2009). 'Contingency and Convergence in Macroevolution: A Reply to John Beatty'. *The Journal of Philosophy* 106: 390–403.
Preston, S. D. and F. B. M. De Waal (2002). 'Empathy: Its Ultimate and Proximate Bases'. *Behavioral and Brain Sciences* 25: 1–72.
Priest, Graham (1981). 'The Argument from Design'. *Australasian Journal of Philosophy* 59: 422–31.
Quine, Willard Van Orman (1953). *From a Logical Point of View. 9 Logico-Philosophical Essays.* Cambridge, Mass.: Harvard University Press.
Quine, Willard Van Orman (1969). *Ontological Relativity and Other Essays.* New York: Columbia University Press.
Quinn, Philip L. (1993). 'The Foundations of Theism Again: A Rejoinder to Plantinga', in Linda Zagzebski (ed.), *Rational Faith.* Notre Dame, IN: University of Notre Dame Press.
Ray, John (1691). *The Wisdom of God Manifested in the Works of Creation.* Eighth Edition. London: printed by William and John Innys (1722).
Reichenbach, Hans (1938). *Experience and Prediction.* Chicago: The University of Chicago Press.
Renan, Ernest (1864). *Vie de Jésus.* Paris: Calmann-Levy.
Reppert, Victor (1989). 'Miracles and the case for theism'. *Philosophy of Religion* 25: 35–51.
Ridley, Mark (2004). *Evolution.* Third Edition. Oxford: Blackwell.
Rowe, William L. (1979). 'The Problem of Evil and some Varieties of Atheism'. *American Philosophical Quarterly* 16: 335–41.
Rowe, William L. (1982). 'Religious Experience and the Principle of Credulity'. *International Journal for Philosophy of Religion* 13: 85–92.
Rowe, William L. (1993). *Philosophy of Religion.* Second Edition. Belmont, Calif.: Wadsworth.
Rowe, William L. (2006). 'Friendly Atheism, Skeptical Theism, and the Problem of Evil'. *International Journal for Philosophy of Religion* 59: 79–92.
Rundle, Bede (2004). *Why There is Something Rather than Nothing.* Oxford: Oxford University Press.
Ruse, Michael (1986). 'Evolutionary Ethics: A Phoenix Arisen'. *Zygon* 21: 95–112.
Ruse, Michael (2006). 'Is Darwinian Metaethics Possible?', in G. Boniolo and G. De Anna (eds.), *Evolutionary Ethics and Contemporary Biology.* Cambridge: Cambridge University Press.
Ryle, Gilbert (1949). *The Concept of Mind.* Harmondsworth: Penguin Books, 1963.
Sanders, E. P. (1993). *The Historical Figure of Jesus.* London: Penguin Books Ltd, published in Penguin Books, 1995.
Sarkar, Sahotra (2007). *Doubting Darwin? Creationist Designs on Evolution.* Oxford: Blackwell Publishing.
Saver, J. L., and J. Rabin (1997). 'The Neural Substrates of Religious Experience'. *The Journal of Neuropsychiatry and Clinical Neurosciences* 9: 498–510.

Schellenberg, J. L. (2005a). 'The Hiddenness Argument Revisited (I)'. *Religious Studies* 41: 201–15.

Schellenberg, J. L. (2005b). 'The Hiddenness Argument Revisited (II)'. *Religious Studies* 41: 287–303.

Schellenberg, J. L. (2006). *Divine Hiddenness and Human Reason.* The 1993 edition with a new preface. Cornell Paperbacks. Ithaca and London: Cornell University Press.

Searle, John R. (1992). *The Rediscovery of the Mind.* Cambridge, Mass.: The MIT Press.

Sextus Empiricus (1933). *Outlines of Pyrrhonism.* With an English translation by the Rev. R. G. Bury. The Loeb Classical Library, vol. 273. Cambridge, Mass.: Harvard University Press.

Silver, David (2001). 'Religious Experience and the Facts of Religious Pluralism'. *International Journal for Philosophy of Religion* 49: 1–17.

Smith, Mark S. (2001). *The Origins of Biblical Monotheism: Israel's Polytheistic Background and the Ugaritic Texts.* Oxford: Oxford University Press.

Smith, Quentin (1991). 'An Atheological Argument from Evil Natural Laws'. *International Journal for Philosophy of Religion* 29: 159–74.

Smith, Quentin (1992). 'The Anthropic Coincidences, Evil and the Disconfirmation of Theism'. *Religious Studies* 28: 347–50.

Smith, Quentin (1998). 'Review article. Swinburne's Explanation of the Universe'. *Religious Studies* 34: 91–102.

Smolin, L. (1992) 'Did the Universe Evolve?' *Class. Qu. Grav.* 9: 173–91.

Sobel, John Howard (2004). *Logic and Theism. Arguments For and Against Beliefs in God.* Cambridge: Cambridge University Press.

Sober, Elliott (1988). 'What is Evolutionary Altruism?' *Canadian Journal of Philosophy,* Supplementary Volume 14: 75–99.

Sober, Elliott (1999). 'Testability'. *Proceedings and Addresses of the American Philosophical Association* 73: 47–76.

Sober, Elliott (2002a). 'Bayesianism – its Scope and Limits'. In Swinburne (BT) *Bayes's Theorem.* Edited by Richard Swinburne. Proceedings of the British Academy, 113. Oxford: Oxford University Press, 2002, pp. 21–38.

Sober, Elliott (2002b). 'What is the Problem of Simplicity?' In H. Keuzenkamp, M. McAleer, and A. Zellner (eds.), *Simplicity, Inference, and Econometric Modelling.* Cambridge: Cambridge University Press, pp. 13–32.

Sober, Elliott (2003). 'The Design Argument'. In N. Manson (2003) (ed.) *God and Design. The Teleological Argument and Modern Science.* London: Routledge, pp. 27–54.

Sober, Elliott (2004). 'The Design Argument'. In W. Mann (ed.), *Blackwell Guide to Philosophy of Religion.* Oxford: Blackwell Publishers, pp. 117–47.

Sober, Elliott (2008). *Evidence and Evolution. The Logic Behind the Science.* Cambridge: Cambridge University Press.

Sober, Elliott, and David Sloan Wilson (1998). *Unto Others: The Evolution and Psychology of Unselfish Behavior.* Cambridge, Mass.: Harvard University Press.

Sosa, Ernest (1984). 'Mind–Body Interaction and Supervenient Causation'. *Midwest Studies in Philosophy* 9: 271–82.

Soskice, Janet Martin (1984). *Metaphor and Religious Language.* Oxford: Oxford University Press.

Spong, John Shelby (1994). *Resurrection: Myth or Reality?* San Francisco: Harper Collins.

Stark, Rodney (2007). *Discovering God. The Origins of the Great Religions and the Evolution of Belief.* San Francisco: HarperOne.

Strauß, David Friedrich (1835/36). *Das Leben Jesu, kritisch bearbeitet.* 2 Vols., Tübingen: C. F. Oslander.

Strawson, P. F. (1959). *Individuals. An Essay in Descriptive Metaphysics.* London: Methuen.

Strawson, P. F. (1966). *The Bounds of Sense. An Essay on Kant's Critique of Pure Reason*. London: Methuen.
Strawson, P. F. (1974). *Freedom and Resentment*. London: Methuen.
Street, Sharon (2006). 'A Darwinian Dilemma for Realist Theories of Value'. *Philosophical Studies* 127: 109–66.
Sudduth, Michael Czapkay (2002). 'Plantinga's Revision of the Reformed Tradition: Rethinking our Natural Knowledge of God'. *Philosophical Books*, issue 43: 81–91.
Suppes, Patrick (1957). *Introduction to Logic*. Mineola, New York: Dover Publications, Inc., 1999.
Swinburne, Richard (BT). *Bayes's Theorem*. Edited by Richard Swinburne. Proceedings of the British Academy, 113. Oxford: Oxford University Press, 2002.
Swinburne, Richard (ChrG). *The Christian God*. Oxford: Oxford University Press, 1994.
Swinburne, Richard (CM). *The Concept of Miracle*. London: MacMillan, 1970.
Swinburne, Richard (CT). *The Coherence of Theism*. Revised Edition. Oxford: Oxford University Press, 1993.
Swinburne, Richard (EG). *The Existence of God*. Second Edition. Oxford: Oxford University Press, 2004.
Swinburne, Richard (EJ). *Epistemic Justification*. Oxford: Oxford University Press, 2001.
Swinburne, Richard (ES). *The Evolution of the Soul*. Revised Edition. Oxford: Oxford University Press, 1997.
Swinburne, Richard (FR). *Faith and Reason*. Oxford: Oxford University Press, 1981 (FRa); Second Edition, 2005 (FRb).
Swinburne, Richard (ICT). *Introduction to Confirmation Theory*. London: Methuen, 1973.
Swinburne, Richard (ITG). *Is There a God?* Oxford: Oxford University Press, 1996.
Swinburne, Richard (PPE). *Providence and the Problem of Evil*. Oxford: Oxford University Press, 1998.
Swinburne, Richard (RA). *Responsibility and Atonement*. Oxford: Oxford University Press, 1989.
Swinburne, Richard (RGI). *The Resurrection of God Incarnate*. Oxford: Oxford University Press, 2003.
Swinburne, Richard (RMA). *Revelation: From Metaphor to Analogy*. Oxford: Oxford University Press, 1992.
Swinburne, Richard (ST). *Space and Time*. Second Edition. London: Macmillan, 1981.
Swinburne, Richard (1976). 'The Objectivity of Morality', *Philosophy* 51: 5–20.
Swinburne, Richard (1983). 'Mackie, Induction, and God'. *Religious Studies* 19: 385–91.
Swinburne, Richard (1996). 'The Beginning of the Universe and of Time'. *Canadian Journal of Philosophy* 26: 169–89.
Swinburne, Richard (2001). 'Plantinga on Warrant'. *Religious Studies* 37: 203–14.
Swinburne, Richard (2003). 'For the Possibility of Miracles', in *Philosophy of Religion: An Anthology*, 4. L. Poyman (ed.). Belmont, CA: Wadsworth Publishing Company.
Swinburne, Richard (2005). 'Second Reply to Grünbaum'. *British Journal for the Philosophy of Science* 56: 919–25.
Swinburne, Richard (2006). 'Return to Sender: Trivial Petitions Don't Warrant a Reply'. *Science and Theology News* 6.
Swinburne, Richard (2007). 'A Simple Theism for a Mixed World: Response to Bradley'. *Religious Studies* 43: 271–7.
Swinburne, Richard (2008). 'What Difference does God make to Morality?' In *Is Goodness without God Good Enough?* Edited by R. K. Garcia and N. L. King. Lanham, MD: Rowman & Littlefield.
Swinburne, Richard (2010). 'God as the Simplest Explanation of the Universe'. *European Journal for Philosophy of Religion* 2: 1–24.

Swinburne, Richard (2011). 'Gwiazda on the Bayesian Argument for God'. *Philosophia* 39: 393–6.
Tabor, James D. (2006). *The Jesus Dynasty: The Hidden History of Jesus, His Royal Family, and the Birth of Christianity*. New York: Simon & Schuster.
Taylor, Charles (2007). 'A Different Kind of Courage'. *The New York Review of Books*, 54.7: 4–8.
Theissen, Gerd, and Annette Merz (1996). *Der historische Jesus. Ein Lehrbuch* (Göttingen: Vandenhoeck & Ruprecht). Translated by John Bowden as *The Historical Jesus: A Comprehensive Guide*. Minneapolis: Fortress, 1998.
Thiselton, Anthony C. (2000). *The First Epistle to the Corinthians. A Commentary on the Greek Text*. Grand Rapids, Michigan/Cambridge, UK: William B. Eerdmans Publishing Company.
Tien, D. W. (2004). 'Warranted Neo-Confucian Belief: Religious Pluralism and the Affections in the Epistemologies of Wang Yangming (1472–1529) and Alvin Plantinga'. *International Journal for Philosophy of Religion* 55: 31–55.
Timmer, John (2006). 'Scientists on Science: Explanatory Power and Predictions'. *Ars Technica*, http://arstechnica.com/journals/science.ars/2006/9/16/5315.
Trivers, R. (1971). 'The Evolution of Reciprocal Altruism'. *Quarterly Review of Biology* 46: 35–57.
Tversky, H. L., and E. J. Marsh (2000). 'Biased Retelling of Events yield Biased Memories'. *Cognitive Psychology* 40: 1–38.
Ussher, James (1650). *Annales veteris testamenti, a prima mundi origine deducti una cum rerum Asiaticarum et Aegyptiacarum chronico, a temporis historici principio usque ad Maccabaicorum initia producto*. London: J. Crook & J. Balier.
Van Bendegem, Jean Paul (2000). 'Analogy and Metaphor as Essential Tools for the Working Mathematician'. In Hallyn (2000) *Metaphor and Analogy in the Sciences*. Dordrecht: Kluwer Academic Publishers, pp. 105–24.
Van Fraassen, Bas C. (1980). *The Scientific Image*. Oxford: Oxford University Press.
Van Inwagen, Peter (1991). 'The Problem of Evil, the Problem of Air, and the Problem of Silence'. *Philosophical Perspectives* 5, *Philosophy of Religion*: 135–65.
Van Inwagen, Peter (2006). *The Problem of Evil. The Gifford Lectures Delivered in the University of St Andrews in 2003*. Oxford: Oxford University Press.
Victor, Jeffrey (1993). *Satanic Panic. The Creation of a Contemporary Legend*. Chicago: Open Court.
Vogelstein, Eric (2004). 'Religious Pluralism and Justified Christian Belief: A Reply to Silver'. *International Journal for Philosophy of Religion* 55: 187–92.
Wagenaar, Willem Albert, and Hans Crombag (2005). *The Popular Policeman and Other Cases*. Amsterdam: Amsterdam University Press.
Wald, Robert M. (1984). *General Relativity*. Chicago, IL: University of Chicago Press.
Ward, Peter D., and Brownlee, Donald (2000). *Rare Earth*. New York: Copernicus Springer-Verlag.
Wasson, R. G. (1961). 'The Hallucinogenic Fungi of Mexico: An Inquiry into the Origins of the Religious Idea among Primitive Peoples'. *Botanical Museum Leaflets*, Harvard University 19(7): 137–62.
Wasson, R. G. et al. (1986). *Persephone's Quest: Entheogens and the Origins of Religion*. Ethnomycological Studies No. 10. New Haven: Yale University Press.
Weisberg, Jonathan (2005). 'Firing Squads and Fine-Tuning: Sober on the Design Argument'. *British Journal for the Philosophy of Science* 56: 809–21.
Weldon, M. S., and K. D. Bellinger (1997). 'Collective memory: Collaborative and Individual Process in Remembering'. *Journal of Experimental Psychology: Learning, Memory, and Cognition* 23: 1160–75.
White, Alan R. (1982). *The Nature of Knowledge*. Totowa: Rowman & Littlefield.

White, Andrew Dickson (1896). *A History of the Warfare of Science with Theology in Christendom*. Buffalo: Prometheus Books, 1993.
Willard, Julian (2003). 'Plantinga's Epistemology of Religious Belief and the Problem of Religious Diversity'. *The Heythrop Journal* 44: 275–93.
Williams, Bernard (1970). 'The Self and the Future'. *Philosophical Review* 79: 161–80.
Williams, D. C. (1966). *Principles of Empirical Realism*. Springfield, IL: Charles C. Thomas.
Williamson, Timothy (2000). *Knowledge and its Limits*. Oxford: Oxford University Press.
Wilson, David Sloan (2002). *Darwin's Cathedral. Evolution, Religion, and the Nature of Society*. Chicago and London: The University of Chicago Press.
Wilson, David Sloan (2007). *Evolution for Everyone. How Darwin's Theory Can Change the Way We Think About Our Lives*. New York: Bantam Dell (Random House, Inc.).
Wilson, David Sloan, and Edward O. Wilson (2007). 'Rethinking the Theoretical Foundation of Sociobiology'. *The Quarterly Review of Biology* 82: 327–48.
Wilson, M., and M. Daly (1997). 'Life Expectancy, Economic Inequality, Homicide, and Reproductive Timing in Chicago Neighborhoods'. *British Medical Journal* 314: 1271–8.
Wittgenstein, Ludwig (1921). *Tractatus Logico-Philosophicus*. With a new edition of the Translation by D. F. Pears and B. F. McGuinness. London: Routledge & Kegan Paul, 1974.
Wittgenstein, Ludwig (1953). *Philosophische Untersuchungen/Philosophical Investigations*. Translated by G. E. M. Anscombe. Oxford: Blackwell.
Wittgenstein, Ludwig (1966). *Lectures and Conversations on Aesthetics, Psychology and Religious Belief*. Edited by C. Barrett. Oxford: Blackwell.
Wittgenstein, Ludwig (1980). *Culture and Value*. Edited by G. H. von Wright in collaboration with H. Nyman, translated by P. Winch. Oxford: Blackwell.
Wohlers, Michael (1999). *Heilige Krankheit. Epilepsie in antiker Medizin, Astrologie und Religion*. Marburg: N.G. Elwert Verlag.
Wolterstorff, Nicholas (2005). 'Religious Epistemology'. Chapter 10 of *The Oxford Handbook of Philosophy of Religion*, edited by William J. Wainwright. Oxford: Oxford University Press.
Wright, N. T. (2003). *The Resurrection of the Son of God*. London: Society for Promoting Christian Knowledge.
Wynn, Mark (1993). 'Some Reflections on Richard Swinburne's Argument from Design'. *Religious Studies* 29: 325–35.
Zangwill, Nick (2004). 'The Myth of Religious Experience'. *Religious Studies* 40: 1–22.
Zimmerman, Dean (2010). 'From Property Dualism to Substance Dualism'. Forthcoming in *The Proceedings of the Aristotelian Society*, Supp. Vol. LXXXIV.
Zuckerman, Ben, and Matthew A. Malkan (1996). *The Origin and Evolution of the Universe*. Boston: Jones and Bartlett Publishers.

Index

Abraham, William 29
accommodation 9, 10
Adler, Jonathan E. 201n.37
Age of Reason, The (Paine) 22
agnosticism 17, 305–6, 316, 343
AIDS epidemic 297
Akaike, Hirotugu 215
Akhenaton, Pharaoh 38, 84
alethic moral objectivism 154, 154n.49
Alhazen 249
Allison, Dale C. 5n.4
Alston, William P. 328–9n.59
altruism 288–9
American Heart Journal 82
analogy 95–7, 103–9
analytic 104
Analytical Behaviourism *see* Logical Behaviourism
Anselm of Canterbury 302–3
anthropic principle 259–60
apologetics xii, 59
Aquinas, Thomas 20–1, 20n.2, 77, 135, 157n.62
arguments *see also* inductive arguments
　abductive 27
　cosmological 146, 221–7, 234–8, *see also* cosmological arguments
　deductive xiii, 221, 222–3
　design 257
　for coherence 107–8, 109–13, 136–9
　from beauty 276–8
　from coincidence 78
　from consciousness 279–81
　from evil 292–302
　from fine-tuning 183, 187, 270–6
　from hiddenness 302–9
　from measurement 281–3
　from moral awareness 286–90
　from order 256
　from providence 290–2
　from religious experience 310–11, 318–29, 346
　inductive xiii, 221, 222–3, 345
　to the best explanation 27, 62–3
Aristotle
　God contemplating his own contemplation 157
　and infinities 225, 225n.9
　and light 249
　model of real knowledge 77
　a monotheistic god 149
　natural theology 19–20, 22
　and necessity 129
　virtuous conduct 156

Armstrong, David 262–3
asteroids 210
atheism 148n.27, 303–5, 309, 318, 335, 341, 343, 346
'Atheism, Theism, and Big Bang Cosmology' (Smith) 244
auxiliary theories 84–5
Ayer, A. J. 25, 34n.10
Aztecs 331

Baal (god) 82
Babbage, Charles 171
background knowledge 147–8, 207–12, 217, 226
bacterial flagellum 269
Balance of Probability, The (Swinburne) 315
Barth, Karl 28, 105
Batteux, Charles 165
Bavinck, Herman 32
'Bayes, God, and the Multiverse' (Swinburne) 211n.63
Bayes' theorem
　application of 191, 203–5, 204n.46, 207, 221, 315
　confirmation theory 313, 340
　cosmological arguments 226, 240, 254
　and likelihoodism 201–4
　Principle of Indifference 246
　Swinburne's form 315, 346
Behe, Michael 269
'Being' 85
beliefs
　basic 32, 70–1
　derived 70–1
　moral 289–90
　properly basic 33–4
Bible, the
　a communication from God 7–8, 58
　Deuteronomy (21.18–21) 12
　Genesis 38
　gospel according to *Mark* 5
　gospel according to *Matthew* 5–7
　incredible stories 8
　Isaiah (45.15) 302
　Kings 18.20–46 82
　Matthew 1.20–21 325
　Paul's letter to the *Corinthians* (15.35–50) 5
　Paul's letter to the *Romans* (I.18–20) 4, 7, 221
　for practical guidance 22
　and theology 3, 7
　Zechariah (Zech.9.9) 180

Big Bang, the
  development of the universe 187, 194–8, 224, 229–30, 275
  and God 238–9, 241–3, 252, 276, 316
  hot theory 125, 183–5, 206, 228, 230, 264
  and Pope Pius XII 238
  singularity 242–6, 248, 250–1, 252–5, 275, see also cosmological arguments; universe the
black holes 136, 210, 271
blood-clotting system 269
bodiless spirits 208, 208n.57
Bohr, Niels 96
Bondi, Hermann 184
Book of Mormon. see Mormon, Book of
Boreas (thermometer) 45–6
brain events 279–83, 285–6
Braithwaite, R. B. 26
Buddhism 80n.4

Calvin, John 22, 36
Calvinism 46–7, 52
Cantor, Georg 224, 226, 233, 242
Carnap, Rudolf 25
Cartesian dualism see substance dualism
causality 165, 223, 232, 236–7, 263, 281
Christ, Jesus
  burden of proof 310, 325
  cognitive dissonance 179–82, 182n.81
  as God incarnate 13–14, 49–50
  kingdom of God 10, 10n.15–16
  resurrection of 5, 5n.3, 170–8, 170n.31/34, 174n.50, see also Christianity; God; Holy Spirit
Christian God, The (Swinburne) 121
Christianity
  Aquinas/Calvin Model 40, 42n.37, 43
  content of revelations 11
  core message 10
  eternal truths of 9, 52
  expansion of 339
  factual religious knowledge 83
  and God 7, 22, 63, 145
  the great central fact 5
  and Greek philosophy 20
  and the Holy Spirit 62
  metaphysics 25
  and natural evil 59
  and neutralizers 54
  original sin 39, 41
  and Pyrrhonian crises 13
  rational/irrational beliefs 59–61
  and the resurrection 181, 181n.79, 344–5
  second belief 51
  and *sensus divinitatis* 58
  and theism/monotheism 56, 59, 63–4
  Trinitarian 121, 154n.51
  truth of 338
C-inductive arguments see inductive arguments

cognitive dissonance xii, 178–82
cognitive interpretation xiv, 86
Coherence of Theism, The (Swinburne) 89, 95, 103–4, 108, 113, 117, 120, 126, 135, 137, 140, 150–1
coherent statements 107, 109
  analytic 104
  synthetic 104
collaborative storytelling 178, 180
collective causality principle 183, 232–3
Collins, Francis 286–9
condition of independence see independence, condition of
condition of non-arbitrariness see non-arbitrariness, condition of
Conee, E. 45n.5
confirmation theory 77
conscience 286
Copernicus, Nicolaus 9–10n.9, 312
coral tiger lilies 57
Cordry, Benjamin S. 278n.82
cosmological arguments
  classification 221–7
  fine-tuning 312
  and God 146, 222, 226–9, 234–8
  God and the Big Bang 238–43, 252–5
  kalam 224
  old universe to God 231–6, 235n.42
  scenarios 227–31, 263–4
  singularity 243–51, 265–6
  synchronic argument 236–8, see also Big Bang, the; universe, the
cosmological constant 270–1
Countable Additivity Principle 246, 247n.85
Craig, William Lane 224–6, 233
creation, the 86–7
Creator, the see God
credal sentences 104–5
Credulity Principle 317–18, 317n.23, 318–21, 321–6, 324n.45/48, 329, 334, 337
Cretaceous-Tertiary mass extinction 295
Critique of Pure Reason (Kant) 24
crypto-atheism 88, see also atheism
cumulative case strategy 256, 279, 311–15, 345
curve-fitting problem 213–15, 214n.72

D_2O see heavy water
Dark Ages 20
dark energy 277
dark matter 277
Darwin, Charles
  acquisition of a moral sense/conscience 155–6
  model of artificial selection 96
  and moral awareness 288–9
  and Swinburne 195
  theory of descent 163, 268–9
  theory of evolution 25
  theory of selection 285

Dawkins, Richard 66, 252
decision tree xiv–xv, 339–40, 342
defence *see* God
Dennett, Daniel 16, 37
Derrida, Jacques 85
Descartes, René
  dualism 98–100, 98n.7/9, 114–15
  first *Metaphysical Meditation* 17–18
  foundationalism 33
  and God's intentions 199
  human organisms as machines 96
  laws of motion/vortex theory 163, 219
  and material entities 129
  and matter 284n.28
  omnipotence/omniscience 250, 250n.104
  and speed of light 249–50
  true universal first principles 77
Descent of Man, The (Darwin) 155, 288
design arguments 257
  global 257
  local 257
de Sitter, Willem 183, 228
Deus absconditus 302
De Waal, Frans 155, 155n.53, 289
diachronic cosmologial scenarios
  others 229n.27
  scenario 1 229–31, 238, 263
  scenario 2 229–31, 263
  scenario 3 229–31, 238, 263
Dialogues Concerning Natural Religion (Hume) 228, 233, 267
dilemma xiv, xv, 8, 62, 80, 84–5, 86–8, 108, 162, 165, 167–8, 265–6, 272, 339–40
Dionysius 157, 157–8n.64
disciples, the 172, 179
diversity, religious *see* religion, diversity
Divine Hiddenness and Human Reason (Schellenberg) 303
divine ineffability 28
doctrina sacra 21
double causality doctrine 184
Drake, Frank 196n.18
Drake equation 196n.18
Drummond, Henry 164
dualism *see* substance dualism
Dummett, Michael 151
Durkheim, Emile 84

Earman, John 167n.20
Eddington, Arthur 204
Eddington lecture 26
eidetic intuition 129
Einstein, Albert 138, 183, 228
Elijah (prophet) 82
Empedocles 249
Enquiry Concerning Human Understanding (Hume) 170
Epicureanism 19
epilepsy 81, 81n.7

episteme 77
Epistemic Justification (Swinburne) 214
Ethics (Spinoza) 22
Euclidean geometry 27
Everitt, Nicholas 261, 263, 299
evidentialism 31, 54, 59, 60, 199–201, 200n.35
evil 59, 292–3, 294–7, 297–302, 307–9
  free-will defence 297–9
  higher-order goods defence 294–6, 296n.74, 305–6
evolutionary altruism 288
Evolution of the Soul, The (Swinburne) 117–18
Existence of God, The (Swinburne) 89, 123, 132, 137, 140, 150, 157, 201, 205, 208, 213, 221, 226, 290, 294, 305–6, 310, 324, 337, 345
explanation 140, 142, 192–4
  complete 193
  diachronic 193, 229
  evolutionary 284–6
  full 192
  personal 142, 146–8, 192–3
  religious 61–4
  scientific 142, 166, 169
  secular 61–4
  synchronic 193
  theistic 177, 181–2
  ultimate 192–9
explanatory power 143–4
externalism xv, 44, 47, 57, 67–70
extrinsic neutralizers
  see neutralizers, extrinsic

faith, leap of 16
Faith and Reason (Swinburne) 72
fall, the 84
falsificationism 77
Feldman, R. 45n.5
Festinger, Leon 178–9
fine-tuning 187, 270–8, 292, 311–12
first-cause argument 223–5
First Mover 19–20
Five Ways (Aquinas) 20–1
Flavius Josephus 172
foundationalism
  classical 33–4
  modern 33
free choice 158, 307
free-will 297–9
Frege, Gottlob 133, 262
Freud, Sigmund 37, 60

Galileo Galilei 208
Ganesha (Hindu-god) 60, 341, 343
Geach, Peter 253
generality problem 44–6
Generatio Spontanea Mundi hypothesis (GSM) 246, 248, 250–1, 252

GIGO principle (Garbage In, Garbage
    Out) 252
Glymour, C. 215
God
  accepting a (fake) death penalty 182
  all knowing 101–2
  an incorporeal being 341
  Aquinas/Calvin Model 36–42, 42n.37
  belief in 55–6
  and the Bible 7–8
  and the Big Bang 238–9, 241–6, 252–5,
    276, 316
  a bodiless person 109–10, 109n.49, 114–9,
    148–9, 148n
  causal theory of proper names for 130–1
  cause of existence 198–9, 199n.30
  and Christianity 7, 22, 53, 58, 63, 145,
    176–7
  C-inductive argument for existence
    of 280–1, 299
  contemplating his own contemplation 157
  and cosmological arguments 146, 222,
    226–9, 234–8
  created the world 231
  and creation of animate beings 158–60,
    195–7, 197n.23, 211, 244–5
  as defined by theism 87–8, 137, 139, 141
  and evil 292–3
  existence argument 66, 191, 203–6, 207–9,
    220, 221, 339, 343–5
  existence of 29–30, 43, 46–8, 53, 61–3,
    144–5, 192–4, 285–6, 315–18
  experience of 311–12
  explanation for biological phenomena 163
  false messages from 10–12
  and fine-tuning 270, 274–5, 277
  and freedom 149, 149n.28, 299–301
  and God-of-the-Gaps 162–6, 169, 344
  hiddenness of 83–4, 84n.11, 302–9
  intentions of 195
  Kingdom of 173
  and the Koran 177
  and the maxima argument 21–2
  metaphorical terms 103
  mind unfathomable 85
  miniessential kind 133–6
  and miracles 166–9, 171–2
  monotheistic 332
  and morality 150–1, 150n.34, 154–6,
    156n.57, 286–7
  and natural evil 59, 292
  necessity of 120–5, 134
  and Newton 25
  no literal description possible 96–7
  not a despotic autocrat 4–5
  in the noumenal world 28
  omniscient/omnipotent/free 250–1
  only existing deity 52, 52n.22, 62
  ontological necessity of 125–30, 127n.28–9
  perceptual experiences of 318–21, 321–30,
    333–7
  a personal ground of being 132
  and person-terminology 135
  physico-teleological argument 23
  and Plantinga 32–4, 33n.8, 36
  properties of 127–30, 127n.28–9
  providential argument 290–2
  and spatial order 266–8
  statement of existence 107–8
  'super-miracle' of a bodily resurrection 175,
    175n.52
  and temporal order 258–61, 259n.13, 265
  and theism 87–8, 104, 161–2, 240
  and theology 3
  threshold for belief of existence 143, 201
  and the universe 140, 150n.32, 184–5,
    186n, 187, 248, see also Christianity;
    Christ, Jesus; Holy Spirit
God Delusion, The (Dawkins) 66
God-of-the-Gaps
  conclusions 340, 344
  cosmological arguments 227, 229, 236,
    238–9, 244
  cul-de-sac of 182
  dilemma of 162–6, 164
  and God 162–6, 169
  order/design arguments 255, 257–8,
    269–70, 276
  and science 184–8
  stratagems to deal with 167
  ultimate explanation/prior probability 191,
    195, 197–8, 207, 211
Gold, Thomas 184
good-father metaphor 298
Gould, Stephen Jay 85, 196–7, 196n.20,
    197n.22
grand strategy 76, 90–1
grave, empty 5, 174
gravity 271
  waves 79
gullibilism 329–30
Gwiazda, Jeremy 314n.13

H_2O see water
Hacker, P. M. S. 110n.51, 112n.57
Hacking, Ian 78, 202
Hamilton, W. D. 288
Hare, Richard 151
Harrison, Jonathan 111
Hartle-Hawking no-boundary conditions 243
heaven 299–302
heavy water 133
Hegel, Georg Wilhelm Friedrich 25
Heidegger, Martin 85, 85n.16
Heller, Mark 45n.5
Hesperus 129–30, 133
Hick, John 15n.25, 28, 125
Hilbert, David 225

Hilbert's Grand Hotel
    (thought experiment) 225
Hinduism 4, 9, 13, 52, 56, 56n.32, 62, 83
Histoire des causes premières (Batteux) 165
historical biblical criticism 7
History of the Warfare of Science with Theology in Christendom, A (White) 11
Holocaust, the 296–7, 298
Holy Spirit
    and Christian beliefs 62
    and God 40–1, 43
    internal instigation 51–3, 58, 60, 68
    and Plantinga 63n.50
    and *sensus divinitatis* 46, 48, 55–6, 62,
    see also Christianity; Christ, Jesus; God
homicide 300–1
Homo sapiens 196
homosexuality 12
Hoyle, Fred 184
Hubble, Edwin 183, 230
humanly free agents 158–9
Hume, David
    and the bodily resurrection 176, 176n.57, 178
    combining words/ascribing properties 105
    cosmological arguments 228, 230, 233–4, 236
    design arguments 267
    God neither omniscient nor omnipotent 250–1
    laws of nature 262–3
    miracles 168n.21, 170–1, 170n.29
    and monotheism 309
    natural history of religion 61–2
    and natural theology 16, 23–7, 23n.5, 68
    use of pain 286
Husserl, Edmund 129, 323
Huygens, Christian 137

independence, condition of 147–8, 147n.22
Indifference Principle 247
Individuals (Strawson) 110
inductive arguments
    C 144, 221–3, 227, 237, 238–41, 252, 258, 266, 270, 275, 277, 279–83, 279n.1, 286–7, 290–3, 294, 299–302, 304, 306–14, 317, 335, 342, 345–6
    P 144–8, 201, 222, 256, 279, 313–14, 315–16, 346
infinity 224–6, 225n.9, 226n.12, 229, 248–50, 253
intelligent design 267n.40, 269
internalism/externalism 44–6, 44n.4, 48, 57, 67–70
interpretation 4, 7, 8, 9, 86–7
intrinsic neutralizers *see* neutralizers, intrinsic
Inverted Lewis Carroll Fallacy 152
Islam
    eternal truths of 9
    factual religious knowledge 83
    and God 52, *see also* Muhammad; Muslims
Israel 38
Is There a God? (Swinburne) 89

James, William 25, 66, 326, 331–2
Jesus Christ *see* Christ, Jesus
Jesus Seminar 174n.50
Jewish Antiquities (Flavius Josephus) 172
John Templeton Foundation 82
Joyce, Richard 156
Judaism 83, 304, 304n.108
Justus of Tiberias 172

kalam cosmological argument 224
Kant, Immanuel 16, 23–8, 77–8, 163n.10, 287, 299
Kaufmann, William J. III 242
Kelly, K. 215
Kenny, Anthony 20, 31n.2, 42n.37, 102, 102n.26, 343–4
Kepler, Johannes 163, 208
Keynes, J. M. 91
Khmer Rouge 298
Kierkegaard, Søren 25, 86
Kim, Jaegwon 98n.9
Koran, the
    content of revelations 11
    contradictions 6, 8–10
    production a miracle 177
    and theology 3
Korsgaard, Christine 151–2n.38
Kretzmann, Norman 29, 157, 157n.62
Kripke, Saul 129–34

Lakatos, Imre 77, 77n.1, 83–4, 89–91
Lancet, The 82
Lane Fox, Robin 338
Language of God, The (Collins) 286
Laplace, Pierre Simon 25, 163, 165
Last Judgement 87
Lateran Council, fourth 231
Laws (Plato) 221
laws of nature 167, 262–3
Leibniz, Gottfried Wilhelm 223, 223n.3, 232–3
Lewis, David 168n.22, 262
light 137
    wave/particle 138
likelihood law (Hacking) 202–4, 202n.41
Locke, John 129
Logical Behaviourism 99
Logical Positivism 99–100
logical probability 142–3n.9
Luther, Martin 13, 22
Lutheran Synod of Missouri 8

Mackie, J. L. 168n.23, 170n.29, 336
Mantra II research 82

Marion, Jean-Luc 85
Mars 208
Marx, Karl 37, 59–60
Mary Magdalene 174–5, 175n.54
mathematics 77, 87–8
Mawson, T. J. 155n.52
measurement 281–3
Melanchthon, Philipp 22
mereological fallacy 209
Merleau-Ponty, Maurice 100
metaphor 95–6
Metaphysical Foundations of Natural Science, The (Kant) 24
Metaphysical Meditations (Descartes) 17
metaphysical necessity 121
metaphysical time 243n.70
metaphysics 20, 24, 25
Metaphysics (Aristotle) 19
Metaphysics of Theism, The (Kretzmann) 157
methods of research 76–80, 81
microreduction 283–4
microscopes 78
MIG-hypothesis 250–1, 250n.106
Mill, John Stuart 296
miniessential kind 131–2
miracles 166–170
 of Christ's resurrection 5, 170–182,
  see also resurrection
Möbius strip 137, 139
'Modernizing the Case for God' (*Time*) 26–7
Mohammed 145, 149, *see also* Muslims
monotheism
 and atheism 343
 compared to polytheism 61, 218
 design argument 267
 fine-tuning 274
 and God 84, 102, 109, 130, 316, 337
 and heaven 301–2
 and Judaism 304
 natural theology 5, 38–9, 56, 60–2
 one necessary being 226
 and theism 140, 146, 246
 truth of 332
 and use of prayer 83
moral access claim 150–2, 154–6, 342
Moral Law 286–7
moral objectivism 151–6, 287
Mormon, Book of 3
Muhammad 7, 331, *see also* Islam; Muslims
multiverse 258–9
Muslims
 God not incarnated 11–12, 49–50
 and God's incarnation in Jesus 177
 Mohammed as God's messenger 145
 and polytheism 6
 sensus divinitatis of 53
 and unbelievers 6,
  see also Islam; Muhammad; Muslims

naive realism concerning perception 57nt.35
Natural Collapse into Nothingness Principle
 (PNCN) 238, 238n.48
natural (or rational) theology
 Aquinas/Calvin Model 36–40, 46–7
 definition 14–15
 downfall/resurrection 25–30
 early history 19–21
 four conditions 15–18
 four dichotomies 67–71
 generality 44–7
 intrinsic neutralizers 54–9
 later history 21–4
 methodological dilemma 80–5
 negative apologetics 59–64
 Plantinga's positive tactic 40–2
 rationality *see* rationality
 reformed objection to 31–4
 refutation of reformed objection 43–4
 religious diversity 48–54
 research methods 76–8
 six reasons 3–5, 8, 10, 11–12
 strategic options 86–9
 strategy 76
 synchronic internalist rationality 71–3
 the tension 89–92
 theory of warrant 34–6
 validation of methods 78–80
Natural Theology (Paley) 25, 145, 163
nebular hypothesis (Kant-Laplace) 163
necessary properties 127–30
necessity 120–5
 de re necessity 129–134
 'necessity B' 129–30, 132
 'necessity D' 125–7
 ontological necessity 125–7
Neoplatonism 157
nested models 216
neutralizers
 of defeaters 49–53, 58
 extrinsic 54–8
 intrinsic 54–8
New Testament 5–6, 12–13, 40, 172–3,
 see also Bible, the
Newton, Sir Isaac 24–5, 162–3, 165, 186, 208, 219, 266
Nietzsche, Friedrich 25, 37, 60, 331
Ninth Bridgewater Treatise (Babbage) 171
non-arbitrariness, condition of 147–8
non-cognitive interpretation xiv, 86–8, 339
non-contingency criterion 122
Non-Overlapping MagisteriA (NOMA) 85
noumenal world 24, 28

Ockham's razor 61, 232, 235, 237, 248, 259
Oppy, Graham 267n.40
original sin 39, 41, 46
Origin of Species, The (Darwin) 25, 96, 163, 268

Pagans and Christians (Lane Fox) 338
Paine, Thomas 22
Paley, William 25, 145, 163, 267n.40
Pascal, Blaise 66, 302–3
Paul, St 4–7, 12, 172–6, 175n.55, 180, 182, 331
Pensées (Pascal) 302
Penzias, Arno Allan 183
personal identity, empiricist theories 114–16
person-terminology 110, 134–5
Phillips, D. Z. 26, 28–9, 86, 339
Philo of Alexandria 172
philosophy, transcendental 24
philosophy of religion xi, xii
   apologetic xii
   critical xii
Phosphorus 129–30, 133
Physics (Aristotle) 19
P-inductive arguments *see* inductive arguments
Pius XII, Pope 184, 198, 206, 224, 238
Planck, Max 138
Planck time 183, 229–31, 241, 243
Plantinga, Alvin 7n.9, 29, 32
   Aquinas/Calvin Model 36–40, 36n.18, 39n, 41, 52n.23, 57–8, 58n.36, 62, 63n.47, 83
   arbitrariness objection 50–1
   belief in God 55–6
   de facto objection 37, 59–61
   defeater deflectors 50n.16
   defeaters 53n.26, 59n.39
   de jure objection 37, 59–61
   explanation of Christian beliefs 63
   global argumentative strategy 32–4, 32–3n
   Great Pumpkin objection 52n.24
   and the Holy Spirit 63n.50
   irrational/rational beliefs 60
   objection to natural theology 88
   objection to Quinn 64n.51
   positive tactic 40–2, 41n.35
   reformed objection to natural theology xv, 31–2, 43, 339
   refutation of reformed objection 43–4
   religious diversity 48, 50
   religious exclusivists 48
   representational theory of perception 57n.35
   secular explanation of religious beliefs 61n.42
   sensus divinitatis 36, 38–9, 53, 55–6
   theory of warrant 34–6, 34n, 35n.16, 39, 40n.33, 42, 44–8, 47n.7, 48n.10, 52–5, 57, 68
Plato 19, 21–2, 34, 221
Pliny the Younger 172
Pollock, John 49
polytheism
   Aquinas, Thomas 21
   broadminded versions of 337
   compared to monotheism 61, 218
   design argument 267

fine-tuning 274
and gullibility 333
and Hinduism 56
and Muslims 6
and Plato 19
plurality of religions 63
religions/any number of gods 246
and theism 217, 240, 248
Popper, Karl 77, 91, 166, 215
Poseidon 333
'positive heuristic' 84
Posterior Analytics (Aristotle) 20, 33–4, 77
Postponed *Parousia* 10
pragmatic test (James) 66
prayer 81–3
   intercessory 82
   petitionary 82–3, 161
predicates
   M 110
   P 110
predictive power 140–50
predictive power dilemma 199
Principe Island 204
Principia (Newton) 162–3
Principle of Credulity *see* Credulity Principle
Principle of Testimony *see* Testimony Principle
Principle of Sufficient Reason *see* Sufficient Reason, Principle of
probability xiii, 142–4, 189, 195–6, 200–3, 204, 212, 256, 272–3, 314, 345
   intrinsic 207–8, 212–3, 216–7, 217–8
   logical 142n.9, 200n.35
   prior 204–7, 208, 212, 245–6
Problem of generality 44–7
proper name 96–7, 118, 130–1, 132
Proslogion (Anselm of Canterbury) 302
psychological egoism 215
psychological terms 97–102, 103
Putnam, Hilary 129–30, 132–3
Pyrrhonian crisis 13–14
Pyrrho of Elis 13
Pythia, the 81, 81n.7, 331

quantum mechanics 138, 186
quantum theory 138, 314
Quine, Willard Van Orman 119n.76
Quinn, Philip 64n.51

rationality 65–75
   adequacy 74–5, 74n.15
   diachronic epistemic 67, 70, 73
   diachronic internalist epistemic 73–4
   epistemic 65–8, 70n.7, 74–5
   external/internal 67–70
   global instrumental 67
   instrumental 65–6, 74
   internalist epistemic 71–2
   synchronic 73
   synchronic epistemic 70–3

rationality (cont.)
    synchronic/evidential epistemic 66
    synchronic internalist epistemic 72
rational theology see natural theology
reductionism 282
Reformation, the 13, 22–3
reformed objection to natural theology
    (Plantinga) 31–2, 43, 339
Reichenbach, Hans 215
relativity
    general theory of (GTR) 239, 314
    theories of 186, 228, 239
relevance condition/criterion 144, 144n.12
reliabilism 44
religion
    beliefs 62, 84–5
    creeds 80
    defined 80
    diversity of 12, 48–54, 63n.46, 68, 332–4
    irrationalist explanation 60–1
    rationality of 66
    Reformed Epistemology of 54
religious experience
    argument 310–11
    avoiding numbers 315–17, 316n.17
    diversity 332–4
    epistemic parity 321–4
    five-fold classification 324–9
    gullibilism 329–30, 333, 335, 337
    presence 334–7
    rationality principles 317–18
    unreliable 331–2
Religious Experience Research Unit, Oxford 326
Renan, Ernest 175n.54
resurrection, the 5, 170–82, 181n.79, 344–5
Resurrection of God Incarnate, The (Swinburne) 176–7, 345
revelation 3, 4
Revelation (Swinburne) 9
Rhees, Rush 26
Rheticus, Georg Joachim 9–10n.9
rigid designators 129
Rissanen, Jorma 215
Roman Catholicism 13–15
Roman Empire 20
Rowe, William L. 168n.23
Rundle, Bede 166n.17, 237, 242
Ruse, Michael 156, 156n.57

Schellenberg, John L. 303–4, 303n.104, 307, 309
Schindler, Oskar 288
schizophrenia 6–7, 331
scientia 77
scientific essentialism 129
Scientific Revolution 22–3
scope of a theory 142, 208–12, 217–9
Searle, John R. 283–4

semantical functionalism 99–100n.15
semantic/syntactic rules 105–6
sensus divinitatis 36–9, 41, 43, 45–8, 52–3, 55–8, 60, 62, 68
Sextus Empiricus 13
simplicity 212–20, 214n.74, 231, 247–51, 252–5, 261–4
Smith, Joseph Jr. 149
Smith, Quentin 244, 265
Smolin, L. 275
Sober, Elliott 215, 289, 343n.11
sola scriptura 22
sooma pneumatikon 5, 174
soul, the 98–9, 98n.7, 118–19, 118n.74, 119n
souls 115, 118, 119, 279, 280, 282, 285
Spinoza, Baruch de 22, 176
S-P-L account (substances-powers-and-liabilities) 263
Spontaneity of Nothingness (SoN) 238n.48
steady-state theory 184
STEP 82–3
Stoicism 19–20
strategy of subsidiary arguments 341
Strawson, P. F. 98n.10, 110n.51
substance dualism 98–9, 98n.7&9, 114–9, 118n.71, 280
Suetonius 172
Sufficient Reason, Principle of 223
Summa theologiae (Aquinas) 20–1
Swinburne, Richard
    analogy 105–6, 106n.39, 108–9
    attributions to God 156–9, 157–8n.64/67
    bare theism 191–2, 191n, 193n.7–8
    between the horns of the dilemma 88–90
    and the Big Bang 239
    bypassing difficulties of quantification 315–17, 316n.17
    the case for theism 340–1
    and coherence 107–8, 107n, 137–9
    completist fallacy 228
    concept of a person 110–13
    cosmological arguments 184–5, 222, 226–7, 229–35, 229n.24, 234n.39, 238–49, 245n, 247n, 252
    creation of humans/demi-gods 159–60, 160n.69, 212, 243–4
    Credulity Principle 319–20, 319n.32, 321–3, 321n, 330–2, 330n
    cumulative case strategy 311–14, 311n.2, 314n.13
    doctrines of accommodation 9–10, 9–10n.13
    doxastic involuntarism 73, 73n.13
    and dualism 99n.13, 115n.63
    and empirical background knowledge 209, 212
    empiricist theory of personal identity 114–19, 117n
    and evil 292–3

evolutionary arguments 284–6
existence of God 29–30, 41, 97, 97n.6, 194, 198–9, 198n.27–8, 199n.39–40, 236
fine-tuning 270–7, 272n.58, 276n.72, 277n.78/80
five-fold classification 324–5, 327–8
and freedom 159, 299–302, 301n.97
free-will defence 297–8
grand strategy 76
hiddenness of God 304–10, 305n
higher-order goods defence 294–5, 294n, 295n
importance of religious knowledge 16
inductive arguments 279–83, 280n.12–13, 281n
internalist epistemic rationality 72–3, 72n.11
and intrinsic probability of a theory 207–11, 209n.58, 210n.61, 213–17
logical probability 142–3n.9,
and metaphorical religious language 103–4, 104–5n
microreduction 283–4
miniessential kind of God 131–4
and miracles 182n.83
modal argument 118n.74
and moral access claim/moral objectivism 150–6, 151–3n, 287–90
omniscient/omnipotent/free God 251–3, 254n
ontological necessity of God 125–30, 126n
order to design arguments 269–70
and Paul's letters 175n.55
perceptual claims 332–7, 332n.75, 334n.78
person-terminology 135–7
properties of/to God 127, 127n.28–9, 130
providential argument 290–1, 290n.51
rationality 74, 74n.16, 317–18, 317n.23, 318n
and the resurrection argument 176, 176–7n
scope of a theory 142n.8
and simplicity 217–19, 219n
temporal order to God arguments 257–66, 257n, 258n.7, 259–60n, 261n.17, 265n.35
and theism 91, 95–6, 161–2, 162n.6, 164, 194–5, 205, 342, 344, 346
theory of justification or warrant 45n.5
thesis that God exists 146, 146n.19–20
threshold value for religious belief 201
too odd/too big for science to explain 166, 168–9, 168n.24, 169n
views on God 120–5, 122–3n
synoptic Gospels, the 172–3, 172n.41, 173n.43
Luke 173, 182
Mark 173–5, 180n.76

Matthew 173–4, 174n.48, 182

Tacitus 172
Tension, The 76, 89, 90, 162, 187, 340
Teresa, Mother 288
Testimony Principle 317–18, 320–1, 323, 329
Theaetetus (Plato) 34
theism
  analogy/coherence 107–9, 134–7, 137–9
  anthropomorphic projections 156–60
  as an ultimate explanation 194–9
  as a theory 95, 96, 136, 140, 144
  and auxiliary theories 83
  bare xiii, 41, 76, 121, 141, 144–6, 148–50, 154, 158, 191–2
  Bayesianism/likelihoodism 201–4
  and Christianity 56, 59, 63
  conclusions 338–46, 342n.9
  cosmological arguments
    see cosmological arguments
  different semantic views 97–103
  elimination of empirical background knowledge 207–12
  evidentialism 31, 60, 199–201, 200n.35
  evil argument 292–3, 294–7, 297–302, 307–9
  evolutionary explanations 284–6
  explaining the resurrection 175–8, 177–8n.64
  fine-tuning 270–8
  freedom 299–302, 299n.85
  free-will defence 297–9
  full explanations 192–4
  and God/bodiless person 87–8, 109–13
  God-of-the-Gaps *see* God-of-the-Gaps
  God's hiddenness 302–10
  God's miniessential kind 130–4
  God's necessity 120–5, 125–7
  higher-order goods defence 294–7, 305–6
  immunization of 161–2
  introduction 95–6
  large-scale phenomena 183–8
  microreduction 283–4
  miracles 166–70, 167n.19
  moral access claim 150–6, 156n.57
  moral truth/awareness 286–90
  and non-cognitivist views 86
  not rational 90–2
  other inductive arguments 279–83
  personal identity 114–19
  and polytheism 217, 240, 248
  predictive power of 140–5, 145–50, 150n.31, 161–2
  prior probabilities 204–7
  properties of God 127–30
  providential argument 290–2
  religious experience *see* religious experience
  resurrection of Jesus 170–5, 170n.31/34

theism (cont.)
  simplicity 212–20, 214n.74, 247–51, 252–5
  spatial order 266–70, 268–9n.43–4
  temporal order 257–66
  theory of the universe 89–90
  ultimate explanation/prior probability 191–2
  use of metaphors 95–6
  uses of words 103–6
Theological-Political Treatise (Spinoza) 22
theology, natural *see* natural theology
Theory of Everything 209
Therapeutic Effects of Intercessory Prayer, Study of (STEP) 82
Thiselton, Anthony C. 174n.48
Thomas *see* Aquinas
three-dimensional notion 210
threshold value 143, 144, 201
Timaeus (Plato) 19
Tooley, Michael 286
Tractatus Logico-Philosophicus (Wittgenstein) 253–4
transcendental philosophy *see* philosophy, transcendental
Trivers, R. 288
true belief 35
truth 13, 151
  criterion of 13
truth-indicator 216
truth-makers 57–8, 57n.34

underdetermination problem 139
universe, the
  age of 38, 86, 232–6
  and the Big Bang 217–18, 221–9
  chaotic 261, 263
  design/chance hypothesis 268n.43
  existence of 183–8, 191–8, 206, 210–12
  expanding 9, 230
  fine-tuning 270–8
  first state of 240–2
  and God 10, 19–20, 23, 83, 88, 123–7, 140, 144, 146, 149, 150n.32, 158, 184–5, 186n, 187, 248
  God intruding in 168
  Godless 260–3, 287
  God's creation 236, 312
  initial first state cause 315
  logically possible 270–4
  and matter 165
  and miracles 166–7
  multiverse 258–9

  possible cosmological scenarios 263–4
  spatial magnitude 277, 277n.81
  theism 89, 145, 147, 161–2
  violent events in 79
  and Vishnu 56, *see also* Big Bang, the; black holes; cosmological arguments
Ussher, Bishop 86–7

validation of methods 78–80
Van Fraassen, Bas C. 27n.12, 144n.12, 214, 214n.73/76
Van Inwagen, Peter 293n.64–5
Varieties of Religious Experience, The (James) 326
Vedas, the 3, 9
violation of laws of nature 167, 167n.20
Vishnu 56
Voetius, Gisbertus 22

wager (Pascal) 66
warrant *see* Plantinga, theory of warrant
Warranted Christian Belief (Plantinga) 37, 39, 41, 48n.10, 51, 54, 59
Watchmaker Argument (Paley) 267n.40
water 129–30, 132–3
Weber, Joseph 79–80
Weltanschauung 10
Wesensschau 129
White, Andrew Dickson 8n.10, 11
whole 233–6
Williams, Bernard 117
'Will to Believe, The' (James) 25
Wilson, David Sloan 289
Wilson, Robert 183
Wittgenstein, Ludwig
  C-inductive argument 280n.13
  conception of philosophy 26–7
  no logically simple propositions 253
  non-cognitive interpretation of religion xiv, 86–8
  psychological concepts 103–4, 119
  rejection of the Cartesian myth 99–101, 101n.21
  religious dilemmas 338, 339
Wolterstorff, Nicholas 85n.16
Wright, N. T. 176n.56, 181n.79, 182n.80

Xenophanes 19

Yahweh *see* God

Zweckrationalität *see* rationality, instrumental

The manufacturer's authorised representative in the EU for product safety is Oxford University Press España S.A. of el Parque Empresarial San Fernando de Henares, Avenida de Castilla, 2 – 28830 Madrid (www.oup.es/en or product. safety@oup.com). OUP España S.A. also acts as importer into Spain of products made by the manufacturer.

www.ingramcontent.com/pod-product-compliance
Ingram Content Group UK Ltd.
Pitfield, Milton Keynes, MK11 3LW, UK
UKHW021136240326
469240UK00020B/153